Sedimentary Petrology

Harvey Blatt
University of Oklahoma

W. H. Freeman and Company
New York

Project Editor: Pearl C. Vapnek
Interior Designer: Ron Newcomer
Cover Designer: Sharon Helen Smith
Production Coordinator: Linda Jupiter
Illustration Coordinator: Richard Quiñones
Compositor: Allservice Phototypesetting Company
Printer and Binder: The Maple-Vail Book Manufacturing Group

Cover photograph: Painted Desert northeast
of Winslow, Arizona. © Kaz Hagiwara

Library of Congress Cataloging in Publication Data

Blatt, Harvey.
Sedimentary petrology.

Includes bibliographies and index.
1. Rocks, Sedimentary. I. Title.
QE471.B653 552′.5 81-22147
ISBN 0-7167-1354-3 AACR2

Printed in the United States of America

3 4 5 6 7 8 9 10 VB 6 5 4 3 2 1 0 8 9 8

Dedicated to my English teachers in
the New York City public school system,
who taught me how to think.

To Bob Folk, who showed
me some things worth thinking about.

And to my dear parents, whose many
sacrifices made everything possible.

Contents

Preface

He that publishes a book runs a very great hazard, since nothing can be more impossible than to compose one that may secure the approbation of every reader.

MIGUEL DE CERVANTES

This book is an introduction to sedimentary petrology: the study of the origin, occurrence, mineral composition, and texture of sedimentary rocks. The description and interpretation of sedimentary structures, environments, and facies are dealt with only briefly, as a more extensive treatment seemed inappropriate in a text centered on rocks and minerals. Thus, structures, environments, and facies are considered only insofar as they affect mineral composition, other lithologic characteristics, and the areal distribution of various rock types.

Because many geology departments separate the study of sedimentary petrology from that of igneous and metamorphic rocks, my publisher suggested this book as a companion volume to *Petrology: Igneous, Sedimentary, and Metamorphic* (W. H. Freeman and Company, 1982), which I coauthored with Ernest G. Ehlers. I have expanded the presentation of the earlier book, making it more suitable for a course devoted exclusively to sedimentary rocks. As in the parent book, the control of sedimentary accumulations by plate movements is stressed.

Nomenclature is basically boring. Nevertheless, students must learn some of the current jargon in order to understand the thought patterns of sedimentary petrologists, and I have tried to make a judicious selection of terminology from the mass presently in use. For example, several dozen classifications of sandstones based on mineral composition have been published. These classifications have been accompanied by new terms whose number and interpretation are rivaled only by the manufacturers of dry cereals and household detergents. I have chosen perhaps 10 percent of the rock names to illustrate petrologic principles. The remainder must be tolerated by advanced professionals but are only a source of confusion for beginners. As an aid for students in their initial encounter with the terminology of sedimentary petrology, a Glossary of more than 300 terms is provided. The Glossary does not include terms normally defined in the mineralogy courses that are prerequisites for a course in petrology.

It has been my experience that field trips are part of all geology curricula, but that adequate field description of rocks is not. Therefore, this book puts considerable emphasis on learning to describe rocks in their natural setting, the Great Outdoors. Many important aspects of rocks occur on a scale too small to be seen without a polarizing microscope or an electron microscope, however, so I have included photomicrographs when necessary. Sometimes it is useful to examine a mountain with a microscope. However, it has been my intent to make the text comprehensible to students not yet familiar with the polarizing microscope.

The emphasis in this book is on the description of sedimentary rocks, rather than on the physics and chemistry of their formation. This separation has not always been possible, but I tried. In instances where observations are interpreted, I hope the distinction between fact and fancy is clear.

The latter part of most chapters describes field examples of the principles explained in the earlier part. Chosen from readily available publications, the examples indicate the types of observations that sedimentary petrologists have found useful, both in the field and in the laboratory, and the types of interpretations that can result. Often these interpretations have economic significance, as has been noted when appropriate. The day when "ivorytowerism" was clearly distinguished from "commercialism" is long past.

This textbook is intended for students in their third or fourth year of a university education in geology. Its suitability can be evaluated only by the teachers and students who use it. I would appreciate constructive criticism from both groups if errors in conceptual or factual material are uncovered.

I would like to thank those individuals who critically reviewed various portions of the manuscript: J. C. Crelling, R. V. Ingersoll, P. C. Lyons, G. V. Middleton, and R. C. Murray. Many thanks are due Marjorie Starr, who did a rapid and superb job of typing the manuscript. Needless to say, none of these individuals is responsible for errors of fact or interpretation that remain in the text.

February 1982 Harvey Blatt

1

The Occurrence of Sedimentary Rocks

Birds of a feather will gather together.

ROBERT BURTON

There are many reasons for studying sedimentary rocks. Sediments and sedimentary rocks are the only source of knowledge about conditions on the Earth's surface before the invention of written language a few thousand years ago. Regardless of whether you are interested in the temperature of surface ocean water during the Jurassic Period, the types of organisms that preceded the evolution of humans, the development of surface topography, or the amount of oxygen in the atmosphere four billion years ago, the source of the information is sedimentary materials. And, in addition to our innate curiosity about esoteric problems such as these, there are the quite practical concerns centering on the existence of our technological civilization. Sedimentary materials contain the coal, petroleum, natural gas, gypsum, sulfur, sand and gravel, uranium, aluminum (in bauxite), iron, and many other substances that permit the standard of living we enjoy in the twentieth century. Shortages of some of these materials have already begun to appear and will certainly become more severe in the coming years.

Pre-Holocene sediments and sedimentary rocks cover 66% of the continental surfaces and probably most of the ocean floor as well. The basic reason for this wide areal extent is the chemical instability of igneous and metamorphic rocks under atmospheric conditions. Rocks and minerals are in equilibrium only under the set of physical and chemical conditions in which they form. Under different conditions they tend to react to reach a new equilibrium state. Igneous and metamorphic rocks form at temperatures and pressures much higher than those at the Earth's surface, and in an environment containing less water, less oxygen, less carbon dioxide, and no organic matter. It is to be expected that such rocks will be unstable and will undergo chemical and physical changes when brought to the surface by tectonic, erosional, or isostatic forces. These changes constitute the process we call *weathering*. Based on Le Châtelier's Principle, we would expect the products of this chemical change to contain more water, more oxygen, more carbon dioxide, and more organic matter than before the change. This expectation is fulfilled, as we shall see in Chapter 2.

Sedimentary rocks consist almost entirely (> 95%) of three types: sandstones, mudrocks, and carbonate rocks. *Sand* is defined as fragmental sediment 2–0.06 mm in diameter; *mud*, smaller than 0.06 mm. *Carbonate rocks* are composed largely of $CaCO_3$ (calcite or aragonite) or $CaMg(CO_3)_2$ (dolomite), with other carbonates (e.g., siderite, magnesite) being rare. As expected, there are transitional rocks that do not fit neatly within the three pigeonholes. For example, *coquina* is a fragmental rock composed of sand-size fragments of fossil shells. It is both a sandstone and a limestone; it is usually included in the limestones. How should we classify a rock composed of subequal amounts of clay and microcrystalline carbonate material (*marl*)? There are no perfect answers to such questions, only agreed-on compromises; sometimes, not even these.

The most abundant sedimentary rocks are the mudrocks, which form 65% of all sedimentary rocks. A moment's reflection about the mineralogy of igneous and metamorphic rocks suggests why mudrocks dominate the average stratigraphic section. Igneous and metamorphic rocks are composed of approximately 20% quartz and 80% other silicate minerals; only the quartz is chemically stable under most surface conditions. The other minerals are unstable when exposed at the surface and are altered to a variety of substances, but mostly to clay minerals. Clay minerals are mud size; hence, mudrocks are the dominant sedimentary rock. The quartz in crystalline rocks is very resistant to chemical attack and occurs chemically unchanged in both mudrocks and sandstones. Based on many thousands of analyses by X-ray, polarizing microscope, and chemical techniques, mudrocks and sandstones have the average detrital mineral compositions listed in Table 1–1. A weighted average shows that the detrital sediment in the sedimentary column consists of 45% clay minerals, 40% quartz, 6% feldspar, 5% undisaggregated rock fragments, and 4% others. About 85% is either clay minerals or quartz, the most stable minerals under surface conditions. Clearly, weathering has been a very effective process through geologic time.

Table 1–1
Average Detrital Mineral Composition of Mudrocks and Sandstones

Mineral composition	Mudrocks, %	Sandstones, %
Clay minerals	60	5
Quartz	30	65
Feldspar	4	10–15
Rock fragments	< 5	15
Carbonate	3	< 1
Organic matter, hematite, and other minerals	< 3	< 1

Figure 1–1
Depositional contact between unaltered Upper Oligocene microcrystalline calcium carbonate ooze and olivine basalt, as seen in core 6.5 cm in diameter, from beneath 335 m of sediment at latitude 8.8°N, longitude 143.5°E. [W. B. Hamilton, U.S. Geol. Surv. Prof. Paper No. 1078, 1979. Photo courtesy W. B. Hamilton, U.S. Geol. Surv.]

Although the areal extent of sedimentary rocks is great, its thickness is not, averaging only 1.8 km on the continents and 0.3 km in the oceanic basins. In part, the thinness of sedimentary cover on the continents results from the definition of the word *sedimentary.* In some areas the base of the sedimentary column is well defined and clearly seen. For example, in mid-continental North America the oldest sediments contain fossils and look in every way like normal mudrocks, sandstones, or carbonate rocks. Directly below them lie granites, gneisses, or schists. The boundary between sedimentary and nonsedimentary can be drawn on the outcrop with a pencil. Similarly, in the deep oceanic basins, sediment such as globigerinid ooze, brown clay, or chert is underlain by basalt, again with a very well-defined contact (see Figure 1–1.)

In many areas, however, the contact between sedimentary and nonsedimentary is gradational, and different geologists would draw the pencil line at stratigraphic positions several hundred meters or more apart. This results from the fact that, with increased burial depth and resultant increased temperature and pressure, the sediment that had largely equilibrated with atmospheric conditions is subjected to an environment so different that it must adjust. But different minerals have different limits of stability; they do not all change at the same depth or at the same rate once the change has begun. For example, some clay minerals are known to grow and/or recrystallize to new clay minerals at a depth of about 3,000 m and a temperature of 80°C. Chert, which is microcrystalline, may recrystallize and coarsen at 150°C; and

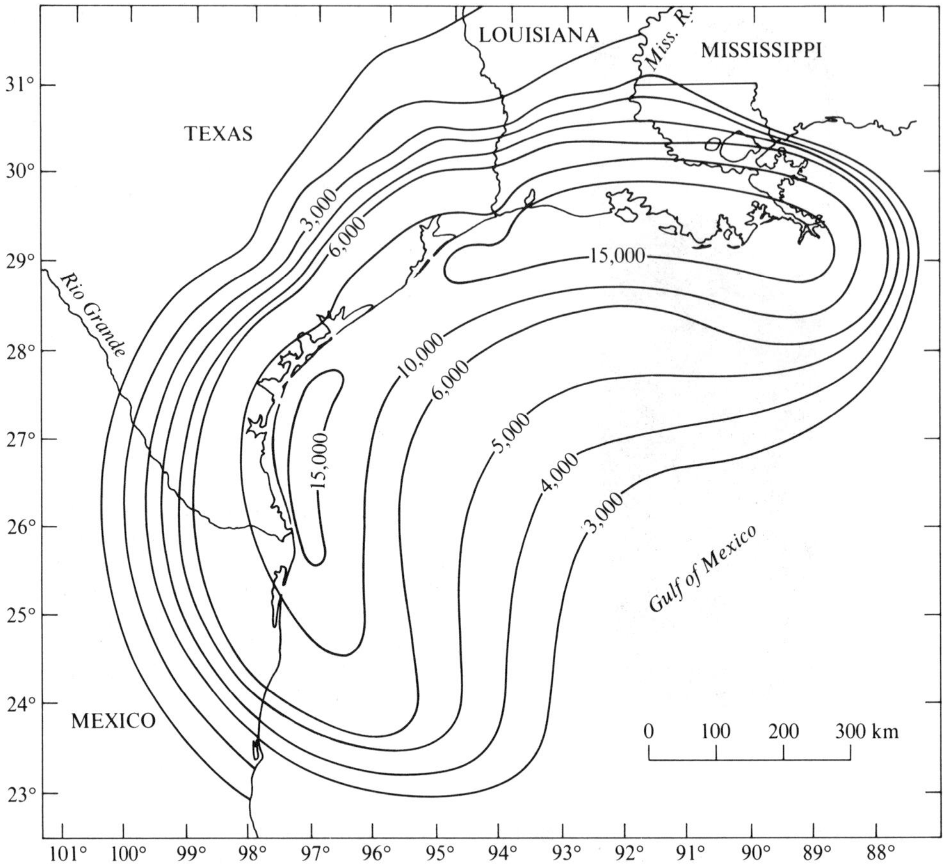

Figure 1–2
Isopach (thickness in meters) map of Cenozoic sediments in Gulf Coast basin. Note irregular contour interval. Approximately 7,000 m of Mesozoic sedimentary rocks underlie Cenozoic rocks of delta region. [G. C. Hardin, Jr., 1962, in E. H. Rainwater and R. P. Zingula, eds., *Geology of the Gulf Coast and Central Texas* (Houston Geol. Soc.).]

feldspar may be dissolved by circulating groundwaters at any depth from the surface downward. Where are we to draw the line between a sedimentary rock and a rock that has been so severely changed by conditions accompanying burial that we believe the term "metamorphic rock" is more appropriate? Standard practice is that each investigator is sovereign in setting the dividing line for his or her rocks. One person's hard shale is another person's slate; a metaquartzite to one investigator may still be a quartz-cemented quartz sandstone to another. In some geographic regions these differences in terminology can seriously handicap communication.

The average thickness of sedimentary rocks on the continents is about 1,800 m but is quite variable, ranging from zero over extensive areas such as the Canadian Shield to more than 20,000 m in some basinal areas such as the Louisiana–Texas Gulf coastal region (see Figure 1–2). The maximum possible thickness is determined by the geothermal gradient in the area, by fluid chemistry, and by the chemical reactivity of the detrital particles. The average temperature at a given depth can vary greatly among geographic-tectonic areas. For example, at a depth of 10,000 m under Pittsburgh, it is about 150°C; under New Orleans, 200°C; under Las Vegas, 260°C; under Los Angeles, greater than 300°C. The mineral content of the sandstones in the sedimentary section can vary from those composed entirely of quartz grains, which are relatively resistant to destruction or recrystallization, to sandstones composed mostly of calcic plagioclase grains and basaltic rock fragments, which alter chemically at very low temperatures.

THE DESTRUCTION OF THE ROCK RECORD

Examination of geologic maps of the continental areas of the Earth reveals that 66% is underlain by sedimentary rocks. The maps also reveal that rocks of more recent geologic periods are more abundant in outcrop than those of older periods and that the decrease in abundance with increasing age follows the same logarithmic law that describes the radioactive decay of elements (see Figure 1–3). The data indicate that

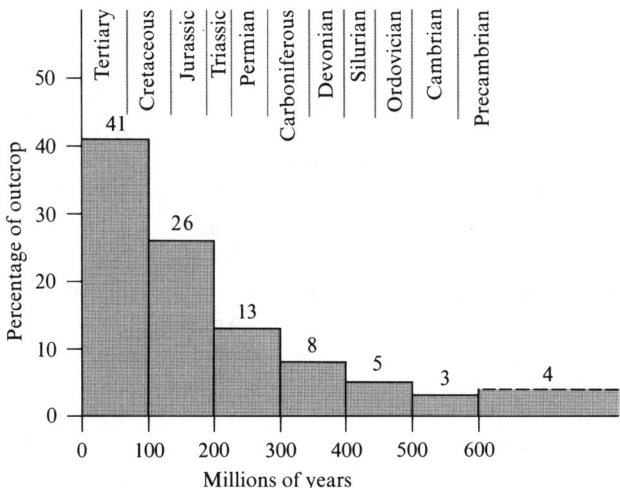

Figure 1–3
Relationship between sedimentary rock age and amount of outcrop area. [H. Blatt and R. L. Jones, 1975, *Geol. Soc. Amer. Bull., 86*.]

half of all outcropping sedimentary rocks are younger than 130×10^6 years, i.e., Cretaceous or younger. As we examine older sedimentary rocks, we have less rock to examine, so that our interpretations must be more generalized. The world of Cambrian time will forever remain more speculative than the world of more recent times.

THE TYPES OF SEDIMENTARY ROCKS

The major types of sedimentary rocks are mudrocks (65%), sandstones (20–25%), and carbonate rocks (10–15%), with all others totaling less than 5%. These world averages vary widely on both a regional and a local scale, however, with some sedimentary basins filled largely with detrital sediments in various proportions and others dominated by nonclastics. For example, the thick Cenozoic section in the Gulf of Mexico basin (Figure 1–2) contains about 90% mudrocks and 10% sandstones, with only minor amounts of nonclastic rocks. The Michigan Basin, in contrast, contains only 18% mudrocks and 23% sandstones, but 47% carbonate rocks and 12% evaporite beds. On the scale of a single stratigraphic section, still greater extremes occur. In later chapters we discuss each of the abundant types of sedimentary rocks (and some of the less abundant ones) in some detail, but it is useful to introduce here their general characteristics.

Mudrocks

Mud particles are defined as grains less than 62 μm in size (silt size is 62–4 μm; clay size is < 4 μm); most are less than 5 μm. Because of this, the particles are easily kept in suspension by even the weakest of currents and can settle and accumulate only in still waters. Many such quiet-water environments exist, in both nonmarine and marine settings: for example, floodplains, deltas, and lakes on the continents; lagoons and areas below wave base in the marine environment.

The thickest accumulations of mudrocks occur in geosynclinal settings, with mudrock thicknesses ranging up to at least 2,000 m in the central Appalachians of Pennsylvania and in the Ouachita Geosyncline in Arkansas. When interbedded with sandstones in the geosynclines, the mudrocks commonly form the bulk of the accumulation, e.g., 56% of a 7,000 m Tertiary section in Indonesia and 61% of a 3,000 m Carboniferous section in the Anadarko Basin of western Oklahoma.

In contrast to the visible siltiness of thick accumulations of geosynclinal mudrocks, those mudrocks deposited in locations such as lakes, abyssal areas far from land, or shallow marine areas below wave base are exceptionally fine-grained—a mixture of clay minerals, organic matter, and very fine-grained quartz. An excellent example is the Chattanooga Shale of Late Devonian age in Tennessee and surrounding states. This unit has a thickness of only 10 m and is a blanket deposit extending over tens of thousands of square kilometers. It is believed to have been deposited in an epicontinental sea in water depths of less than 30 m and is composed of about 20–25% quartz,

25–30% clay and mica, 10% alkali feldspar, 10–15% pyrite, 15–20% organic matter, and 5% miscellaneous constituents. The quartz grains are nearly all of fine silt and clay size and are distributed as thin laminae within the black organic matter and clay-mineral mixture. Petrologically identical sediment can accumulate also in the deep ocean, as is presently occurring in the numerous small fault basins off the coast of southern California. What are the basic controls of mudrock occurrence and composition? We consider this question in more detail in Chapter 3.

Sandstones

Sand accumulates in areas characterized by relatively high kinetic energies, i.e., environments of rapidly moving fluids. Examples of these environments include desert dunes, beaches, marine sandbars, river channels, and alluvial and submarine fans. However, some sands deposited on the shallow seafloor are subsequently carried down to great depth in the sediment–water mixtures called *turbidity currents.* As a result, coarse-grained sediments can occur in quiet water.

Many common sites of sand accumulation are elongate, such as beaches and rivers, but in the geologic record the sands deposited in these environments are commonly sheetlike in character. This difference results from the displacement of the depositional site through time; e.g., a beach migrates inland during a marine transgression, resulting in a slight increase in the thickness of the sand body but an extreme increase in its width. It is, however, possible for a sand-dominated beach–dune complex to exist at the same geographic locality for a long period of time. This could occur in the tectonic setting of a slowly subsiding basin, resulting in pure sand deposits hundreds of meters thick with relatively narrow areal extents. The Cambro-Ordovician quartz sands of the western United States may be an example of this.

As in the case with mudrocks, environment of deposition can commonly be related to mineral composition. Sands deposited in loci of highest kinetic energy, such as beaches and desert dunes, tend to be more quartz-rich than the sandbars of sluggish rivers. This occurs because of the relative ease of breakage and elimination of cleavable minerals, such as feldspar, or foliated fragments, such as shale or schist. However, it is not a good idea to base an environmental interpretation on detrital mineral composition. Some fluvial sands contain more than 90% quartz, and many modern beaches contain high percentages of feldspar and rock particles. Mineral composition is not a good environmental indicator, as we discuss further in Chapter 4.

Carbonate Rocks

Modern carbonate sediments are composed almost entirely of the hard parts of marine organisms, and there is every reason to believe this has been true of carbonates throughout Phanerozoic time. Because of the great chemical reactivity of calcium carbonate, however, most carbonate particles are recrystallized sometime after depo-

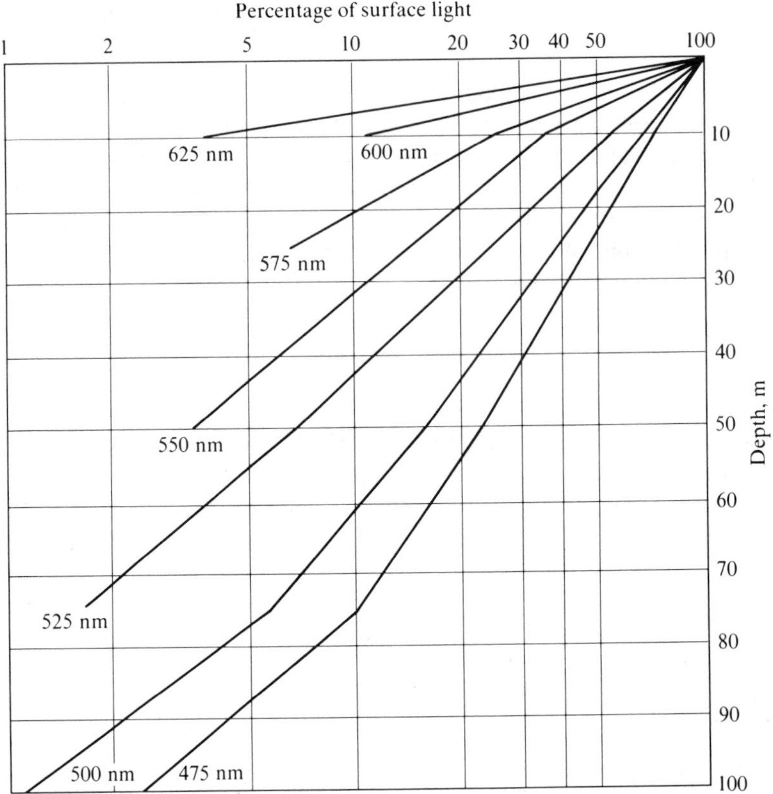

Figure 1–4
Depth to which light of various wavelengths will penetrate clear ocean water. In coastal waters, penetration is commonly only 1–10% of that in clear water. Plants can photosynthesize in a wide variety of wavelengths and intensities; e.g., red algae prefer red light. <400 = ultraviolet, 400–500 = blue, 500–600 = green, 600–700 = red, >700 = infrared.

sition so that their organic origin is not always evident. This is particularly true of the microcrystalline particles that form the bulk of ancient limestones.

Because they are organic in origin, the abundance of carbonate particles is tied to the occurrence of the phytoplankton at the base of the food chain; and the phytoplankton, in turn, are tied to the depth of penetration of light into seawater. If there is no light, there can be no photosynthesis and no phytoplankton. The depth of penetration of light in seawater is shown in Figure 1–4, and it is apparent that most wavelengths are absorbed at very shallow depths. Because of this, most organisms live within 10 m of the sea surface. However, the carbonate material generated in these shallow waters need not be deposited there. Planktonic, carbonate-shelled organisms such as *Globigerina* are abundant in the open ocean and may settle to the deeper ocean floor to accumulate (see Figure 1–5). Few of these deep-sea carbonates appear

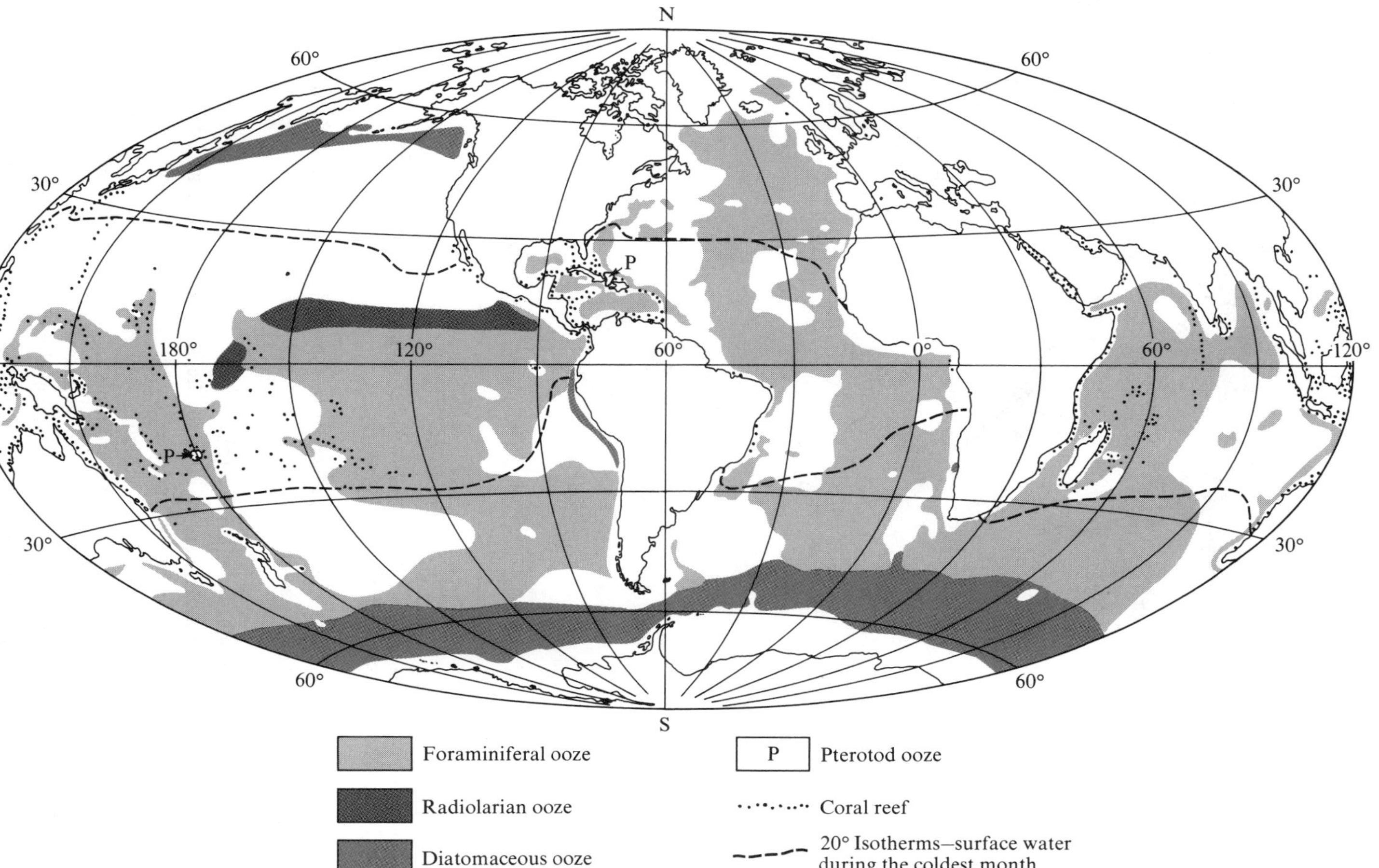

Figure 1–5
Distribution in oceanic surface sediment of globigerinid and pteropod ooze (calcareous) and diatomaceous–radiolarian ooze. Also shown are 20° isotherms in surface water during coldest month of year and distribution of shallow-water coral–algal reefs. Reefs occur almost exclusively between 20° isotherms. Calcareous oozes have a more widespread latitudinal distribution. [G. Dietrich, 1963, *General Oceanography* (New York: Wiley).]

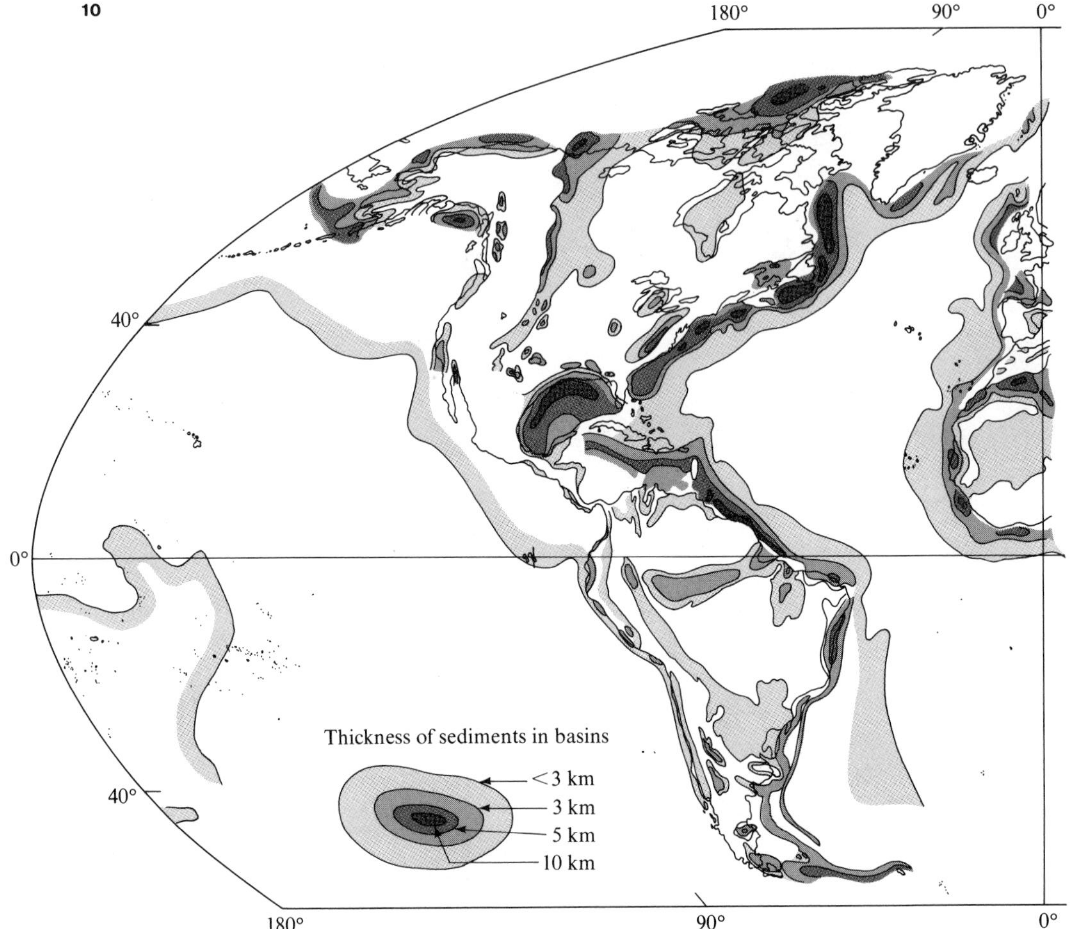

Figure 1–6
Sediment thickness in relation to continental margins. [W. C. Kruger, Jr., 1977, *Thickness of Sediments* (Tulsa, OK: Amoco Production Company).]

in the stratigraphic record because the shells dissolve in the cold waters of the deep ocean. As a result, most carbonate rocks we see in the stratigraphic record are of shallow marine origin.

Sand-size fragments of carbonate-shelled organisms are not difficult to recognize and identify in thin sections of limestones, at least to the level of phylum and class (see Chapter 7). Sometimes even the genus and species can be specified, and rather detailed reconstructions of depositional environments are possible using these and other data. For example, analyses of the relative amounts of the different isotopes of oxygen present in the calcite of unrecrystallized shell material can be used to determine the temperature of the water in which the organism lived. Organisms are very sensitive to their environments and, because of this, limestones can be gold mines of information about the shallow marine waters of the geologic past.

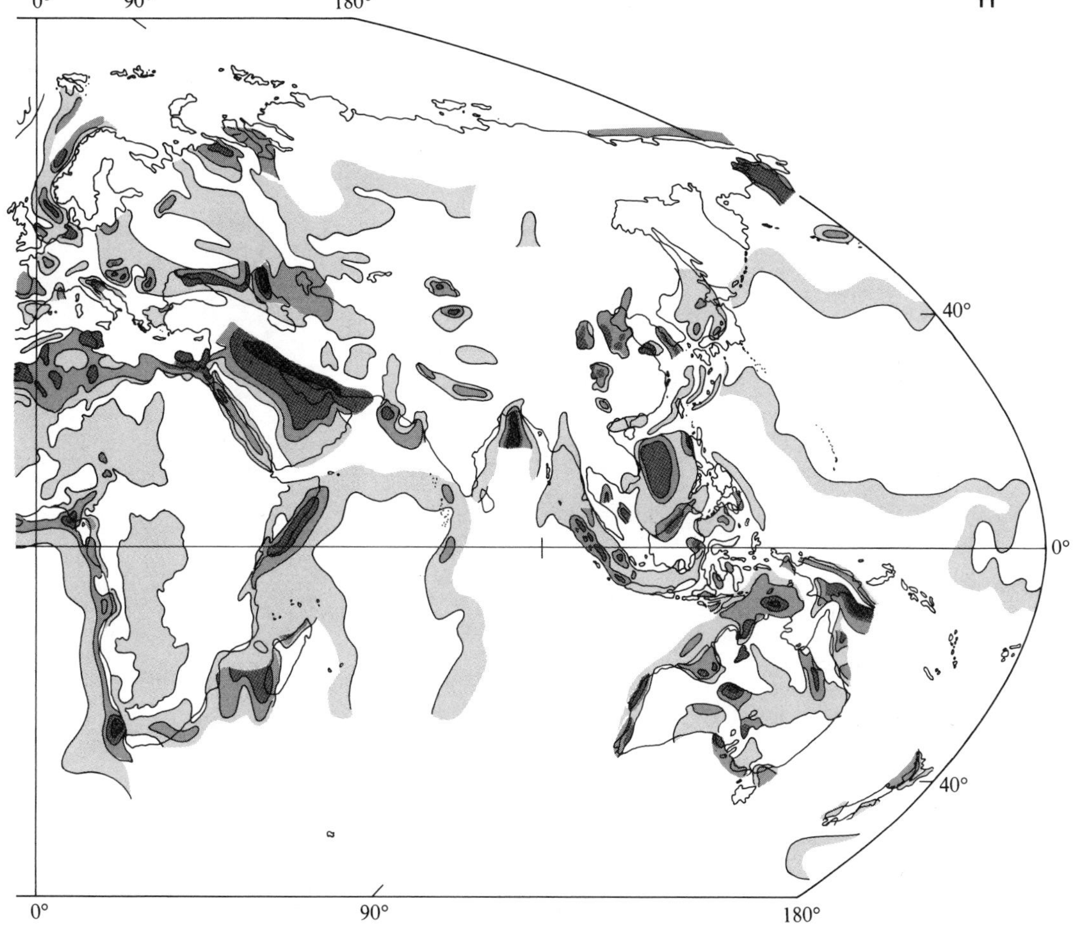

DEPOSITIONAL BASINS, GEOSYNCLINES, AND PLATE TECTONICS

Based on stratigraphic data accumulated during the past hundred years, it is clear that the majority of preserved sedimentary rocks is marine. There are several reasons for the dominance of marine sediments:

1. The light sialic material that forms a large proportion of continental masses is limited in volume, with the result that continental areas constitute only about 30% of the Earth's surface. The marine areas are sediment traps that cover 70% of the Earth's surface.

2. Because of the movement and subduction of oceanic crust and mantle at the edges of some continental blocks, topographically depressed areas (*trenches*) with adjacent easily eroded highland areas exist at some continental margins (see Figure 1–6).

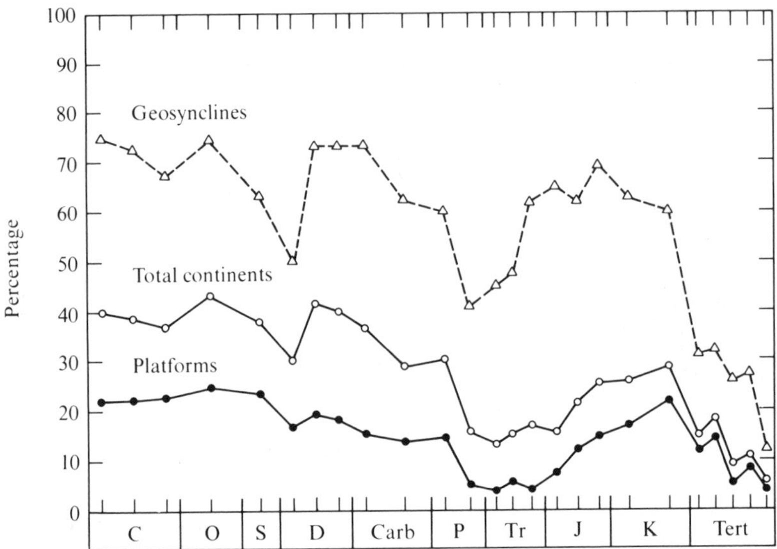

Figure 1–7
Variation during Phanerozoic time in percentage of continental blocks covered by marine sedimentary rocks. [Ronov et al., 1980.]

3. There has been a pronounced tendency through time for broad areas of the continental blocks to be invaded by shallow marine waters (*epicontinental seas*). The resulting shallow-water marine sediments are laterally extensive on the craton, although usually they are thin. Figure 1–7 shows the extent of these epicontinental seas through geologic time based on extremely detailed paleogeographic and facies maps. Several things are obvious from the graph:

a. Holocene time is characterized by a nearly complete absence of marine waters on the continental blocks.

b. During the past 570 million years there have been three major advances of the sea onto the continent (Early Ordovician, Late Devonian, and Late Cretaceous) and three withdrawals (Siluro-Devonian, Permo-Triassic, and Late Tertiary). The tectonic explanation of these marine inundations and regressions is not certain but is thought to result from changes in the volume of the oceanic rift system (Donovan and Jones, 1979). Increases in volume of the oceanic ridges result in a decrease in volume of the ocean basins and subsequent flooding of the low-lying parts of continents. Calculations indicate that this mechanism is adequate to cause flooding on the scale seen in the geologic record.

4. Continental deposits are, by definition, formed above sea level and hence are subject to removal should the rate of accumulation fall behind the erosion rate. It is no accident that stratigraphic sections on the continental blocks contain many unconformities, while deeper marine or oceanic sections are more complete.

As our understanding of the Earth's history has grown, it has become increasingly clear that there is a close relation between sediments and tectonics. Beyond the obvious relation between uplift and erosion and subsidence and accumulation, it has become clear that the sedimentary assemblages of some areas or times differ markedly from those of other places and times and that these differences are in some way related to the architecture of the basin and/or the stage in its evolution and filling. Prior to the advent of the theory of plate tectonics in the 1960s, the ruling idea used to explain the thickest accumulations of marine sediments was the geosynclinal theory. *Geosynclines* were thought of as areally extensive, usually linear, structural depressions located most commonly at the edges of continents. In North America the "type locality" of the geosyncline was the Appalachian trend along the eastern margin of the United States and extending northeastward into the maritime provinces of Canada.

The idealized geosyncline was divided into two parallel segments, termed miogeosyncline and eugeosyncline. In general, *miogeosynclinal terranes* are characterized by clear-cut depositional contacts with the underlying continental basement and a stratigraphic section whose marine sediments were all deposited in shallow water (fossils) and lacked interstratified volcanic rocks or volcanic detritus. The sediments grade landward into continental, cratonic sediments rich in quartz. As a first approximation, the miogeosyncline can be interpreted as a thick accumulation of strata on the margin of a continent.

In contrast, *eugeosynclinal terranes* are characterized by equivocal contact relationships with the continental basement, an abundance of volcanic rocks and detritus within the sedimentary sequence, and a conspicuous absence of evidence of shallow-water deposition. Eugeosynclinal strata apparently form in the deeper part of an oceanic basin adjacent to the shallow-water miogeosynclinal strata and in an area of active volcanic activity. In many cases of geosynclinal deposits, the boundary between the miogeosyncline and the eugeosyncline (if exposed) is marked by a tectonically chaotic mixture of very large fragments of older sedimentary and crystalline rocks, some several kilometers in length, set in a pelitic matrix. This intensely disturbed sedimentary rock unit is called a *mêlange*. In the Appalachian region, the miogeosynclinal rocks are exposed in the Valley and Ridge Province; the eugeosynclinal rocks to the east, in New England and the maritime provinces of Canada.

As indicated in Figure 1–8, dating of Precambrian rocks indicates that the continental block has grown during geologic time by the accretion of successive geosynclinal deposits. Pettijohn et al. (1973, p. 545) estimate that 82% of sediments by volume in North America are geosynclinal accumulations (eugeosynclines, 59%; miogeosynclines, 23%) and only 18% are cratonic. Figure 1–9 shows the extent of accretion onto the Precambrian nucleus of the United States during Phanerozoic time. Also shown is the position of the United States with respect to the equator during each geologic period, and it is apparent that the area was located within 30° of the equator during the entire Paleozoic Era. This implies a continual tropical to subtropical climate, and climate is an important control of the mineral composition of sediments (see Chapter 2).

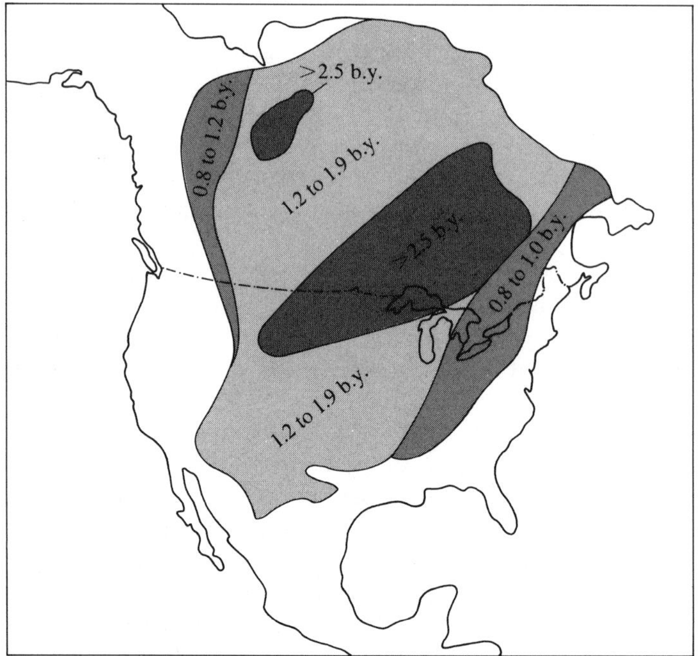

Figure 1–8
Generalized map of ages of Precambrian rocks in North America, illustrating continental "nuclei" and apparent enlargement of nuclei through time. [R. W. Ojakangas and D. G. Darby, 1976, *The Earth Past and Present* (New York: McGraw-Hill).]

Modern theories concerned with the origin of large sedimentary basins and their accumulations of sediment center on plate movements. According to the theory of plate tectonics, the settings of basins can be described with reference to three fundamental factors:

1. The type of lithosphere that serves as substratum for the basin (oceanic, transitional, or continental).
2. The proximity of the basin to a plate margin.
3. The types of plate junctions nearest the basin.

The term *geosyncline* is rapidly passing into disuse but may still be useful to indicate a geographically extensive area in which the accumulation of sediment is much thicker than in surrounding areas.

The use of the theory of plate tectonics as a framework of thought precludes the possibility of a neat catalog of basin types. The key factors in basin evolution are the

Figure 1–9
Present outline of conterminous United States, showing accretion to Precambrian craton during Phanerozoic time. Also shown is current best estimate of location of conterminous United States in relation to the equator during each Phanerozoic period. During Cretaceous and Tertiary times the United States was located more than 2,000 km north of the equator because of a consistent drift northward of North American plate. 1,100 km = about 10° of latitude.

types of plate interactions and settings, but the order in which they may be arranged in space and time can vary within wide limits. A great variety of developmental schemes can be accommodated within the theory of plate tectonics. The evolution of sedimentary basins is incidental to the formation and consumption of lithosphere. The major perturbations of a stable and level surface of the Earth are related to the opening of oceanic basins, accompanied by the rifting and fragmentation of continental blocks, and to the closing of oceanic basins, accompanied by the collision and assembly of continental blocks.

For this reason, it is most meaningful to catalog the major types of sedimentary basins according to their location and origin as determined by the interplay between tectonics and sedimentation:

1. *Oceanic basins.* These are the areas of deposition (*depocenters*) underlain by oceanic lithosphere; for example, the bulk of the Pacific Ocean basin.

2. *Grabens along continental margins.* These depocenters are formed by rifting along the margins of stable continental crust in association with the formation of an oceanic basin. Most such grabens are oriented parallel to the sides of the oceanic basin (Tertiary of Spitsbergen [Atkinson, 1962], and most of the Triassic basins along the eastern coast of the United States), but others can be oriented at a large angle to the coastline, in a reentrant (an *aulacogen;* Burke, 1977).

3. *Arc-trench system basins.* Numerous and distinct types of sedimentary basins can develop in association with a plate margin that is being consumed beneath island arcs or continental margins. Modern examples rim the Pacific Ocean basin. An ancient example of sedimentary rocks deposited in one of these types of basins is the Great Valley sequence of Late Mesozoic age in California (Dickinson and Seely, 1979).

4. *Suture-belt basins.* These are basins that develop in areas where continental blocks are in crustal collision. Examples of such basins are found in the Himalayan Mountain belt and in the Ouachita belt (Carboniferous) in Arkansas and Oklahoma (Graham et al., 1976).

5. *Intracontinental basins.* These depocenters seem to be formed independently of tectonic activity at plate margins and at unpredictable locations within the continents.

Oceanic Basins

The principal settings of oceanic facies controlled by tectonic relations are, as shown in Figure 1–10:

1. Bathymetric highs of ridge crests at divergent plate junctions where layered igneous rocks are formed along the trends of the spreading centers.

2. Where the oceanic substratum gradually subsides as it cools in moving away from spreading centers.

3. Deep basins beneath which the thermal contraction of the lithosphere is essentially complete.

The pelagic sediment that covers the basaltic ocean crust has a stratigraphy and a facies relationship that reflect changing water depths. Near the upper part of the ridge the water depth is commonly less than about 4,000 m, so that pelagic shells of

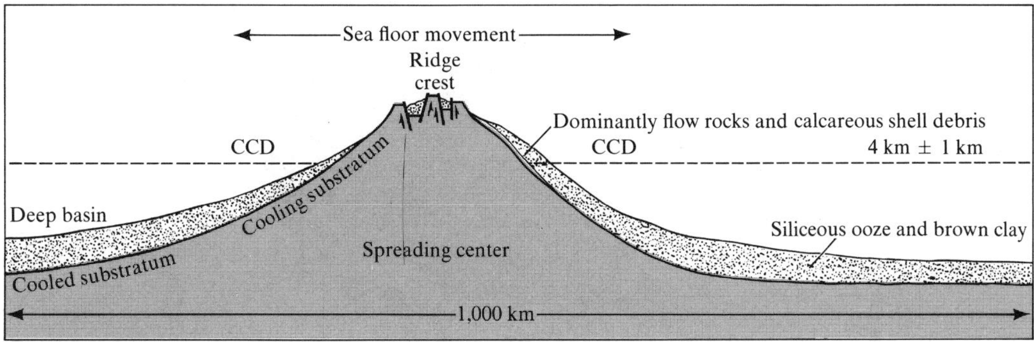

Figure 1–10
Idealized cross section of an oceanic spreading center, showing accumulations of ponded and peripheral pelagic sediment and volcanic rocks in extensional fault basins at ridge crest and on flanks above carbonate-compensation depth (CCD) and siliceous ooze and brown clay overlying mafic rocks to the sides of the spreading center. Turbidites may be intercalated with siliceous ooze and brown clay near continental edges. Total sediment thickness far from continental margins may be a few hundred meters.

calcium carbonate can accumulate. Below this depth, called the *carbonate-compensation depth,* the degree of undersaturation of the water with respect to calcite or aragonite is so great that shell accumulation is not possible (see Chapter 7). Lower on the rise flanks, in the deep basins and in cold waters, siliceous shells or brown clay accumulates. If the depositional site is sufficiently near a landmass, continental sediment may be carried into the deep oceanic basin by turbidity currents (see Chapter 4) and be interbedded with the calcareous and siliceous oceanic deposits.

Rifted Continental Margins

Rifted continental margins form in pairs when a divergent plate junction (such as the site of the Mid-Atlantic Ridge) forms within a continent (Pangea). The series of rifts on the western side of the Atlantic are shown in Figure 1–11; an equivalent set occurs in North Africa (Schlee, 1980). Grabens can form singly when magmatic arcs are rifted away from the margins of continental blocks by spreading behind the arcs. The Sea of Japan owes its origin to the rifting of the Japanese islands from the Asian mainland.

The first stage of rifting is thermal arching, typically associated with the extrusion of lavas rich in sodium and potassium. The balance between the rate of accumulation of such volcanics and the rate of erosion of the thermal arches they crown is uncertain; but when erosion predominates, uplifted terranes of granitic basement are prominent as sediment sources.

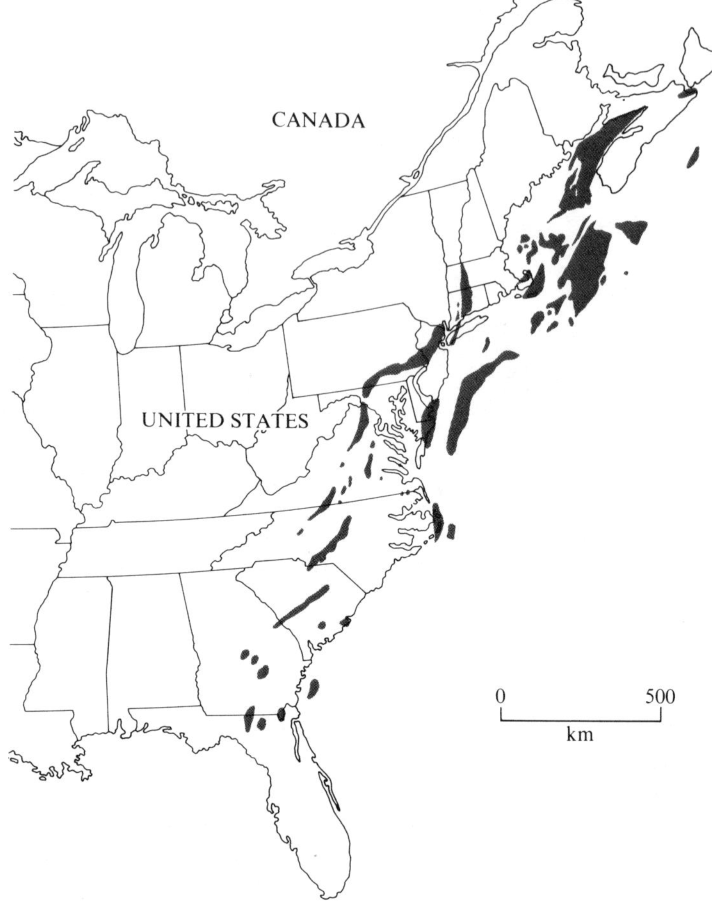

Figure 1–11
Map of eastern margin of United States and Canada, showing location of known Triassic fault basins (grabens) that originated about 30 million years before North Atlantic Ocean came into existence along site of Mid-Atlantic Ridge. Similar grabens occur in Triassic rocks of northwestern Africa. [F. B. Van Houten, 1977, *Amer. Assoc. Petroleum Geol. Bull., 61.*]

When sufficient crustal extension affects the arched region, rift valleys begin to form as graben (see Figure 1–12). Probably these develop first within the domal uplifts, but later extend as an essentially continuous branching network along the full trend of the rift belt. In the rift valleys, continental redbeds are interbedded with volcanics that continue to erupt through the growing system of crustal fractures. Broad regions to either side of the eventual zone of rupture between the separating continents can be affected by the extensional faulting. For example, the Triassic basins of the Appalachian region, which are filled with nonmarine sediment and volcanic flow rocks, lie 250–500 km inland from the present continental slope; the slope can be taken as marking roughly the line of Jurassic continental separation.

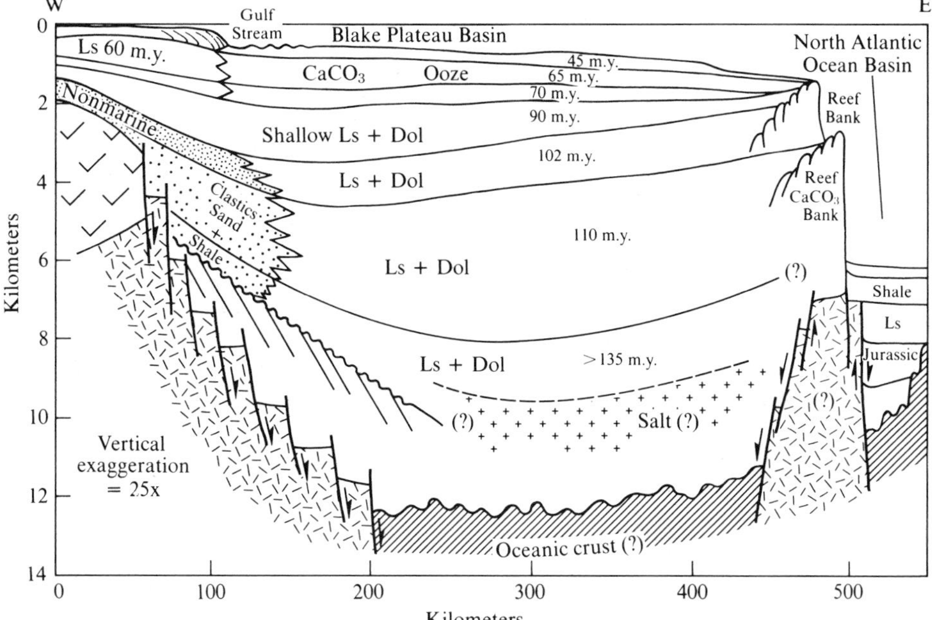

Figure 1–12
Diagrammatic cross section of Blake Plateau east of Florida, revealing underlying sediment accumulation to be a highly faulted graben of Triassic age, now 12 km deep. Formation of graben preceded opening of North Atlantic Ocean by 30 million years. Note salt deposit at base of section. Most of Gulf of Mexico basin also has salt immediately above Paleozoic basement rocks. Analogous grabens of Triassic age occur in North Africa, the other side of the rifted Paleozoic landmass. Ls = limestone, Dol = dolomite. [R. E. Sheridan, 1976, in M. H. Bott, ed., *Sedimentary Basins of Continental Margins and Cratons* (New York: Elsevier).]

As continued crustal separation induces subsidence along the zone of incipient continental rupture, the floors of the main rift valleys become partially or intermittently flooded to form proto-oceanic gulfs. Restricted conditions in these basins may promote the deposition of evaporites in suitable climates. For example, thicknesses of 5–7.5 km of evaporites are present in the subsurface beneath parts of the Red Sea. Extensive evaporites several thousand meters thick are known also from coastal basins on both sides of the Atlantic Ocean, where they apparently are correlative and represent dismembered portions of the same elongate trend of evaporite basins (see Figure 1–13).

Subsequent evolution of the rifted region is marked by the formation at the interface between continental and oceanic crust of a wedge of marine and nonmarine strata built upward to form an isostatically balanced continental terrace. The terrace develops on continental crust and extends to the slope break at the shelf edge, from which the continental slope leads down to deep water where turbidites accumulate along the edge of oceanic crust. The distance from the thin edge of the shelf-sediment wedge to the top of the slope is perhaps 100–250 km. Examples of such sediment wedges are the Lower Paleozoic carbonate–shale sections in the Appalachian and Cordilleran miogeosynclines.

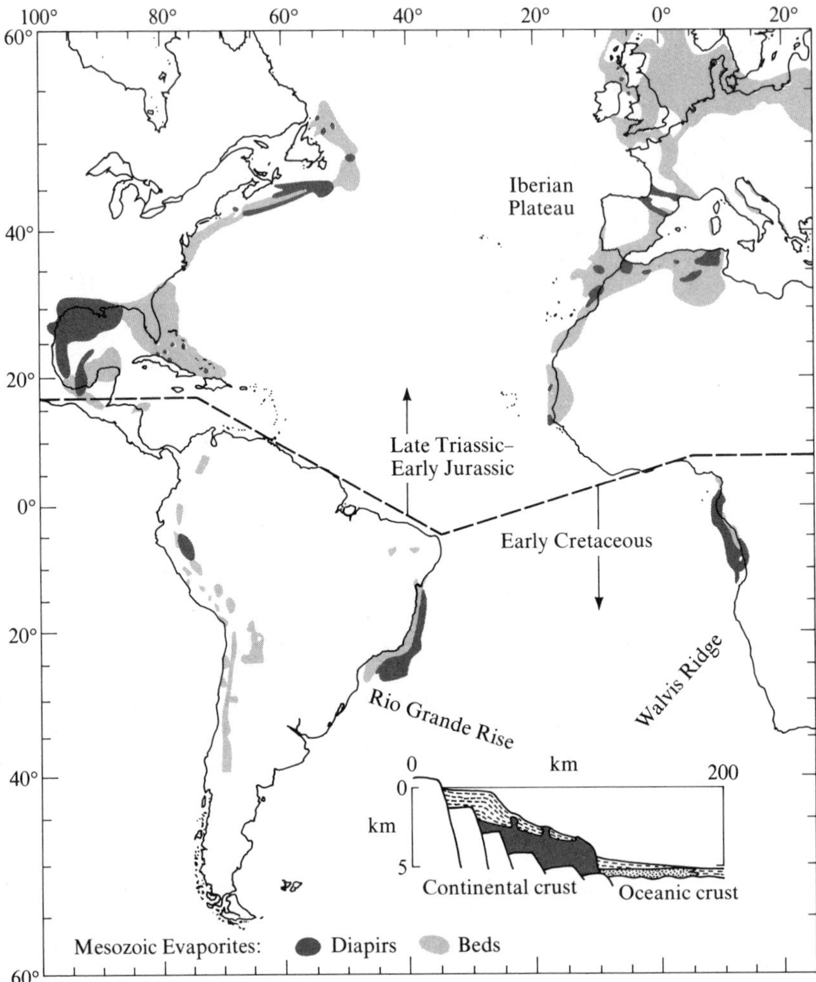

Figure 1–13
Distribution of Mesozoic evaporites and their relationship to continental margins of Atlantic Ocean. [K. O. Emery, 1977, Amer. Assoc. Petroleum Geol. Short Course Notes No. 5, *Geology of Continental Margins*.]

Arc-Trench System Basins

Arc-trench systems are a characteristic geologic expression of convergent plate junctions and are the depositional sites of most of the eugeosynclinal suite of sedimentary rocks. The system consists of five morphotectonic elements (see Figure 1–14), each of which may accumulate a different sedimentary assemblage:

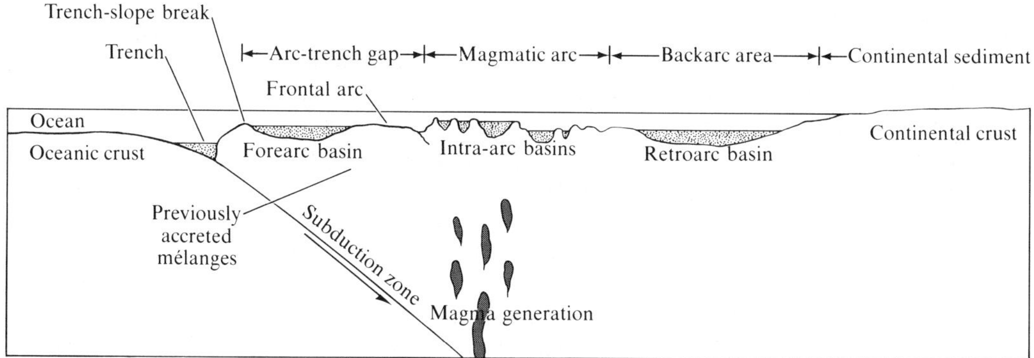

Figure 1–14
Generalized sketch of arc-trench system along convergent continental margin, showing spatial relationships and nomenclature of plate tectonics and related sedimentary basins. Marine sediment accumulations in basins are stippled.

1. The *trench,* a bathymetric deep, floored by oceanic crust.
2. The *subduction zone.*
3. The arc-trench gap, a belt within which a *forearc* (or frontal arc) *basin* may occur between the trench and the magmatic arc.
4. The magmatic arc, within which *intra-arc basins* may occur.
5. The backarc area, within which may lie either an *interarc basin* floored by oceanic crust and separated from the rear of the arc by a normal fault system or a *retroarc* (or backarc) *basin* floored by continental basement and separated from the rear of the arc by a thrust fault system. (A retroarc basin is retro with respect to the arc. The basin may also be termed a *foreland basin* because of its position with respect to the continent.)

In each of these morphotectonic settings, sedimentation, volcanism, and plutonic intrusions occur contemporaneously, although not necessarily at the same site.

Trench Sediment

On the trench floor, variable thicknesses of turbidites are deposited above the oceanic sediment layers. Transport by turbidity currents within a trench is mainly longitudinal along the trench axis, although the initial entry of sediment into the trench may occur along the inner wall as well as from the ends of the trench. The volume of sediment within the trench at a given time reflects the balance between the rate of supply and the rate of plate consumption into the subduction zone.

Forearc Basins

Immediately landward of the top of the trench lie forearc basins, which overlie older, deformed orogenic belts or perhaps oceanic or transitional crustal material. Forearc basins receive sediment mainly from the extensive nearby arc structures, where not only volcanic rocks but also plutonic and metamorphic rocks exposed by uplift and erosion may serve as sources. Sources may also include local uplands along the trench-slope break or within the arc-trench gap itself. There may be little transfer of sediment into the subduction zone from the forearc basins; frequently, the forearc basins seem to completely override the subduction zone. The Great Valley sequence of California is an example of a forearc-basin deposit.

By inference from the bathymetry of modern forearc basins and from the sedimentology of older sequences inferred to have been deposited in similar settings, forearc basins may contain a variety of facies. Shelf and deltaic or terrestrial sediments, as well as turbidites, may occur in different examples. The local bathymetry is controlled by the elevation of the trench-slope break, the rate of sediment delivery to the forearc basin, and the rate of basin subsidence. Various facies patterns occur in different basins.

Intra-Arc Basins

The sedimentary strata in modern intra-arc basins include turbidite aprons of volcanic debris shed backward from the rear sides of magmatic arcs. These turbidite wedges rest almost directly on the igneous oceanic crust with few or no intervening pelagic deposits present. Landward from the intra-arc spreading centers, sedimentation varies markedly.

Retroarc Basins

The sedimentary record of retroarc basins includes fluvial, deltaic, and marine strata as much as 5 km thick deposited in terrestrial lowlands and epicontinental seas along elongate, cratonic belts between continental margin arcs and cratons. Sediment dispersal into and across retroarc basins is from highlands on the side toward the magmatic arc and from the craton toward the continental side. The Sea of Japan is a modern example of an extensional retroarc basin. Ancient examples of compressional (thrust-faulted) retroarc basins are the Upper Cretaceous basins of the interior and Rocky Mountain region of North America.

Suture-Belt Basins

Suture belts form along lines of continental collision and, therefore, involve both continental and oceanic lithosphere. The Himalayan Mountain range along the site of suturing of the Indian subcontinent to the Eurasian continent is an example of a suture zone. Because such zones involve two continental masses and an intervening oceanic area, the zones contain deformed examples of all the various types of sedimentary sequences discussed in connection with oceanic basins, rifted-margin prisms, and arc-trench systems. In addition, the collision process can give rise to sedimentary basins located at the suture zone. These basins are generated by depression of the continental block by partial subduction. The sediment fill is clastic debris of continental origin, characteristically wedges of fluvial and deltaic strata.

Intracontinental Basins

Basins within the continental craton are difficult to explain in terms of activities at plate margins (Sleep, 1980). Examples of such basins include the Michigan Basin with a sediment accumulation 3,000 m thicker than its geographic surroundings and the Williston Basin in North Dakota and Montana, with 4,500 m more than its surroundings. The basins may have resulted from aborted continental rifting, local cooling in the asthenosphere, downbowing of the crust near convergent plate boundaries, or causes not now recognized.

The term *craton* is used generally to refer to tectonically "passive" parts of a continent, typically formed of Lower to Middle Precambrian igneous rocks and metamorphosed sedimentary rocks and overlain by essentially flat-lying Upper Precambrian or younger sedimentary rocks. Cratons consist of former geosynclinal deposits that have been accreted to earlier continental nuclei. A craton tends to be an area of positive relief; portions of it are generally exposed (shield area) even during times of maximum continental submergence. Cratonic sedimentary rocks are thin but laterally extensive (platform sediments), and contain many unconformities (see Figure 1–15). Both nonmarine and shallow marine deposits can occur, and the sandstones

Figure 1–15
Generalized cross section of cratonic Lower Paleozoic rocks in mid-continental North America. Carbonate rocks are shallow-water, fossiliferous limestones and dolomites containing many small reefs; sandstones are composed almost entirely of fine- to medium-grained quartz grains cemented by quartz and calcite. [P. E. Potter and W. A. Pryor, 1961, *Geol. Soc. Amer. Bull., 72.*]

are highly quartzose because of repeated and intense abrasion. Some of these well-rounded and uniformly sized quartz sands may be carried by river systems long distances into bordering mobile belts.

SUMMARY

Sedimentary rocks are composed almost entirely of mudrocks (65%), sandstones (20–25%), and carbonate rocks (10–15%). Clay minerals and detrital quartz grains form about 85% of the mineral grains in these rocks. The thickness of sedimentary rocks on the continents ranges from zero over extensive areas such as the Canadian Shield and the Siberian Shield to more than 20,000 m in the deepest parts of some geosynclinal areas; the lower limit is set by the local geothermal gradient and the susceptibility of the minerals to recrystallization.

Mudrocks are composed of 60% clay minerals that, because of their small grain size, can accumulate only in areas of low kinetic energy. Sandstones dominate in areas of high kinetic energy such as beaches and desert dunes. The occurrence of carbonate rocks is controlled primarily by the depth of penetration of light into the sea, so that most carbonate rocks accumulate within a few tens of meters of the sea surface.

The location and size of depositional basins are controlled by continental drift and plate tectonics. Five distinct areas of structurally controlled accumulation of sediments can be recognized: (1) oceanic basins, (2) rifted continental margins, (3) arc-trench systems, (4) suture belts, and (5) intracontinental basins. Numerous examples of each type are known from both modern and ancient examples. The mineral composition of the sediments in each type of basin is determined by its location with respect to a continental margin, the nature of the underlying crustal material, and the types of plate junctures nearest the basin.

FURTHER READING

Atkinson, D. J. 1962. Tectonic control of sedimentation and the interpretation of sediment alternation in the Tertiary of Prince Charles Foreland, Spitsbergen. *Geol. Soc. Amer. Bull., 73,* 343–364.

Bathurst, R. G. C. 1975. *Carbonate Sediments and Their Diagenesis,* 2nd ed. New York: Elsevier, 658 pp.

Burk, C. A., and C. L. Drake (eds.). 1975. *The Geology of Continental Margins.* New York: Springer-Verlag, 1,009 pp. A collection of 71 articles by various authors covering all aspects of the origin and development of continental margins.

Burke, K. 1976. Development of graben associated with the initial ruptures of the Atlantic Ocean. *Tectonophysics, 36,* 93–112.

Burke, K. 1977. Aulacogens and continental breakup. In F. A. Donath, F. G. Stehli, and G. W. Wetherill (eds.), *Ann. Rev. Earth Planet. Sci., 5,* 371–396. Palo Alto, CA: Annual Reviews, Inc.

Conant, L. C., and V. E. Swanson. 1961. *Chattanooga Shale and Related Rocks of Central Tennessee and Nearby Areas.* U.S. Geol. Surv. Prof. Paper No. 357, 91 pp.

Conybeare, C. E. B. 1979. *Lithostratigraphic Analysis of Sedimentary Basins.* New York: Academic Press, 555 pp.

Dickinson, W. R. 1974. Plate tectonics and sedimentation. In W. R. Dickinson (ed.), *Tectonics and Sedimentation.* Soc. Econ. Paleontol. Mineral. Spec. Pub. No. 22, pp. 1–27.

Dickinson, W. R., and D. R. Seely. 1979. Structure and stratigraphy of forearc regions. *Amer. Assoc. Petroleum Geol. Bull., 63,* 2–31.

Donovan, D. T., and E. J. W. Jones. 1979. Causes of world-wide changes in sea level. *Jour. Geol. Soc., 136,* 187–192.

Firstbrook, P. L., B. M. Funnell, A. M. Hurley, and A. G. Smith. Not dated but probably 1980. *Paleooceanic Reconstructions,* 160-0 Ma. Washington, DC: National Science Foundation, 41 pp.

Graham, S. A., R. V. Ingersoll, and W. R. Dickinson. 1976. Common provenance for lithic grains in Carboniferous sandstones from Ouachita Mountains and Black Warrior Basin. *Jour. Sed. Petrology, 46,* 620–632.

Hallam, A. 1981. *Facies Interpretation and the Stratigraphic Record.* San Francisco: W. H. Freeman and Company. 291 pp.

Hsü, H. J. 1973. The odyssey of a geosyncline. In R. N. Ginsburg (ed.), *Evolving Concepts in Sedimentology.* Baltimore: Johns Hopkins University Press, pp. 66–92.

Jenkyns, H. C. 1980. Cretaceous anoxic events: from continents to oceans. *Jour. Geol. Soc., 137,* 171–188.

Pettijohn, F. J., P. E. Potter, and R. Siever. 1973. *Sand and Sandstone.* New York: Springer-Verlag, 618 pp.

Ronov, A. B., V. E. Khain, A. N. Balukhovsky, and K. B. Seslavinsky. 1980. Quantitative analysis of Phanerozoic sedimentation. *Sedimentary Geol., 25,* 311–325.

Schlee, J. S. 1980. *A Comparison of Two Atlantic-Type Continental Margins.* U.S. Geol. Surv. Prof. Paper No. 1167, 21 pp.

Scotese, C., et al. 1979. Paleozoic base maps. *Jour. Geol., 87,* 217–277.

Shaw, D. B., and C. E. Weaver. 1965. The mineralogical composition of shales. *Jour. Sed. Petrology, 35,* 213–222.

Sleep, N. H. 1980. Platform basins. In F. A. Donath, F. G. Stehli, and G. W. Wetherill (eds.), *Ann. Rev. Earth Planet. Sci., 8,* 17–34. Palo Alto, CA: Annual Reviews, Inc.

Sloss, L. L., and R. C. Speed. 1974. Relationships of cratonic and continental margin tectonic episodes. In W. R. Dickinson (ed.), *Tectonics and Sedimentation.* Soc. Econ. Paleontol. Mineral. Spec. Pub. No. 22, pp. 98–119.

Thompson, S. L., and E. J. Barron. 1981. Comparison of Cretaceous and present Earth albedos: implications for the causes of paleoclimates. *Jour. Geol., 89,* 143–167.

Wilson, R. C. L., and C. A. Williams. 1979. Oceanic transform structures and the development of Atlantic continental margin sedimentary basins—a review. *Jour. Geol. Soc., 136,* 311–320.

2

The Formation of Sediment

In time the Rockies may crumble, Gibraltar may tumble—
They're only made of clay.

IRA GERSHWIN

Sediment originates at the Earth's surface because of the chemical and, to a lesser extent, mechanical instability of igneous and metamorphic rocks under atmospheric conditions. The variety of new substances produced by the chemical alteration depends on both the surface conditions (e.g., amount of water, temperature, availability of gaseous oxygen) and the chemical composition of the minerals being altered. For example, calcite cannot be produced by the alteration of orthoclase because there is no calcium in orthoclase. If calcite is to form, a calcium-bearing mineral such as plagioclase or hornblende must occur in the rock to supply calcium ions. In most igneous and metamorphic rocks a variety of minerals of greatly differing chemical composition is present, so that a variety of sedimentary minerals is produced at most geographic sites.

The disintegration and decomposition of exposed igneous and metamorphic rocks to form materials more stable under surface conditions is controlled very largely by plants, animals, and bacteria. Although chemical alteration can occur in a completely sterile environment as long as water is present, the rate of mineralogic change is much slower than in the presence of living organisms. Plants first colonized the land surface during the Silurian Period and doubtless were preceded by bacteria. It is fair to say that no sedimentary rock in the existing stratigraphic column has formed free of organic influences, although the degree of influence probably increased greatly about 420 million years ago. All sedimentary accumulations contain organic matter. Although living matter forms under atmospheric conditions, it is composed of highly organized constituents that can be maintained only during the life of the organism, so that sediments contain carbonaceous material in all stages of decomposition (weathering). In both modern and ancient sediments the entire range of altered tissue is found, from nearly intact tissue through partially decomposed parts of the plant structure to free amino acids and finally to essentially pure, free carbon (anthracite coal or graphite).

CHEMICAL WEATHERING

The degree of influence that chemical weathering has on the destruction of igneous and metamorphic rocks and on the formation of sediments is determined basically by climate. Climate controls the temperature, availability of water, and extent of plant cover. The plant cover, in turn, controls the acidity of the soil water and the intensity of leaching through its release of organic acids and carbon dioxide gas during metabolism and decay. Plant decay in a temperate climate typically causes the percentage of carbon dioxide in soil gas to be 10–100 times greater than the percentage in the atmosphere. This gas combines with the soil moisture to form the carbonic acid that is important in the leaching process.

Granite

If we examine a new roadcut in granite in a humid temperate area, one thing we notice immediately is the freshness of the outcrop face. The nearly vertical surface has a micro-jagged relief, and a view through a hand lens reveals the sharp edges and corners of the individual crystals. Cleavage faces in minerals such as feldspars and hornblende are shiny and clear; magnetite is coal-black and unstained; and the concentric, semicircular ridges of the conchoidal fracture of quartz are clearly visible. There is no evidence of life visible on the outcrop surface.

A return to the same cliff face a few months or years later reveals a different scene. The near-vertical face is partly covered by vegetation. A mixture of brown and black stain has spread over the outcrop. Examination of the surface with a hand lens reveals no sharp, jagged corners on crystals, but instead softer and duller crystal surfaces. Fragments of granite lying on the ground must be broken with a rock hammer to obtain a surface clean and clear enough to see cleavages or the true colors of the minerals. The wavy fracture surfaces of quartz crystals now lack the sharp edges of the original conchoidal fractures. Surficial staining is widespread.

What has caused these changes? The most obvious sites of chemical alteration occur along crystal boundaries, twin composition surfaces, cleavage planes, and fractures in the rock. Apparently, water has percolated along cracks and weakly bonded surfaces, creating a trail of clay minerals and iron oxides in its wake. Some minerals are more altered than others. Hornblende, biotite, and plagioclase are in worse condition than are orthoclase and muscovite crystals; quartz shows few ill effects. More detailed study of this outcrop and others in which a variety of mineral types is present

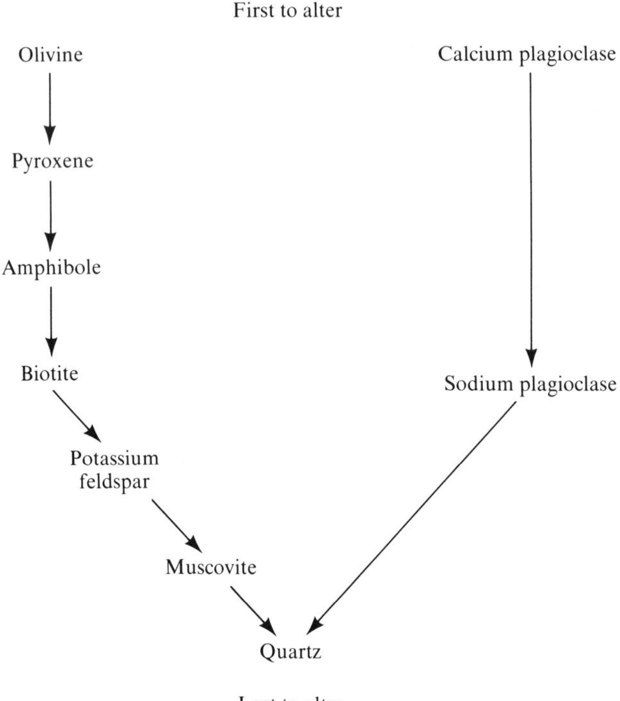

Figure 2–1
Relative rates of chemical alteration during weathering, based on field and laboratory data.

reveals that a consistent ordering is present in the degree of chemical alteration (see Figure 2–1). It is apparent that the sequence is exactly the same as the sequence of crystallization of minerals from a basaltic magma–Bowen's Reaction Series. In weathering, however, augite does not turn into hornblende; nor does hornblende alter to biotite. As we will see, all three are converted during weathering into the same materials; only the rates of conversion differ.

The earliest fairly thorough field study of mineralogic changes in granitoid rock during weathering was made by Goldich in 1938. Goldich studied mineralogic and chemical changes in a granitic gneiss, two diabases, and an amphibolite by examining both the unaltered rocks and their weathering products. Based on his observations, Goldich established the weathering stability series shown in Figure 2–1, and these results have since been duplicated in many other weathering profiles. Goldich also determined the relative rates of loss of elements from the rocks he studied. Most rapid losses were recorded for sodium and calcium, then potassium and magnesium; weathered residues were relatively enriched in water, titanium, aluminum, and silicon. The explanation for the relative mobility of these elements is the type of bond they form with oxygen in the mineral crystal structure. Only the titanium, silicon, and aluminum form bonds with oxygen that are dominantly covalent in character. The other four elements form bonds that are predominantly ionic, and ionic bonds are more easily broken by the force of the dipolar molecules of which water is composed. (The elements in evaporite minerals are also joined by ionic bonds, which is the reason for their great solubilities.)

Goldich observed that the orthoclase crystals contained a potassium-rich phyllosilicate mineral (sericite or illite) and inferred that it had formed by alteration of the orthoclase. We can write this reaction as follows:

$$\underset{\text{orthoclase}}{3\ KAl\ Si_3O_8} + 2\ H^+ + 12\ H_2O \rightarrow \underset{\text{muscovite}}{KAl_3\ Si_3O_{10}(OH)_2} + \underset{\text{soluble silica}}{6\ H_4SiO_4} + 2\ K^+$$

(Sericite and illite are not chemically well-defined substances but are approximately the composition of muscovite with slightly less potassium.) Goldich also noted that the most abundant clay mineral formed from the orthoclase was kaolinite. This can be produced by continued alteration of the muscovite:

$$\underset{\text{muscovite}}{2\ KAl_3\ Si_3O_{10}(OH)_2} + 2\ H^+ + 3\ H_2O \rightarrow \underset{\text{kaolinite}}{3\ Al_2\ Si_2O_5(OH)_4} + 2\ K^+$$

It appears, then, that the weathering of potassium feldspar produces a clay mineral of some kind, silica in solution, and potassium ions. We can write the reaction for the weathering of plagioclase feldspar as

$$\text{albite} + H_2O + H^+ \rightarrow \text{sodium montmorillonite} + H_4SiO_4 + Na^+$$

$$\text{anorthite} + H_2O + H^+ \rightarrow \text{calcium montmorillonite} + H_4SiO_4 + Ca^{2+}$$

In fact, both field observations and laboratory experiments reveal that all the abundant minerals (except quartz) in igneous rocks alter during weathering to (1) phyllosilicate minerals, (2) silica in solution, and (3) alkali and alkaline earth cations (see Table 2–1). This common pattern is violated only by iron released from ferromagnesian minerals. The iron in ferromagnesian minerals is present mostly in the ferrous form; but on release from the crystal structure, it oxidizes immediately to the very insoluble ferric ion, which hydrates and precipitates in place as $Fe(OH)_3$. For example,

$$\underset{\text{augite}}{(Ca,Na)(Mg,Fe,Al)(Si,Al)_2O_6} + H^+ + H_2O \rightarrow$$

$$\text{sodic and calcic montmorillonite} + H_4SiO_4 + Na^+ + Ca^{2+} + Fe(OH)_3$$

This brown, amorphous substance subsequently dehydrates to the red mineral hematite, Fe_2O_3:

$$2\ Fe(OH)_3 \rightarrow Fe_2O_3 + 3\ H_2O$$

Table 2-1
Summary of Weathering Reactions of Common Minerals in Igneous Rocks

Input		Output	
Mineral	Others	Phyllosilicate or clay mineral	Others
Potassium feldspar	H_2O, H^+	Illite, sericite (muscovite)	Dissolved silica Potassium ions
Sodium feldspar	H_2O, H^+	Sodium montmorillonite	Dissolved silica Sodium ions
Calcium feldspar	H_2O, H^+	Calcium montmorillonite	Dissolved silica Calcium ions
Pyroxenes Amphiboles Biotite	H_2O, H^+	Calcium–sodium montmorillonite	Dissolved silica Calcium ions Sodium ions Magnesium ions Ferric hydroxide (precipitate)
Olivine	H_2O, H^+	Serpentine (antigorite + chrysotile)	Magnesium ions Ferric hydroxide (precipitate)

Bustin and Mathews (1979) studied the effects of the alteration of biotite in granite on the destruction of the rock. They found biotite alteration to be the most notable mineralogical change in the rock. Fresh biotite is black, shiny, and dense; but with increasing weathering and chemical alteration, the color changes progressively through dark yellow-brown to golden yellow. In large flakes the alteration of biotite results in a zoning of color from a golden-yellow rim to a dark-brown core. In thin section, leaching of iron from the biotite is visible along cleavage planes and at crystal edges (see Figure 2–2), and the amorphous character of the iron oxide on biotite surfaces is plainly visible with a scanning electron microscope (see Figure 2–3). With the removal of iron and potassium from its edges, the biotite exfoliates (see Figure 2–4) and is converted into a randomly interstratified biotite-vermiculite and pure vermiculite (pure Mg–mica), as revealed by X-ray diffraction studies. The physical expansion of the biotite causes microfractures to form in the granite, greatly acceler-

Figure 2–3
Scanning electron photomicrograph of (010) face of partially exfoliated biotite crystal, showing development of apparently noncrystalline lumps of iron oxide. [Bustin and Mathews, 1979. Photo courtesy R. M. Bustin.]

Figure 2–2
Thin-section photomicrograph showing development of iron oxide along biotite surfaces and at edges, frayed (exfoliated) edges of crystal, and development of microfractures filled with iron oxide in adjacent quartz crystal. [Bustin and Mathews, 1979. Photo courtesy R. M. Bustin.]

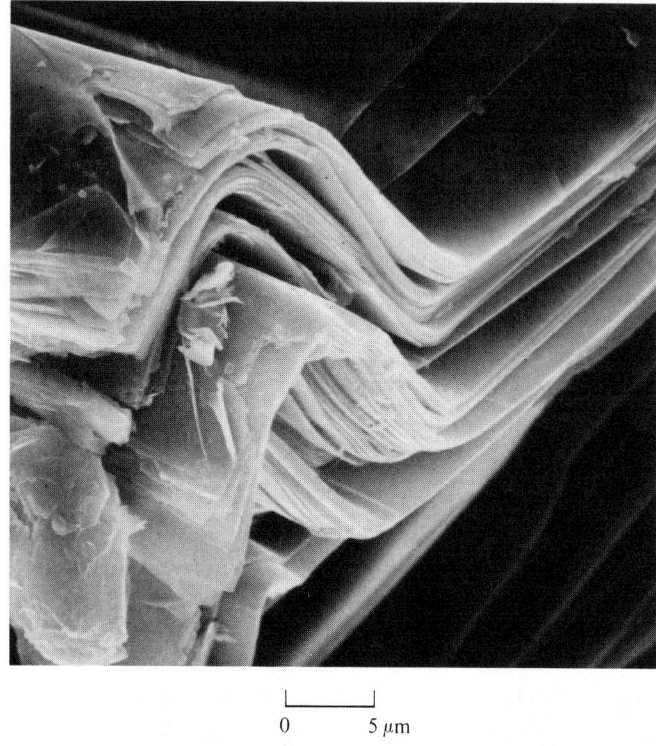

Figure 2–4
Scanning electron photomicrograph of partially exfoliated biotite, showing expansion along cleavage planes and accompanying folding of crystal edges. [Bustin and Mathews, 1979. Photo courtesy R. M. Bustin.]

ating its decay by percolating meteoric water. In more advanced stages of decomposition, the biotite exfoliates rather completely and is converted through extensive leaching into kaolinite and gibbsite (see Figure 2–5). Table 2–1 summarizes the products generated by weathering of the common silicate minerals.

Basalt

The weathering of basalt follows the same pattern as the weathering of granite. The elemental composition of the input minerals determines completely the types of substances produced as output. The only differences between the weathering of granite and that of basalt are (1) the rate of alteration of the input minerals and (2) the proportions of the different substances produced.

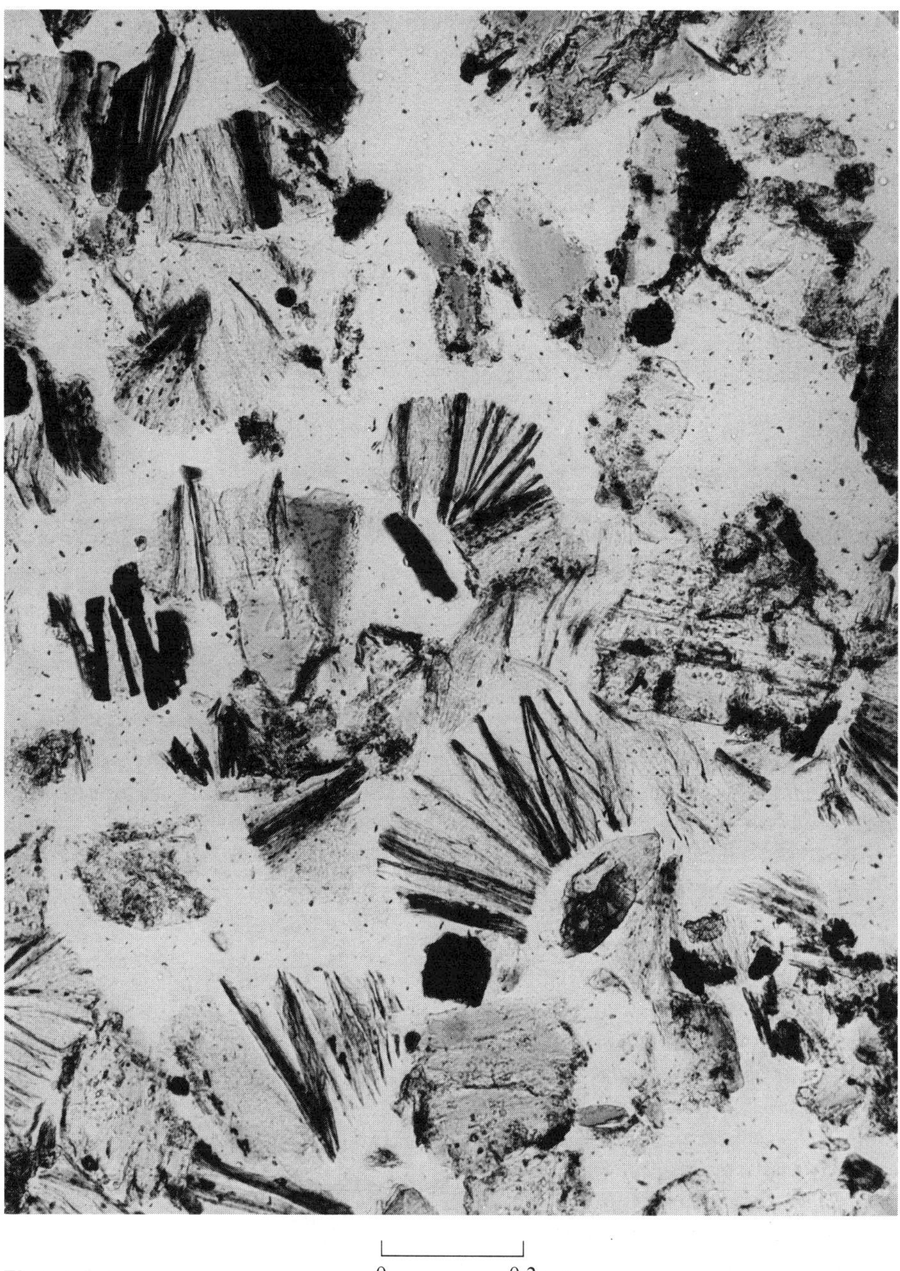

0 0.2mm

Figure 2–5
Thin-section photomicrograph of grains from naturally disintegrated gabbro in northeastern Scotland, showing kaolinite–gibbsite vermiforms (light-colored) developing from iron-stained biotite. [Basham, 1974. Photo courtesy I. R. Basham.]

Mafic rocks such as basalt and andesite are altered chemically at a much faster rate than is granite for several reasons:

1. Calcic plagioclase and pyroxenes contain less silicon than do alkali feldspars and biotite, so that fewer of the strong silicon–oxygen bonds are present to resist chemical attack.

2. Chemical attack occurs along surfaces. Crystals in basalt are smaller than those in granite; thus, they have higher surface/volume ratios.

3. Volcanic flow rocks chill rapidly on leaving the site of eruption and, as a result, commonly have a glassy groundmass as well as cracks due to thermal contraction. Glass is amorphous, and amorphous substances alter much faster than crystalline ones because the ions and ionic groupings in them are disorganized and hence bonded less strongly.

Because mafic rocks such as basalts and andesites differ a great deal in composition from granites, we can anticipate a parallel difference in the proportions of output substances. Basalts will produce no illite because they lack potassium, much more calcic montmorillonite and hematite because they are calcium- and iron-rich rocks, more calcium and magnesium and less silica in solution.

Vegetation

Inorganic chemical weathering on the Earth has been minimal since the evolution of land plants about 420 million years ago. There is little doubt that the presence of plants has a pronounced effect on weathering processes. Plants evolve both oxygen and carbon dioxide during their metabolism; and when they decay after death, they produce additional carbon dioxide from the breakdown of their organic tissues. Perhaps of even greater importance is the fact that plant decay generates organic acids, and the structure of these organic molecules enables them to act as chelating agents. *Chelation* involves the holding of an ion, usually a metal, within a ring structure. The ion is held (sequestered) by more than one chemical bond (see Figure 2–6) and is completely removed from direct contact with the soil water. The water does not "see" the magnesium ion in Figure 2–6 and hence is able to dissolve additional magnesium from a magnesium-bearing mineral, increasing the apparent solubility of the mineral. All ions can be chelated: silicon, aluminum, iron, magnesium, calcium, sodium, and potassium. Smaller, more highly charged ions are chelated more strongly. Chelating agents can extract ions from otherwise insoluble solids and enable the transport of ions in chemical environments where they normally would be precipitated.

A graphic illustration of the effect of vegetation on the rate of weathering was provided by Jackson and Keller (1970) in a field and laboratory investigation in Hawaii. They studied the weathering of lava flows of known composition and age, some parts of which were colonized by the lichen *Stereocaulon vulcani* and parts of

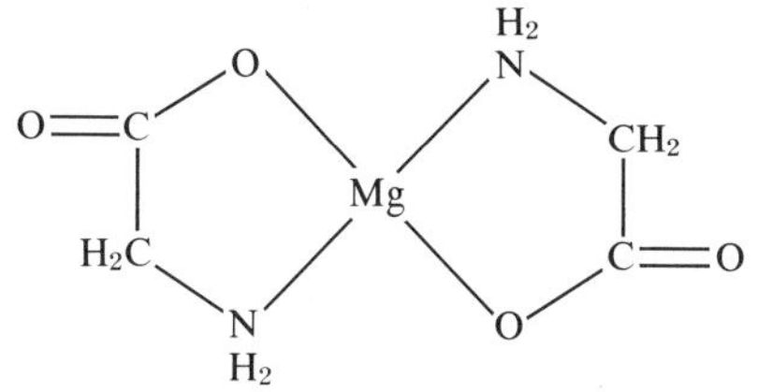

$$
\begin{array}{c}
\text{Chelate ring structure: } O{=}C(-O-)(-CH_2-NH_2-)\ \text{Mg}\ (-NH_2-CH_2-)C(=O)(-O-)
\end{array}
$$

Figure 2–6
Chelation of magnesium ion by two molecules of glycine, $H_2N \cdot CH_2 \cdot COOH$.

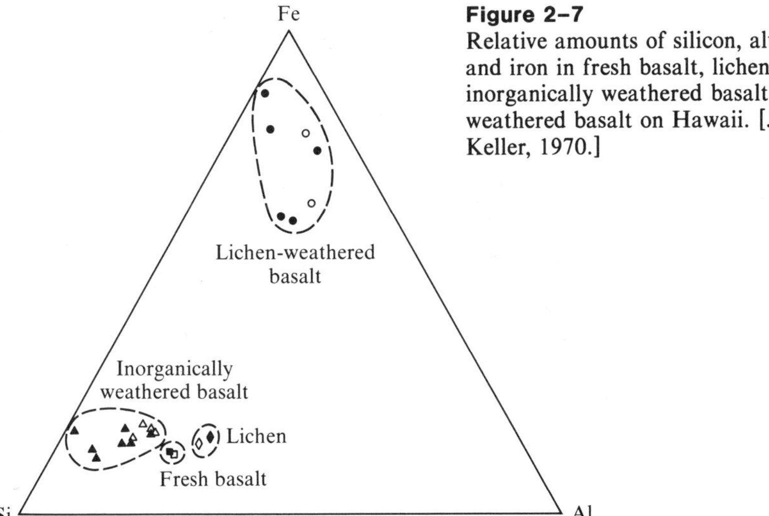

Figure 2–7
Relative amounts of silicon, aluminum, and iron in fresh basalt, lichen tissue, inorganically weathered basalt, and lichen-weathered basalt on Hawaii. [Jackson and Keller, 1970.]

which were bare. They found that the weathering crust (i.e., soil) on the colonized rock was 10–100 times thicker than on the bare rock. The crust was particularly depleted in silicon compared to the fresh tholeiitic basalt parent rock (see Figure 2–7) because this element is especially susceptible to being chelated by organic molecules; silicon has a high ionic charge and small size compared to the other ions in the basalt. The lichen-free soil contains 5–10 times more silicon than does the soil formed by the lichens. Both the common occurrence of dissolution of quartz near coal beds and the presence of abundant silicified wood in some formations testify to the strong tendency toward chemical interaction between organic molecules and silicon.

Many of the organic compounds dissolved in soil water and groundwater have this ability to chelate. In fact, a large proportion of some economically important elements (e.g., copper and manganese) is transported in streams in chelate compounds, to be precipitated when the organic matter decomposes. Primitive plants such as lichens are particularly adept at snatching metal ions out of bare rock, although seed plants are able to perform this feat as well. Thin skins of lichen can commonly be seen as the first colonists on the floors of granite quarries.

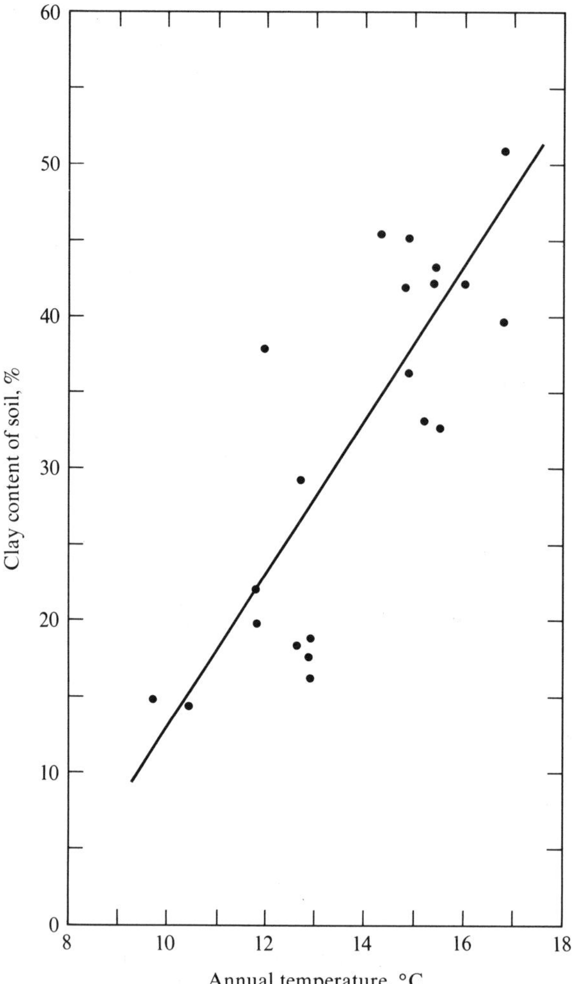

Figure 2–8
Clay content of residual soils developed in diorite and gabbro along north–south line from New Jersey to Georgia, as function of mean annual temperature. Mean annual rainfall is fairly constant at 100–125 cm. Clays are mostly montmorillonite in north and kaolinite in south. [H. Jenny, 1941, *Factors in Soil Formation* (New York: McGraw-Hill).]

Temperature

Temperature plays an important role in chemical weathering. Increasing temperatures promote weathering by greatly accelerating the rate of leaching and formation of clay minerals (see Figure 2–8). In addition, high temperatures permit the formation of substances that do not appear under cooler conditions. In this regard, it must be remembered that the latitudinal positions of the continents have varied greatly through geologic time (see Figure 2–9). In addition, variations in solar radiation can cause climatic changes independently of latitude (Dury, 1971), as appears to have

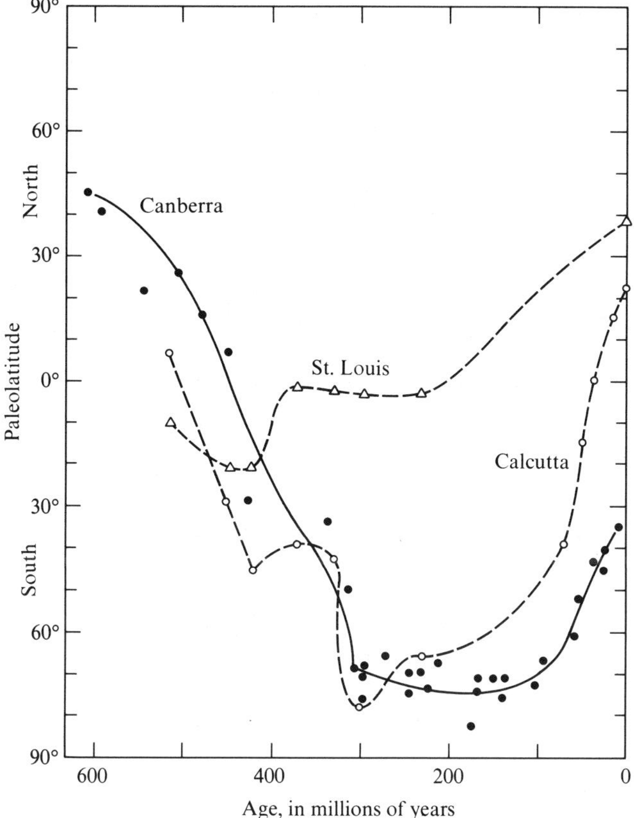

Figure 2–9
Phanerozoic paleomagnetic latitudes of Canberra, Australia; Calcutta, India; and St. Louis, United States. Solid circles, open circles, and triangles are data points for best-fit curves shown. [W. B. Hamilton, U.S. Geol. Surv. Prof. Paper No. 1078, 1979; and data from C. Scotese et al., *Jour. Geol., 87,* 1979, and P. Tapponnier, *Scientific American, 236,* 1977.]

been the case during Pleistocene time. It is noteworthy that bauxite pebbles of probable Early Tertiary age have been found in Pleistocene sediments in Massachusetts (42°N), and Tertiary laterite profiles are preserved beneath lavas in Ireland (53°N). Certainly, neither bauxite nor laterite could form at these latitudes today, and the amount of latitudinal continental drift during the Tertiary is inadequate to account for the presence of these tropical soils.

SUBMARINE "WEATHERING"

We normally think of weathering as occurring, by definition, at the air–rock interface. However, essentially the same process can occur on the seafloor at the water–rock interface, where it is usually termed *submarine alteration.* When it occurs near rift zones in the oceanic basins, the alteration is called *hydrothermal* in recognition of the fact that the liquids and gases issuing from the rifts are at temperatures of several hundred degrees Celsius. There has been only one study of undersea "weathering" of a granitoid rock, a granodiorite at the continental margin off the coast of southern California. It was found that the alteration of the rock was very similar to the alteration that would have occurred in contact with the atmosphere. The alteration products were iron oxide, sericite, montmorillonite, and kaolinite—as would be expected in a rock composed of quartz, orthoclase, plagioclase, and ferromagnesian minerals. An assemblage of plants and animals was growing attached to the undersea rock wall, and it probably played a part in the alteration process.

Because of the existence of spreading centers and fracture zones throughout the world oceans, basalt either is exposed on the seafloor or lies at a shallow depth beneath the water–sediment interface over much of the ocean floor. In either case, the basalt is subject to chemical alteration by circulating seawater. The most abundant new mineral produced is a potassium montmorillonite. The montmorillonite is the expected clay mineral from the weathering of a plagioclase–augite rock; the potassium is supplied by the ocean water. Also produced are small amounts of analcite ($NaAlSi_2O_6 \cdot H_2O$) and phillipsite ($(\frac{1}{2}K, Na, Ca)_3\ Al_3Si_5O_{16} \cdot 6H_2O$. Both of these minerals seem reasonable as results of the interaction between basalt and seawater.

Certain areas of the seafloor are notably hotter than others, presumably because of the presence of magma at shallow depth below the water–rock (sediment) interface. The seafloor rocks and the water in them are warmed, which results in the movement of hot water through the fractured rocks or sediments above. Cold seawater is drawn down into the sediment to establish a convective pattern, as in a pot of water heating on a stove. The hot water reacts strongly with the seafloor basalt and takes into solution a variety of minor elements present in the rock. The rising hot water is cooled and diluted by the addition of cold seawater, resulting in the precipitation of elements such as copper, nickel, cadmium, and others as sulfides in fractures. Some economic mineral deposits, such as the copper deposits in Cyprus, may have formed in this manner.

Metalliferous deposits of economic value formed in association with ocean-floor spreading centers and fracture zones need not be confined to vein occurrences. An excellent example of blanket-type deposits is provided by the deposits forming today on the floor of the Red Sea. The minerals in the Red Sea are disseminated as sulfides in a widespread sediment blanket. As the rising solutions cool in contact with seawater, they spread along the seafloor, combine with sulfide ions either from the brine itself or from bacterial reduction of the sulfate in seawater, and are deposited in the enclosing detrital or carbonate sediments. Such deposits are known from the Miocene of Japan and are suspected as the explanation of metalliferous deposits in other areas as well.

The manganese nodules that cover some parts of the floor of the Pacific Ocean result from enrichment of oceanic bottom water in manganese from basalts and rift-zone fluids. The nodules contain an abundance of other valuable metals as well, and numerous proposals for mining the nodules have been made in recent years.

THE NATURE OF CLAY MINERALS

Clay minerals are the most common mineral group in sedimentary rocks, totaling about 45% by weight or volume. They are very small particles, commonly less than 1 μm in size (because of their defect structures), and are the major constituent in clay-size sediment. In addition, they are the only abundant minerals in detrital rocks that are not inherited from igneous or metamorphic parents. The particular variety of clay mineral formed reflects the mineral weathered to produce it, but the clay-mineral structure is generated in the sedimentary environment.

The mechanism by which the phyllosilicate sheet structure of clays is produced from the tectosilicate structure of a feldspar or the inosilicate structure of an amphibole or pyroxene is unknown. The scale of the transformation is too small to be traced by thin-section studies and can be followed only by using an electron microscope in combination with detailed chemical analysis. But even with these techniques, there are great difficulties. Microchemical analysis of a particle less than 1 μm in size is not easy, even with the most sophisticated modern instruments. Figure 2–10 shows the appearance of kaolinite flakes produced on a feldspar surface; there are no visible

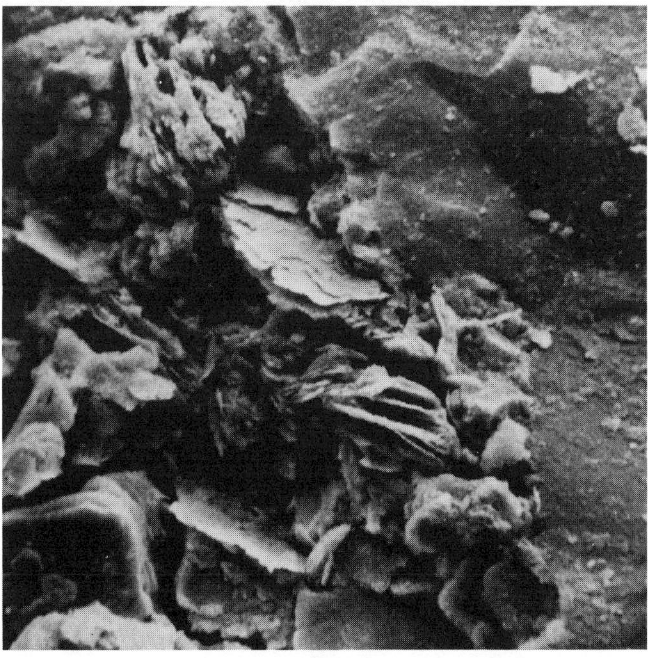

0 1 μm

Figure 2–10
Contact of orthoclase crystal and sheaves of kaolinite in weathering horizon of Butler Hill Granite (Precambrian), Missouri. Photo taken with scanning electron microscope. [Keller, 1978. Photo courtesy W. D. Keller.]

intermediate steps. At the scale of either a petrographic microscope or a scanning electron microscope, we see a sharp contact between the tectosilicate structure of feldspar and the sheet structure of a clay mineral. More sophisticated chemical techniques reveal the existence of a leached layer at the exterior of the feldspar that is depleted in potassium ions, but the crystal structure of the layer is still that of feldspar.

THE STRUCTURE OF CLAY MINERALS

The major clay minerals are kaolinite, montmorillonite, and illite. Nearly all clay-bearing sedimentary rocks contain more than one type of clay mineral, and this fact, coupled with their very small size and similar optical properties, requires the use of X-ray diffraction techniques to identify them. These techniques are standardized and provide unequivocal identification of the various kinds of clay minerals. In clay mixtures, however, the estimation of relative percentages of each clay mineral is only semiquantitative. Estimates to within 10% are acceptable for most purposes.

The crystal structure of clay minerals is the same as that of micas: sheeted-layer structures with strong bonding (covalent) within each sheet and among the sheets, but weak bonding (hydrogen bonds and van der Waals bonds) between the adjacent two- or three-sheet layers. The weak bonding between layers permits not only the excellent cleavage of clays but also the adsorption of metallic cations and organic substances on clay-mineral surfaces. This latter factor is important in chemical reactions that occur in weathering horizons.

The clay-mineral structure contains two types of sheets. One is composed of tetra-

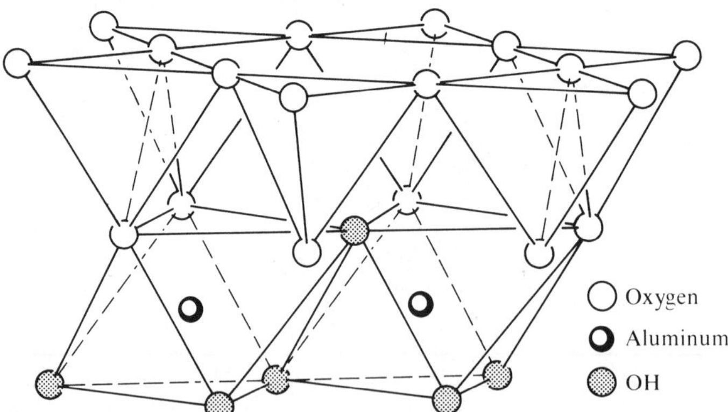

Figure 2–11
Diagrammatic sketch of structure of kaolinite with tetrahedral sheet bonded on one side to octahedral sheet. [C. S. Hurlburt, Jr., and C. Klein, 1977, *Manual of Mineralogy,* 19th ed. (New York: Wiley).]

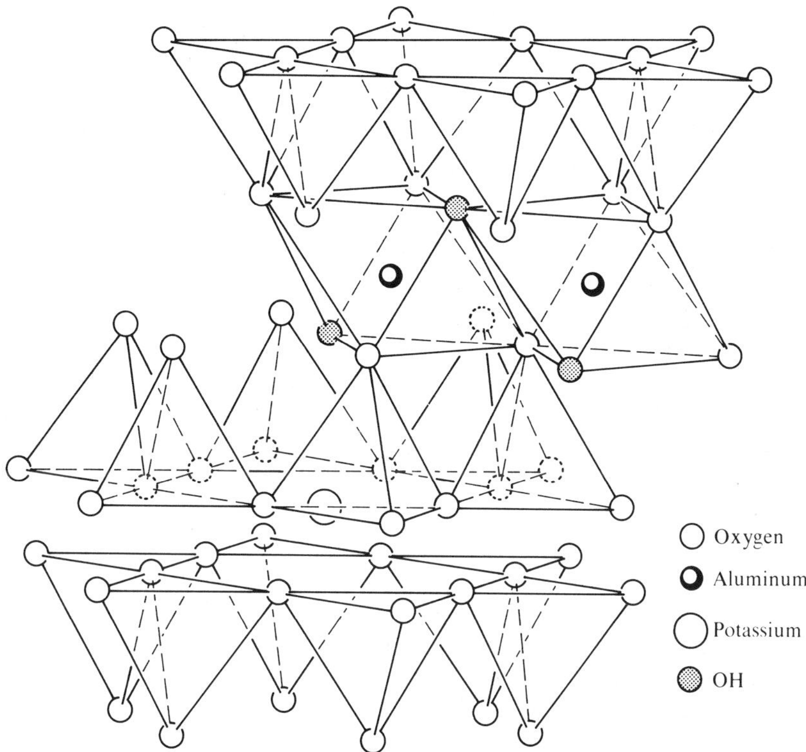

Figure 2–12
Diagrammatic sketch of muscovite (illite) structure and position of required interlayer cation. [C. S. Hurlburt, Jr., and C. Klein, 1977, *Manual of Mineralogy,* 19th ed. (New York: Wiley).]

hedrally coordinated Si–O and Al–OH groups in which three of the four oxygen atoms of each group are shared with adjoining groups. The cations in this sheet are at least 50% silicon atoms; the proportion of aluminum is greatest in illite and least in kaolinite. The second type of sheet is composed of Al–OH or Mg–OH groups in octahedral coordination, with ferrous iron sometimes substituting for magnesium. A single flake of kaolinite consists of one Si–O tetrahedral sheet and one Al–OH octahedral sheet (see Figure 2–11) with essentially no cation substitution in either sheet. All other clays (and micas) are composed of three sheets: an octahedral sheet sandwiched between two tetrahedral sheets (see Figure 2–12). There is abundant substitution within all three sheets: aluminum for silicon in the tetrahedral sheets, magnesium and ferrous iron for aluminum in the octahedral sheet. These substitutions cause charge imbalances within the sheets (Al^{+3} versus Si^{+4}; Mg^{+2} and Fe^{+2} versus Al^{+3}) that are rebalanced by the adsorption of metallic cations on the surfaces of each clay flake (see Figure 2–12). The cations adsorbed are those available in soil waters: the potas-

Table 2–2
Abundant Phyllosilicates in Rocks and Weathering Horizons, Showing Their Idealized Compositions[a]

Name	Chemical formula	Comment
	Two-sheet structure	
Kaolinite	$Al_2Si_2O_5(OH)_4$	Almost no substitution
Antigorite	$Mg_3Si_2O_5(OH)_4$—platy	Forms from serpentine—no Al is present
Chrysotile	$Mg_3Si_2O_5(OH)_4$—fibrous	Forms from serpentine—no Al is present
	Three-sheet structure	
Pyrophyllite	$Al_2Si_4O_{10}(OH)_4$	Almost no substitution
Montmorillonite	$Al_2Si_4O_{10}(OH)_2 \cdot xH_2O$	Mg may partly replace Al; interlayer Na and Ca present
Muscovite (illite)	$KAl_2(AlSi_3O_{10})(OH)_2$	In illite, Mg, Fe partly replace octahedral Al; interlayer K present
Talc	$Mg_3Si_4O_{10}(OH)_2$	
Vermiculite	$Mg_3Si_4O_{10}(OH)_2 \cdot xH_2O$	
Phlogopite	$KMg_3(AlSi_3O_{10})(OH)_2$	
Biotite	$K(Mg,Fe)_3(AlSi_3O_{10})(OH)_2$	
Chlorite	$Mg_5Al(AlSi_3O_{10})(OH)_8$	

[a]The Si_2O_5 or Si_4O_{10} grouping (± Al substituting for Si) is the tetrahedral sheet. The other cations and hydroxyl are the octahedral sheet.

sium that was released from orthoclase, the sodium and calcium that were released from plagioclase and ferromagnesian minerals. Illite prefers potassium; montmorillonite prefers sodium and calcium. These relationships are summarized in Table 2–2.

CLAY MINERALOGY AND CLIMATE

The relationship between the composition of the mineral being weathered and the composition of the initial clay mineral produced is clear. Orthoclase is a potassic mineral and produces a potassic clay (illite). Plagioclase, amphiboles, and pyroxenes are sodic and calcic and produce sodic or calcic clay (montmorillonite). But what can we expect to occur as the initially formed clay minerals are attacked by the meteoric waters passing over them after they form? Will they be transformed into different clay minerals? What will happen if the intensity of weathering increases because of a climatic change; for example, increased temperature caused by changes in solar radiation as during recent interglacial epochs or increased temperature resulting from drifting of the continent from higher to lower latitudes as during the Permo-Triassic breakup of the supercontinent Pangaea?

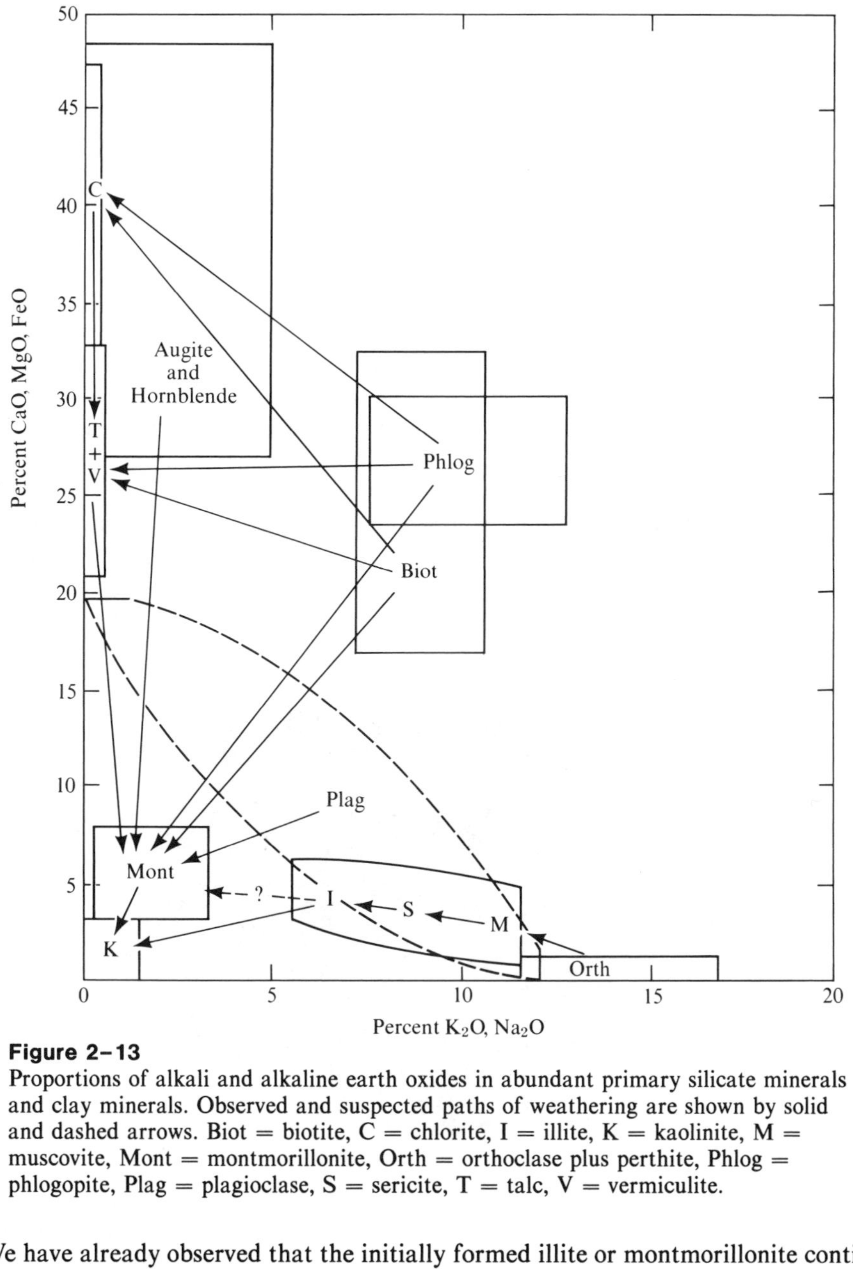

Figure 2–13
Proportions of alkali and alkaline earth oxides in abundant primary silicate minerals and clay minerals. Observed and suspected paths of weathering are shown by solid and dashed arrows. Biot = biotite, C = chlorite, I = illite, K = kaolinite, M = muscovite, Mont = montmorillonite, Orth = orthoclase plus perthite, Phlog = phlogopite, Plag = plagioclase, S = sericite, T = talc, V = vermiculite.

We have already observed that the initially formed illite or montmorillonite continues to lose potassium, sodium, or calcium as weathering proceeds. The reason this occurs is that the K^+, Na^+, and Ca^{2+} initially released from orthoclase, plagioclase, and ferromagnesian minerals are carried away in solution. Consideration of chemical equilibrium tells us that if the products of a reaction are removed, more of them will be produced. Eventually, a clay mineral containing only aluminum and silicon as cations will remain: kaolinite. The weathering paths of the primary silicate minerals are shown in Figure 2–13. Average weathering residues should contain more montmorillonite than illite, because the total volume of plagioclase plus ferromagnesian minerals in crystalline rocks is greater than the volume of orthoclase plus sodium-rich plagioclase. In a warm, humid climate with good drainage, such as might occur in South Carolina or Louisiana, kaolinite should be (and is) the most abundant clay mineral produced.

Laterite

Suppose the intensity of weathering becomes extreme, as in the Amazon River Basin or southeast Asia: extremely high rainfall and good drainage, coupled with very high mean annual temperature. Under these conditions, even the kaolinite is unstable and silica is leached from the clay mineral to leave an aluminous residue of amorphous and crystalline $Al(OH)_3$, the mineral gibbsite. Such a soil residue is called either an aluminous laterite or bauxite (see Figure 2–14) and is the world's major source of aluminum. The major economic deposits are located in Jamaica, Surinam, Guinea,

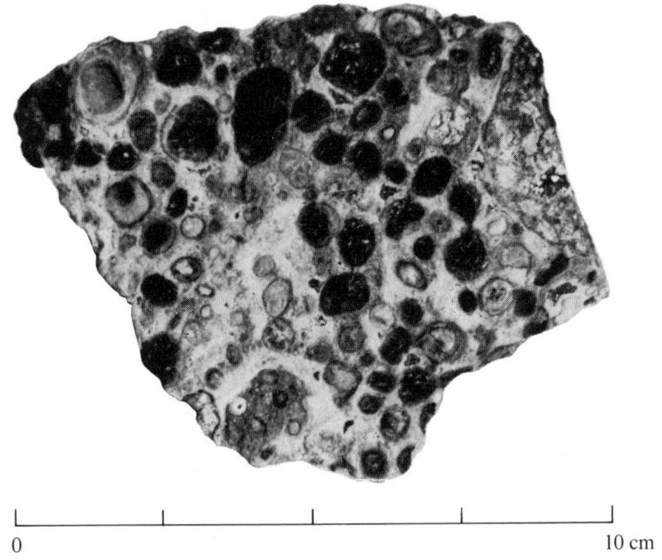

Figure 2–14
Specimen of bauxite, showing characteristic pisolitic structure formed by accretion of hydrated aluminous compounds around nuclei. [A. C. Tennissen, 1974, *Nature of Earth Materials* (Englewood Cliffs, NJ: Prentice-Hall). Photo courtesy A. C. Tennissen.]

and Australia. The Australian deposits are residual from a warmer and more humid climate during the Tertiary Period.

The behavior of iron during weathering is complicated by the fact that it can occur in the sedimentary environment in more than one state of oxidation. Most iron in ferromagnesian minerals is in the reduced state because of a general deficiency of gaseous oxygen in the environments of formation of igneous and metamorphic rocks. In an aerated weathering environment the iron is converted to the oxidized form to precipitate as ferric hydroxide. Once precipitated, it cannot be reduced easily and accumulates. Such iron-rich crusts are best developed on basalts because basalts contain three times as much iron as granites (12% versus 4%); soils developed on basalt in tropical climates are ferruginous laterites and are commonly tens of meters thick. These residual soils also contain aluminum and titanium oxides and hydroxides, the other superstable weathering residues. A complete gradation in composition occurs between aluminous and ferruginous laterites.

MECHANICAL WEATHERING

Mechanical weathering is of trivial importance compared to chemical weathering because of the extraordinary dissolving power of even the smallest amount of dipolar water molecules. Even in the driest of deserts there is a significant amount of water in the air (i.e., the humidity is not zero), and in the cool predawn hours this moisture condenses on the desert sand. Field observations reveal much flaking of fallen Egyptian monuments near the air–sand interface, very little at the air–monument interface. The ancient inscriptions are decomposed and illegible on the monument surface partly buried in the sand, legible on the upper surface of the monument. Thin sections of crystalline rocks exposed in desert areas always show the formation of films of iron oxide and/or clay along crystal boundaries or mineral cleavages. Desert sand originates primarily because of chemical decomposition, not mechanical abrasion by wind-transported older sediment.

The most important cause of mechanical weathering is frost-wedging, which is visible on the upper parts of mountainous areas in wet, temperate climates. When water freezes, it expands about 9% in volume, generating a force more than adequate to spring loose flat slabs of rock. Surfaces at higher elevations of the Rocky Mountains and Sierra Nevada are often covered by these slabs. The water seeps into cracks during the warm daylight hours and freezes at night, loosening the slabs. This process is repeated every day for much of the year and is very effective.

The importance of mechanical weathering is that it makes many small fragments from a single larger one, greatly increasing the surface area of the rock mass. Chemical activity operates fastest on exposed surfaces, so that the nighttime activity of the H_2O molecules paves the way for the more destructive activities of the liquid H_2O molecules the following day.

SUMMARY

Sediment is formed because of the instability of igneous and metamorphic rocks at the Earth's surface. The crystalline rocks alter because they formed at temperatures of 200–1000°C, at fairly high pressures, and generally in the presence of relatively small amounts of water and oxygen gas, compared to the amounts available at the surface. The presence of living organisms at the surface not only accelerates the formation of sediment but also makes possible chemical reactions not possible in purely inorganic systems.

The residue of the chemical-alteration process under surface conditions is enriched in ferric iron oxide, in hydrated substances, and in aluminosilicate minerals depleted in alkali and alkaline earth cations. The most abundant aluminosilicate mineral group in sediments is clay, which forms about 45% of all minerals in sedimentary materials. The type of clay that is most abundant depends during early stages of weathering on the mineral composition of the parent rock, but in later stages entirely on the climate. Illite and montmorillonite appear initially, to be succeeded in humid temperate climates by kaolinite. In humid tropical areas gibbsite and ferric iron oxides form the final residue of the original igneous or metamorphic rock.

FURTHER READING

Baas Becking, L. G. M., I. R. Kaplan, and D. Moore. 1960. Limits of the natural environment in terms of pH and oxidation-reduction potentials. *Jour. Geol., 68,* 243–284.

Basham, I. R. 1974. Mineralogical changes associated with deep weathering of gabbro in Aberdeenshire. *Clay Minerals, 10,* 189–202.

Bustin, R. M., and W. H. Mathews. 1979. Selective weathering of granitic clasts. *Canad. Jour. Earth Sci., 16,* 214–223.

Degens, E. T., and D. A. Ross (eds.). *Hot Brines and Recent Heavy Metal Deposits in the Red Sea.* New York: Springer-Verlag, 600 pp.

Dixon, J. B., and S. B. Weed (eds.). 1977. *Mineral in Soil Environments.* Madison, WI: Soil Science Society of America, 948 pp.

Dury, G. H. 1971. Relict deep weathering and duricrusting in relation to the palaeoenvironments of middle latitudes. *Geog. Jour., 137,* 511–522.

Ehlmann, A. J. 1968. Clay mineralogy of weathered products of river sediments, Puerto Rico. *Jour. Sed. Petrology, 38,* 885–894.

Eswaran, H. 1979. The alteration of plagioclases and augites under differing pedo-environmental conditions. *Jour. Soil Sci., 30,* 547–555.

Fry, E. J. 1927. The mechanical action of crustaceous lichens on substrata of shale, schist, gneiss, limestone, and obsidian. *Ann. Bot., 41,* 437–460.

Goldich, S. S. 1938. A study in rock weathering. *Jour. Geol., 46,* 17–58.

Huang, W. H., and W. C. Kiang. 1972. Laboratory dissolution of plagioclase feldspars in water and organic acids at room temperature. *Amer. Mineral., 57,* 1849–1859.

Jackson, T. A., and W. D. Keller. 1970. A comparative study of the role of lichens and "inorganic" processes in the chemical weathering of recent Hawaiian lava flows. *Amer. Jour. Sci., 269,* 446–466.

Keller, W. D. 1978. Kaolinization of feldspar as displayed in scanning electron micrographs. *Geology, 6,* 184–188.

Keller, W. D. 1979. Bauxitization of syenite and diabase illustrated in scanning electron micrographs. *Econ. Geol., 74,* 116–124.

Kerr, R. A. 1978. Seawater and the ocean crust: the hot and cold of it. *Science, 200,* 1138–1141, 1187.

Smale, D. 1973. Silcretes and associated silica diagenesis in southern Africa and Australia. *Jour. Sed. Petrology, 43,* 1077–1089.

Syers, J. K., and I. K. Iskandar. 1973. Pedogenic significance of lichens. In V. Ahmsdjian and M. E. Hale (eds.), *The Lichens.* New York: Academic Press, pp. 225–248.

3

Mudrocks

Though the mills of God grind slowly,
yet they grind exceeding small.
Though with patience He stands waiting,
with exactness grinds He all.

FRIEDRICH VON LOGAU

Although they form approximately two-thirds of the stratigraphic column, mudrocks are poorly understood and inadequately studied. Few sedimentary petrologists have chosen to study mudrocks because of several difficulties:

1. Mudrocks are composed mostly of clay minerals that absorb water easily and in large amounts, so that mudrocks become plastic and flow downslope readily. Hence they form valleys rather than hills. Cliffs formed of mudrock are not common.
2. Mudrocks are extremely fine-grained. In outcrop there is little that can be described other than color and the presence or absence of fissility. In thin section many of the rock constituents cannot be resolved because of their small size, the intermixing of the different clay minerals, and the common occurrence of opaque hematite stain or organic matter.
3. It is almost impossible to distinguish between quartz and feldspar in such small grains unless the feldspar is twinned. But because large feldspar grains tend to weather and break along twin composition surfaces, most silt-size feldspar grains are untwinned.
4. Because of their small size and sheeted structure, clay minerals are easily and frequently altered after deposition. The original clays have very often been recrystallized and/or changed into clays of a different chemical composition.

Is there any hope? Can we salvage two-thirds of the stratigraphic column from the scrap heap of petrology? Recent advances in analytical techniques suggest that the answer to these questions is "yes." Scanning electron microscopy permits magnifications of at least 50,000×, in contrast to the polarizing microscope with its limitation of about 500×. New wet-chemical techniques permit the isolation of quartz and feldspar from the mass of clay minerals, so that the feldspars can be studied petrographically. The development of the electron microprobe about 1950 and its greatly increased use by sedimentary petrologists during the past few years have revolutionized our ability to make microchemical analyses of tiny grains in sediments. It is now possible to make quantitative chemical analyses of areas only 1 μm in diameter—the size of individual clay flakes.

In addition, there is now a much greater interest in the organic matter in mudrocks by the major oil companies. How are the organic tissues of microscopic organisms converted into petroleum and natural gas? It seems likely that the decade of the 1980s may well be "the age of mudrocks" in terms of the amount of study they receive and the increase in our understanding of these enigmatic rocks.

FIELD OBSERVATIONS

As is the case for all rocks, an adequate field description of a mudrock should include information about the texture, structure, and mineral composition. It is harder to do this for mudrocks than for sandstones or carbonate rocks because of the fine grain size of mudrocks.

Textures

Texture is defined as the size, shape, and arrangement of the grains or crystals in a rock. Individual grains cannot always be seen in a mudrock with only a hand lens, but despite this handicap it is still possible to make a semiquantitative estimate of the ratio of silt to clay. The method is to nibble a bit of the rock between the teeth to determine whether the rock is gritty (see Figure 3–1). Mudrocks are formed almost entirely of quartz and clay. Quartz is gritty; clay is slimy. If you sense no abrasion of your teeth, the rock contains more than two-thirds clay minerals; and clay minerals are the bulk of the clay-size particles. If grit is sensed, perhaps there are between two-thirds and one-third clay minerals. If there are less than one-third clay minerals, enough quartz silt is present to be seen with the hand lens.

The shape of the grains in a mudrock cannot be determined in outcrop because of their small size. This is not a problem, however, because we can safely infer that the quartz grains in mudrocks are quite angular. Particles with diameters less than about 60 μm travel almost entirely in suspension, being small enough to be suspended by even weak currents. This means that they are not abraded by impacts with other grains and will be as angular after transport as when they began as fragments from a crystalline rock. Clay minerals, on the other hand, have a shape determined by their crystal structure. Most commonly, they are shaped like a sheet of paper, although this cannot be seen with a hand lens.

Figure 3–1
Geologist hungry for knowledge determining grain-size distribution of mudrock sample from Hennessey Formation (Permian), central Oklahoma. [Photo courtesy S. Bock.]

Structures

Fissility and Lamination

The major sedimentary structures visible in outcrop are fissility and lamination. *Fissility* is a property of a mudrock that causes it to break along thinly spaced planes parallel to bedding (see Figure 3–2) and to the orientation of the sheetlike clay flakes. The existence of fissility depends on many factors, only one of which is the abundance of clay minerals. Mudrocks with identical percentages of clay can differ greatly in fissility because of differences in the perfection of orientation of the clay flakes. In outcrop the reasons for the lack of parallelism of clays cannot be determined, but observations made in modern muddy environments suggest several possible explanations.

All small particles have large surface/volume ratios and, as a result, have a strong tendency to adhere when they collide during transport or when settling in water. This tendency is increased by salinity and the presence of organic matter in the water.

Figure 3–2
Well-developed fissility in Antrim Shale (Mississippian), Alpena County, Michigan. Hand lens is about 2 cm in diameter. [R. V. Dietrich and B. J. Skinner, 1979, *Rocks and Rock Minerals* (New York: Wiley). Photo courtesy R. V. Dietrich.]

Maximum possible adhesion is attained when the salinity has increased to about 2,000 ppm and changes very little above this value. Therefore, as particles enter more saline water, they clump into aggregates containing randomly oriented clay-mineral flakes and very fine-grained quartz. This process is termed *flocculation*. The modal size of the floccules consistently lies in the silt range (4–62 m) and is controlled by the initial size of the constituent grains (see Figure 3–3). The coarser floccules are concentrated in the areas of more rapidly moving currents such as tidal inlets. The correlation between floccule size and dynamic conditions is not surprising in view of the dependence of grain size on dynamic conditions (see Chapter 4). The result of flocculation is to impair the development of fissility in a mudrock. Clearly, a mudrock formed of clay–quartz lumps rather than neatly oriented individual flakes will not be fissile.

Bottom-dwelling organisms in the depositional environment also affect the development of fissility. As these organisms scavenge through the fresh mud for organic matter, they swallow great amounts of clay; and when the clay passes through the alimentary canal of an organism, it is formed into aggregates: more clumps. Also, as the organism burrows through the mud, it destroys any clay-flake parallelism formed during the settling of individual clay flakes to the bottom (see Figure 3–4). Studies of modern mud environments suggest that the rate of burrowing and mealtime clump formation is frequently greater than the rate of deposition of the mud particles.

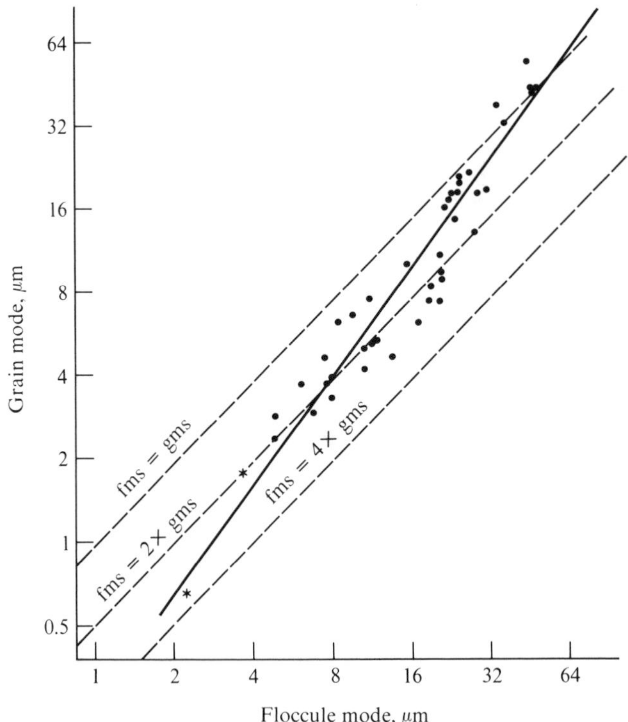

Figure 3–3
Relationship between floccule modal size (fms) and grain modal size (gms) in artificially (*) and naturally flocculated sediment samples. Natural floccule sizes fall in silt range, 4–62 μm. [Modified from K. Kranck, 1973, *Nature, 246*. Reprinted by permission. Copyright © 1973 Macmillan Journals Limited.]

Figure 3–4
Photograph of sawed hand specimen from Sonyea Group mudrock (Devonian), New York, showing bioturbation of partly lithified (broken, flat fragments), fine-grained sediment. Intense burrowing is evident in center of specimen. [C. W. Byers, 1974, *Sedimentology, 21*. Photo courtesy C. W. Byers.]

Perhaps it is more meaningful to ask how mudrocks can ever be fissile than why many are not. Fissility in some mudrocks may be produced during diagenesis of the clay minerals (see below).

Lamination refers to parallel layering within a bed. By definition, a bed is thicker than 1 cm; a lamina, thinner than 1 cm (see Figure 3–5). Lamination can have many origins that are related both to variations in current strength during deposition of the layer and to changes in composition of the sediment deposited. For example, a bed of mudrock may contain laminae of black organic matter, zones of green color within a

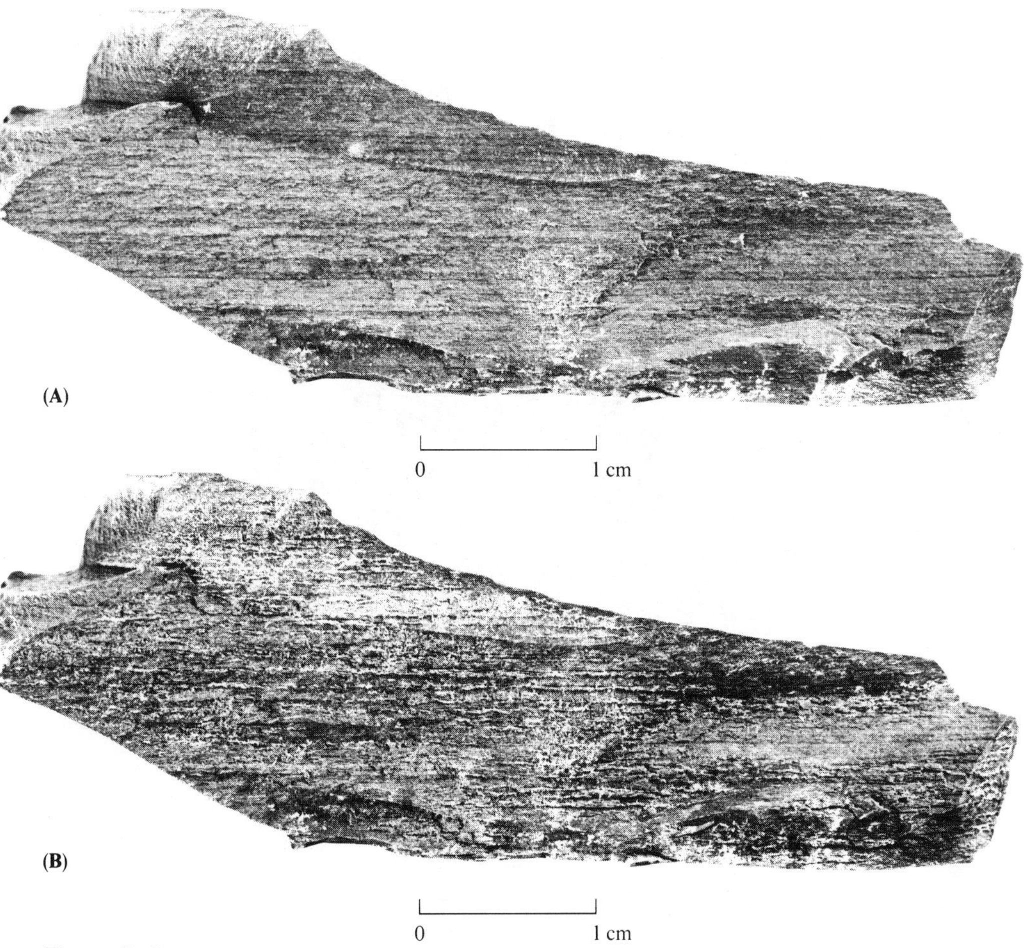

Figure 3–5
Marine shale in Yorkshire Coal Measures (Carboniferous), England, showing laminations thought to be varves. Dark and light laminae are caused by presence and absence of lenticles composed of organic matter and clay minerals. Laminae are 0.4–0.8 mm thick. In **B** water has been added to sawed rock surface, causing porous lenticles to swell in dark laminae. [D. A. Spears, 1969, *Jour. Sed. Petrology, 39*. Photos courtesy D. A. Spears.]

dominantly red unit, or microplacering of quartz silt within an otherwise clayey mudrock. Each of these features can give unique information about oxidation/reduction reactions at the site of deposition, changes in these conditions after burial, or variations in current strength with time during deposition of the bed. All departures from randomness and homogeneity should be recorded at an outcrop. Perhaps their meaning is unclear at the moment, but "The Truth" may emerge on reflection in the laboratory or office.

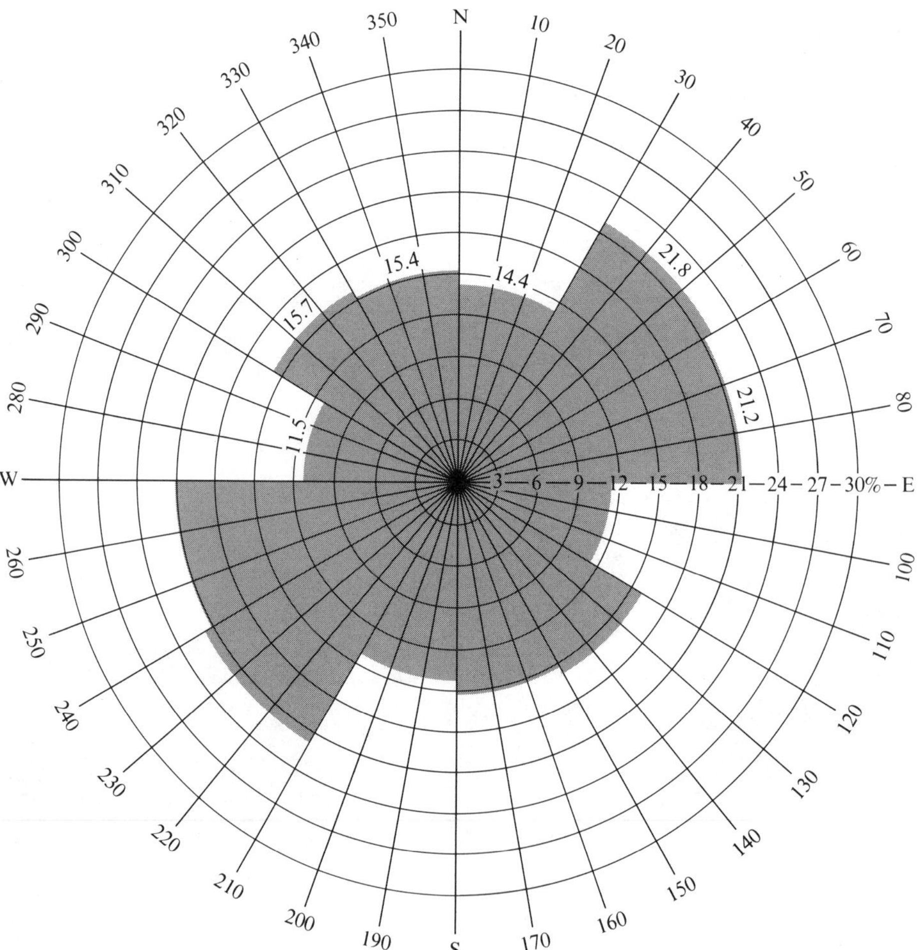

Figure 3–6
Rose diagram (circular histogram) of orientations of 316 carbonate concretions in Upper Devonian mudstones, New York. [After Colton, 1967.]

Concretions

Although concretions are much more frequent in sandstones than in mudrocks, such structures occur in all types of sedimentary rocks. They generally are inequant in shape, and the direction of elongation indicates the direction of maximum permeability of the sediment during diagenesis. An interesting example of the use of inequant concretions in mudrocks is provided by Colton (1967). He measured the direction of elongation of 316 carbonate concretions 1 cm to 5 m in length in shale and mudstone of Late Devonian age in northwestern New York and found the distribution shown in Figure 3–6. Colton interpreted the preferred orientations in the NE–SW and NW–SE directions as indicating a fabric formed in the mud by the action of the marine bottom currents. Because mudrocks are impermeable to water after they have been compacted, the calcium carbonate must have been deposited from the waters within a very brief period after the mud was deposited (see Figure 3–7), before compaction eliminated permeability. Probably this occurred within a few meters of the water–sediment interface at the Devonian seafloor.

Figure 3–7
Inequant calcareous concretion in dark-colored shale (Cretaceous), Magdalena Valley, Colombia. Note curvature of shale bedding around concretion, indicating that concretion formed during very early diagenesis, before mud was compacted. [L. G. Weeks, 1953, *Jour. Sed. Petrology, 23.*]

Colors

The colors of mudrocks fall almost entirely into two groups: gray-black and red-brown-yellow-green. The gray-black shades reflect the presence of 1% or more of free carbonaceous material, which, in turn, reflects deposition in an oxygen-deficient or reducing environment. In well-oxygenated water the concentration of gaseous oxygen is 10^{-4} mol/L. To obtain conditions sufficiently reducing for the black free carbon to accumulate requires a decrease to about 10^{-6} mol/L. This implies a lack of circulation of aerated water in the depositional environment to prevent complete oxidation of the organic tissues to carbon dioxide plus water. It is important to note that lack of circulation is not related to depth of water. Most of the ocean floor at depths of many thousands of meters is kept well oxygenated by the cold bottom currents that originate at the surface in polar regions and circulate throughout the oceans. There are many shallow areas of the ocean floor that are stagnant, such as the lagoons between the Texas Gulf Coast and the offshore barrier bars. Some of the geographically most extensive black mudrocks are known to have been deposited in water only a few tens of meters in depth, e.g., the Chattanooga Shale noted in Chapter 1.

The best-studied example of a large, modern stagnant environment is the Black Sea, which has an average depth of 3,700 m but no measurable oxygen below 200 m. Circulation of oxygen-rich water from the Mediterranean Sea is prevented by a rock barrier at the southwestern end of the Black Sea (Bosporus) that rises to within 40 m of the surface.

The red-brown-yellow-green color grouping reflects the presence or absence of ferric oxide (red), hydroxide (brown), or limonite (yellow) as colloidal particles among the clay-mineral flakes. Hematite (Fe_2O_3) is red; goethite [FeO(OH)] is brown. Only a few percent of hematite are sufficient to give a deep-red color to a mudrock. If these compounds are absent, the true green color of most clay minerals shows through; illite, chlorite, and biotite are green. In many red mudrocks, ovoid or tubular green spots are present, reflecting the reduction of ferric iron adjacent to a plant root or other bit of organic matter and the removal of the ferrous ions in groundwater.

In color descriptions the colors of exposed sediments should be recorded from a fresh, dry surface showing a typical texture. The colors of modern marine sediments and other subaqueous deposits should be described immediately upon obtaining a sample because of rapid color changes resulting from desiccation, oxidation, and bacterial activity. In any event, the condition of the surface being described and the lighting conditions should be noted.

Color can be specified in the field by reference to one of several easily available color charts. The two most widely used charts are those published by the Geological Society of America and by the Munsell Color Company. The charts consist of a booklet of cardboard pages on which are glued pieces of colored paper about 15 cm^2. These colors cover the range seen in sedimentary rocks and should be used in descriptions of stratigraphic sections and hand specimens.

Nomenclature

There is no uniform usage of the many terms that refer to detrital sediment finer than sand size. Such terms as "shale," "argillite," "argillaceous," and "clayey" mean different things to different people. The terminology we will adopt (see Table 3–1) is simple, usable in fieldwork, and consistent with whatever detailed laboratory studies may subsequently follow the outcrop descriptions. Table 3–2 is an example of a good field description of mudrock units in the New Albany Shale (Devonian) in Kentucky.

LABORATORY STUDIES

Laboratory investigations of mudrocks tend to concentrate on mineralogy and chemistry rather than on texture, probably because *some* textural observations can be made in the field, but almost no mineralogic determinations are possible without sophisticated laboratory instruments.

Laboratory Techniques

The most valuable techniques and instruments for studying mudrocks are (1) X-ray diffraction, (2) sodium bisulfate fusion, (3) electron microprobe, (4) scanning electron microscope, (5) polarizing microscope, and (6) radiography of rock slabs.

Table 3–1
Classification of Mudrocks

Ideal size definition	Field criteria	Fissile mudrock	Nonfissile mudrock
> ⅔ silt	Abundant silt visible with hand lens	Silt–shale	Siltstone
> ⅓ < ⅔ silt	Feels gritty when chewed	Mud–shale	Mudstone
> ⅔ clay	Feels smooth when chewed	Clay–shale	Claystone

Source: H. Blatt et al., 1980, *Origin of Sedimentary Rocks,* 2nd ed. (Englewood Cliffs, NJ: Prentice-Hall).

X-Ray Diffraction

This method is used for mineral identification and is particularly useful when the particles are very small (e.g., clays) so that the petrographic microscope is not effective. The sample is powdered, mounted on a glass slide, and bombarded with X-rays.

Table 3–2
Measured Stratigraphic Section of New Albany Shale West of Berea, Kentucky, Illustrating Good Field Description of Mudrock Beds. (The numbers and letters in parentheses, e.g., 5YR 2/1, refer to colors on the rock color chart published by the Geological Society of America. The numbers 11 through 15 are the beds as measured from the base of the sequence. Beds 1–10 were omitted by the authors of the table.)

	Thickness, m
Quaternary (?)	
15. Soil, olive-gray (5Y 4/1) to yellowish-gray (5Y 8/1), containing near base, quartzite pebbles, siliceous geodes, and phosphatic nodules derived from nearby and underlying bedrock units; weathers to grassy flat. Erosional contact.	1.5+
Devonian:	
New Albany Shale (incomplete):	
14. Shale, brownish-black (5YR 2/1), weathers light gray (N7), with some iron oxide and sulfide stain on fractures and bedding planes along with rosettes and prisms of selenite 3–5 mm long; silt in discontinuous laminae 1–2 grains thick; pyrite in discoidal concretions and disseminated grains; phosphate in nodules that are round to amoebiform, 2–3 cm thick, elongate, some more than 13 cm across, brownish-gray (5YR 4/1) and brownish-black (5YR 2/1), weather yellowish-gray (5Y 8/1) on surface, with earthy luster and rough fracture; fossils include *Tasmanites* spores and a vitrain layer 3 mm thick 1.5 m above base. Top at or near contact with grayish-green shale of basal Borden Formation seen in outcrop less than 1.6 km to the west. Basal contact placed at lowest phosphate nodule.	2.7
13. Shale, brownish-black (5 YR 2/1) to grayish-black (N2), like unit above, except contains no phosphate nodules. 2.4 m above base is zone of pyrite concretions 30 cm in diameter and 1.3 cm thick, concentrated along bedding planes; shale weathers to fissile, brittle flakes and plates as much as 120 cm in diameter; *Tasmanites* abundant; possible fish scale, 1 mm across near top of unit; silt laminae increase in abundance downward; tough and dense, with subconchoidal fracture where fresh. Sharp basal contact.	5.5
Three Lick Bed:	
12. Shale, greenish-gray (5GY 5/1–4/1), weathers yellowish-gray (5Y 8/1), clayey, subconchoidal fracture on joints; partings coated with limonite and sulfate stain. Sharp contact with underlying unit.	0.2
11. Shale, black (N1) to grayish-black (N2) and brownish-black (5YR 2/1); weathers light gray (N7), with iron oxide and sulfate stain; silt laminae 1–2 grains thick, commonly about 5 mm to 1 cm apart; brittle flakes litter outcrop; burrowed in upper 30 cm; burrows filled with greenish shale from overlying unit?; discontinuous cone-in-cone limestone layer, dark-gray (N5), 30 cm thick, about 180 cm below top (sampled); basal contact sharp.	0.4

Source: Potter et al., 1980.

The X-rays are diffracted by planes of atoms in the crystal structure, and a tracing is produced on a paper chart (see Figure 3–8). The chart (diffractogram) is an x–y plot of diffraction angle versus intensity of diffracted radiation; it reveals the interplanar spacings and, in turn, the type of mineral. This is the best method for identifying the various types of clays in a rock. It is sometimes used in conjunction with differential thermal analysis, a technique in which the sample is heated in a furnace to determine the temperatures at which compositional and structural changes occur. Different temperatures are characteristic of different minerals.

Sodium Bisulfate Fusion

The purpose of this technique is to isolate the quartz and feldspar grains from the mass of clay minerals and other substances in the mudrock. Pea-size fragments of the sample are fused over a Bunsen burner in sodium bisulfate. The only materials that

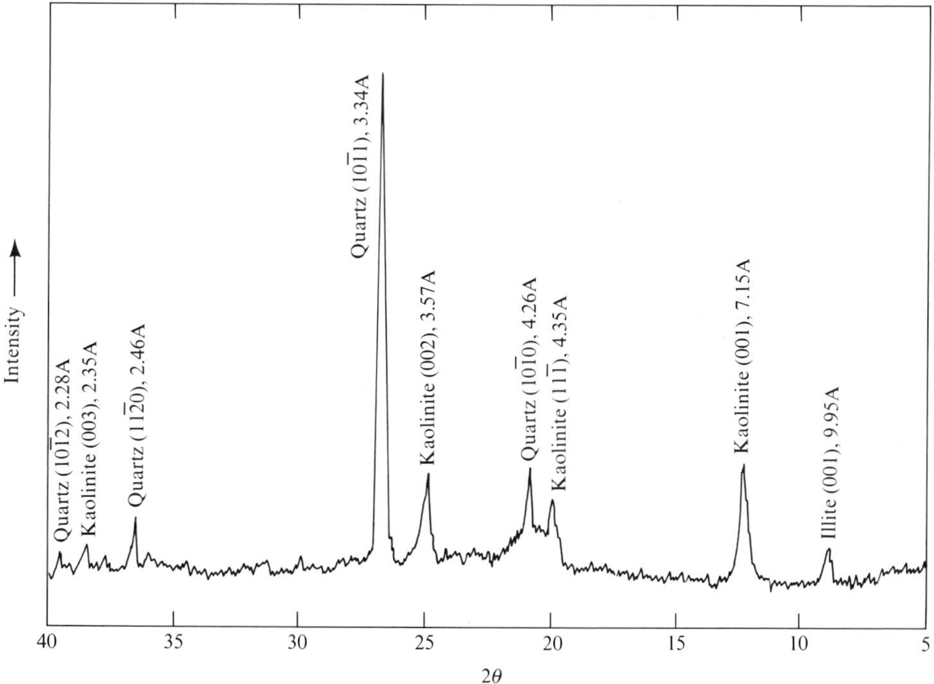

Figure 3–8
X-ray diffraction pattern of unoriented mixture of one-third each quartz, kaolinite, and illite. Differences in peak height among the three minerals result from combined effect of differences in crystal structures and orientation of grains on glass slide. Labels show lattice plane that produces particular peak and distance between these lattice planes. θ = angle between incident beam and plane within crystal. Peaks occur when relation $n\lambda = 2d \sin \theta$ is satisfied. n = integral number, λ = wavelength of X-rays, d = interlayer spacing.

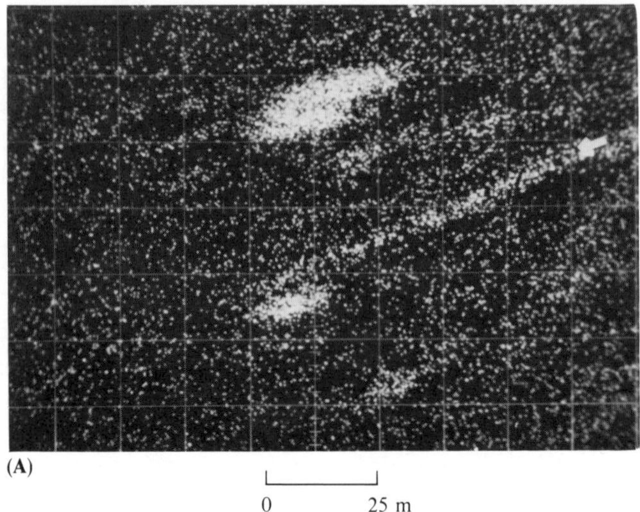

Figure 3–9
Electron microprobe scans of Levis Shale (Ordovician), Quebec, Canada. (**A**) Distribution of iron (white dots).

survive are quartz and feldspars, which are almost completely unaffected in either composition or grain size. These grains can then be analyzed petrographically.

Use of the Electron Microprobe

This instrument provides a means of determining the chemical composition of very small volumes at the surface of polished thin sections or grain mounts. An electron beam is focused on the area of interest, which can be as small as 1 μm in diameter. The impact of the beam on the sample causes X-rays to be emitted whose wavelengths are characteristic of the elements present in the area hit by the beam. The intensity of the X-rays reveals the concentration of the element (see Figure 3–9). This technique is sufficiently sensitive to determine concentrations of trace elements, as well as major elements, in the sample.

Scanning Electron Microscopy

The scanning electron microscope is used for both textural and mineralogic determinations. A piece of the rock, perhaps a centimeter in diameter, is coated under vacuum with a gold–palladium mixture. The coated specimen is then bombarded by electrons, which are scattered by the gold–palladium coating to produce the detailed topography of the fragment. Magnifications of 50,000× with excellent resolution

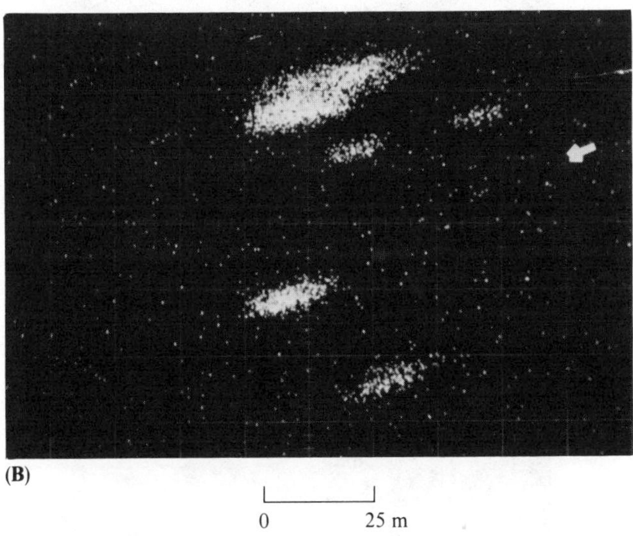

(**B**) Distribution of sulfur. This patch of micro- to cryptocrystalline pyrite is not visible with only a polarizing microscope. Large pyrite crystal is 30 μm long. Arrows point to layer of iron-rich chlorite. [R. Siever and M. Kastner, 1972, *Jour. Sed. Petrology, 42.* Photos courtesy M. Kastner.]

and great depth of field are obtained easily (see Figure 3–10), and enlargements of 100,000× are possible with somewhat diminished but quite usable resolution. X-ray attachments to the scanning electron microscope are commonly used and permit at least semiquantitative analyses of the elements in the sample.

Polarizing Microscopy

Although it is possible in theory, with perfect imaging and monochromatic light, to resolve grains as small as 0.2 μm, in practice it is not possible to achieve better than about 1 μm. This is about the average size of the particles in a mudrock, so that much of the rock appears in thin section as an irresolvable, birefringent mass of clay minerals. Most mudrocks are mixtures of different clay species, and their optical properties are sufficiently similar to make distinctions among them impossible unless individual flakes can be isolated. The most appropriate use of the light microscope in mudrock studies is to study textural features that can be seen in thin section but not in outcrop, such as small-scale cross bedding, structures produced by organic burrowing, microdesiccation features, or other structures that give insights into the history of the rock (see Figure 3–11).

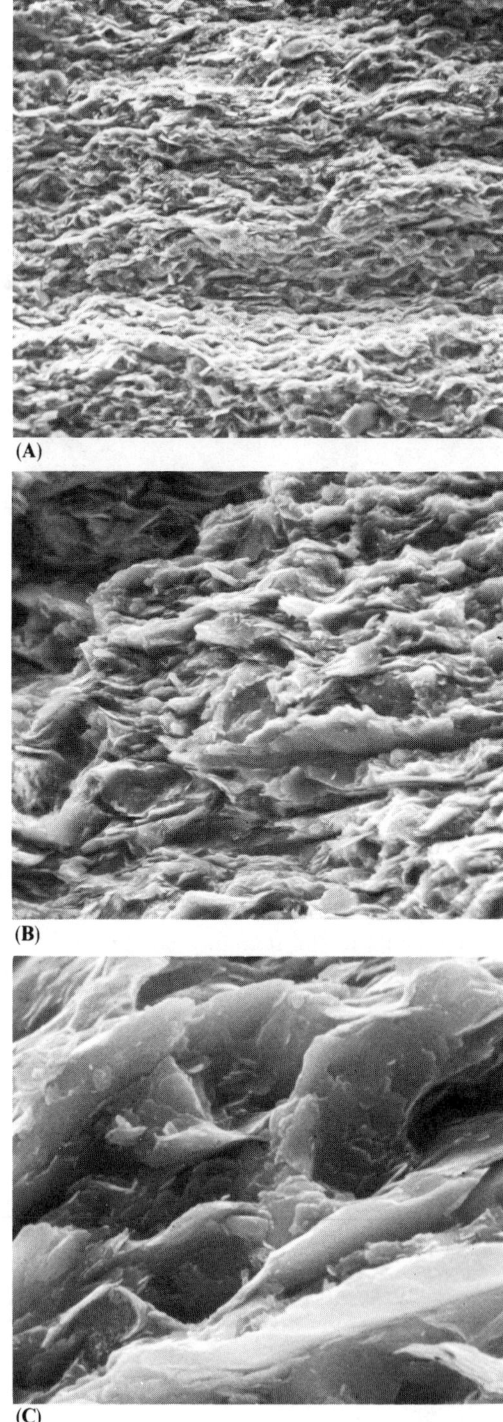

Figure 3–10
Scanning electron micrographs of Perche Shale (Devonian), north of Las Cruces, New Mexico. Fissility clearly results from parallelism of clay-mineral flakes. (**A**) 300×. (**B**) 1,000×. (**C**) 5,000×. [Photos courtesy K. P. Helmold.]

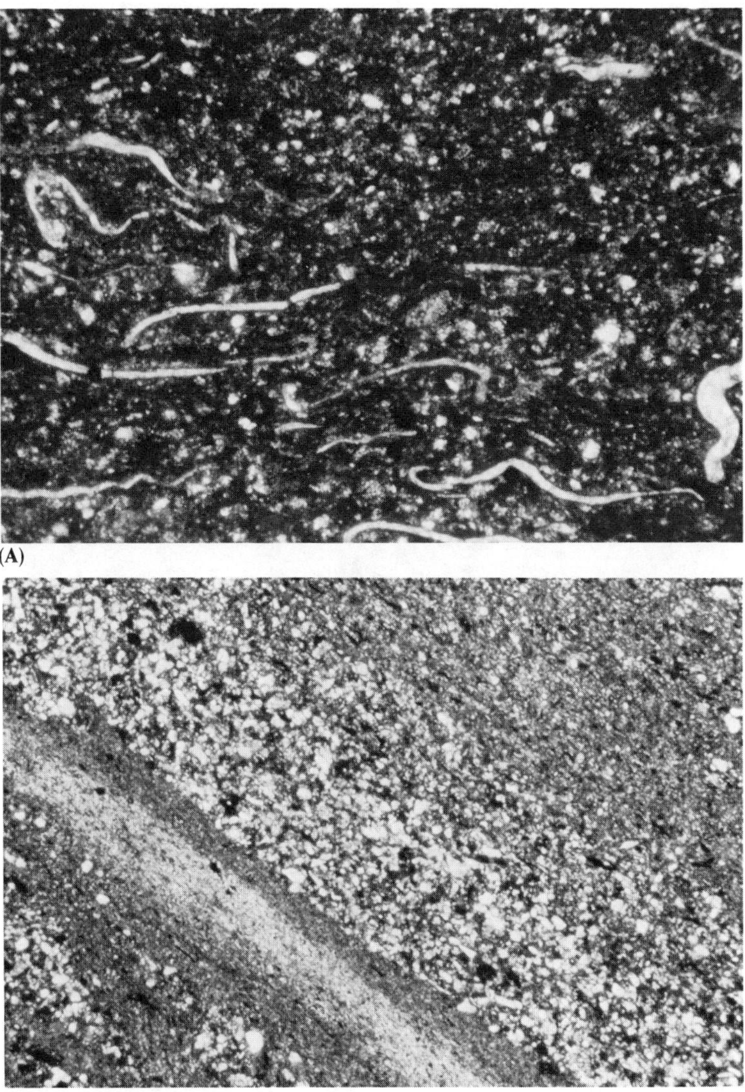

Figure 3–11
Thin-section photomicrographs. (**A**) Quartz silt, clay, calcite, and organic matter in Ordovician mudstone near Waynesville, Ohio. Abundant elongate trilobite fragments in lower part are oriented parallel to bedding, as are bits of dark organic matter and many of the quartz grains. Field diameter is 2 mm. [Photo courtesy D. S. Brandt.] (**B**) Well-laminated shale, Gammon Shale (Cretaceous), Bowman County, North Dakota. Clay plates are oriented 45° to crossed polars. Pronounced preferred orientation of clay minerals is evident in lower left third (difference in color). Muddy quartz silt lamina in center grades upward into more clayey layer. Field diameter is 3.4 mm. [Photo courtesy D. L. Gautier.]

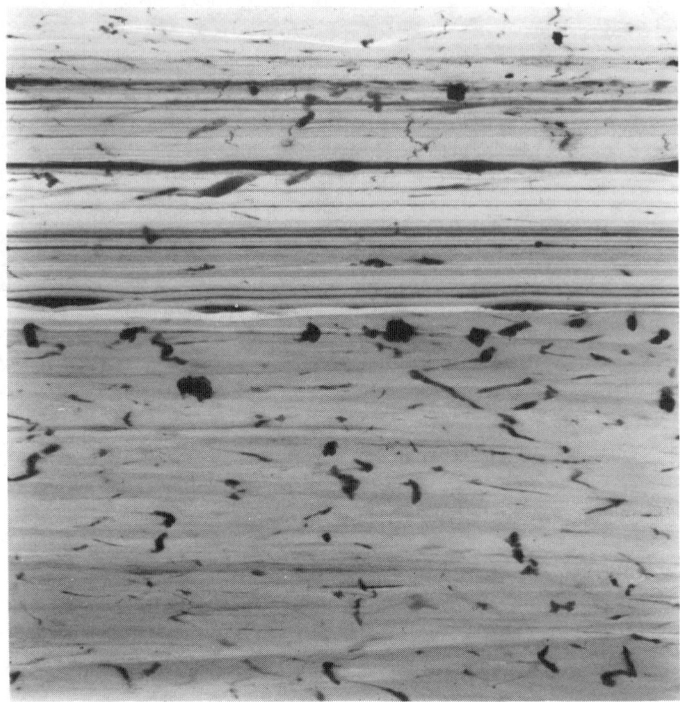

Figure 3–12
Radiograph of section of subsurface core in Selmier Shale (Devonian-Mississippian), Illinois, showing interbedded, thickly laminated black shale (upper half) and indistinctly bedded gray shale (uppermost centimeter and lower half). Lower contact of laminated shale is sharp, but upper contact is gradational and several burrows extend into top of laminated bed. Burrows and many of the laminations cannot be seen without radiography. Width of core is 10 cm. [Cluff, 1980. Photo courtesy R. M. Cluff.]

Radiography of Rock Slabs

Many mudrocks appear structureless in outcrop (or in the small chip of rock present in thin section) but may contain sedimentary structures too subtle to be visible with the naked eye. These can be made visible by the use of radiography (see Figure 3–12). In this technique, a slab of mudrock 15 cm long, 10 cm wide, and 0.5 cm thick is cut perpendicular to the bedding and placed directly on X-ray film. A photograph is taken with either a medical, a dental, or an industrial X-ray unit. Previously obscure features such as root tubules, organic burrows, slump features, and subtle laminations are clearly visible because textural and mineralogic variations in a rock affect the ability of the X-rays to penetrate the rock.

Composition

Use of the techniques described and others is beginning to reveal the range in the mineral composition of mudrocks and its variation with depositional environment and diagenetic history.

Clay Minerals

The most abundant mineral group is the clays (principally illite, montmorillonite, and kaolinite)—a result of the great chemical stability of a sheeted Si–O, Al–OH crystal structure (see Chapter 2). Although the abundance of clays in the stratigraphic column is not surprising, the relative abundance of the three main types is. Available data, based largely on X-ray diffraction studies during the past 50 years, indicate that the relative abundances change markedly with the age of the mudrock (see Figure 3–13). Modern muds in temperate climates contain about twice as much montmorillonite as illite (see Figure 3–14), as would be expected because of the greater abundance in crystalline rocks of plagioclase and ferromagnesian minerals compared to potassium feldspar. Kaolinite in temperate climates is a poor third in

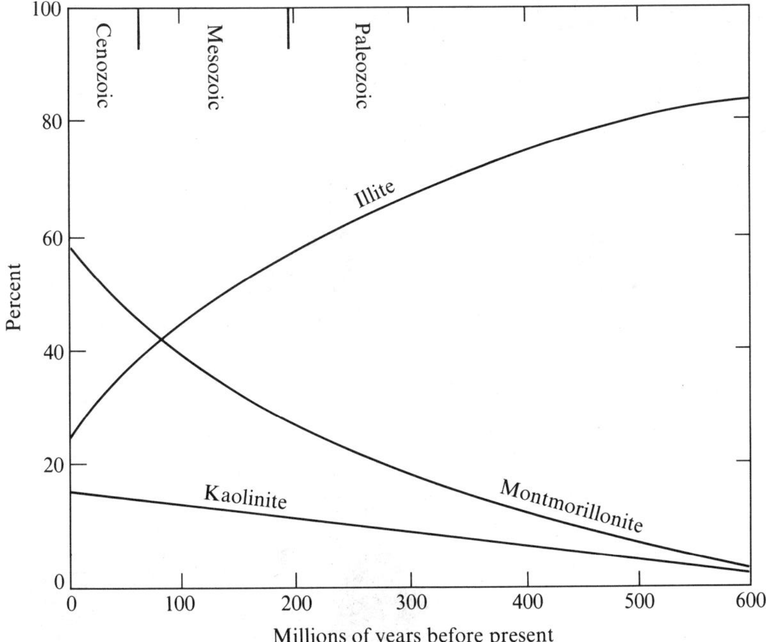

Figure 3–13
Generalized relative abundances of major groups of clay minerals in Phanerozoic mudrocks.

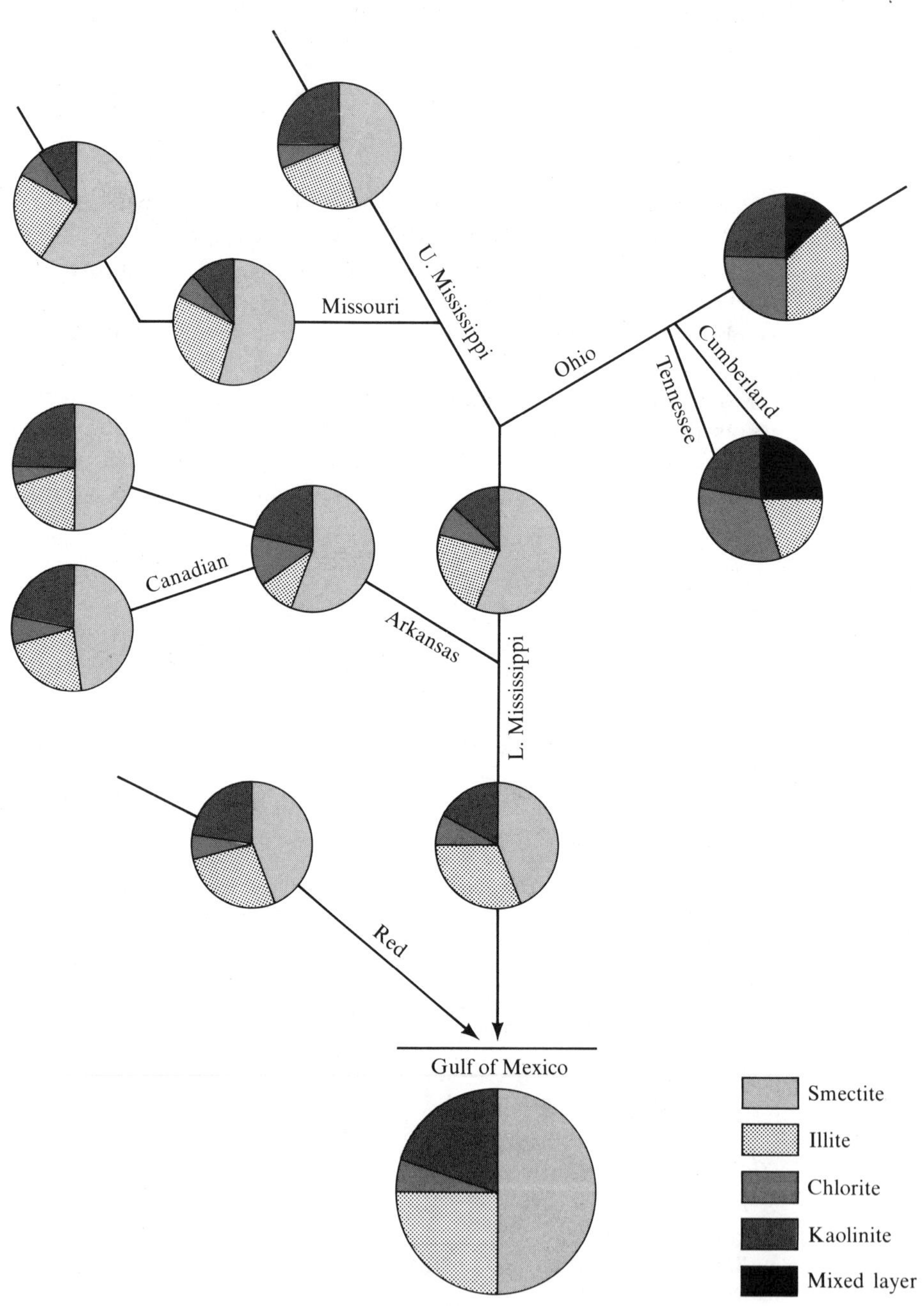

U. Mississippi
Missouri
Ohio
Cumberland
Tennessee
Canadian
Arkansas
L. Mississippi
Red
Gulf of Mexico
Smectite
Illite
Chlorite
Kaolinite
Mixed layer

abundance because it requires a hot, wet climate to form in large amounts. The bulk of the Earth's land area at present is in the cool, temperate climatic belt that favors the stability of montmorillonite and illite.

As we go back in time, however, it is clear that relative abundances change, with illite becoming overwhelmingly dominant in Paleozoic mudrocks, forming as much as 80% of all clay minerals. Several explanations for this change have been suggested:

1. Continual increase in basaltic volcanic activity through Phanerozoic time, thus increasing the relative abundance on the Earth's surface of plagioclase and ferromagnesian minerals, the parent materials of montmorillonite. This would imply, however, that progressively lesser amounts of granite and gneiss (which are rich in potassium feldspar) have been exposed through geologic time. Field evidence does not support this hypothesis. The percentage of sand-size feldspar, which must originate largely in granites and gneisses, is at least as great in Mesozoic and Cenozoic sandstones as in older ones. This suggests, if anything, an increase in silicic crystalline rocks with time rather than a decrease.

2. Change in the biologic controls of chemical weathering. Plant life first colonized the land surface during the Silurian Period, about 420 million years ago. The great abundance of coal beds of Carboniferous age suggests that this spread of land plants might be responsible for the change in clay mineralogy through time. Plants require at least as much potassium ion for their nutrition as they do calcium and sodium combined. Perhaps the increase in the amount of plant life on the land surface has resulted in an increasing removal of potassium from developing soils, making it less possible for illite to form from the weathering of potassic feldspar. Present data concerning the illite/montmorillonite ratio through time are inadequate to determine whether the ratio changed sharply in the Late Paleozoic. Also, no one knows whether the spread of plants over the land surface required 100 million years, 1 million years, or possibly only 100 years.

3. The relative proportions of the different clay minerals are strongly dependent on climate and the weathering regime it imposes on rocks, and climate varies as a function of latitude. Perhaps progressive drifting of continents through geologic time has generated the progressive change in the proportions of clay minerals we find in ancient rocks. For this explanation to be acceptable, we would need to demonstrate that temperate climates, which favor illite formation, were more common during Paleozoic time than they are today. The positions of the continents during Paleozoic time, however, seem to have been in higher or lower latitudes, on the average, than at present (see Chapter 9). For much of the Paleozoic the southern continents were

---

Figure 3–14
Pie diagrams showing relative amounts of different clay minerals in Mississippi River, its major tributaries, and Gulf of Mexico. Smectite is the group name for expandable clay minerals. Montmorillonite is the most abundant member of the group. [P. E. Potter et al., 1979, *Bull. Centre Recherches*. Pau-SNPA, 9.]

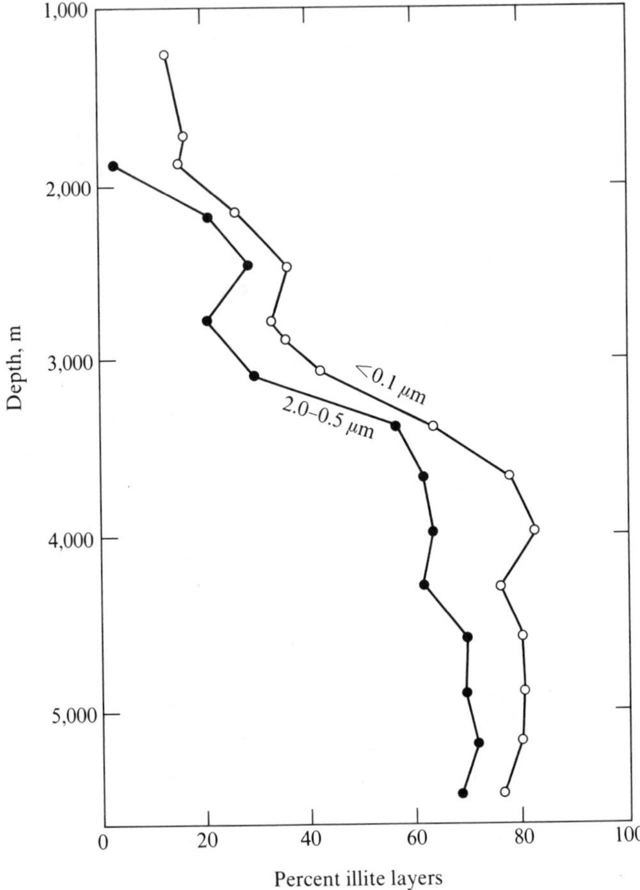

Figure 3–15
Proportion of illite layers in interlayered illite–montmorillonite clay as function of depth in Oligocene-Miocene sediment in Gulf Coast borehole. Conversion to illite is faster in finer-grained (hence more reactive) clay. [J. Hower et al., 1976, *Geol. Soc. Amer. Bull., 87.*]

centered near the South Pole as the supercontinent Gondwanaland, and northern continents appear to have been closer to the equator than at present. In addition, variation in the latitudinal positions of the continents does not explain the change through time in the abundance of montmorillonite. The relative abundance of illite and montmorillonite in any geologic period is a function of the chemical composition of the rock being weathered (silicic rocks → illite; mafic rocks → montmorillonite), not climate. Both minerals are characteristic of temperate climates.

4. The change through time toward increasing illite abundance could also be caused by diagenetic processes. During the past 20 years, many data have accumulated that suggest postdepositional processes as the major cause of the increase in the illite/montmorillonite ratio with time. In the United States, the studies have concentrated on changes in clay-mineral composition in Late Tertiary sediments in the Gulf Coast basin. Similar results have been obtained from other areas of the world, such as the Niger Delta in West Africa and the Rhine graben area of West Germany. In each area, there occurs an abundance of clay minerals composed of alternating illitic and montmorillonitic sheets. Such crystals are most abundant at shallow depth; but as sample depth increases, the montmorillonitic sheets in the crystals are converted into illitic sheets (see Figure 3–15). Suggested sources of the required potassium ions include orthoclase grains in the mudrocks and potassium ions in the connate waters trapped in the rocks. The conversion to illite is largely complete at a depth of 4,000 m and at a temperature of approximately 100°C. The chemical change can be idealized by the reaction

$$\text{Ca and Na montmorillonites} + Al^{3+} + K^{+} \rightarrow \text{illite} + Si^{4+} + Mg^{2+} + Fe^{2+} + Na^{+} + Ca^{2+}$$

The released silica may crystallize as chert in the mudrocks; the calcium may precipitate in calcite, dolomite, and ankerite; the magnesium, in dolomite; the iron, in ankerite or hematite. The fate of the sodium is unknown; it may leave the mudrock, escaping upward to the seafloor.

Quartz

Based on studies using both X-ray techniques and sodium bisulfate fusion, the average amount of quartz plus chert in mudrocks is 30% ± 3%. More than 95% of this is single crystals of quartz. The mean size of this sediment is about 6 ϕ (15 μm) and is $\frac{1}{8}$ fine and very fine sand size, $\frac{6}{8}$ silt size, and $\frac{1}{8}$ clay size; all grains are angular. The amount of quartz in a mudrock is correlated with the grain size of the quartz, with a lesser amount of quartz implying a smaller mean size of the grains (see Figure 3–16). This relationship seems quite reasonable from the viewpoint of sediment-transport processes: the very weak currents that allow low-density floccules of silt and clay size to settle cannot have transported relatively high-density grains such as quartz to the site of deposition.

The origin of the quartz and chert is varied. Some of it results from the chipping of larger igneous and metamorphic quartz grains during transport in streams or by wind or during pounding by the surf on beaches and barrier bars. Some of it is quartz released as silt-size grains from slates, phyllites, and low-grade schists. And some of it may be secondary, having crystallized as very fine quartz or chert from the silica released during illitization of interlayered illite–montmorillonite after deposition of the mud. No method presently exists to distinguish among these alternatives.

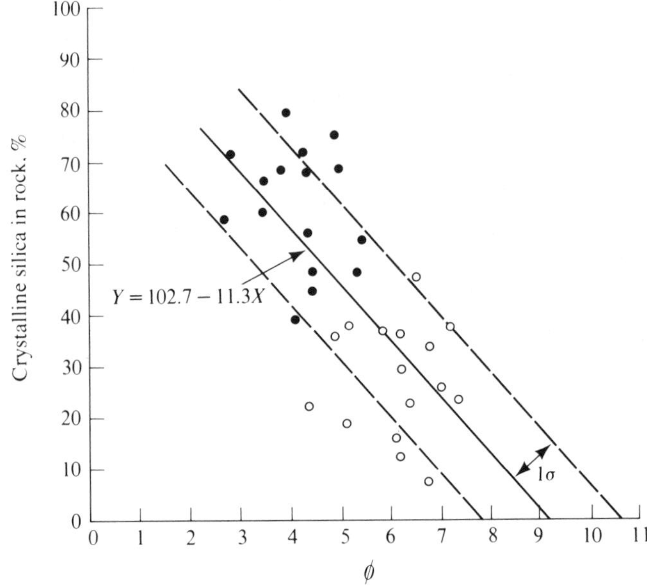

Figure 3-16
Correlation between percentage of crystalline silica and mean size of crystalline silica in interbedded mudrocks and sandstones. The solid line is the line of best fit to the data points; $\pm 1\ \sigma$ lines enclose two-thirds of the points. $\phi = -\log_2$ mm. Open circles = mudrock. Solid circles = sandstone. Correlation coefficient = -0.68. [Blatt and Schultz, 1976.]

Feldspar

Feldspar grains average about 5% ± 2% of the average mudrock. In order to study effectively the mineralogy of grains present in such small amounts, we must first isolate and concentrate them. The sodium bisulfate fusion technique can be used for this purpose. As noted previously, this technique destroys the part of the mudrock that is neither quartz nor feldspar, so that the feldspar percentage is increased in the residue. The grains may be examined by a variety of techniques, such as the polarizing microscope or electron microprobe. Such studies are only now beginning.

Carbonate Minerals

Carbonate material is almost as abundant in mudrocks as are feldspars but is even more poorly understood. Presumably, most of it is calcite, rather than dolomite, siderite, or other carbonate mineral, but this assumption is based entirely on the fact that calcite is the dominant carbonate in sandstones. There is some evidence that the ratio of dolomite to calcite increases in older mudrocks, as is true of carbonate rocks (see Chapter 7).

The origin of the carbonate in mudrocks may be organic, inorganic, or a combination of the two. Perhaps it is particulate material derived from the breakdown of shell material. Perhaps it is an inorganic precipitate formed during compaction from brines generated by the concentration of formation waters. No one has attempted to resolve these uncertainties.

Other Substances

Other materials commonly reported to occur in mudrocks in small amounts include organic matter, pyrite, and hematite. Of these, organic matter is probably most abundant and certainly is the most important from either an academic or an economic point of view. Mudrocks contain about 95% of the organic matter in sedimentary rocks, although the amount in the average mudrock is probably less than 1%. The black matter in these rocks is much darker in color and more carbonaceous than living organic matter because of the processes that formed it. These processes depend fundamentally on the way by which the living tissue was decomposed after death. If gaseous oxygen is present, the tissue is changed into $CO_2 + H_2O$, leaving no residue. The CO_2 either is dissolved in soil water or escapes into the atmosphere. If gaseous oxygen is absent, the tissue is inefficiently and only partially decomposed, leaving a residue of black substances. These organic residues are the raw material from which petroleum is produced by natural processes. That is, mudrocks are the source rocks in which our major energy resource is formed; subsequently, the oil is squeezed into the sandstones and carbonate rocks, where we find it.

The environmental conditions that cause the reduction of the organic matter toward free carbon also cause the reduction of other substances in the water. For example, ferric iron is changed to the ferrous form, and sulfur is reduced from a valence of +6 in sulfate ion (SO_4^{2-}) to −2 (S^{2-}). These two reduced species then combine to produce amorphous FeS and pyrite, FeS_2. Many black mudrocks contain pyrite.

Hematite is a very common pigmenting material in mudrocks; possibly it is more abundant in mudrocks than in sandstones. As noted previously, iron atoms in ferromagnesian minerals exist mostly in the reduced state, but oxidize immediately on contact with the atmosphere. The substance formed initially is brown goethite.

The granitic igneous rocks, gneisses, and schists that contain ferromagnesian minerals form deep in the crust and reach the surface through tectonic and isostatic uplift and erosion. Hence, they occur initially in highland areas, mountain ranges such as the Sierra Nevada or ancestral Rockies. Precipitation in the mountains causes the release of the iron atoms so that they may oxidize. Some of the atoms combine with hydroxyl ions to form colloidal goethite; some are adsorbed onto clay-mineral surfaces to balance charge deficiencies; some are adsorbed onto organic matter in the soil; some are adsorbed onto the amorphous materials that form about

20% of the soil solids. The oxidized iron atoms are then transported to lowland areas where they are deposited. Colloidal material, clay minerals, and organic matter are hydraulically equivalent, in that they are all very fine-grained and will therefore settle together when water velocity decreases. In interbedded mudrock–sandstone sequences it is common to find the mudrocks redder than the sandstones.

Some iron atoms may be added to those that arrive at the depositional site. After deposition, iron may be leached from the octahedral layer of clay-mineral structures or from detrital ferromagnesian minerals to contribute additional intensity to the red coloration of the fine-grained sediment (see Chapter 6).

Bentonite

Bentonite is a special and important variety of mudrock defined by the type of material from which it formed. It is composed almost entirely of montmorillonite and colloidal silica produced as the alteration product of glassy volcanic debris, generally a tuff or volcanic ash. In a pure bentonite the only other minerals present are clearly volcanic in origin, such as euhedral brown biotite, idomorphic zircon, or relict glass shards (fragments). The montmorillonite is normally the calcic variety, reflecting the fact that the parent material is basaltic or andesitic. Sodic and potassic bentonites are known, however, and contain, in addition to the essential montmorillonite, euhedral sanidine and quartz grains with the high-temperature crystal habit (beta-quartz outlines). Paleozoic bentonites generally are composed of mixed-layer illite–smectite with illite layers predominating, a result of diagenesis of originally pure smectite. Bentonites frequently are interbedded with impure tuffs.

A bed of bentonite is the result either of a single eruption (e.g., Mount St. Helens, Washington, May 18, 1980) or of several eruptions within a very brief period, perhaps a few years. As a result, bentonite beds normally are less than 50 cm thick. Stratigraphic and sedimentologic studies have demonstrated regularities in the distribution patterns of these beds consistent with their mode of origin. The thickness of the bed and the size of the unaltered fragments in it decrease logarithmically with distance from the volcanic vent (Fisher, 1964). These fragments also tend to be graded within the bed, with heavier ones near the base and lighter ones near the top. From a stratigraphic point of view, the important facts about bentonite beds are that they are widespread (see Figure 3–17) and that each bed defines a "timeline" in the geologic section, or as close to an infinitely thin, synchronous surface as it is possible to get in geologic materials. In a basin where there are pronounced variations in sedimentary facies, such beds often provide the only means of stratigraphic correlation. In marine ash deposits the precision of the correlation can be decreased by the burrowing and sediment-mixing activities of benthonic organisms.

It has been suggested that volcanic ash deposits might be used in the construction of maps showing planetary wind-circulation patterns and for determining continental-plate rotations relative to the pole (Eaton, 1964). Present-day deposits faithfully reflect tropospheric and lower stratospheric circulation patterns, and preliminary

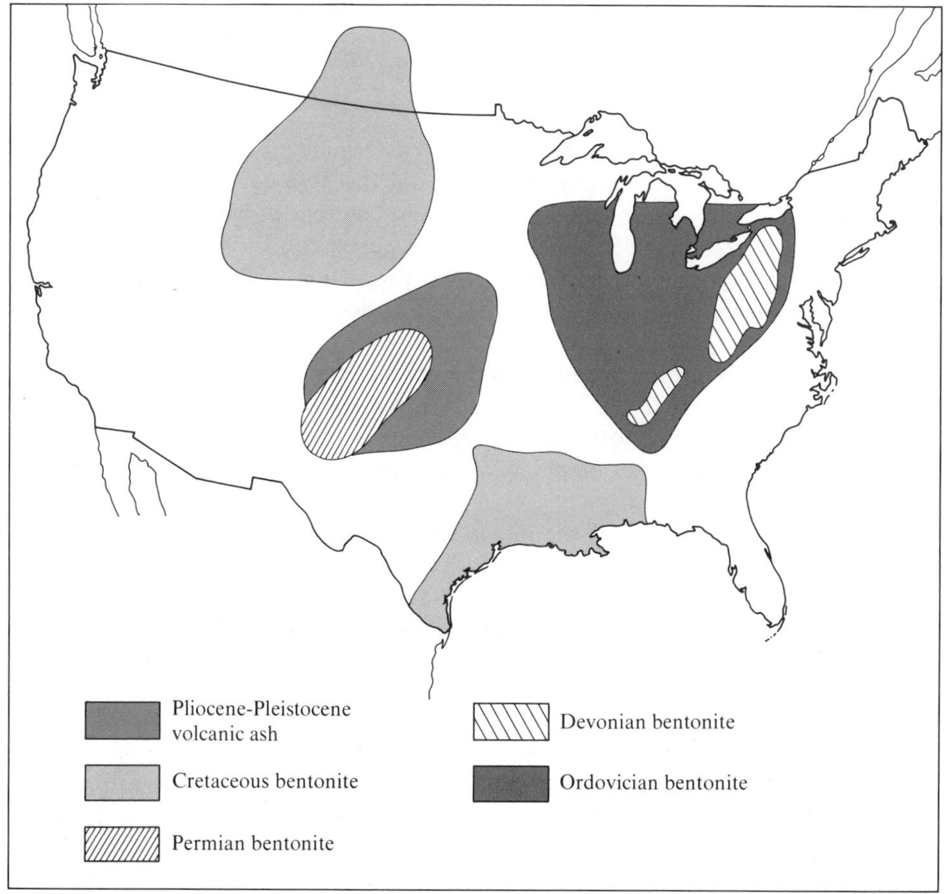

Figure 3–17
Distribution of volcanic ash and bentonite in conterminous United States. Each of the areas outlined includes a succession of beds of volcanic ash-fall material. [G. P. Eaton, 1978, in R. W. Fairbridge and J. Bourgeois, eds., *Encyclopedia of Sedimentology.* Copyright © 1978 by Dowden, Hutchinson & Ross, Inc., Stroudsburg, Pa. Reprinted by permission of the publisher.]

studies of Cretaceous bentonites support the view that the upper-air circulation pattern during the Cretaceous Period was essentially like that of the present (Slaughter and Early, 1965).

From an economic point of view, bentonites are important as bases for the drilling mud used in oil exploration. Unfortunately, the swelling characteristic that makes them desirable in these muds causes them to form very unstable slopes in outcrop. Construction projects on mudrocks that contain montmorillonite require great care. From a general academic point of view, bentonites are important because they are among the rare mudrocks that provide a clear description of the rocks from which they formed.

SOURCE AREAS

One of the tasks of sedimentary petrology is to determine *paleogeography* and *provenance* (source area). From which directions did the mud come? Did the drainage basin from which the muds were derived contain outcrops of igneous and metamorphic rocks, or were only older sedimentary rocks exposed? If crystalline rocks were exposed, of what types were they and in what proportions? The aim is to construct paleogeologic maps for earlier periods that are as accurate and detailed as geologic maps of present outcrop patterns.

The chief clue to upstream geology is the mineral composition of the downstream sediment accumulation. If the sediment contains kyanite, it means that high-grade aluminous schists were exposed in the drainage basin. Similarly, the presence of labradorite implies gabbro or amphibolite; strongly abraded quartz sand means that older sandstone was present.

How can an examination of the mineral composition of mudrocks contribute to paleogeographic reconstruction? The major constituent of mudrocks is clay minerals. But, as we have seen, the intensity of weathering and diagenetic alteration are at least as important as determinants of clay mineralogy in a mudrock as are the crystalline rocks from which the clays developed. Kaolinite can be formed from any aluminosilicate mineral. Both kaolinite and montmorillonite are often changed to illite after burial. Because of this, the character of the original crystalline rocks probably is determined better by minerals other than clays. If we are to obtain source-area information from mudrocks, we will have to concentrate on the other one-third of the mudrock: the part composed of quartz, feldspars, and miscellaneous accessory minerals. Unfortunately, however, only one such study has yet been published, and it was concerned only with the average size distribution of quartz grains in mudrocks. The usefulness of mudrocks for provenance studies is unknown.

ANCIENT MUDROCKS

Paleocurrent Indicators

It is much more difficult to determine transport directions in mudrocks than in sandstones. The coarser-grained rocks commonly contain such obvious current indicators as cross bedding, ripple marks, or flute casts (see Chapter 4). Mudrocks, on the other hand, usually lack these features except in the coarser siltstones. The strengths of currents from which muds are deposited are weak, and directional structures are correspondingly obscure or absent. With effort, however, directional indicators can be found; three types have been uncovered:

1. Progressive decrease in quartz grain size and percentage in the downcurrent direction.

2. Alignment of inequant quartz grains parallel to the depositing current.
3. Alignment of inequant fossils.

Blaine Formation (Permian), Western Oklahoma

The Blaine Formation is 35–40 m thick and consists of three shale units separated by three gypsum beds. It is of Middle Permian age and outcrops in a continuous north–south band extending for more than 600 km through Kansas, Oklahoma, and Texas (see Figure 3–18). It is present as well in the subsurface to the west and has been penetrated during drilling for petroleum and natural gas in the Anadarko Basin. Existing paleogeographic and stratigraphic studies establish that the nearest Permian land area lay to the east. Because of the brief period of time represented by Blaine

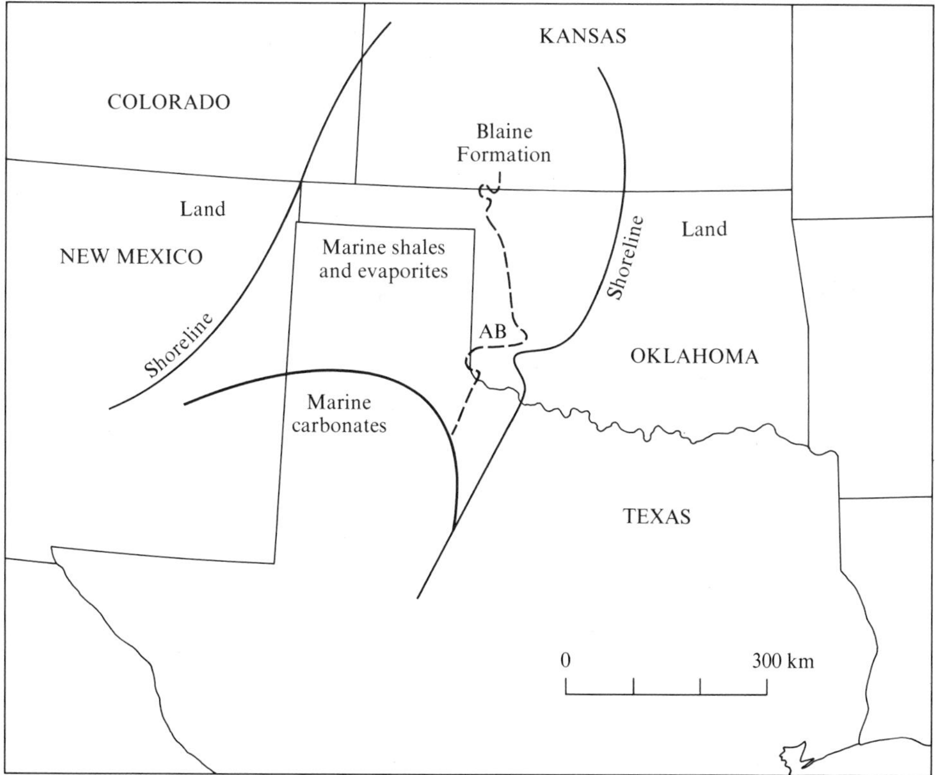

Figure 3–18
Paleogeography and principal facies in relation to outcrop pattern of Blaine Formation in Middle Permian basin of southwestern United States. AB = location of Anadarko Basin. [Modified from K. S. Johnson and R. E. Denison, 1973, *Geol. Soc. Amer. Guidebook No. 6.*]

deposition (perhaps a few million years), the position of the shoreline to the east of the outcrop belt can be assumed constant. Because of the thinness of the Blaine and the presence of three correlatable gypsum marker beds within the unit, stratigraphic position during a sampling program can be determined rather precisely.

Blatt and Totten (1981) collected 89 surface samples of the Blaine in western Oklahoma to determine whether there exists a decrease in mean grain size and percentage of quartz with increasing distance from the Middle Permian shoreline and whether the decrease is regular enough to be useful in studies of ancient marine shales. The quartz grains were isolated from the shale by the sodium bisulfate fusion technique, and size distributions were determined by sieving. X-ray analyses revealed that the fusion residue contained only 5% feldspar.

Figure 3–19 shows the relationship between the mean grain size of quartz in the Blaine Formation and the distance from the shoreline, or, more precisely, from the line offshore where sands grade into muds. It is clear that a strong relationship exists; grain size decreases as distance from the sand–mud line increases. The correlation between the two variables is highly significant statistically. Analysis reveals that there is less than one chance in a hundred that a trend this well defined could exist by chance alone ($r = .71$). It is noteworthy that the graph predicts the mean grain size of quartz at the sand–mud line to be 4.75 ϕ, a coarse silt. This seems quite reasonable for the transition point offshore from sand to muddy sediment.

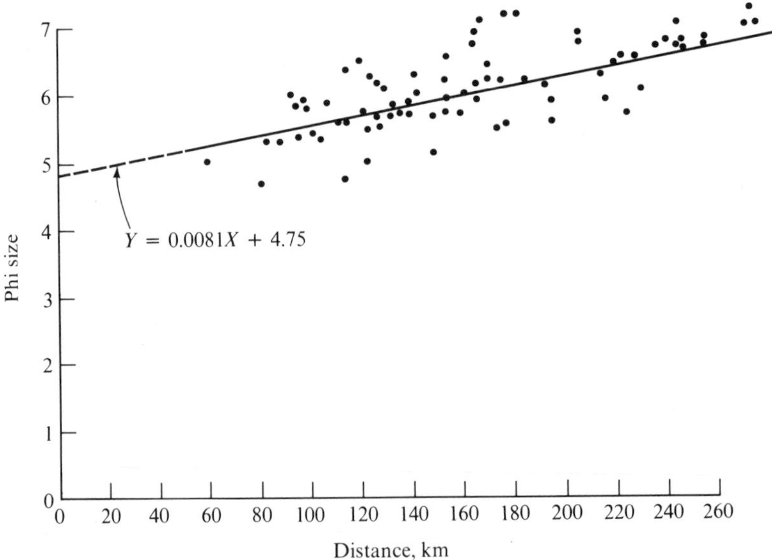

Figure 3–19
Relationship between mean grain size of detrital quartz and distance from sand–mud line offshore of beach in Blaine epicontinental sea. [Blatt and Totten, 1981.]

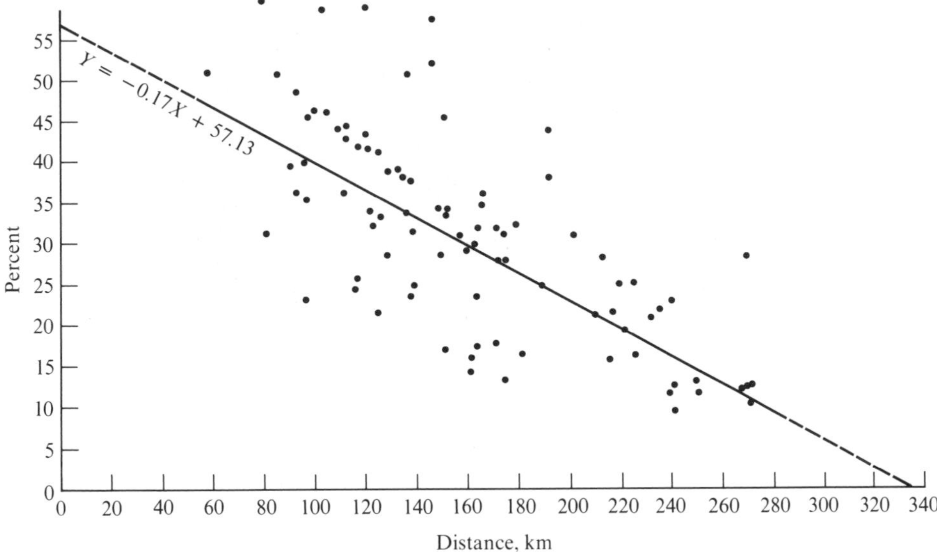

Figure 3–20
Relationship between percentage of detrital quartz and distance from sand–mud line offshore of beach in Blaine epicontinental sea. [Blatt and Totten, 1981.]

Figure 3–20 shows the effect of increasing distance from shore on the percentage of detrital quartz in Blaine Formation sediment. Once again, a clear relationship is evident ($r = .70$). The percentage of quartz decreases as distance from the sand–mud line increases; the percentage of clay increases correspondingly. Extrapolation of the line of best fit to the sand–mud depositional boundary in the Blaine sea indicates the quartz/clay ratio to be about 1/1.

Gowlaun Member (Silurian), Western Ireland

The Gowlaun Member is a continental slope/deep-sea fan deposit several hundred meters thick composed of graded, laminated mudrocks overlying graded, fine-grained turbidite sandstones. Piper (1972) collected compass-oriented samples from the Gowlaun Member to determine how the laminated silts and muds were deposited and, as part of the investigation, determined the direction of current flow during deposition of the beds. Figure 3–21 is a thin-section photomicrograph of the Gowlaun. All the grains larger than 20 μm in a 2 mm wide strip of thin section normal to the bedding were measured, and the results are shown in Figure 3–22. The figure shows that the finer silt grains are present throughout the 16 mm section of rock examined. There is, however, an upward decrease in the size of the largest silt grains and an upward decrease in the proportion of silt in the rock (fewer data points toward the top of the section).

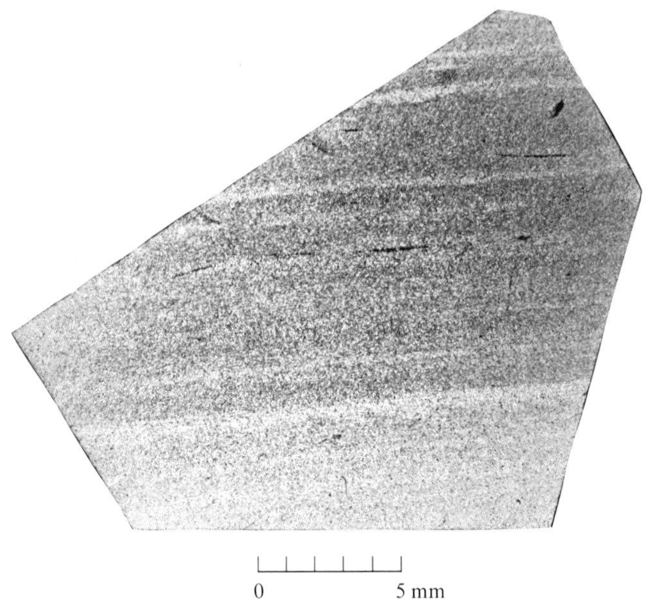

Figure 3–21
Thin-section photomicrograph of graded, laminated, silty mudstone, Gowlaun Member (Silurian), western Ireland. [Piper, 1972. Photo courtesy D. J. W. Piper.]

Thin sections cut parallel to bedding planes in the Gowlaun Member show a preferred orientation of silt grains in silt laminae and an apparently random orientation in mud laminae (see Figure 3–23). Presumably, the difference results from the fact that the proportion of silt in a mudrock lamina reflects the strength of the current from which the particles were deposited. Coarser and/or more abundant silt reflects stronger currents and a better chance of producing a preferred orientation of inequant quartz grains. The thicker silt laminae contain 20–65% silt coarser than 10 μm; the thinner silt laminae and clay-rich laminae contain only 0.3–2% silt grains coarser than 10 μm.

Chattanooga Shale (Devonian), Northeastern Tennessee

Current directions can also be determined in mudrocks through the use of inequant fossils such as graptolites, cephalopods, gastropods, ostracods, brachiopods, and plant spores. An excellent example is given by Jones and Dennison (1970). Figure 3–24 illustrates orientation diagrams (360° bar graph) for six fossil taxa measured on bedding surfaces of the Chattanooga Shale (Devonian) in northeastern Tennessee. It is apparent that strong preferred orientation is present in most of the taxa, although

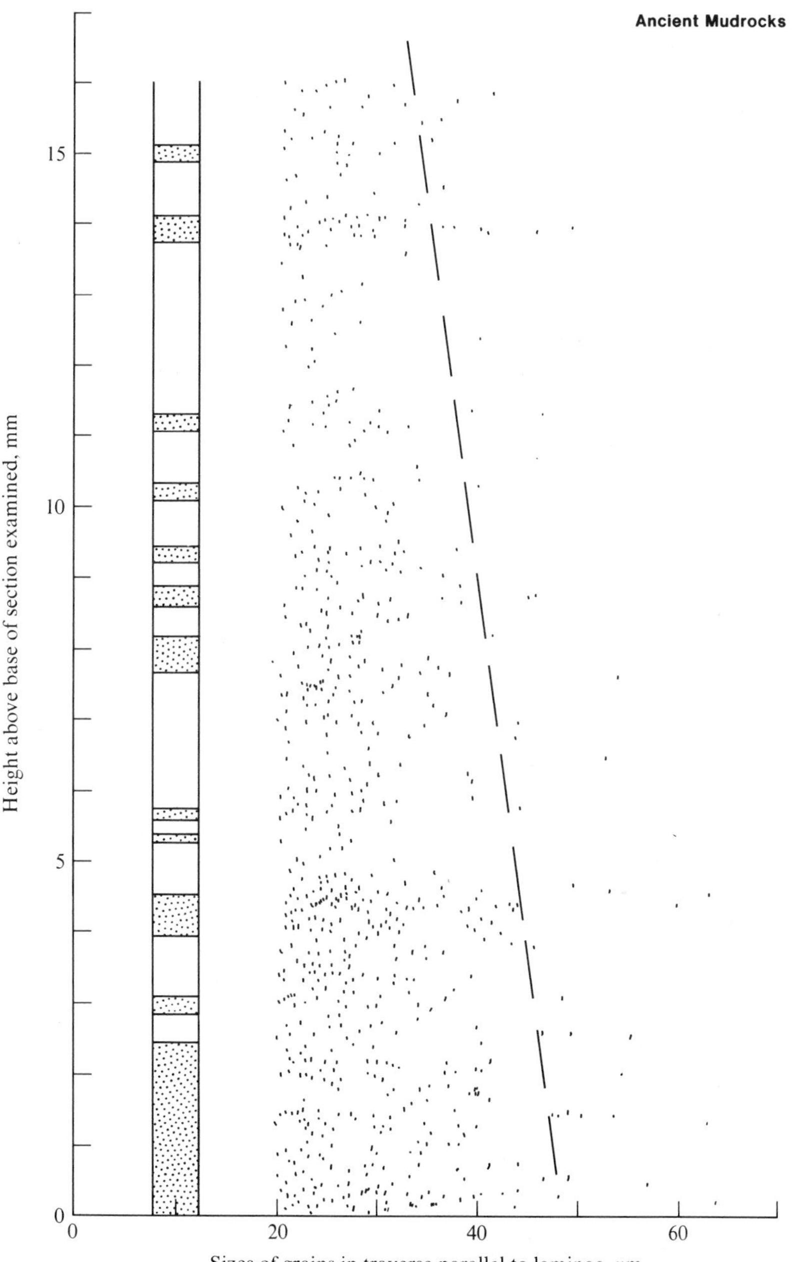

Figure 3–22
Variation of quartz grain size in section 16 mm thick in thin section of Gowlaun Member. Column on left indicates main silty laminae. Dashed line is line of best fit for four coarsest grains per 1 mm vertical interval within 16 mm section examined. [Piper, 1972.]

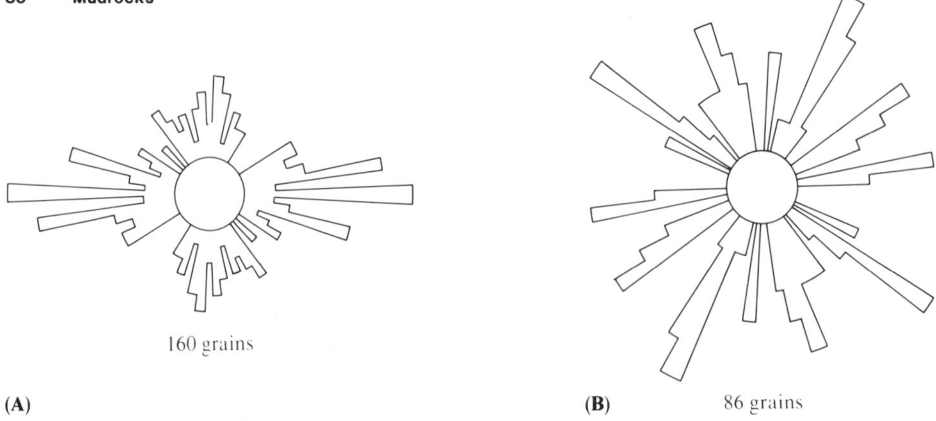

Figure 3–23
Orientation of silt-size quartz grains in Gowlaun Member. (**A**) Silt lamina. (**B**) Clay-rich lamina. Strong preferred orientation is evident in silt-rich lamina; orientation is random in clay-rich lamina. [Piper, 1972.]

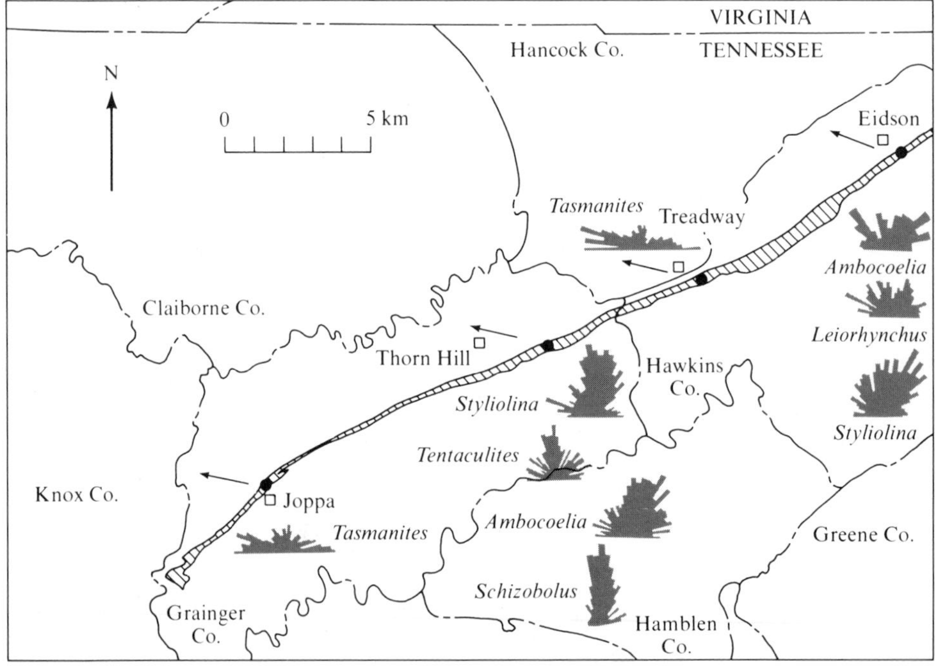

Figure 3–24
Paleocurrent map based on preferred orientations of fossils in Chattanooga Shale (Devonian) in northeastern Tennessee. Ruled area is outcrop of shale. *Ambocoelia, Schizobolus,* and *Leiorhynchus* are brachiopods elongate normal to hinge line. Direction measured in field was hinge-line orientation, so that direction or orientation shown on map appears to be 90° from paleocurrent direction. [Jones and Dennison, 1970.]

the relationship between the shape of the organisms and the direction of current flow is not always what we might predict. For example, although the long axis of *Tasmanites* (an elongate plant spore) parallels the current direction as determined from cross bedding in interbedded sandstones, the orientations of the cone-shaped fossils such as *Tentaculites* and *Styliolina* are normal to the current direction. This results from the small size of the cones, which causes them to roll downcurrent rather than assume a stable position pointing into the current. Larger cone-shaped fossils such as the Mesozoic straight-shelled cephalopods do orient themselves parallel to the flow direction of the depositing current. We can conclude that the orientations of fossils must be used with some care.

Petrology

There have been very few studies of the mineral composition and petrology of ancient mudrock units, and even fewer that have tried to relate the petrology of a specific formation to depositional environment. Two of these few are the investigations by Scotford (1965) and Weiss et al. (1965) of shales of Cincinnatian age (Late Ordovician) in Ohio, Indiana, and Kentucky.

The Cincinnatian Series in the area studied is about 200 m thick and outcrops in an oval-shaped band 40 km wide around the Jessamine Dome on the Cincinnati Arch (see Figure 3–25). It consists of shales and fossiliferous limestones deposited in one of the many shallow epeiric seas that covered parts of North America during Paleozoic time. The mudrocks range from 50% to more than 90% of the series.

Nearly all the siltstone and shale beds are laminated and often interfinger with either the edges or the surfaces of the interbedded limestones. Whole valves or bits of skeletal material scattered on a parting plane in shale may increase in amount laterally into a distinct bed of limestone, with the lamination in the argillaceous beds bending to conform to the shape of the limestone layers. Apparently, the carbonate sediment was lithified before the muds, and the limestones served as rigid units during mud compaction. Some of the silty mudrocks are ripple-marked or cross-laminated but not the fissile units (shales). Presumably, this indicates that the clay-mineral floccules were too small in size (< 10–$20\ \mu m$) to permit this type of current-generated structure to develop.

The mudrocks are medium- to fine-grained silts with less than 3% sand. No variation in grain size is present either laterally or vertically within the series, but size variation sometimes occurs from the middle to the boundaries of individual mudrock beds; coarser sizes tend to occur near the contacts with adjacent limestone beds. Perhaps this reflects a correlation between increased current strength and an increased abundance of carbonate-secreting organisms on the shallow seafloor during Cincinnatian time.

Most of the mudrocks contain some calcite in the form of both shell fragments and micrite of uncertain origin. The bulk of the mudrocks, however, is clay minerals (72%

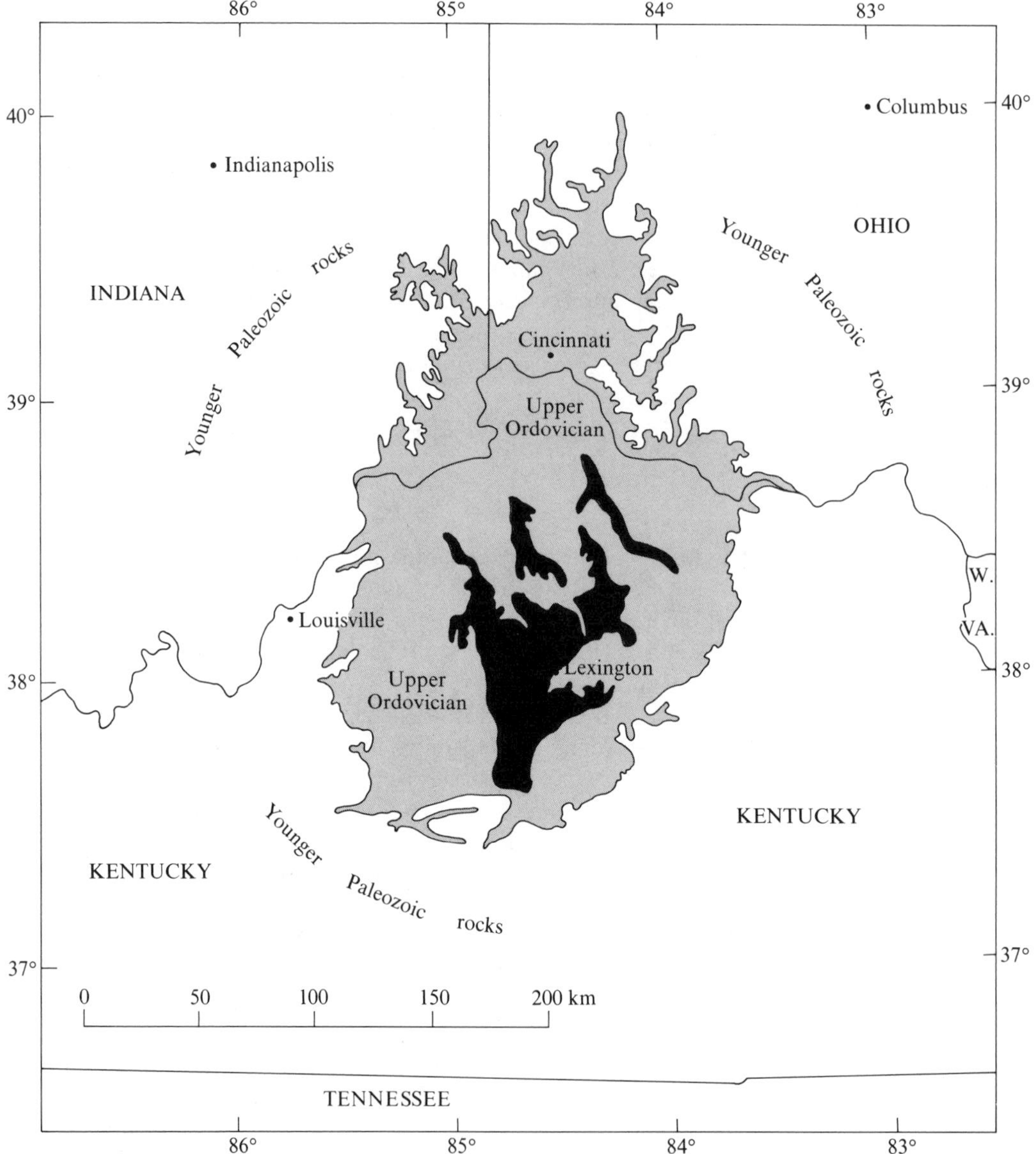

Figure 3–25
Geologic map of Jessamine Dome area, showing Cincinnatian (dark color) outcrop pattern.

of the noncarbonate fraction), mostly illite but with lesser amounts of chlorite. Angular, chemically etched, silt-size quartz forms almost all the rest of the insoluble fraction; trace constituents include orthoclase, plagioclase, biotite, and a few stable, detrital heavy minerals such as garnet and zircon. Authigenic pyrite occurs in some mudrocks. The uniformity and simplicity of mineral composition and fine grain size throughout the Cincinnatian mudrocks indicate a uniformity of source terrane, specifically an extensive area with very low relief and no nearby significant exposures of crystalline rocks. Based on present knowledge of Early Paleozoic paleogeography and tectonics in the mid-continental area, the nearest crystalline rocks were those of the Canadian Shield in eastern Canada.

Many different types of mineralogic and element analyses were made of the Cincinnatian mudrocks, and most showed no trends with distance from the axis of the Cincinnati Arch. Environmental conditions apparently were static. Statistical treatment of the data provided no evidence for the existence of the north–south trending Cincinnati Arch in Ordovician time. Detailed study of the Clarksville Member of the series, however, did reveal numerous trends suggesting the presence of an east–west topographic high. For example, there were consistent north–south increases in the percentages of sand and silt, a decrease in the amount of clay, and a linear increase in the silt/clay ratio. The variations were interpreted to result from changes in depositional environment during Clarksville time. The increase in detrital grain size indicates that current strengths were greater toward the south, suggesting shoaling toward an area of more abundant organic growth, particularly of carbonate-shelled organisms.

It is clear that analyses of the nonclay fraction of mudrocks deposited in epeiric seas can yield important paleogeographic information. The early stages of growth of an underwater structural or topographic feature can be detected many millions of years before the structure is emergent and sheds coarser clastic debris.

MODERN MUD ENVIRONMENTS

As noted earlier, there exist many environments in which muds can accumulate. These environments can be either nonmarine or marine, shallow or deep, aerated or stagnant. The goal in the remainder of this chapter is to describe a few muddy depositional settings and to interpret the mineral composition of the muds in terms of source area, climate, and water circulation.

Santa Clara River, California

There have been few studies of the total mineral composition of river muds, although numerous investigations have been made of clay-mineral compositions in various fluvial environments. One exception is a study of soil and river mud in the Santa Clara drainage basin, located about 80 km northwest of Los Angeles. In order to establish a genetic sequence of mineralogic changes, samples were collected from the bedrock (which is approximately 75% sedimentary and 25% igneous plus metamorphic), *in situ* soil, riverbed, suspended-stream sediment, and continental-shelf sediment in the area where the river empties into the Pacific Ocean.

The drainage basin is 105 km in length from west to east, 60 km in maximum width, and about 4,100 km^2 in area. The stream gradient is rather steep in the mountainous headwaters, averaging 30 m/km, but is only 3 m/km on the coastal plain. The steepest tributaries have gradients of 60 m/km. Annual rainfall averages 60 cm in the headwaters and 40 cm on the coastal plain. Mean annual temperature is 10°C.

In the Santa Clara drainage basin the mud is derived almost entirely from disaggregation of the sedimentary rocks and the poorly developed soils. Less than 10% of the mud is obtained directly from the igneous and metamorphic rocks in the basin. In mud coarser than about 15 μm, the mineral composition clearly reflects the source rock from which it came. Rocks exposed in the upper 20 km of the basin are nearly all igneous and metamorphic, about two-thirds granite plus gneiss and about one-third anorthosite; small outcrops of schist also occur in this area. The ratio of plagioclase to quartz reflects these facts (see Figure 3–26). It is consistently greater than 3/1 in the upper 20 km and reaches a value of 7/1 in the heart of the crystalline rock area. Farther downstream, where outcrops are largely friable Tertiary sandstones and

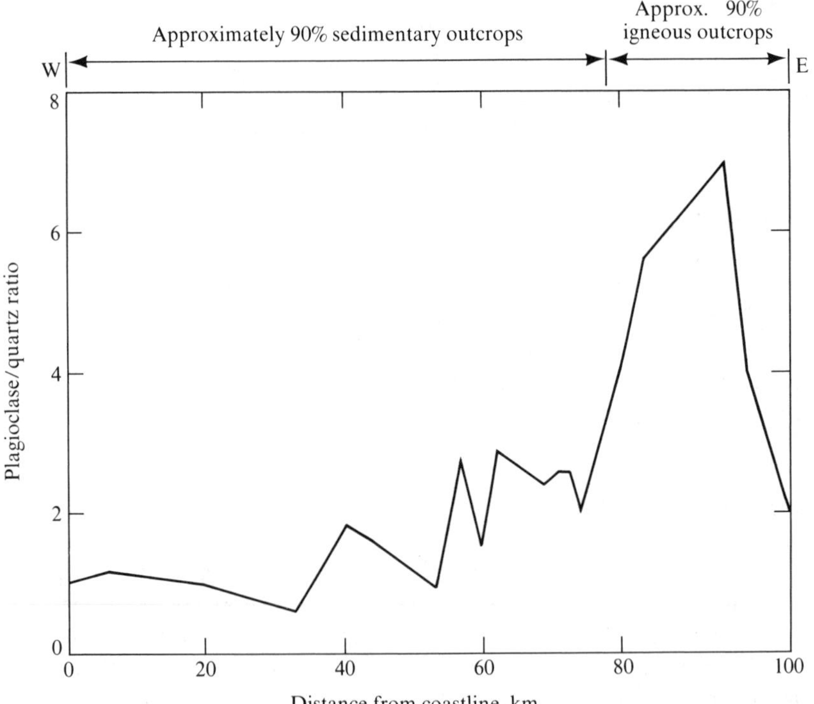

Figure 3–26
Relationship between distance of transport and ratio of plagioclase to quartz in silt fraction of sediment in Santa Clara River drainage basin. [Data from Fan, 1976.]

mudrocks, dilution rapidly reduces the plagioclase/quartz ratio to 1/1. Abrasion during transport may be a factor in the loss of plagioclase feldspar downstream, but the nearly instantaneous decrease in the ratio as the stream leaves the area of crystalline rocks suggests that the change in source is responsible rather than abrasion. Also, silt grains travel in suspension rather than traction, and few shattering impacts with other grains are possible. Without impacts during transport, there can be little mechanical reduction in grain size.

In the accessory-mineral suite of the 62–15 μm size fraction, there occur blue-green hornblende, actinolite, and sphene from the metamorphic rocks; rounded zircon, apatite, and epidote from the older sedimentary rocks. Ilmenite is the only distinctive mineral from the igneous outcrops, particularly from the anorthosite.

In the size fraction finer than about 15 μm, clay minerals begin to dominate the detrital mineral fraction. These minerals occur most commonly in aggregates, rather than as separate clay flakes. Probably, the aggregated structure was produced in the soil during weathering, but it is possible that part of the aggregation (flocculation) was generated during stream transport. As would be expected from the wide variety of source rocks in the drainage basin, the aggregates consist of montmorillonite, illite, kaolinite, and mixed-layer montmorillonite–illite; that is, all varieties of clay minerals are represented.

Eastern Mediterranean Sea

The Mediterranean Sea is a nearly landlocked basin enclosed on the north and east by Europe and the Middle East, on the south by Africa, and on the west by the juncture of Africa and Europe at the Strait of Gibraltar. The floor of the basin, where not covered by sediment, is oceanic basalt. The Mediterranean is an oceanic basin that formed in Late Tertiary time immediately south of the former position of the Tethys seaway. Most of the Tethys was destroyed at the close of the Mesozoic Era in the plate collisions that gave birth to the Alps, the Himalayas, and other mountain ranges that stretch in a west–east band across southern Europe, Turkey, Iran, Afghanistan, northern India, and southwestern China.

Although the Mediterranean Sea is nearly landlocked, its location on the northern edge of the low-latitude arid belt prevents it from becoming stagnant like the neighboring Black Sea (see below). Evaporation at the surface of the Mediterranean Sea causes the oxygen-rich surface water to become more saline and dense and to sink. The dense water flows westward out of the basin along the seafloor through the Strait of Gibraltar, which has a sill depth of 320 m. Less saline Atlantic Ocean water flows in near the surface above the denser Mediterranean water to replenish the basin water mass. Fresh, light water also is supplied from southern Europe and the shallow, aerated part of the Black Sea, through the Bosporus and the Dardanelles.

The topography and bathymetry of the Mediterranean area divide the basin into a western and an eastern section. The eastern section has a maximum water depth

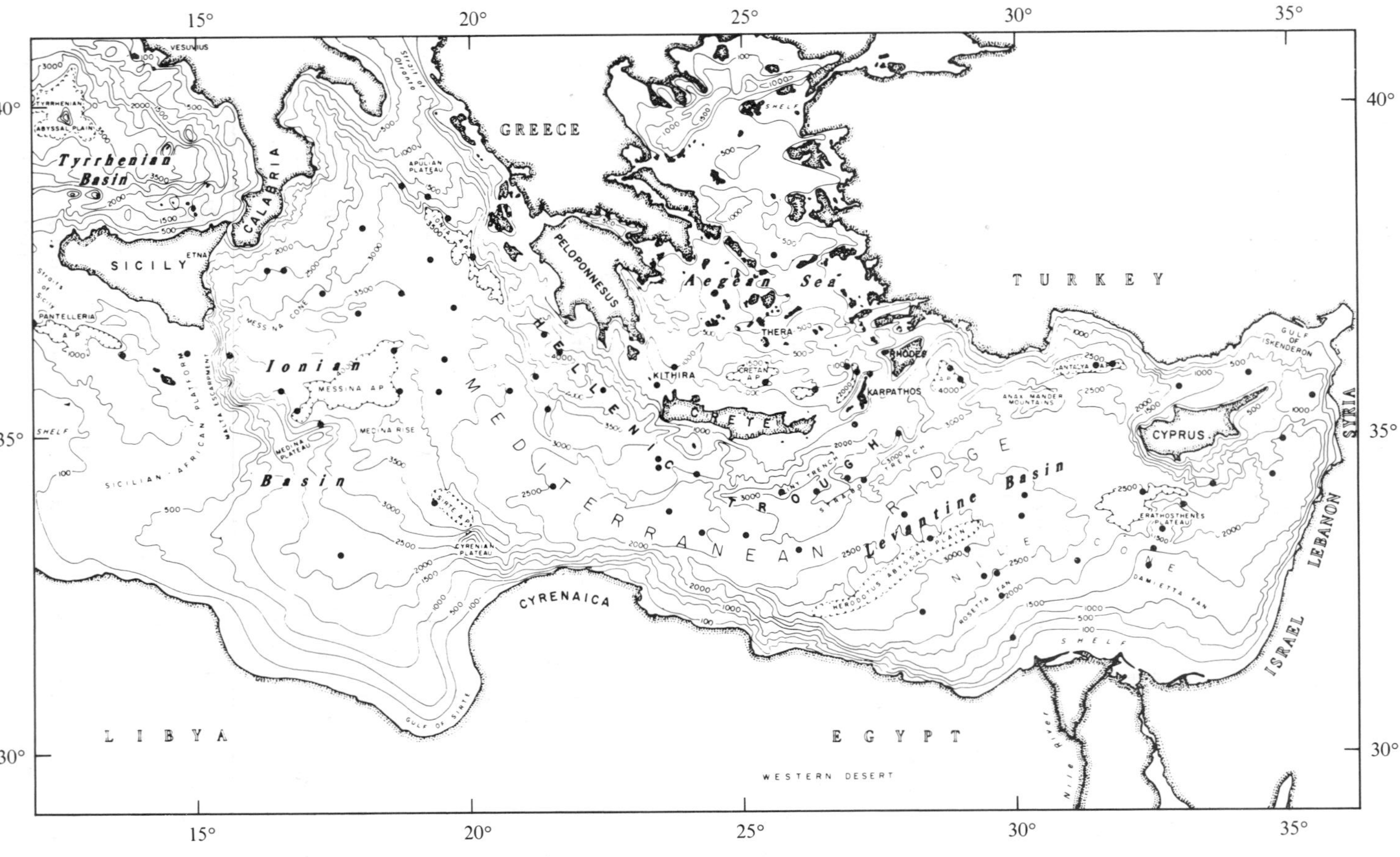

Figure 3–27
Bathymetry in eastern Mediterranean basin. Contours in meters. The dots are data points for Figures 3–28 and 3–29. [K. Venkatarathnam and W. B. F. Ryan, 1971, *Marine Geol., 11.*]

greater than 4,000 m and an average depth of about 2,000 m (see Figure 3–27). The distribution of bottom sediments, sands and muds, depends on bottom morphology. In areas with broad shelves, sands predominate; where the shelves are narrow, so are the belts of sand accumulation, and muds occur not far from the shoreline. The distribution of sediment grain sizes is also affected by climatic conditions. In the relatively humid zone of the eastern Mediterranean (Lebanon, Israel), mud production is high and sands are restricted to depths shallower than 50–100 m; muds cover most of the continental slope. In the more arid zone west of the Nile Delta, sands are prominent to outer-shelf depths of 100–250 m before being overwhelmed by muds. These observations are consistent with the facts that chemical weathering produces mud and that chemical weathering is minimal in arid regions.

The Nile River was the dominant supplier of sediment to the eastern Mediterranean basin until 1970, when the Aswan High Dam was completed 900 km upstream from the mouth of the Nile. Although much of the sandy material is derived from the Arabian Desert, the muds originated largely in the humid tropical upper reaches of the Nile where chemical alteration of Cenozoic basaltic volcanic rocks results in a steady stream of montmorillonitic and kaolinitic clays. As would be expected, the Nile muds also are markedly enriched in iron (7.8%) and titanium (1.4%). Many small globules of ferric hydrate and acicular rutile are present. These distinctive mineralogic characteristics enable the Nile River muds to be traced as they enter the eastern Mediterranean basin and are distributed by the currents.

Circulation of surface water in the eastern basin is counterclockwise, and the effect of this on the muds that poured from the Nile Delta is clearly shown by the distribution of montmorillonite (see Figure 3–28). It dominates the clay-mineral suite along the entire eastern Mediterranean coastline, from the Nile to southwestern Turkey.

Chlorite and illite, on the other hand, are generated in soils of higher-latitude areas with less intense conditions of chemical weathering. Such conditions occur on the northern side of the eastern Mediterranean basin and are clearly reflected in the marine muds west of the Turkish coastline, in the waters off the coast of southern Greece and Italy (see Figure 3–29). The control of clay-mineral assemblages by source rock, climate in the source area, and current pattern in the marine depositional basin is exceptionally clear in the eastern Mediterranean. The same controls are almost certainly present in the nonclay mineral fraction of the muds, but this fraction has not been studied.

Black Sea

The best-studied example of a modern depositional basin dominated by reducing conditions is the Black Sea, which was formerly a part of the Mesozoic Tethys seaway. Drilling in the Black Sea has not penetrated crystalline basement; but based on regional geology and seismic data, the floor underlying the basin is believed to be formed of metamorphosed sedimentary rocks. Sediments above the basement are 8–14 km in thickness and range from Cretaceous to Holocene in age.

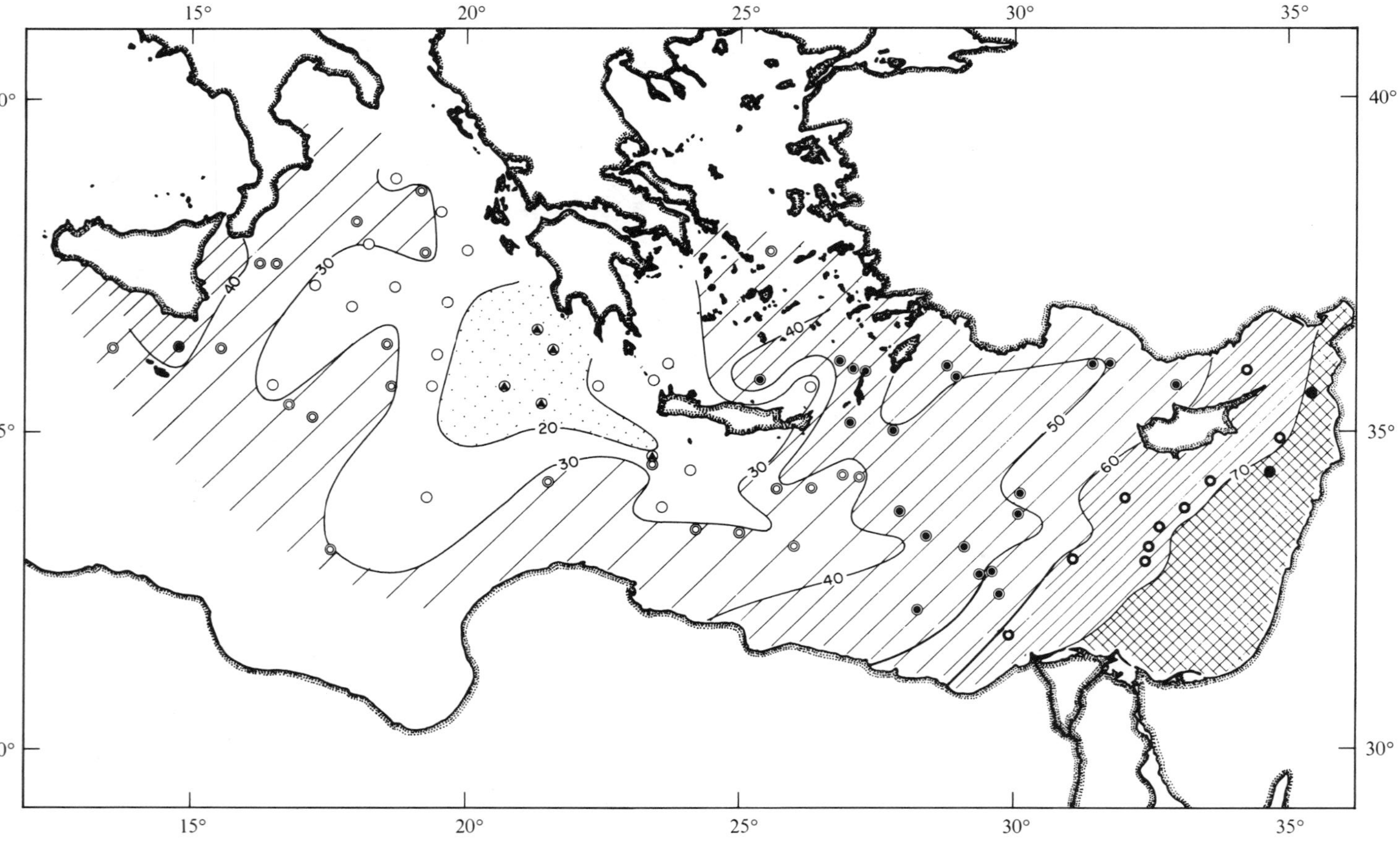

Figure 3–28
Percentage of montmorillonite in <2 μm fraction of clay-mineral assemblage of eastern Mediterranean basin. The circles are sampling sites. [K. Venkatarathnam and W. B. F. Ryan, 1971, *Marine Geol., 11.*]

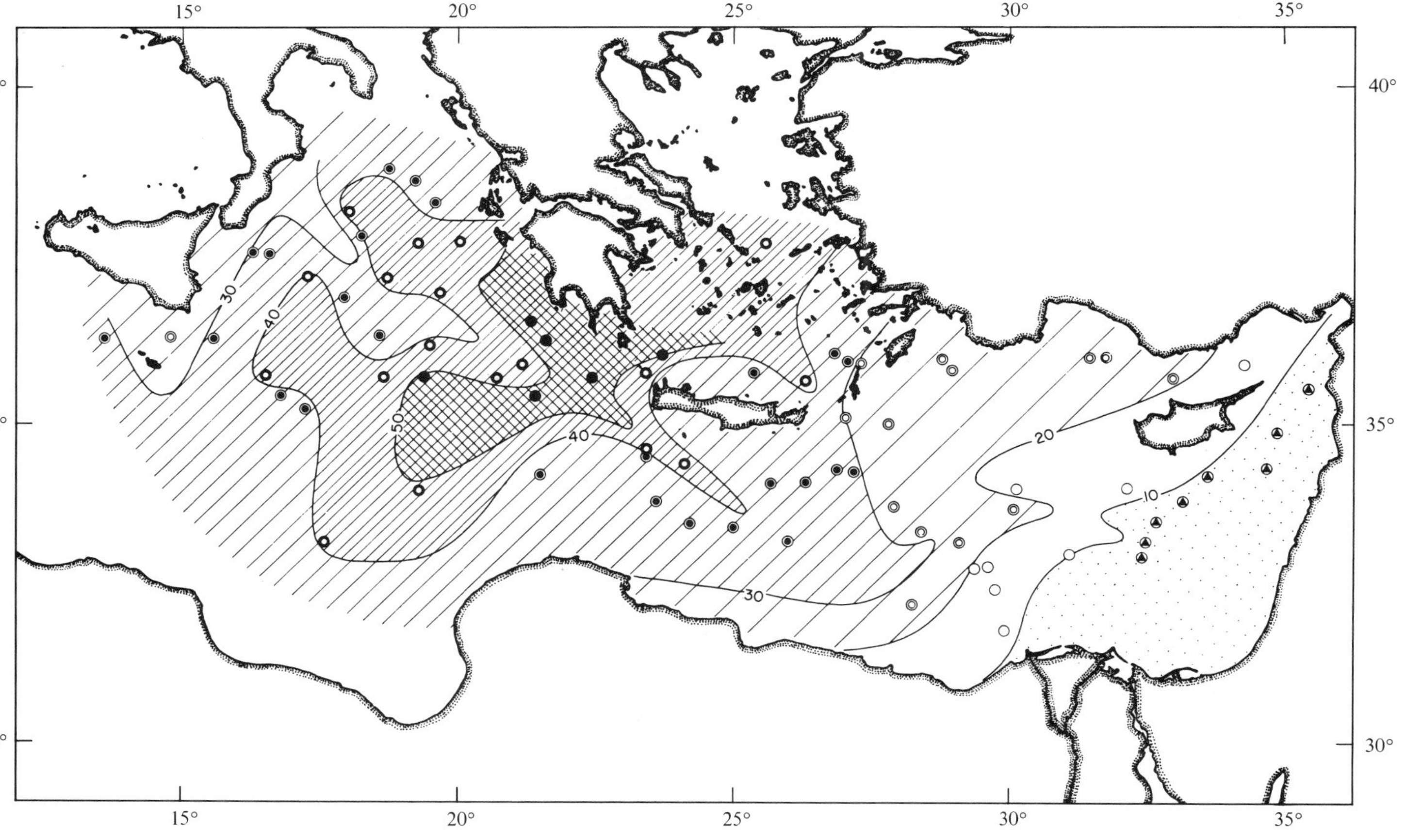

Figure 3–29
Percentage of illite in <2 μm fraction of clay-mineral assemblage of eastern Mediterranean basin. The circles are sampling sites. [K. Venkatarathnam and W. B. F. Ryan, 1971, *Marine Geol., 11.*]

Figure 3–30
Bathymetric chart of Black Sea. Note change in contour interval at 200 m and

Submarine slopes are quite steep in most nearshore areas, normally 4–6° and in places 12–14° (see Figure 3–30). Gentle slopes exist only along the Ukrainian and Romanian coasts on the northwestern side of the basin. Average water depth is 1,200 m, but depths as great as 2,260 m are known.

The water mass in the Black Sea is dominated by reducing conditions that resulted

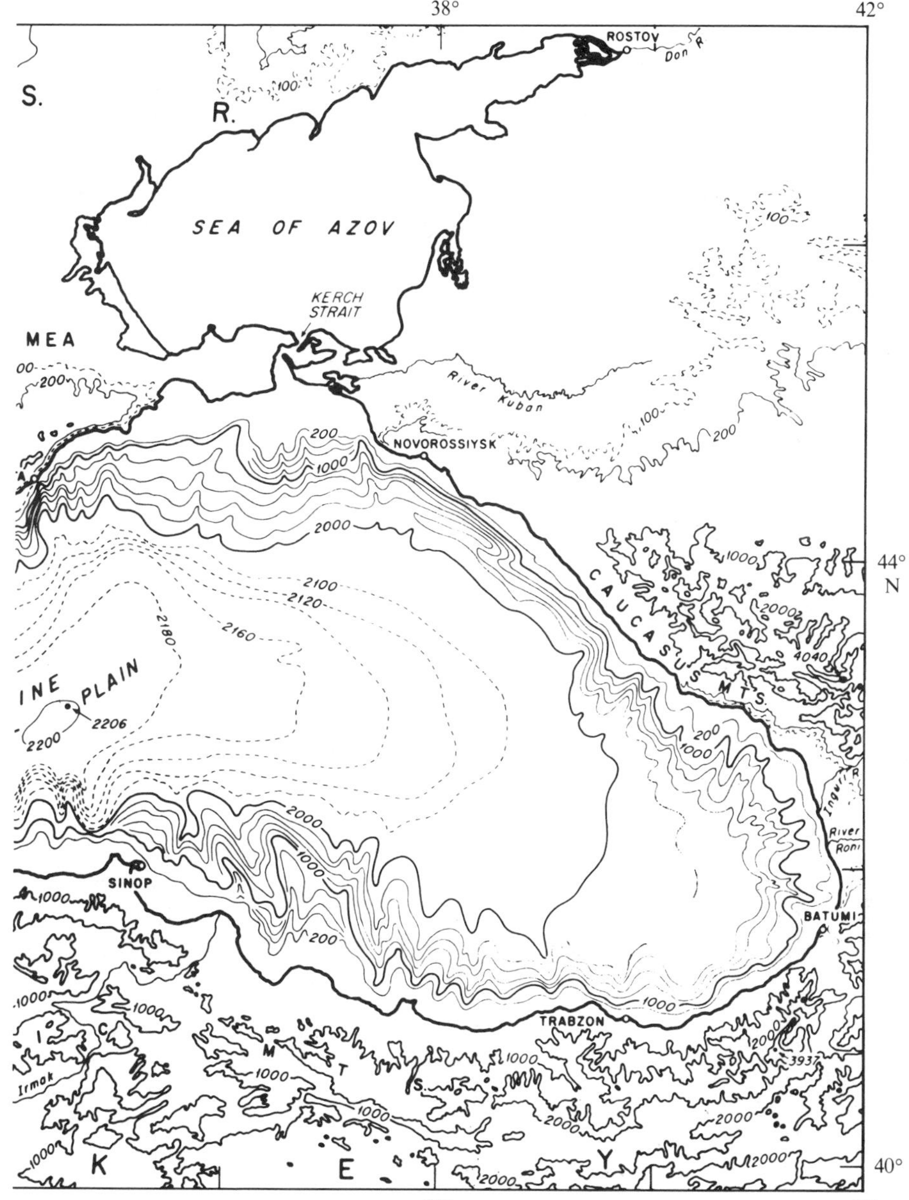

2,000 m. [D. A. Ross et al., 1974, Amer. Assoc. Petroleum Geol. Mem. No. 20.]

from the isolation of its water from the Mediterranean Sea about 7,000 years ago. At that time, circulation through the Bosporus Strait at the southwestern end of the Black Sea became very restricted as the undersea Bosporus ridge rose tectonically to within 40 m of the water surface. The ridge prevented renewal of Black Sea bottom water by oxygen-rich Mediterranean Sea water. Below a depth of 50–70 m the oxygen content of the water decreases steadily, and below 180 m there is no measurable amount of oxygen dissolved in the water (see Figure 3–31). Few organisms can survive when the content of dissolved oxygen is below 0.2 mol/L (0.5 ml/L).

The deep-water sediments deposited during the last 25,000 years in the Black Sea are distinguished by three sedimentary units (see Figure 3–32). The upper unit is a microlaminated layer, about 30 cm thick, consisting mainly of the remains of the coccolithophore *Emiliania huxleyi*. This alga lives in the upper, aerated part of the water mass. The microlaminations consist of alternating light layers which contain at least 40% white coccolith remains, and darker layers, which contain fewer coccoliths

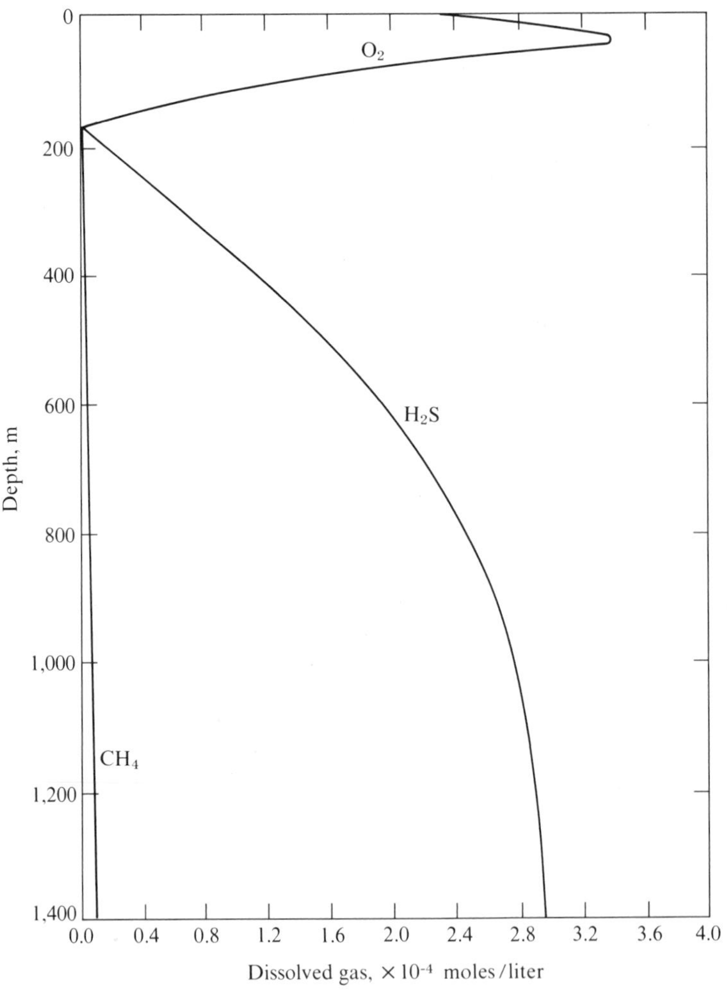

Figure 3–31
Typical profile of dissolved-gas concentrations in Black Sea. Oxygen maximum at 30 m results from metabolism of phytoplankton.

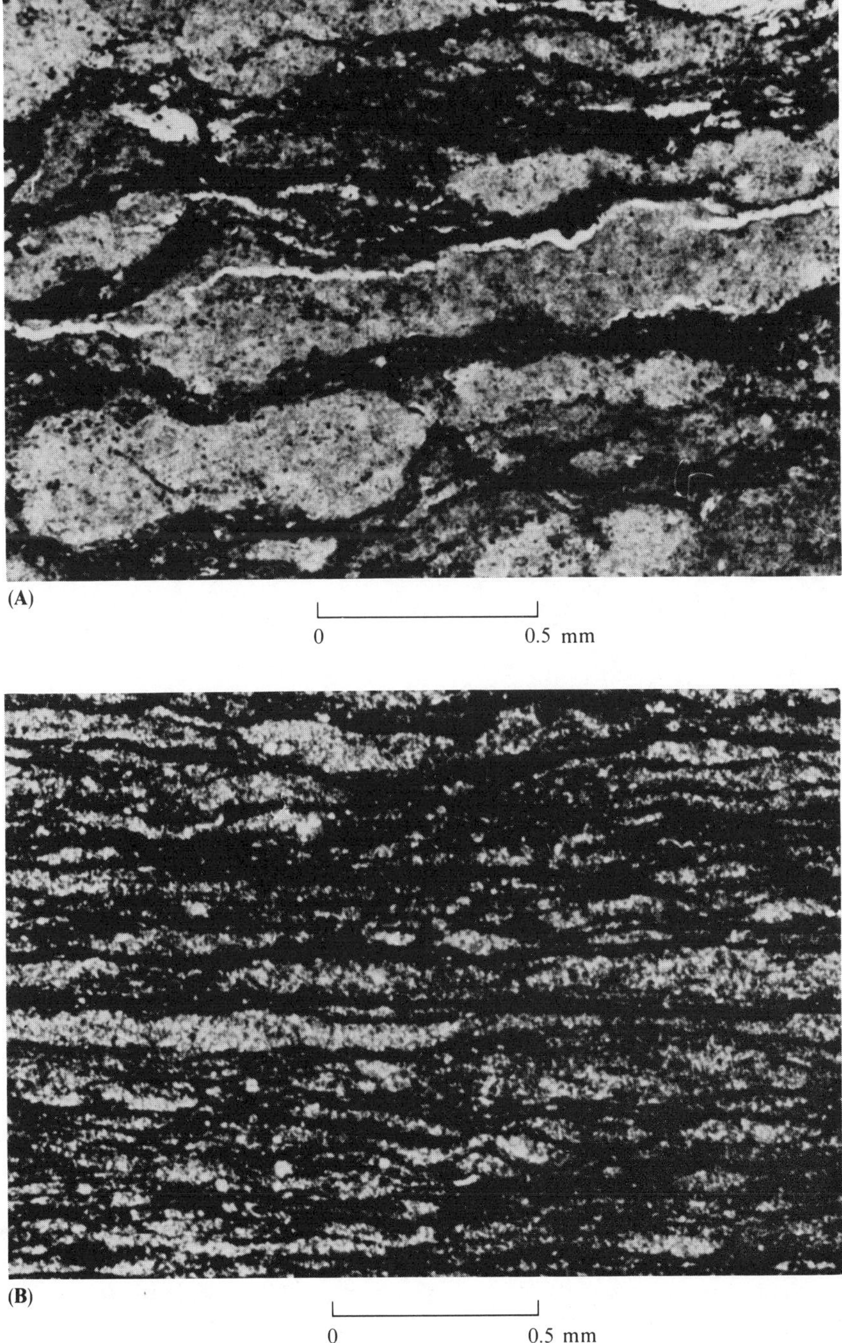

Figure 3–32

Photomicrographs of typical sediments from Black Sea. (**A**) Upper unit: coccolith mud and interstratified carbonate layers composed of coccoliths and terrigenous material. (**B**) Middle unit: layers rich in organic matter alternating with terrigenous silt layers.

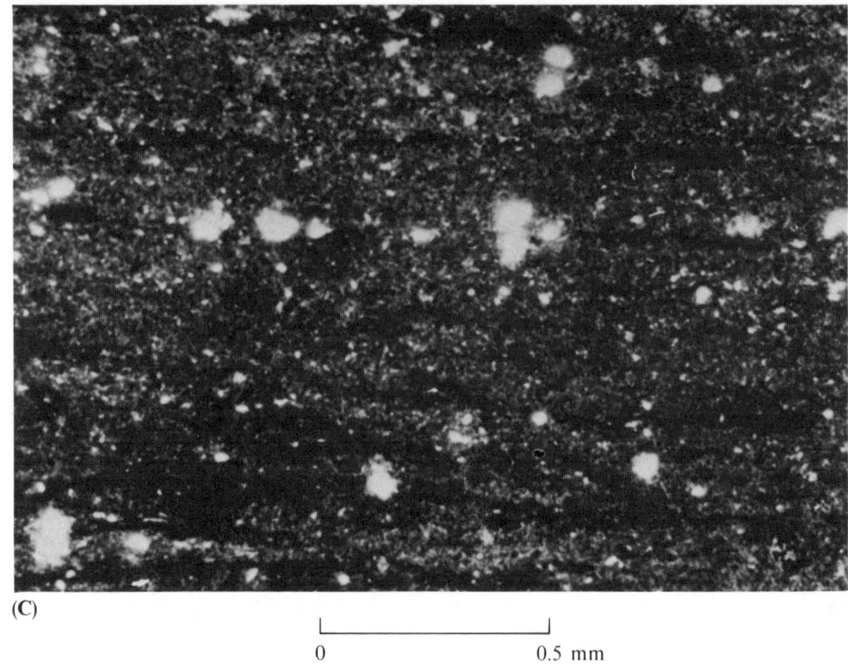

Figure 3–32 *(continued)*
(C) Lower unit: silt with thin layers of organic material. [G. Müller and P. Stoffers, 1974, Amer. Assoc. Petroleum Geol. Mem. No. 20. Photos courtesy G. Müller.]

and significantly more black organic matter. Each lamina is only 0.1–0.2 mm thick and required 30–60 years to accumulate, based on radiocarbon dating. The base of this unit has been dated at 3,000 years before the present.

Underlying the upper unit with sharp discontinuity is a *sapropel* (an ooze composed of plant remains) about 40 cm thick. It is vaguely laminated by light-colored skeleta of coccoliths or dinoflagellates and dark organic matter that is so well preserved that details of the plant cellular structures can be studied with the electron microscope. The organic matter in this unit is sometimes as much as 50% of the dry sediment weight. The 40 cm of sediment in this unit accumulated during the period 7,000–3,000 years before the present. The accumulation rate of 40 cm in 4,000 years is the same as that of the upper unit, whose formation required 3,000 years for 30 cm of sediment.

Underlying the sapropel with sharp discontinuity is a unit of uncertain thickness whose base is thought to be about 25,000 years old. It is microlaminated like the two upper units but contains less than 1% organic carbon. The dark laminae are formed by concentrations of unstable ferrous sulfides that quickly oxidize when the sediment cores are opened. In places, sandy and sometimes graded layers are present in this unit, indicating occasional slumping of sediment from shallow areas on the fringes of the Black Sea basin.

The three sedimentary units can be correlated over most of the Black Sea. Individual laminae, some as thin as 1 mm, can be correlated from one end of the basin to the other, a distance of about 1,000 km. This indicates extremely uniform depositional conditions during the past 25,000 years. Mud forms at least 90% of the inorganic fraction of the units almost everywhere, and the bulk of the mud is clay over at least half the basin floor. Organic carbon is an important component of bottom sediment everywhere, averaging about 2% with values of 4–5% being common (see Figure 3–33). The distribution of the clay-mineral species (see Figure 3–34) reflects the source rocks and climate surrounding the basin: illite and chlorite from silicic granites, gneisses, and schists in the western Caucasus and in the Danube River drainage basin; noticeable amounts of kaolinite from sedimentary rocks in the Danube River drainage; and montmorillonite from Eocene volcanic rocks along the northern Turkish coast. Quartz and feldspars (see Figure 3–35) are relatively minor constituents of Black Sea sediment because of the extremely fine-grained character of the sediment. As in most modern muds and ancient mudrocks, quartz is considerably more abundant than feldspar, except along the northern Turkish coast where the mafic volcanics that supply the montmorillonite clay also supply abundant feldspar. In this nearshore area it is common to find feldspar more abundant than quartz.

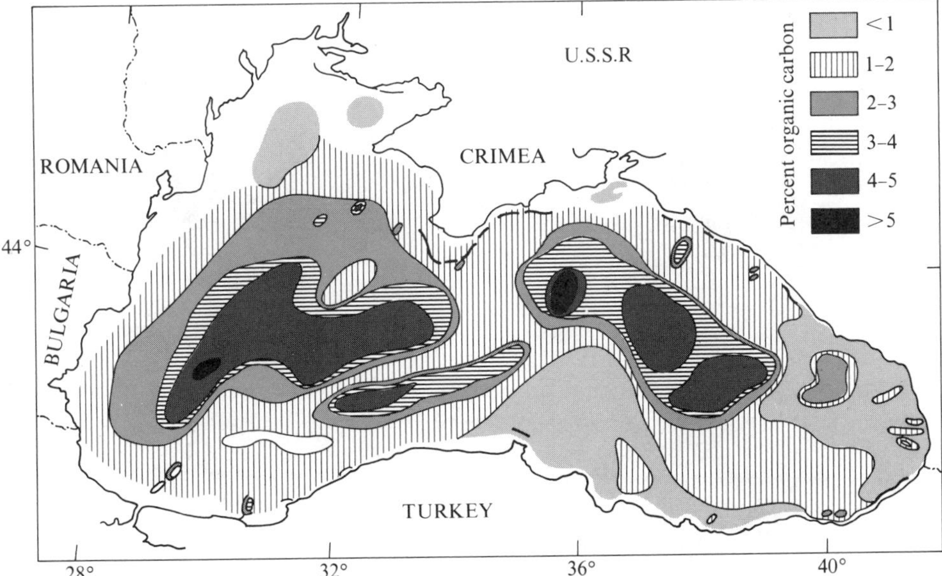

Figure 3–33
Content of organic carbon in modern Black Sea sediments. Heavy black line indicates areas not covered by modern sediments. [K. M. Shimkus and E. S. Trimonis, 1974, Amer. Assoc. Petroleum Geol. Mem. No. 20.]

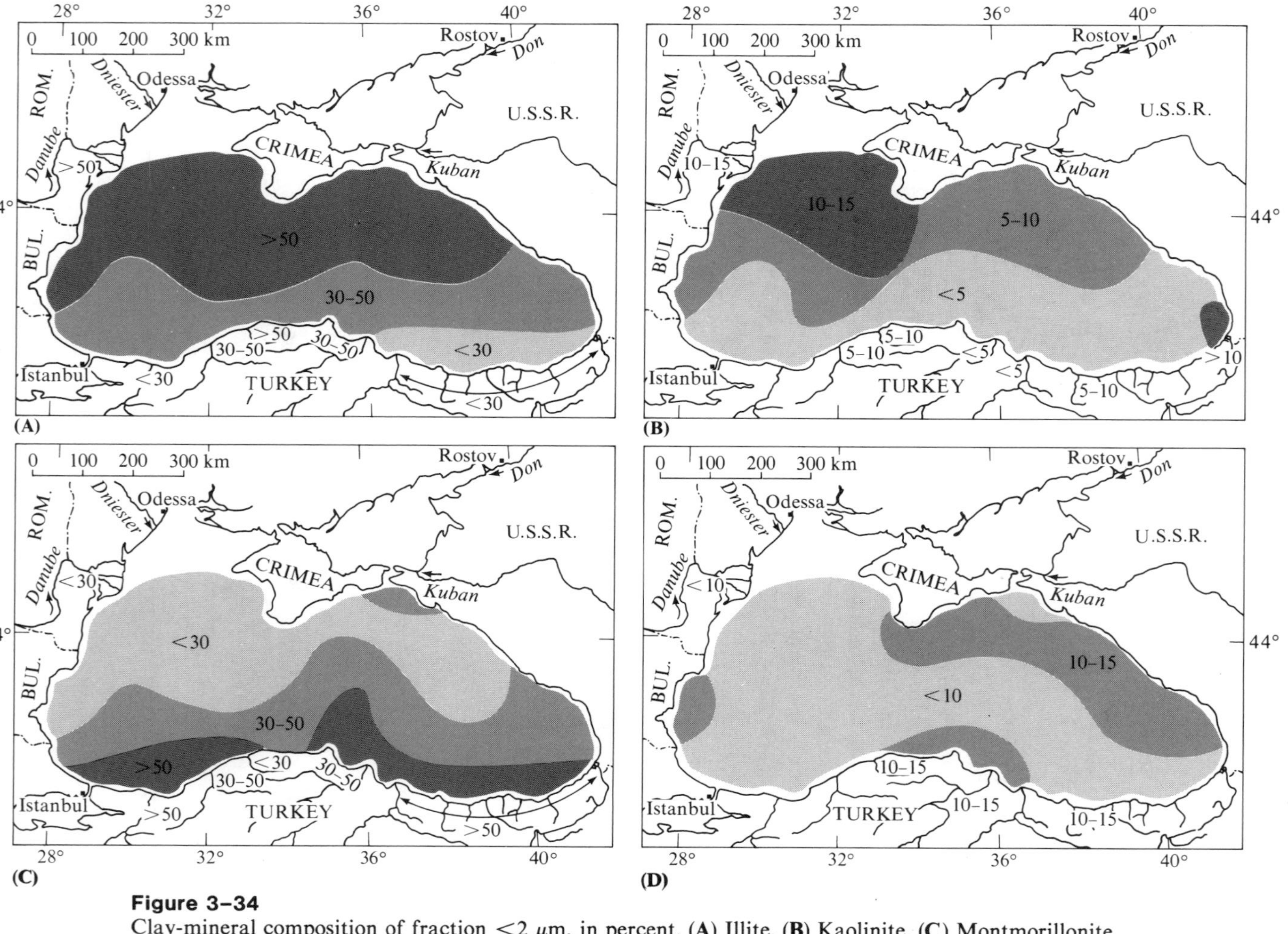

Figure 3–34
Clay-mineral composition of fraction <2 μm, in percent. (**A**) Illite. (**B**) Kaolinite. (**C**) Montmorillonite. (**D**) Chlorite. [G. Müller and P. Stoffers, 1974, Amer. Assoc. Petroleum Geol. Mem. No. 20.]

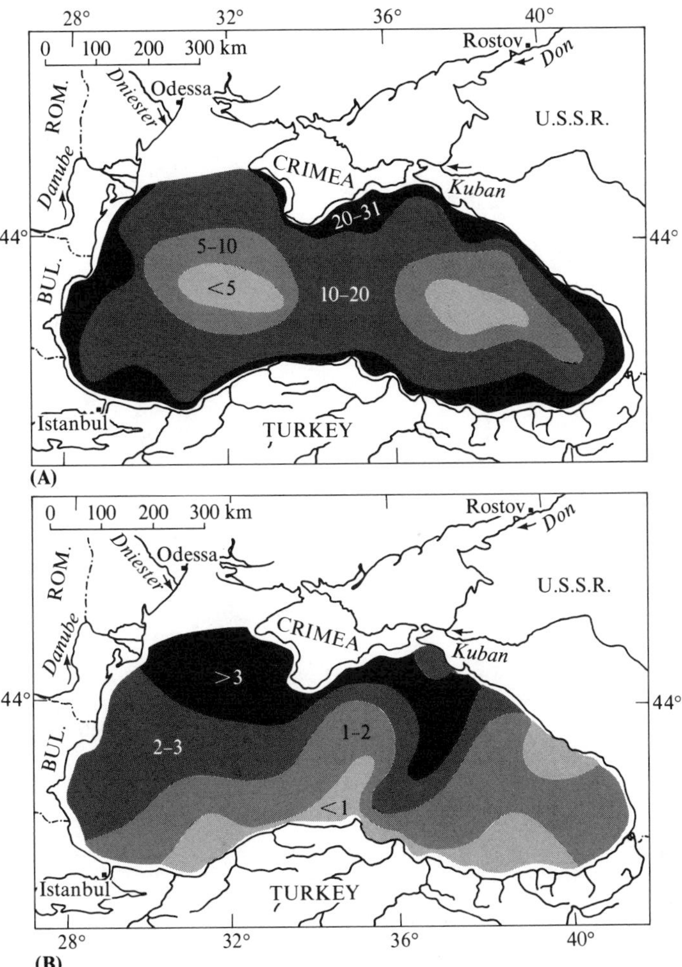

Figure 3–35
Quartz and feldspars in upper unit of Black Sea bottom sediment. (**A**) Percent quartz. (**B**) Quartz/feldspar ratio. [G. Müller and P. Stoffers, 1974, Amer. Assoc. of Petroleum Geol. Mem. No. 20.]

Organic matter has very high adsorptive capacities for trace elements—a fact long recognized by agricultural experts. Because of this, we would anticipate a correlation between the abundances of organic matter and the various minor elements in Black Sea muds. This is indeed the case (see Figure 3–36). Ancient mudrocks are also notable for their rich content of trace elements. In phosphatic mudrocks, for example, the value of the uranium obtained as a byproduct of industrial phosphate processing may exceed the value of the phosphate.

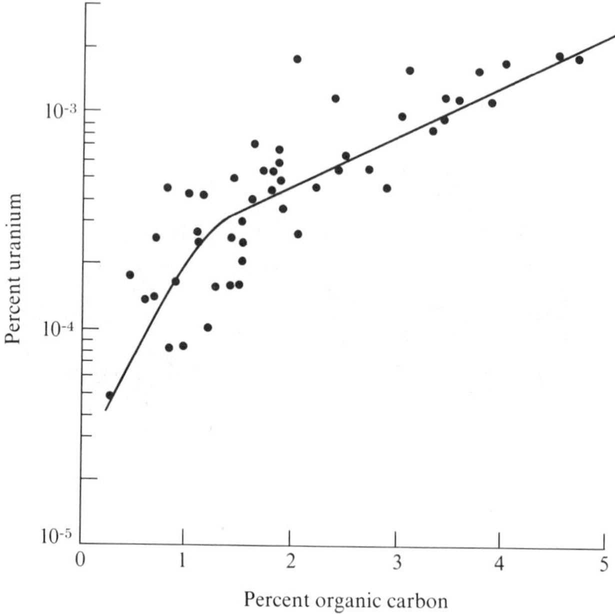

Figure 3-36
Correlation between amounts of organic carbon and uranium in Black Sea. [E. M. Emelyanov, in Stanley, 1972. Copyright © 1972 by Dowden, Hutchinson & Ross, Inc., Stroudsburg, Pa. Reprinted by permission of the publisher.]

Atlantic Ocean

The largest areally definable marine basins are the major oceanic basins: the Pacific, Atlantic, and Indian. During the past 15 years, since the advent of plate tectonics, interest in the sedimentology and petrology of oceanic sediments has increased significantly. Our understanding of both silicate and carbonate materials has improved as a result of the newer investigations. Our concern here is with the silicate materials, particularly the muds. We focus our attention on the Atlantic basin, but analogous descriptions can be given of sediments in the other oceans.

Clay minerals are the dominant mineral group on the floor of the Atlantic Ocean, as in most other depositional areas characterized by nonturbulent water. All of the major clay groups are present: gibbsite, kaolinite, montmorillonite, illite, and chlorite. This is what we would anticipate because of the very large size of the drainage area supplying the sediment and the wide range of rock types being weathered. Further, we would expect to find a close relationship between the climate on the adjacent land surface and the type of clay supplied to the basin. Chlorite, which is most susceptible to destruction, should be most common where chemical weathering is least effective:

in polar regions where water is perpetually frozen and therefore chemically inactive. This expectation is fulfilled. In the South Atlantic, for example, the abundance of chlorite among the clay minerals increases from less than 5% in equatorial waters to an average of 10% at 30°S and 20% at 50°S (see Figure 3–37). The same trend toward increasing amounts of chlorite is present north of the equator.

The reverse gradient in abundance is shown by kaolinite (see Figure 3–38). In low latitudes it forms 25–50% or more of the clay-mineral assemblage but decreases to less than 5% at 50° north and south of the equator. An even more intense gradient is shown by gibbsite, which is the "ultimate clay" in terms of the intense weathering required to produce it. Using the abundance of illite for comparison, we find that the

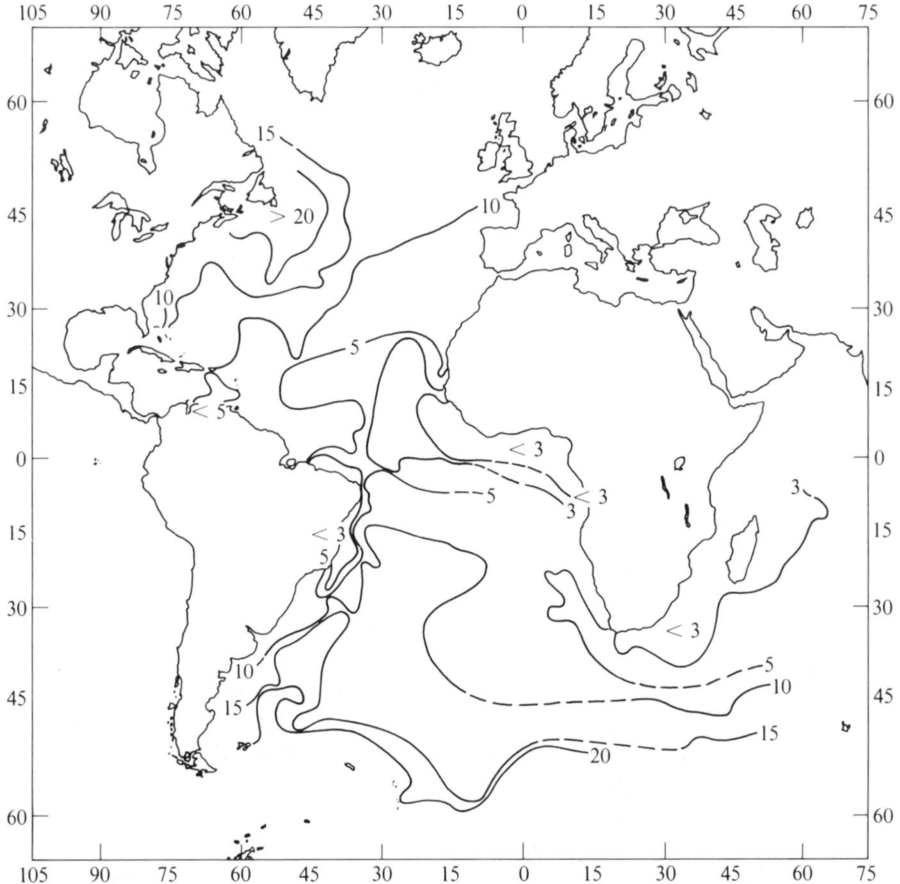

Figure 3–37
Percentage of chlorite in <2 μm size fraction of Atlantic Ocean basin sediment. [Biscaye, 1965.]

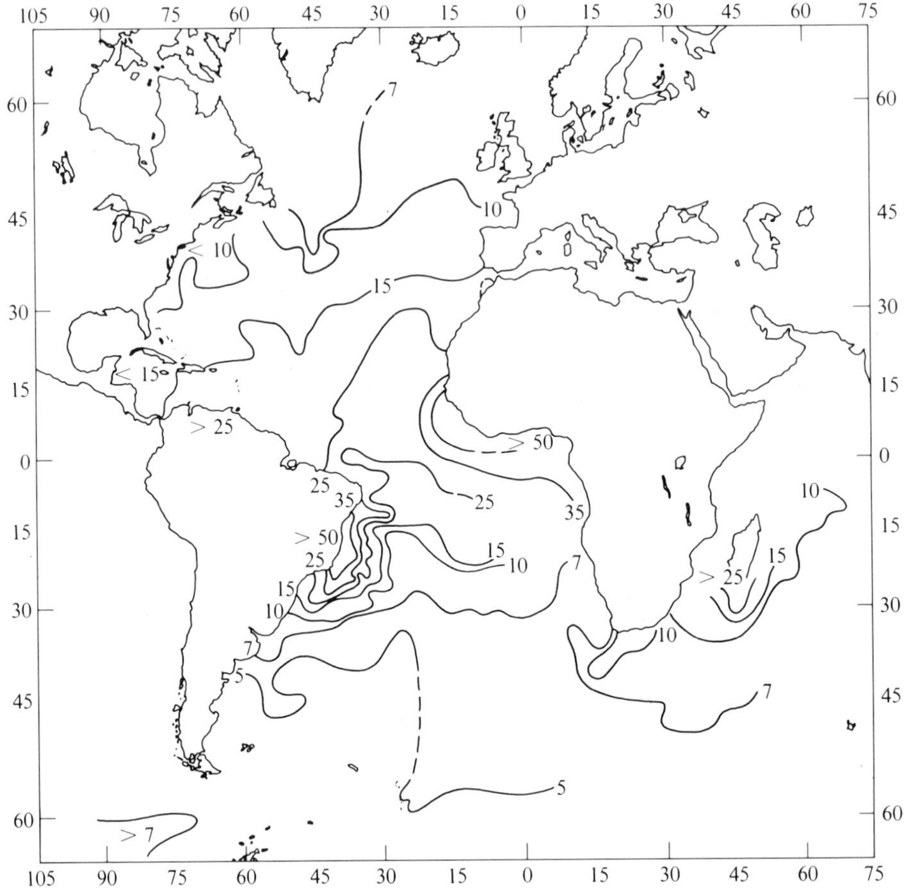

Figure 3–38
Percentage of kaolinite in <2 μm size fraction of Atlantic Ocean basin sediment. [Biscaye, 1965.]

ratio of gibbsite to illite decreases from values greater than unity at the equator to zero at latitudes higher than 30–40°. That is, there is no gibbsite detectable in the oceanic mud when the climate on the adjacent landmass has a mean annual temperature less than approximately 10°C.

The climatic control of mineral composition shown by clay-mineral distributions in the Atlantic Ocean basin is duplicated by the nonclay sediment fraction. An illustration is the amount of the chemically unstable amphibole group in the mud fraction (see Figure 3–39). The amphibole/illite ratio varies systematically from 0.05/1 near the equator to greater than unity in polar regions, paralleling the occurrence of chlorite. Similarly, feldspar is rare in equatorial Atlantic muds but abundant in polar muds.

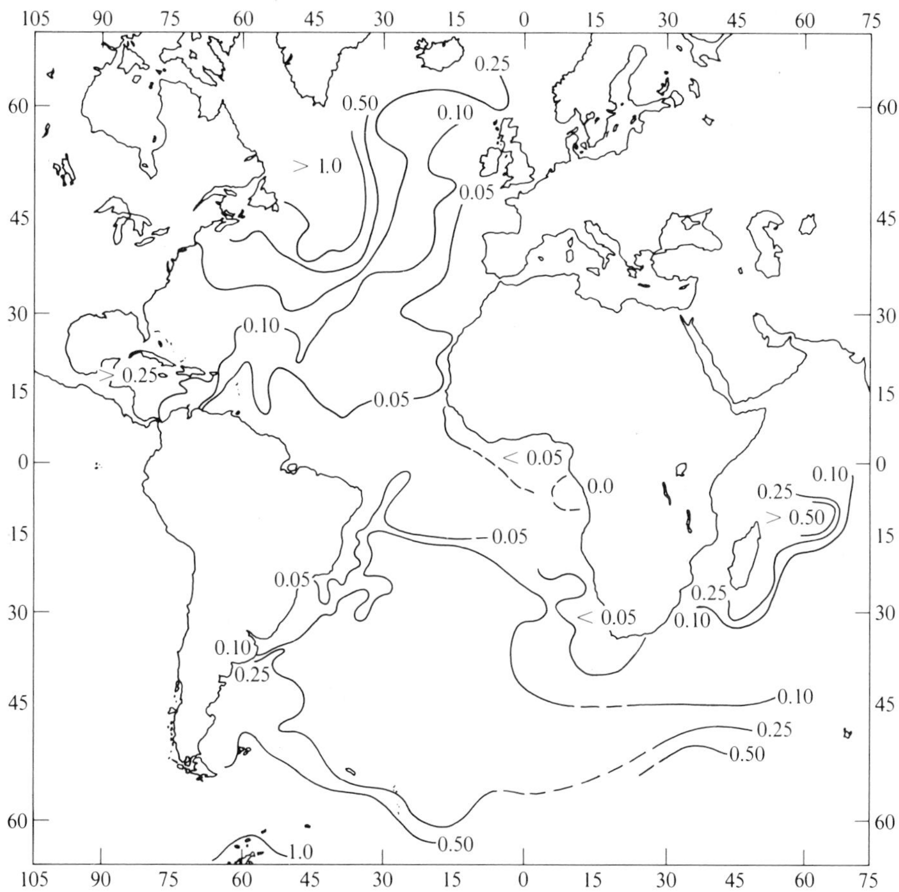

Figure 3–39
Amphibole/illite ratio in 2–20 μm size fraction of Atlantic Ocean basin sediment. [Biscaye, 1965.]

The clear control of detrital mineral composition in the oceans by the climate on adjacent land surfaces reflects not only the importance of climate but also the *un*importance of ocean water as a modifying influence. Ocean water is a very concentrated solution of ions, containing 35,000 ppm dissolved solids in contrast to only 120 ppm in river water. Because of this difference, ocean water has the chemical potential to convert cation-poor clays such as gibbsite and kaolinite to cation-rich clays such as illite and chlorite. The fact that there is no evidence for these changes in the pattern of clay distribution in the ocean indicates that the process of change is too slow to be noticeable in time periods of a few million years. Many long cores of sediment have been retrieved from the oceanic basins during the past decade of drilling by JOIDES cruises (Joint Oceanographic Institutions for Deep Earth Sampling). They, too, reveal no evidence for change due to the reaction of detrital clay with ocean water. Apparently, such change requires higher temperatures to be effective. As noted in the discussion of ancient mudrocks, field evidence suggests that temperatures of at least 50°C are needed. Temperatures at the ocean floor are about 2°C.

SUMMARY

Mudrocks are the most abundant type of sedimentary rock, although their abundance in outcrop may be less than that of sandstones. Mudrocks are not well understood because their aphanitic texture and the difficulty of disaggregating them make them hard to study.

The relative proportion of fissile to nonfissile mudrocks is not known. Fissility can be produced during deposition, but the rapid rate at which burrowing organisms on the seafloor destroy clay-flake parallelism suggests that a large proportion of the fissility seen in ancient mudrocks has been produced during diagenesis.

Mudrocks contain most of the Earth's buried organic matter, so that gray–black mudrocks are common. Black organic matter can be preserved only in an oxygen-deficient environment, and in such an environment advanced forms of life such as burrowing organisms cannot exist. As a result, black mudrocks are more likely to be fissile than red ones, which form and are preserved under oxygenated conditions.

Clay minerals form 60% of muds and mudrocks. Illite forms about 25% of modern clay-mineral suites but increases in abundance with increasing age to perhaps 80% in Lower Paleozoic mudrocks. Both field observations and geochemical experiments indicate that the cause of the change is the progressive change of montmorillonite and mixed-layer montmorillonite–illite clay to pure illite during diagenesis. The reaction is initiated at temperatures as low as 50°C but is not completed until temperatures above 200°C are attained. Detrital silt-size quartz forms most of the remainder of the typical mudrock.

Mudrocks can be used in provenance studies of modern muds; but because the character of the clay-mineral suite changes during diagenesis, only the nonclay fraction can be used in provenance studies of ancient mudrocks. Quartz, feldspar, and accessory minerals present in mudrocks can be used for this purpose.

FURTHER READING

Arthurton, R. S. 1980. Rhythmic sedimentary sequences in the Triassic Keuper Marl (Mercia Mudstone Group) of Cheshire, northwest England. *Geol. Jour., 15,* 43–58.

Biscaye, P. E. 1965. Mineralogy and sedimentation of recent deep-sea clay in the Atlantic Ocean and adjacent seas and oceans. *Geol. Soc. Amer. Bull., 76,* 803–832.

Blatt, H., and D. J. Schultz. 1976. Size distribution of quartz in mudrocks. *Sedimentology, 23,* 857–866.

Blatt, H., and M. W. Totten. 1981. Quartz in mudrocks as an indicator of distance from shoreline, Blaine Formation (Permian), western Oklahoma. *Jour. Sed. Petrology, 51,* 1259–1266.

Cluff, R. M. 1980. Paleoenvironment of the New Albany Shale Group (Devonian-Mississippian) of Illinois. *Jour. Sed. Petrology, 50,* 767–780. See also *51,* 1027–1031.

Colton, G. W. 1967. Orientation of carbonate concretions in the Upper Devonian of New York. U.S. Geol. Surv. Prof. Paper No. 575B, pp. 57–59.

Cubitt, J. M. 1979. *Geochemistry, Mineralogy and Petrology of Upper Paleozoic Shales of Kansas.* Kansas Geol. Surv. Bull., *217,* 117 pp.

Curtis, C. D. 1980. Diagenetic alteration in black shales. *Jour. Geol. Soc., 137,* 189–194.

Degens, E. T., and D. A. Ross (eds.). 1974. *The Black Sea—Geology, Chemistry, and Biology.* Amer. Assoc. Petroleum Geol. Mem. No. 20, 633 pp.

Demaison, G. J., and G. T. Moore. 1980. Anoxic environments and oil source bed genesis. *Amer. Assoc. Petroleum Geol. Bull., 64,* 1179–1209.

Eaton, G. P. 1964. Windborne volcanic ash: a possible index to polar wandering. *Jour. Geol., 72,* 1–35.

Englund, J.-O., and P. Jørgensen. 1973. A chemical classification system for argillaceous sediments and factors affecting their composition. *Geologiska Föreningens Förhandlingar, 95,* 87–97.

Fan, P.-F. 1976. Recent silts in the Santa Clara River drainage basin, southern California: a mineralogical investigation of their origin and evolution. *Jour. Sed. Petrology, 46,* 802–812.

Fisher, R. V. 1964. Maximum size, median diameter, and sorting of tephra. *Jour. Geophys. Res., 69,* 341–355.

Folk, R. L. 1962. Petrography and origin of the Silurian Rochester and McKenzie Shales, Morgan County, West Virginia. *Jour. Sed. Petrology, 32,* 539–578.

Hallam, A., and M. J. Bradshaw. 1979. Bituminous shales and oolitic ironstones as indicators of transgressions and regressions. *Jour. Geol. Soc., 136,* 157–164.

Jones, M. L., and J. M. Dennison. 1970. Oriented fossils as paleocurrent indicators in Paleozoic lutites of southern Appalachians. *Jour. Sed. Petrology, 40,* 642–649.

Kranck, K. 1975. Sediment deposition from flocculated suspensions. *Sedimentology, 22,* 111–123.

McBride, E. F. 1974. Significance of color in red, green, purple, olive, brown, and gray beds of Difunta Group, northeastern Mexico. *Jour. Sed. Petrology, 44,* 760–773.

Matalucci, R. V., J. W. Shelton, and M. Abdel-Hady. 1969. Grain orientation in Vicksburg loess. *Jour. Sed. Petrology, 39,* 969–979.

O'Brien, N. R., K. Nakazawa, and S. Tokuhashi. 1980. Use of clay fabric to distinguish turbiditic and hemipelagic siltstones and silts. *Sedimentology, 27,* 47–61. See also E. Azmon, Discussion. *28,* 733–735.

Piper, D. J. W. 1972. Turbidite origin of some laminated mudstones. *Geol. Magazine, 109,* 115–126.

Potter, P. E., J. B. Maynard, and W. A. Pryor. 1980. *Sedimentology of Shale.* New York: Springer-Verlag, 306 pp.

Ross, C. S. 1955. Provenance of pyroclastic materials. *Geol. Soc. Amer. Bull., 66,* 427–434.

Scotford, D. M. 1965. Petrology of the Cincinnatian Series shales and environmental implications. *Geol. Soc. Amer. Bull., 76,* 193–222.

Shaw, D. B., and C. E. Weaver. 1965. The mineralogical composition of shales. *Jour. Sed. Petrology, 35,* 213–222.

Slaughter, M., and J. W. Earley. 1965. *Mineralogy and Geological Significance of the Mowry Bentonites, Wyoming.* Geol. Soc. Amer. Spec. Paper No. 83, 116 pp.

Spears, D. A. 1969. A laminated marine shale of Carboniferous age from Yorkshire, England. *Jour. Sed. Petrology, 39,* 106–112.

Stanley, D. J. (ed.). 1972. *The Mediterranean Sea: A Natural Sedimentation Laboratory.* Stroudsburg, PA: Dowden, Hutchinson, and Ross, 765 pp.

Tourtelot, H. A. 1962. *Preliminary Investigation of the Geologic Setting and Chemical Composition of the Pierre Shale, Great Plains Region.* U.S. Geol. Surv. Prof. Paper No. 390, 74 pp.

Tourtelot, H. A. 1979. Black shale—its deposition and diagenesis. *Clays and Clay Minerals, 27,* 313–321.

Vandenberghe, N. 1976. Phytoclasts as provenance indicators in the Belgian septaria clay of Boom (Rupelian age). *Sedimentology, 23,* 141–145.

Weiss, M. P., W. R. Edwards, C. E. Norman, and E. R. Sharp. 1965. *The American Upper Ordovician Standard.* VII: *Stratigraphy and Petrology of the Cynthiana and Eden Formations of the Ohio Valley.* Geol. Soc. Amer. Spec. Paper No. 81, 76 pp.

4

Conglomerates and Sandstones: Textures and Structures

It requires a very unusual mind to undertake the analysis of the obvious.

ALFRED NORTH WHITEHEAD

Conglomerates and sandstones form 20–25% of the stratigraphic column and have received much more than 25% of the attention of sedimentary petrologists. This has occurred for several reasons. The grains are coarse enough to be seen easily, and sandstones typically contain textures and structures that can be described and photographed, such as pebble elongation, grain rounding, cross bedding, ripple marks, and graded bedding. These can be diagnostic of transport mechanism or depositional environment. Also, sandstones supply about half of the world's production of petroleum and natural gas, in addition to the bulk of the commercial uranium ores. The volume of publications that deal with sandstones is sufficiently large to require three chapters in this text for adequate treatment. In this chapter we consider textures and structures. The following chapter deals with detrital mineral composition; the succeeding one, with diagenetic effects.

FIELD OBSERVATIONS

Textures

Grain Size, Sorting, and Skewness

A wide variety of textural features can be seen and described from outcrops of conglomerates and sandstones. Perhaps the most fundamental is grain size (see Table 4–1)—both the average size and the variety of sizes in the rock. Describing the size of

Table 4-1
The Standard Grain-Size Scale for Clastic Sediments[a]

	Name	Millimeters	Micrometers	ϕ
		4,096		−12
GRAVEL	Boulder			
		256		−8
	Cobble			
		64		−6
	Pebble			
		4		−2
	Granule			
		2		−1
SAND	Very coarse sand			
		1		0
	Coarse sand			
		0.5	500	1
	Medium sand			
		0.25	250	2
	Fine sand			
		0.125	125	3
	Very fine sand			
		0.062	62	4
MUD	Coarse silt			
		0.031	31	5
	Medium silt			
		0.016	16	6
	Fine silt			
		0.008	8	7
	Very fine silt			
		0.004	4	8
	Clay			
		↓	↓	↓

[a]As devised by J. A. Udden (1898) and C. K. Wentworth (1924). The ϕ scale (Krumbein, 1934) was devised to facilitate statistical manipulation of grain-size data and is commonly used. $\phi = -\log_2$ mm.

grains is not so simple as it first appears. The size of a sphere is its diameter (or a function of its diameter such as volume), but detrital grains are not spheres. When dealing with gravel-size grains, which typically depart widely from a spherical shape, the appropriate definition of "size" is not at all obvious. Should it be the longest dimension of the grain, an intermediate length, or an estimate of grain volume obtained by multiplying a long, intermediate, and short axis of the grain? There is no perfect answer to this question. The standard procedure is to use an intermediate grain length as the size. This may or may not be clearly related to factors such as transportability, settling velocity, or mass. Grains of gravel size can be measured in the field with a meter stick graduated in centimeters.

Sand-size grains are more equant in shape than most gravel particles so that the definition of size is a less serious question, although there does exist a sizable volume of publications discussing the problem. It is difficult, at least initially, to calibrate your eyes precisely enough to distinguish among grain lengths of 0.5, 0.25, and 0.125 mm. Therefore, it is useful to construct a simple grain-size comparator for field use. Two pieces of thin cardboard, a hole puncher, transparent tape, and some laboratory-sized sand grains are all that is needed (see Figure 4–1). With this aid it is not difficult to estimate average grain size and the distribution of sizes in a sandstone. Some detrital rocks have two pronounced sizes (*modes*), and this should be mentioned in the description of grain size. For example, the bed may be a mixture of coarse sand and very fine-grained sand with only a small amount of sediment in the intermediate sizes (see Figure 4–2).

The distribution of sizes is normally given as the range in phi units that includes approximately two-thirds of the grains. This range is twice the standard deviation. For example, if the mean size is 2.0 ϕ and two-thirds of the grains have sizes between 1.5 ϕ and 2.5 ϕ, then the standard deviation is 0.5 ϕ. ***Standard deviation*** is the accepted measure of the "sorting" of the sediment (see Figure 4–3), and sorting values tend to differ among different sedimentary environments. (We return to this point later in the chapter.)

An additional characteristic of a size distribution commonly used is *skewness,* or lopsidedness. Most grain-size distributions are symmetrical; that is, there is as much sediment in each size finer than the average as there is in each size coarser than the average. Some distributions, however, are all skewed up. For example, if the sediment

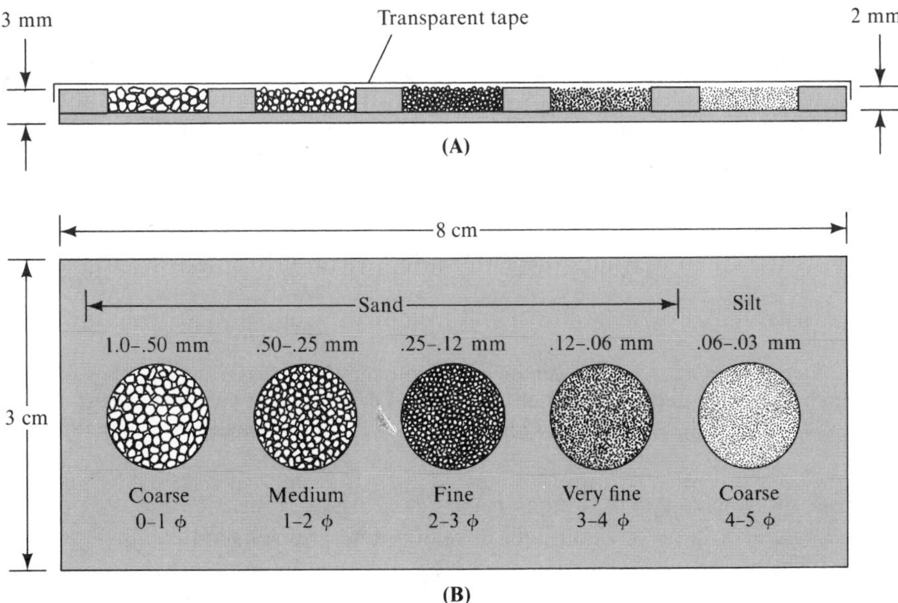

Figure 4–1
Sketch showing construction of simple grain-size comparator for field use. (**A**) Side view. (**B**) Top view.

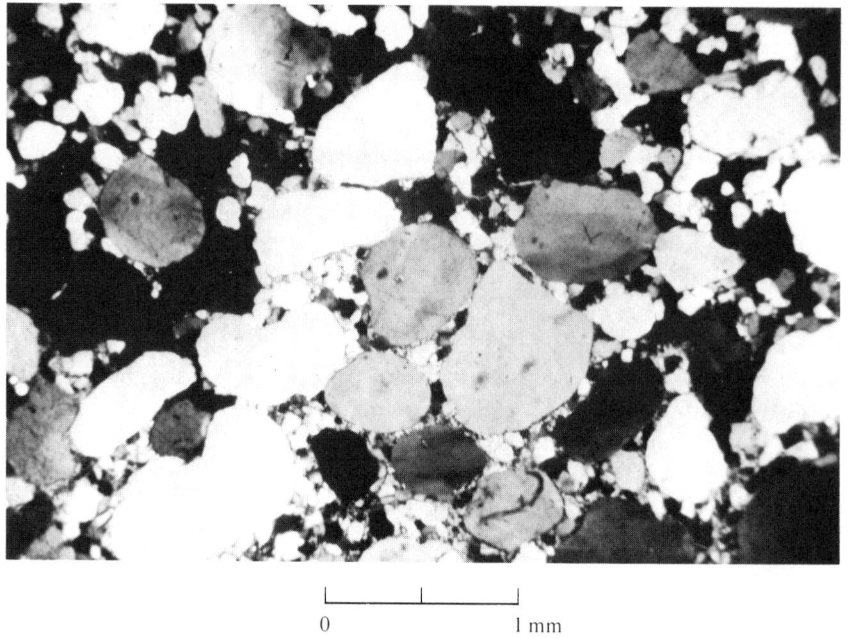

Figure 4–2
Thin-section photomicrograph of bimodal pure quartz sandstone, Bar River Formation (Precambrian), near Blind River, Ontario, Canada. [J. Wood, 1973, Geol. Assoc. Canada Spec. Paper No. 12. Photo courtesy J. Wood.]

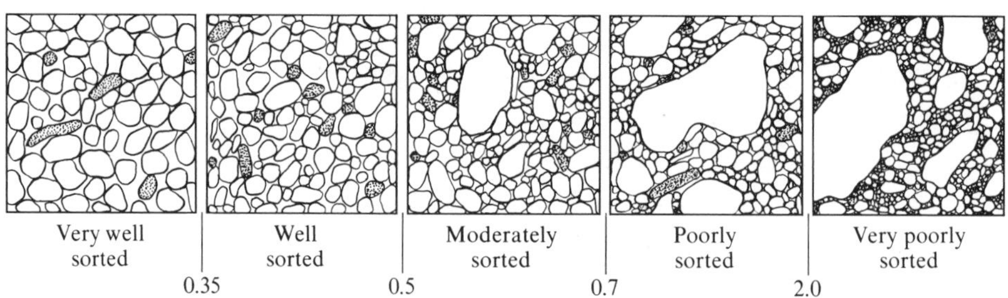

Figure 4–3
Classification of degrees of sorting as seen through square hand lens. Silt- and clay-size sediments are indicated by fine stipple. Values of standard deviation that divide each class of sorting are also shown. [R. R. Compton, 1962, *Manual of Field Geology* (New York: Wiley).]

contains 80% pebbles and 20% fine-grained sand, the size distribution is said to be *positively skewed.* If the sediment contains 80% fine-grained sand and 20% pebbles, the size distribution is said to be *negatively skewed.* In most conglomerates and sandstones the determination of skewness is best left for laboratory studies.

After the average grain size and the variety of sizes have been estimated, the sediment must be named, as a shorthand to make communication easier. As noted earlier, there is general agreement concerning a scale of grain size (Wentworth scale,

Table 4–1) and agreement as well for degrees of sorting (see Figure 4–3). There is, unfortunately, no agreement among geologists concerning a method of naming mixtures of different sizes. For example, analysis of typical usage reveals that field geologists are likely to call a deposit a conglomerate even if gravel forms less than 50% of the rock. The presence of only 10% or 20% pebbles or cobbles creates a dominant impression when observed in outcrop. Nevertheless, it is clearly desirable to have a uniform usage. One system in common use is shown in Figure 4–4; many others have been published and are used. All are equal in merit, and the choice among them is entirely arbitrary.

Grain Shape

The shapes of grains in conglomerates and sandstones vary widely from those shaped like spheres to those that approximate a disk or a rod. Two aspects of grain shape can be seen either in outcrop or in a hand specimen: sphericity and roundness. *Sphericity* is defined by the relative equidimensionality of three mutually perpendicular axes through the grain. If the axes are approximately equal, the grain is spherical; if two of the axes are noticeably longer than the third, the grain is disk-shaped or platy; if two axes are noticeably shorter than the third, the grain is rod-shaped or elongate (Sneed and Folk, 1958). Pebbles formed of material with relatively isotropic mechanical properties, such as quartz, tend to be either rods (in rivers) or disks (on beaches), although shape overlap is common between the two environments (Dobkins and Folk, 1970). Strongly nonisotropic pebbles have a sphericity controlled largely by the structure of the fragment. For example, phyllite fragments, because of their foliation, are platy irrespective of either their mechanism of transport or their environment of deposition.

Grain *roundness* is controlled by grain size, hardness, and environments of transportation and deposition. As a general rule, particles coarser than 5–10 mm are nearly always rounded (Mills, 1979); those 0.1–5 mm may be either round or angular; and those smaller than 0.1 mm are nearly always angular. These trends exist because of the mechanism by which grains become round after release from crystalline rocks. Roundness is produced on grains by impacts with other grains during movement; larger grains impact with more force. Grains larger than 5–10 mm are transported mostly by rolling or sliding on the stream bottom or beach surface, resulting in relatively rapid loss of sharp corners on the grains. Grains 0.1–5 mm (essentially the sand sizes) may travel either mostly along the stream bottom or mostly in the hopping pattern termed *saltation*. In the latter case, the grains spend most of their travel time within the body of the moving fluid rather than in contact with other solid particles and, as a result, are rounded very slowly, if at all. It is believed by most sedimentologists that sand-size quartz grains, because of their hardness, cannot be rounded during stream transport irrespective of the distance of transport—quartz gravel, yes; but quartz sand, no. Sand-size quartz grains can be rounded only in environments characterized by higher average kinetic energies, such as beaches, offshore marine barrier bars, tidal channels, or desert dunes.

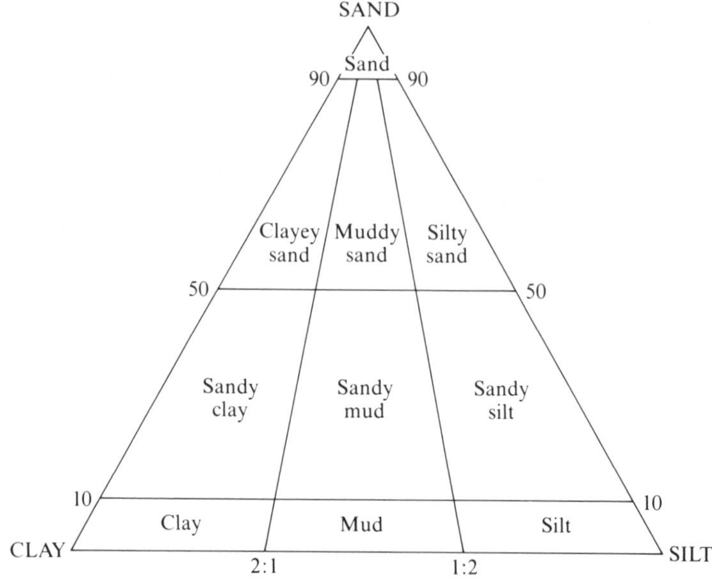

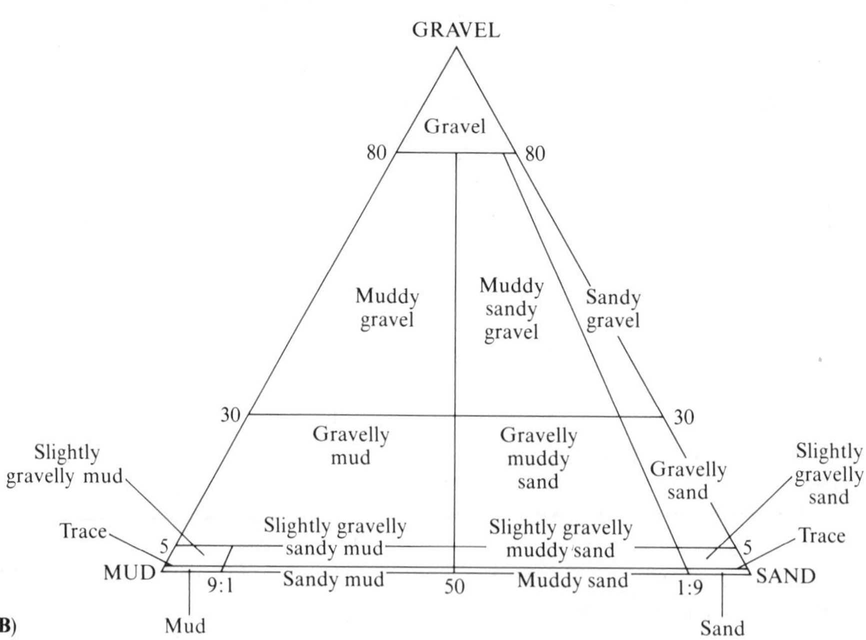

Figure 4–4
Triangular classification of grain sizes in detrital rocks. (**A**) If no gravel is present, this triangle is used. (**B**) If gravel is present, this one is used. Note emphasis given to even trace amount of gravel. [R. L. Folk, 1954, *Jour. Geol., 62*. Reprinted by permission of The University of Chicago Press. Copyright © 1954 by The University of Chicago.]

Silt-size quartz grains are rarely rounded because they travel almost entirely in suspension, cushioned by water or air. Deposits composed entirely of silt are uncommon because both silt and clay are so easy to keep suspended in a fluid that they travel together and are not segregated from each other as effectively as gravel is from sand. Mudstones and mudshales are more common in the geologic column than are mudrocks dominated by either silt sizes or clay sizes. An exception is the sediment known as *loess,* a unit composed almost entirely of silt-size sediment and found only rarely in pre-Quaternary deposits. The source of the silt is fine-grained loose sediment on floodplains (such as in the lower Mississippi River valley) or in desert areas (such as the Gobi Desert in Mongolia). The silt is preferentially picked up by winds and deposited mostly within a few tens of kilometers from the source. Clay-size sediment is uncommon in loess deposits because clay-size grains are composed mostly of clay minerals, which are more difficult to remove from the floodplain because their flat surfaces adhere. When blown out of desert areas, clays are so easily kept in suspension that they circle the Earth and are scattered widely over its surface.

Grain roundness is more precisely determined in thin section than in a hand specimen, but the precision is low in either event. The usual method for evaluating roundness is to compare the grain outline with a standard set of photographs (see Figure 4–5) but, unfortunately, there seem to be "angular people" and "round people." Some people simply see grains as being more angular or more round than other people see them. In the extreme cases of grain angularity and roundness, such as differentiating a round, coarse, sand-size grain on a beach from an angular one just fallen from a granite, there is no problem. But for most grains in most sandstones the differences in roundness among the grains are too small to be determined reliably with a visual estimate.

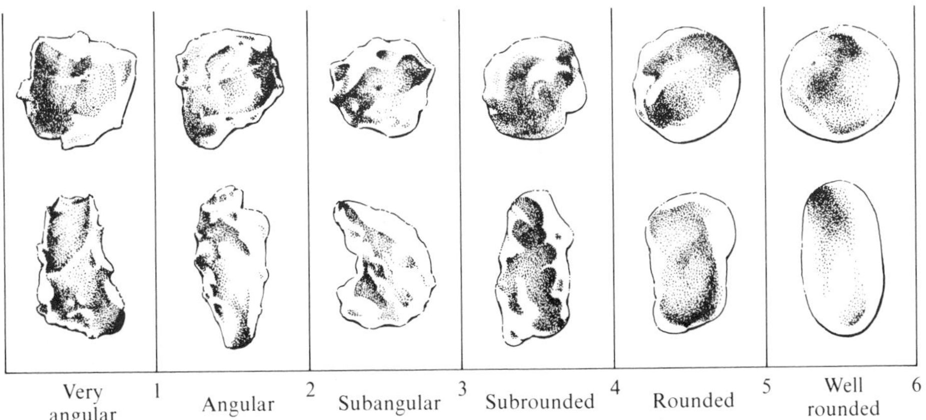

Figure 4–5
Terminology for degree of roundness of detrital grains with hand lens. The numbers assigned to each roundness class permit calculation of mean roundness and standard deviation.
[M. C. Powers, 1953, *Jour. Sed. Petrology, 23.*]

Ehrlich and Weinberg (1970) have developed and applied a method of determining grain roundness that is based on a mathematical technique called Fourier analysis. This technique measures shape by resolving a grain's two-dimensional outline into a series of shape harmonics: the lower-order harmonics measure overall shape (sphericity), and the higher-order harmonics measure small-scale irregularities (roundness). The Fourier method appears to have much potential but has not yet been widely applied.

Textural Maturity

Following release from a source rock, the texture of a detrital sediment is modified in stages during transport in a stream. Within a few minutes the clay and most of the silt are washed out and carried downstream. Within a few hours or days the stream currents separate different sizes of sand and gravel, and the best sorting possible in the environment is achieved. As the sand is transported from the upstream areas, the grains are abraded (except for quartz) and rounded. This process may be either rapid or slow, depending on the mechanical resistance of the grains. For example, carbonate sand would round rapidly; feldspar, more slowly; quartz, not at all. This sequence of clay removal, sorting, and rounding is termed *textural maturity* (see Figure 4–6;

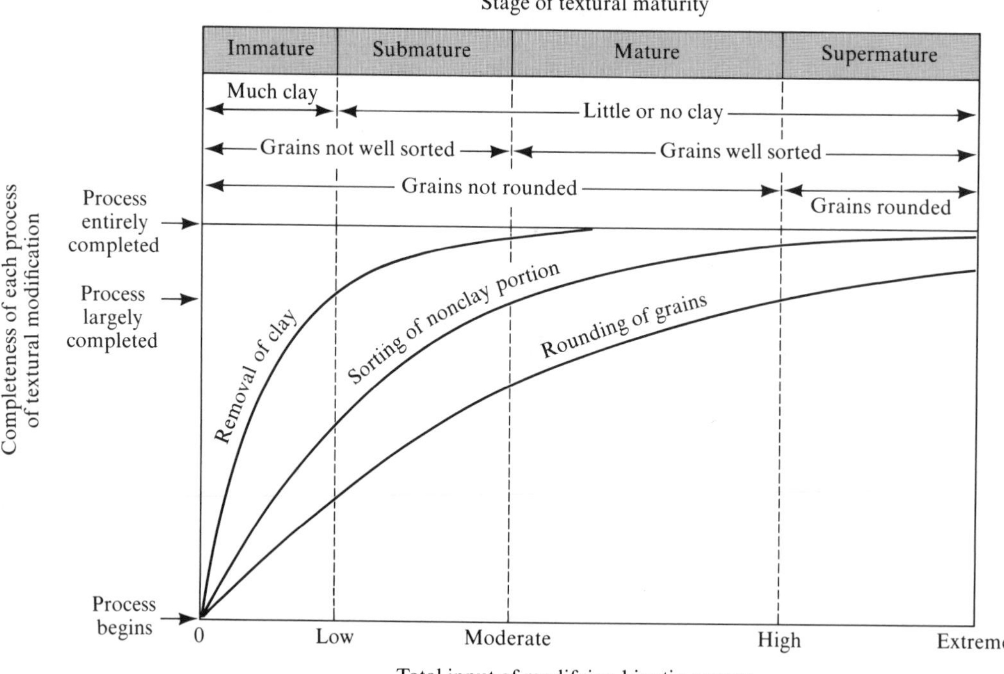

Figure 4–6
Textural maturity of sands as function of input of kinetic energy. [Folk, 1951.]

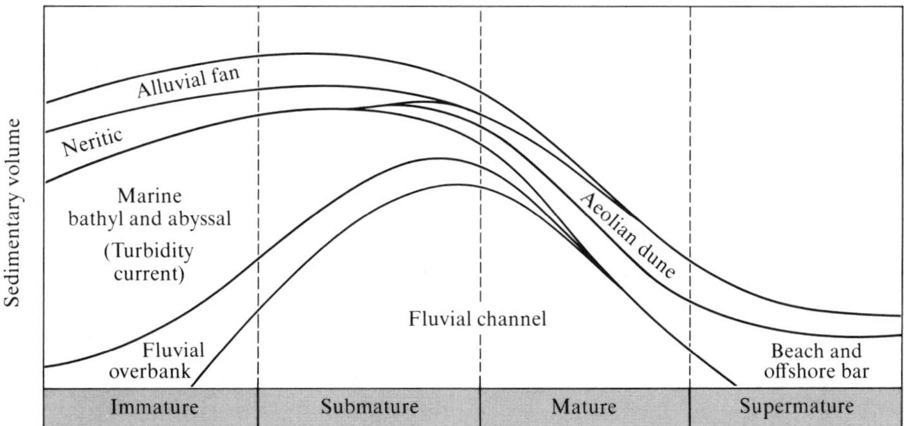

Figure 4–7
Relationship among sedimentary volumes, environments of deposition, and textural maturity. Diagram is based on perceptions of sedimentologists from field studies; adequate numerical data do not exist.

Folk, 1951)—a key concept in the analysis of sandstone units. There are two problems in its application, however, one of which, the control of quartz roundness by environment as well as by transport distance, has already been noted. The second problem is that clay matrix is produced after burial in sandstones that contain chemically unstable lithic fragments. (We consider this problem in more detail in Chapter 6.)

Textural maturity is determined by current strengths at the site of deposition. Environments such as beaches and desert dunes are characterized by high kinetic energies; alluvial fans and offshore marine areas, by lower ones. It follows, therefore, that a correlation should exist between depositional texture and depositional environment. The correlation is far from perfect, as Figure 4–7 shows; so it is not wise to assign an environment of deposition to a sandstone on the basis of textural maturity alone. In fact, no single characteristic of a lithologic unit is an infallible guide to depositional environment.

Colors

The color of a rock is an obvious characteristic that should be described in the field because it can sometimes be used effectively for tracing sedimentary units. The common colors of rocks are red, brown, and yellow, which result typically from the presence of ferric oxide cement (hematite); and gray–black, which reflects the presence of free carbon (organic matter). Colorless rocks, such as quartz-cemented quartz sandstones, contain neither ferric oxide nor free carbon. Local variations in the color of a rock can result from weathering phenomena; for example, a bit of organic matter in a sandstone can react chemically with surrounding hematite cement to reduce the ferric iron to the ferrous state so that the red color vanishes. Conversely, weathering at the outcrop of a colorless rock can release ferrous iron from minerals such as hornblende and biotite, oxidize the iron, and produce hematite.

Dispersal Pattern and Transport Distance

Many field studies have been made of the effects of topography, climate, and other variables on the generation of detrital sediment. It is clear from these studies that relief is the most important factor; rate of erosion increases very rapidly with increasing relief (see Figure 4–8). Thus, most detrital particles originate in highland areas such as the Alps, North American Rocky Mountains, or Himalayas.

Because of the great relief, stream gradients are relatively high in mountainous regions (sometimes as high as 1°, 17.4 m/km), the ability of the stream to transport large particles (*stream competence*) is correspondingly high, and coarse particles can be moved downstream. As the stream leaves the mountains, however, the stress of the water on the streambed decreases with the slope, competence decreases, and coarse particles can no longer be moved. Many studies of conglomerates have used this concept to infer distance of transport from a mountain front (see Figure 4–9). It is also possible to apply it to sandstones (see Figure 4–10); although as grain size

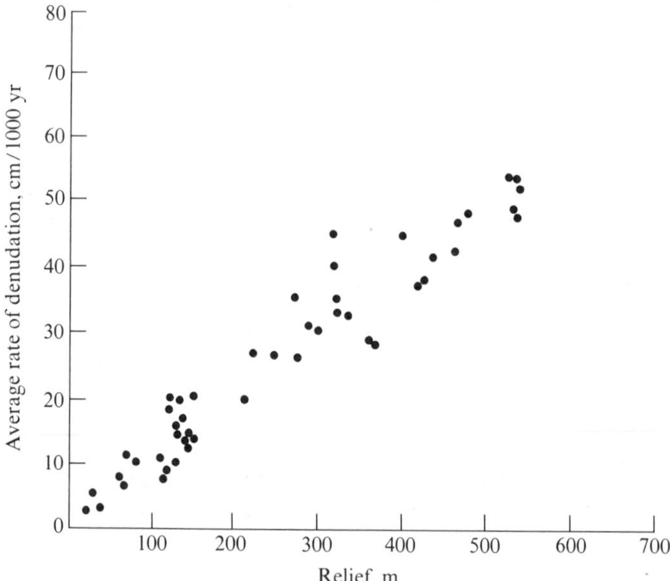

Figure 4–8
Average rate of denudation as function of relief on andesitic stratovolcano of Holocene age in northeastern Papua. [B. P. Buxton and I. McDougall, 1967, *Amer. Jour. Sci., 265.*]

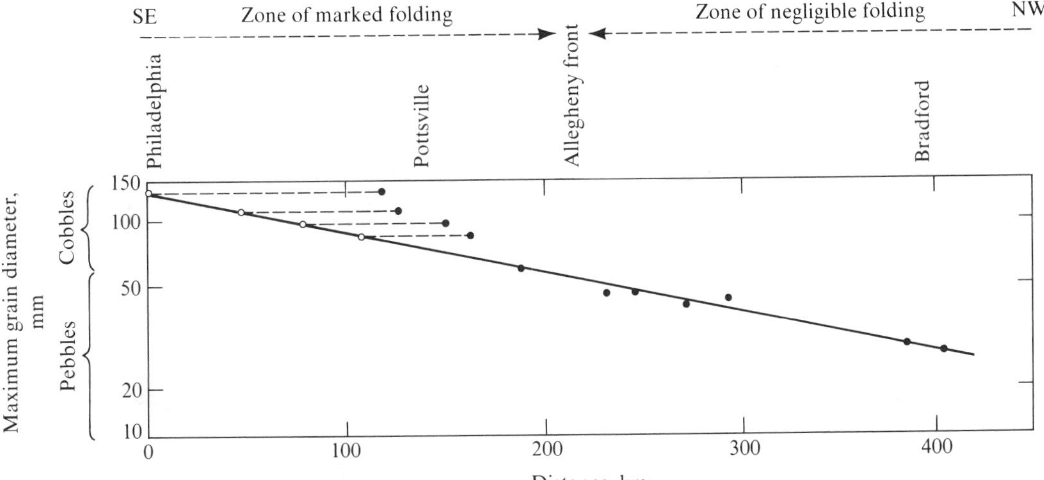

Figure 4–9
Relationship between distance of stream transport and maximum grain size in Pottsville conglomerates (Pennsylvanian), Pennsylvania. Source of gravel is thought to be in vicinity of Philadelphia, based on existing structural and deformational patterns. Solid circles = observed data. Open circles = presumed prefolding position of beds based on extrapolation from observed data points. [Pelletier, 1958.]

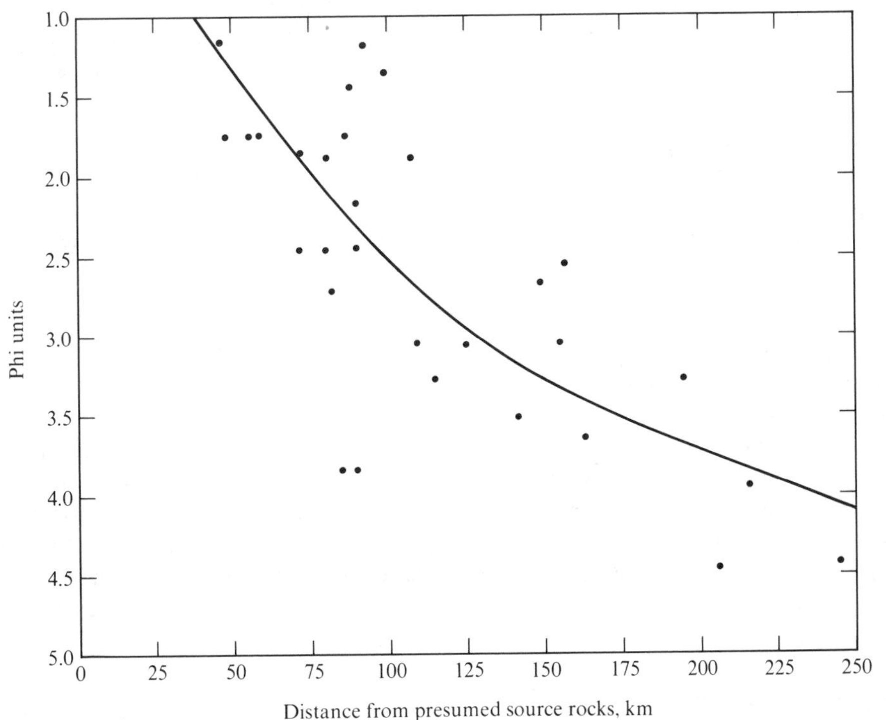

Figure 4–10
Relationship between distance of transport and maximum grain size (long dimension in thin section) in Ludlowville Formation (Devonian), western New York. Line of best fit to data points is shown. [Towe, 1963. Reprinted by permission of The University of Chicago Press. Copyright © 1963 by The University of Chicago.]

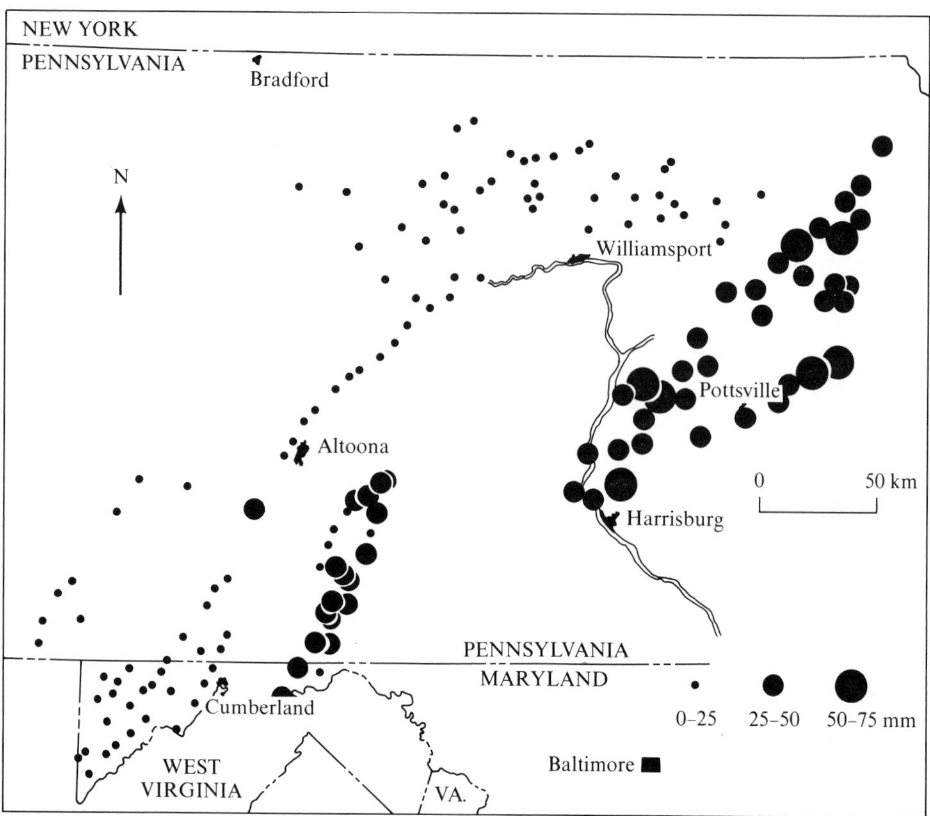

Figure 4–11
Areal distribution of pebble size in Pocono Formation (Mississippian), Pennsylvania and Maryland. [Pelletier, 1958.]

decreases, it becomes easier for local variations in stream character to obscure the larger-scale trend toward size decrease in the downstream direction; that is, the scatter of the data points increases.

Decrease in grain size can be seen areally when a lithologic unit is traced laterally from its source area. An example of the decrease in maximum pebble size in the Pocono Formation (Mississippian) in Pennsylvania and Maryland is shown in Figure 4–11. In a finer-grained detrital sequence, probably the easiest way to detect an areal decrease in grain size with distance from source is through the sand/mudrock ratio (see Figure 4–12), although a contour map of maximum grain size in the formation should yield the same result.

Bedding Structures

Structures are larger-scale features than textures and can be formed in many ways, such as by changes in depositional conditions, by erosion, or by diagenetic accentuation of an originally gradational boundary. Perhaps as many as 100 distinct sedimen-

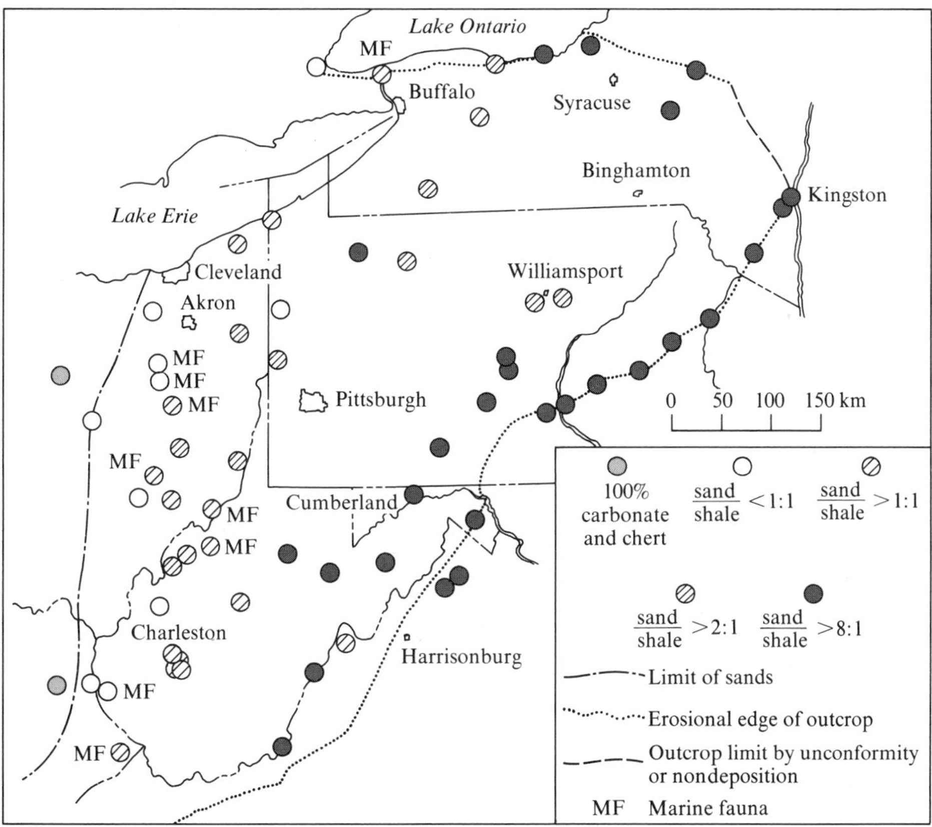

Figure 4–12
Sandstone/mudrock ratio map of Tuscarora Formation (Silurian) and equivalent beds in part of northeastern United States. [Yeakel, 1962.]

tary structures have been described and photographed in outcrop during the past century. They can be grouped as shown in Table 4–2. Detailed discussions of many of them can be found in most modern sedimentology texts, such as Blatt et al. (1980), Pettijohn and Potter (1964), and Potter and Pettijohn (1977).

The most common sedimentary structure is bedding, the subplanar discontinuity that separates adjacent layers of rock. At the outcrop, we describe both its external geometry and internal character, as indicated in Table 4–2. The distinction between *bedding* and *lamination* may be arbitrary so that, for consistency, it is a good idea to establish a nongenetic system for defining these terms (see Table 4–3). When measurements of the thicknesses of large numbers of beds are made in a stratigraphic unit, it is common to find that the thicknesses cumulate as a straight line when plotted on semilog graph paper (see Figure 4–13). The relationship stems from the interaction between the average current velocity and the size of sediment being transported and deposited, but the exact cause is not understood.

Table 4–2
Classification of Primary Sedimentary Structures

Bedding, external form
1. Beds *equal* or *subequal* in thickness; beds laterally uniform in thickness; beds continuous
2. Beds *unequal* in thickness; beds laterally uniform in thickness; beds continuous
3. Beds *unequal* in thickness; beds laterally variable in thickness; beds continuous
4. Beds *unequal* in thickness; beds laterally variable in thickness; beds discontinuous
Bedding, internal organization and structure
1. Massive (structureless)
2. Laminated (horizontally laminated; cross-laminated)
3. Graded
4. Imbricated and other oriented internal fabrics
5. Growth structures (stromatolites, etc.)
Bedding plane markings and irregularities
1. On base of bed:
a. Load structures (load casts)
b. Current structures (scour marks and tool marks)
c. Organic markings (ichnofossils)
2. Within the bed:
a. Parting lineation
b. Organic markings
3. On top of bed:
a. Ripple marks
b. Erosional marks (rill marks; current crescents)
c. Pits and small impressions (bubble and rain prints)
d. Mud cracks, mud-crack casts, ice-crystal casts, salt-crystal casts
e. Organic markings (ichnofossils)
Bedding deformed by penecontemporaneous processes
1. Founder and load structures (ball-and-pillow structures, load casts)
2. Convolute bedding
3. Slump structures (folds, faults, and breccias)
4. Injection structures (sandstone dikes, etc.)
5. Organic structures (burrows, "churned" beds, etc.)

Source: F. J. Pettijohn, P. E. Potter, and R. Siever, 1973, *Sand and Sandstone* (New York: Springer-Verlag).

Table 4-3
Terminology for Distinguishing Between Bedding and Lamination

Term	Criterion
Very thickly bedded	> 1 m
Thickly bedded	30–100 cm
Medium bedded	10–30 cm
Thinly bedded	3–10 cm
Very thinly bedded	1–3 cm
Thickly laminated	0.3–1 cm
Thinly laminated	< 0.3 cm

Source: R. L. Ingram, 1954, *Geol. Soc. Amer. Bull., 65.*

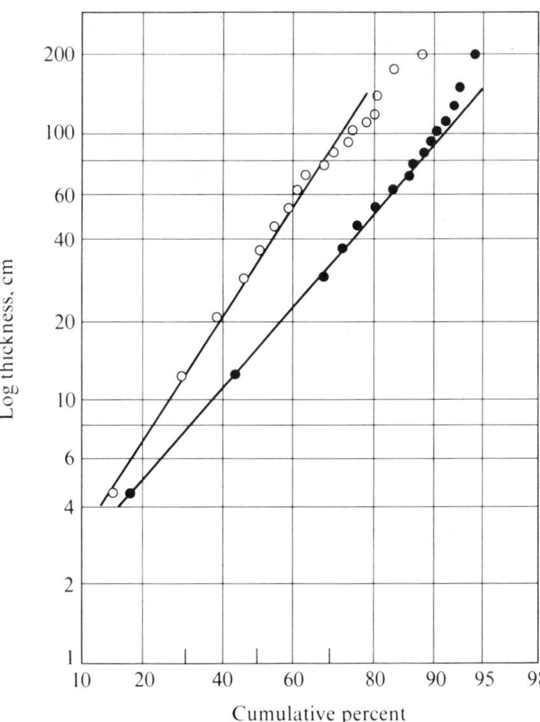

Figure 4-13
Illustration of log normal distribution of bed thicknesses of turbidite sandstone (Cretaceous), Chile, based on measurements of 271 beds. Solid circles = sandstone beds. Open circles = shale beds. [K. M. Scott, 1966, *Amer. Assoc. Petroleum Geol. Bull., 50.*]

Figure 4–14
Thinly interbedded sandstone and shale overlain by very thickly bedded uniform sandstone, Moehave Formation (Triassic), Utah. Cause of radical change in bedding characteristics is unknown. [Pettijohn and Potter, 1964. Photo courtesy F. J. Pettijohn.]

A bedding surface testifies to a change of some sort during deposition of the sediment, but often the nature of the change is uncertain. Many of these changes represent *diastems,* brief periods of time during which no sediment was deposited, none was eroded, and no rock record exists at that location. When sedimentation resumed, some factor had changed so that a discontinuity was created (see Figure 4–14).

Bedding surfaces are commonly generated by contrasting textures (see Figure 4–15), for example, the contrast between the very coarse texture of a conglomerate and the finer texture of an overlying sandstone. What mechanism produced the boundary between the coarser- and the finer-grained layers? Was it a rapid decrease in stream-current velocity, a lateral shift in channel position, or the transgression of an offshore marine sandbar over a gravel beach? The causes of bedding surfaces are

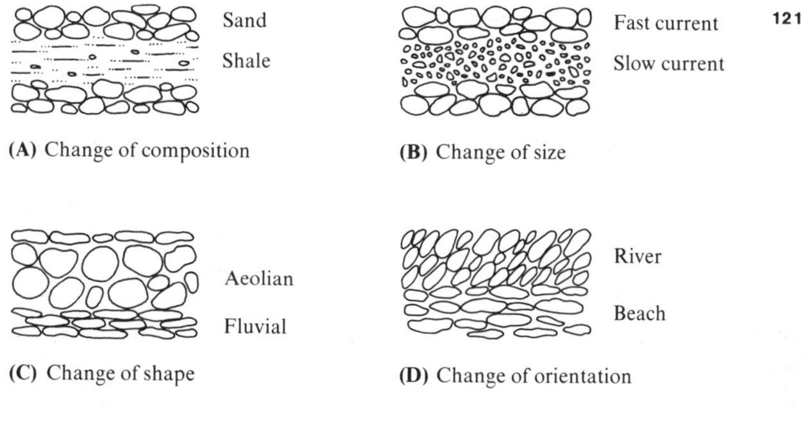

(A) Change of composition (B) Change of size

(C) Change of shape (D) Change of orientation

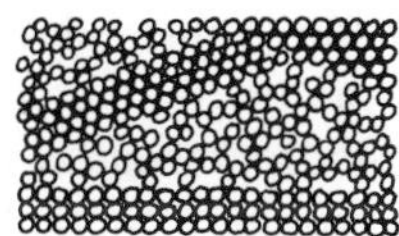

(E) Change of packing

Figure 4–15
Idealized examples of some types of textural changes sensed by naked eye as bedding. [J. C. Griffiths, 1961, *Jour. Geol., 69*. Reprinted by permission of The University of Chicago Press. Copyright © 1961 by The University of Chicago.]

many, and commonly they are not understood. Their existence, however, means that *something* must have changed during the interval of time represented by the surface. Perhaps a change in water chemistry had caused incipient cementation of the earlier sediment, perhaps there had been a change in water temperature resulting from a change in solar radiation, or perhaps a change in stream-current velocity had occurred as a result of a spring flood on April 7 of the early Middle Devonian. It is worth remembering that most of the time elapsed since the Earth was formed is represented in the geologic record by bedding surfaces, not by rocks. Hence, bedding surfaces and bedding in general deserve careful examination.

An example of an analysis and interpretation of bedding that produced results relevant to paleogeography is the study of Tertiary rocks in Spitsbergen by Atkinson (1962). The rock sequence consists of 2,000 m of detrital conglomerates, sandstones, and shales, which occur with varying bed thicknesses and varying order of superposition. Statistical analysis of these changes revealed that conglomerates were preferentially overlain by sandstones, and sandstones by shales. An analysis of the ratios of thicknesses of beds in these cycles suggested short-term meteorologic variations as the cause of the observed preferences in lithologic association. Larger-scale lithologic sequences also were found and were attributed to variation in the balance between the rate of upward movement of the source area and the rate at which erosion reduced the resulting topographic relief.

Graded Bedding

Graded beds are layers of detrital sediment marked by a gradation in grain size from base to top, normally coarser at the base and finer at the top (see Figure 4–16). The beds range in thickness from less than a centimeter to several meters; typically, thicker graded units show more change in grain size between base and top than thinner units do. Most graded detrital rocks are deposited by turbidity currents, mixtures of gravel, sand, mud, and water that move downslope from shallow depth in lakes or in the sea toward the basin floor. As the turbidity current slows toward the foot of the continental slope or lake center, the coarser grains in the sediment–water mixture are deposited first to form the base of the graded bed, with the finer grains of silt and clay following in sequence. As the mixture continues outward on the submarine fan on the deep basin floor somewhat depleted in coarser grains, a lateral change develops until at the *distal* (outer) end of the fan only the finest sediment remains (see Figure 4–17). Thus, the ratio of sandstone beds to mudrock beds in the vertical section of the fan reveals whether you are dealing with a *proximal* (upfan) section or a distal section. Turbidity currents contain varying proportions of sand and mud, so it is not possible to establish a general sand-bed/mudrock-bed ratio to distinguish upfan

Figure 4–16
Outcrop of Minnitaki Group (Archean), Ontario, Canada, showing graded bedding with very coarse sand at base and muddy sediment at top. Total bed thickness is about 6 cm. [Photo courtesy R. G. Walker.]

from midfan from downfan. Individual graded beds can commonly be traced laterally in the field for several kilometers in a cross section of an ancient submarine fan (see Figure 4–18).

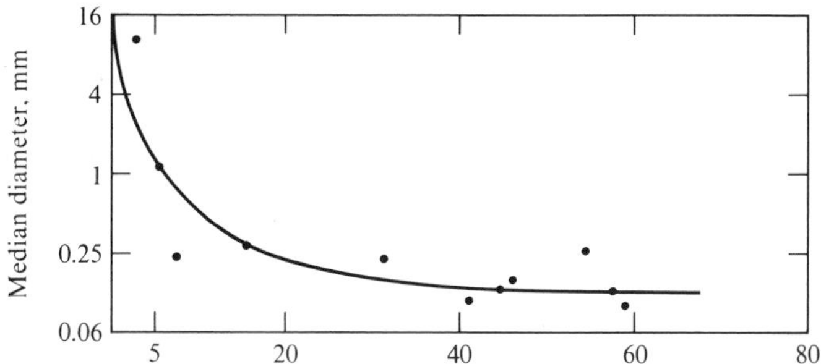

Figure 4–17
Downfan decline of grain size in surface samples from modern turbidite submarine fans. [F. J. Pettijohn et al., 1973, *Sand and Sandstone* (New York: Springer-Verlag).]

Figure 4–18
Cross section of medial part of submarine fan (Miocene), valley of Santerno River, northern Italy, showing long, even beds of graded-bedded sand deposited by sporadic turbidity currents interbedded with pelagic shale. [F. J. Pettijohn et al., 1973, *Sand and Sandstone* (New York: Springer-Verlag). Photo courtesy P. E. Potter.]

Destruction by Bioturbation

Many types of mobile organisms live on the seafloor, and a large proportion of them feed by burrowing through freshly deposited sediment looking for edible organic matter. This activity destroys bedding and lamination. Burrows are recognized in cross section by the contrast between the texture or mineral composition of the burrow fill and that of the host rock. If burrows are sufficiently abundant, only vestiges of the original bedding may remain; in extreme cases complete homogenization is produced within thick sand units (see Figure 4–19).

A large variety of biogenic structures has been described (Frey, 1975), and many are useful as paleoenvironmental indicators. In the sea the major factor determining environment is depth, and because of this fact the great variety of trace fossils (burrowings, feeding trails, etc.) can be grouped into a relatively small number of depth-controlled communities. The major environments and some of their typical traces are shown in Figure 4–20. In general, shallow waters are characterized by vertically oriented structures; and deep waters, by traces along bedding planes, because organisms in shallow waters seek to modify the extreme variations in environmental conditions that characterize the nearshore zone. At great depth the conditions

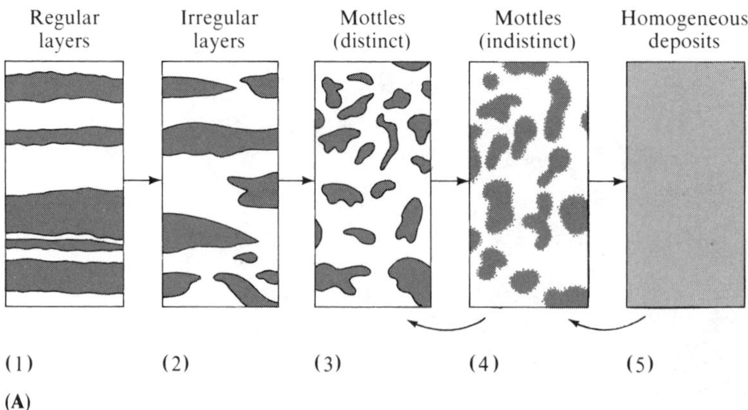

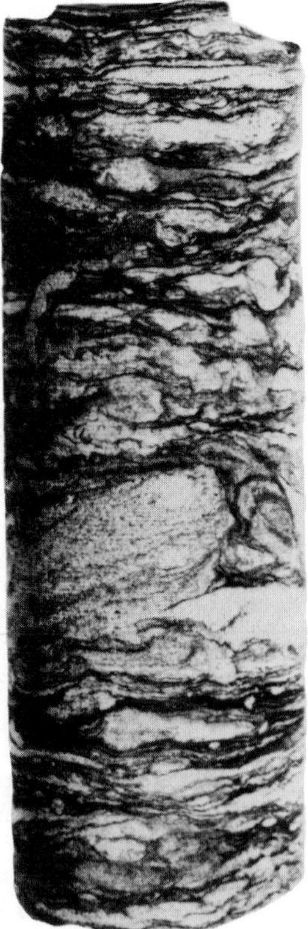

Figure 4–19
(**A**) Diagrammatic sketch showing progressive destruction of bedding by burrowing organisms (1–5). It is also possible for burrowing within a homogeneous sand to result in mottled sand when burrows are filled with material not originally present in the sand (arrows). [D. G. Moore and P. C. Scruton, 1957, *Amer. Assoc. Petroleum Geol. Bull., 41*.] (**B**) Bedding of siltstone (light) and organic-rich shale (dark) disrupted by burrowing of benthonic marine organisms, Tidewater Formation (Pennsylvanian), Kentucky. Diameter of core is 6 cm. [Photo courtesy P. E. Potter.]

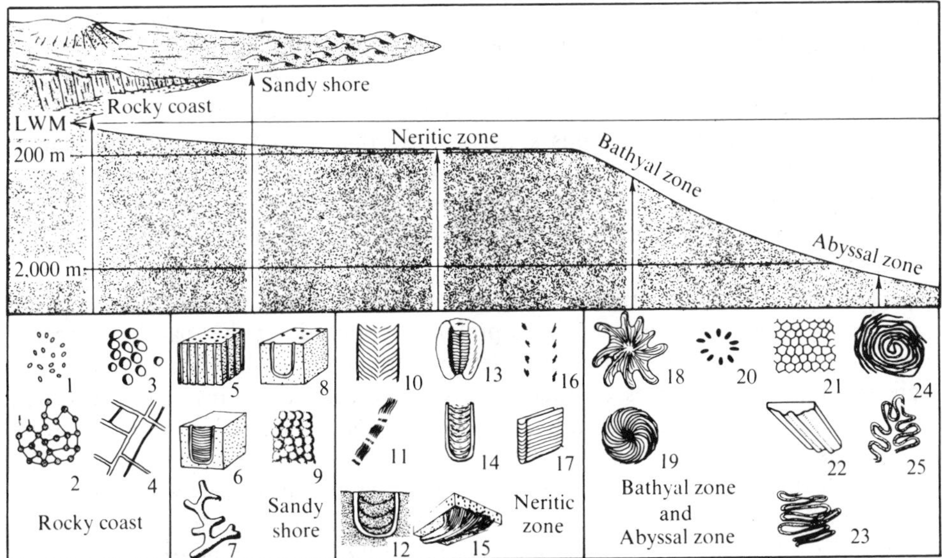

Figure 4–20
Summary diagram of the most common marine facies and depth-related trace fossils. (1–4) All borings. (5) *Skolithus.* (6) *Diplocraterion.* (7–8) Other deep burrows. (9) Detail of wall of *Ophiomorpha* burrow. (10) *Cruziana.* (11, 13, 16) Other trilobite traces. (12, 14, 15, 17) Feeding burrows, with weblike "spreiten" between vertical parts of U tubes. (18) *Zoophycus.* (19) Another complex feeding burrow. (20) Radiating trace. (21) *Paleodictyon.* (23, 24, 25) Typical meandering or spiral surface traces. (22) Enlarged portion of one section of meandering bilobate grazing trail, preserved as sole mark. [T. P. Grimes, 1975, in R. W. Frey, ed., *The Study of Trace Fossils* (New York: Springer-Verlag).]

are essentially constant, and there is no need for organisms to burrow vertically into the sediment. In addition, it is pointless for an organism at great depth to burrow into the sediment for organic matter to eat. The rate of deposition of the fine sediment at depth is so slow that it can be eaten before it is buried.

Directional Structures

A large variety of sedimentary structures can indicate the direction of current flow in sandstone beds. Some of them occur within beds, such as imbrication and cross bedding; others occur on the surfaces of beds, such as ripple marks or flutes. Measurement of the orientation of such structures in fluvial or marine sediments can define the *paleoslope* or regional depositional dip. For example, measurements of cross-bedding directions in mid-continental United States in fluvial sandstones younger than Middle Cambrian age consistently indicate that streams flowed toward the south (Potter and Pryor, 1961). That is, the drainage direction for the past 500 million years has been toward the Gulf of Mexico. Many geophysicists and students of plate tectonics believe that the Mississippi embayment is an aulacogen that originated in Precambrian time and has been reactivated several times since then. The bounding faults of the aulacogen strike approximately north–south, accounting for the direction of flow of the lower Mississippi River and its ancestors.

The structures discussed in this section, those that give directional information useful in paleogeographic reconstructions, are (1) cross bedding, (2) ripple marks, (3) imbrication, (4) flute casts, (5) groove casts, and (6) parting lineation.

A large number of other sedimentary structures has been described from outcrops of sandstones and conglomerates, and most of them can be used to interpret some aspect of depositional or diagenetic conditions. For example, in conglomerates the long axes of gravel particles can have a preferred orientation either parallel or transverse to the paleoflow direction. Parallel orientation is produced by rapid mass-transport processes, such as characterize turbidity currents; transverse orientation occurs when the gravel particles are transported by the rolling action typical of rivers and beaches.

In the field the general rule to follow is that anything that departs from isotropism and homogeneity has the potential to supply useful information about the origin of the sediment. Careful observation and recording either in notebook or on film often prove invaluable in the office or laboratory. "Every little squiggle has a meaning all its own."

Cross Bedding

Perhaps the most familiar of sedimentary structures in sand and silt deposits is cross bedding; it is rare in conglomerates. A vast literature exists concerning its origin and interpretation. High-angle cross bedding is produced largely by the avalanching of sand or coarse silt down the *lee* (downstream) slope of wavelike structures such as ripples and dunes. Therefore, cross beds dip downcurrent. The initial angle of dip is controlled by the angle of repose of loose sediment in air or water (about 25–35°), but in ancient rocks the dip angle is more likely to be 15–20° because of compaction following burial. Cross bedding at a low angle (<10°) is produced during deposition by accretion on sloping surfaces, such as on the *stoss* (upstream) slope of wavelike structures or on the face of a beach. In tide-dominated environments where the current direction is constant for six hours before reversing, a distinctive herringbone pattern can be produced (see Figure 4–21). It is also possible, however, that one direction of tidal flow dominates to the extent that sands deposited during the preceding tidal cycle are eroded and only one direction of flow is indicated by the resulting deposit.

Ripple Marks

The first structures formed on a bedding surface when the current becomes fast enough to transport sand are ripple marks. They are usually asymmetrical in cross section, with the downstream slope steeper than the upstream slope (see Figure

Figure 4–21
Herringbone cross bedding in Winchell Creek Member, Great Meadows Formation (Ordovician), New York, underlain by planar lamination and overlain by ripple forms. Bar = 15 cm. [S. J. Mazzulo, 1978, *Jour. Sed. Petrology, 48*. Photo courtesy S. J. Mazzulo.]

Figure 4–22
Cuspate ripple marks on thin-bedded sandstone, Caseyville Formation (Pennsylvanian), Illinois. Current flowed from left to right. [Pettijohn and Potter, 1964. Photo courtesy F. J. Pettijohn.]

4–22). This type of ripple is formed by frictional stresses between the moving fluid and the sandy bed when there is continued movement of the fluid in the same general direction. The fluid may be air or water. Desert wind currents, stream currents, marine longshore currents, turbidity currents, or deep-sea currents all produce rippled sand surfaces. Hence, ripples occur on sand dunes, stream bottoms, and at all

Figure 4–23
Imbrication in conglomerate (Jurassic), southwestern Oregon. Scale is parallel to bedding, and imbrication is clearly visible just under scale. Current flowed from left to right. Imbrication is generally more pronounced in modern sediments than in ancient rocks because of effect of compaction during burial. [R. G. Walker, 1977, *Geol. Soc. Amer. Bull., 88*. Photo courtesy R. G. Walker.]

depths in the sea. In some cases there may be differences in either the scale or the form of the ripples in different environments because of variations in the fluid dynamics of ripple formation in different environments. Oscillatory movement of fluid (i.e., back-and-forth movement rapidly repeated) such as in some lakes produces symmetrical ripples that cannot be used to indicate the downstream direction.

Imbrication

The stacking of grains with their flat surfaces at an angle to the major bedding plane is termed imbrication. It is seen most clearly in gravels and conglomerates (see Figure 4–23) but is present in sands as well, as has been demonstrated by examination of thin sections cut perpendicular to bedding but parallel to the direction of current flow as defined by other criteria. In conglomerates the imbrication is developed best where the pebbles and cobbles are in direct contact (i.e., not much sand is present) and the gravel particles are platy in shape. The gravel dips upstream because this is the stable position; gravel oriented in the opposite direction tends to be overturned by the moving water.

Flute Casts

These asymmetrical structures are found on the sole (bottom) of some sandstone beds (see Figure 4–24), particularly in environments where turbidity currents have been common. For this reason, flute casts have become associated with deep-water depos-

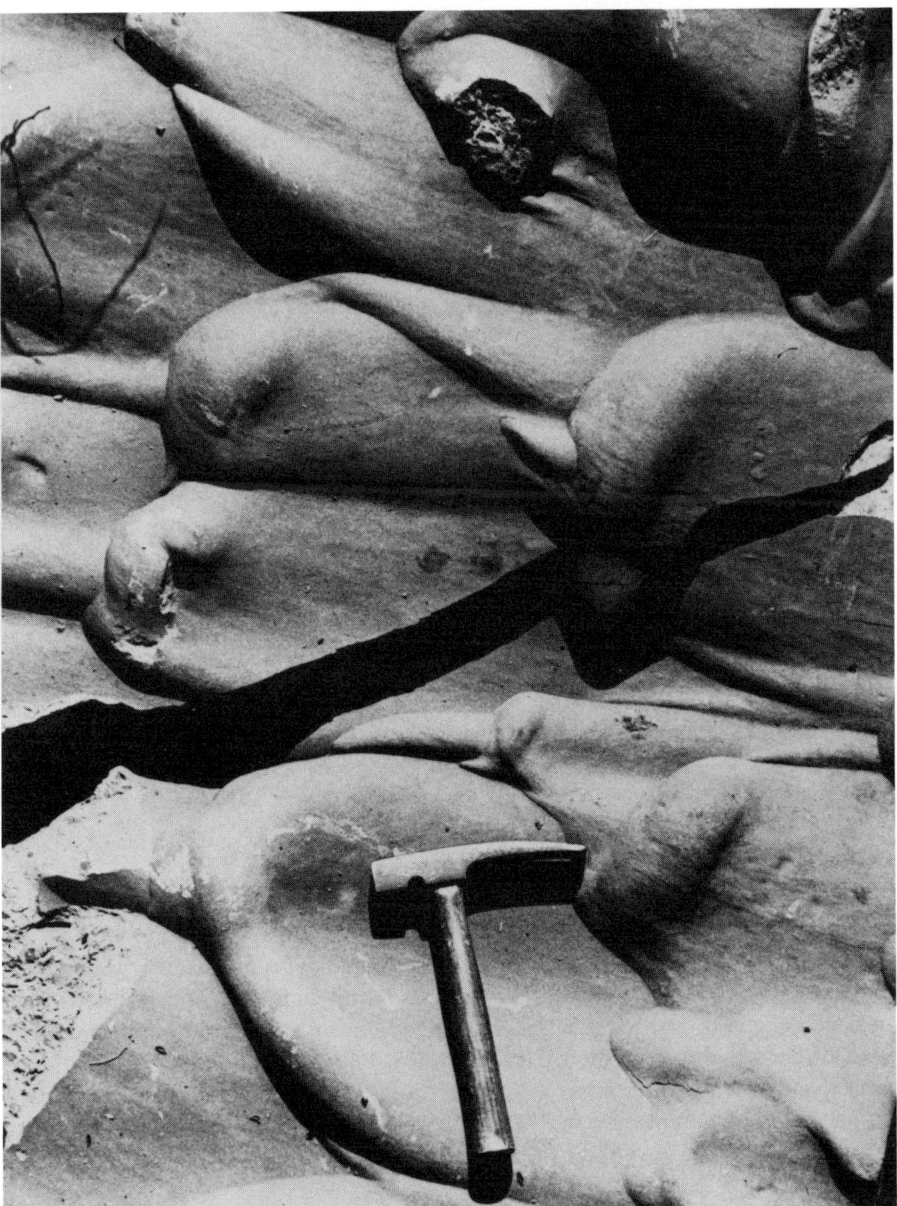

Figure 4–24
Large, closely spaced, and sometimes overlapping flute casts on base of sandstone bed, Smithwick Formation (Pennsylvanian), Burnett County, Texas. Current flowed from left to right. [Pettijohn and Potter, 1964. Photo courtesy F. J. Pettijohn.]

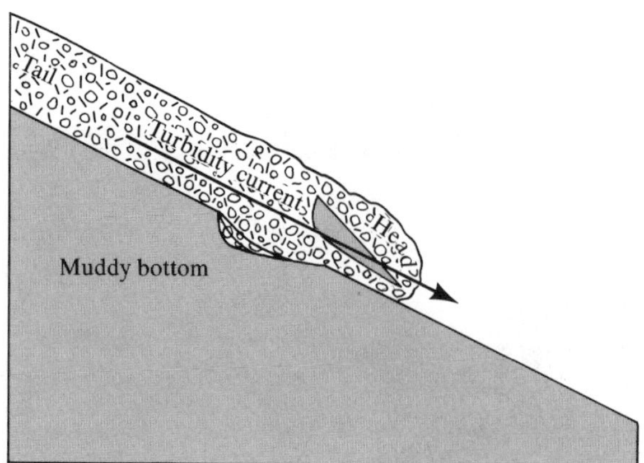

Figure 4–25
Schematic cross section illustrating formation of flute by bottom scour from head of turbidity current on seafloor. Cross-bedded sand from tail of current has begun to fill in flute. Current is normally many orders of magnitude larger than the flute.

its, although turbidity currents are not restricted to the deep oceanic basins. The scooplike structure of the flute is produced by the erosion of freshly deposited mud on an underwater slope. A turbidity current flows over the mud, and turbulent eddies form at the current–mud interface at the head of the current, resulting in the erosion of a spoon-shaped gob of mud with the deepest part of the spoon upcurrent (see Figure 4–25). As the tail of the turbidity current passes over the spoon, fine-grained sand is deposited in it. This sand is subsequently seen as a downward projection on the base of the sandstone that overlies the mud in the stratigraphic record. In outcrops, normally the mud (now mudrock) is weathered away to reveal the oriented flutes.

Groove Casts

Groove casts are long, straight structures on sandstone soles and are thought to be produced by tools being dragged along a soft, muddy surface (see Figure 4–26). An embossed variety of groove cast is the *chevron mark,* composed of continuous, open V marks arranged to form a straight ridge, the V forms closing downcurrent. Related types of current-direction indicators include structures termed *prod marks, bounce marks,* and *brush marks,* all of which are produced by a tool of some sort impacting with a muddy sediment bottom as the tool moves downcurrent. All are illustrated by Pettijohn and Potter (1964).

Parting Lineation

This type of directional structure is found on the bedding surfaces of some thin-bedded sandstones (see Figure 4–27). The structure appears as subparallel, very faint, flat, linear grooves and ridges of very slight relief and is known from both fluvial and turbidite sandstones. The long axis of sand grains parallels the grooves and ridges.

Figure 4–26
Groove casts (a), prod marks (b), and brush marks (c) on base of sandstone bed, Gardeau Formation (Devonian), near Danville, New York. Current flowed from bottom to top. [Pettijohn and Potter, 1964. Photo courtesy F. J. Pettijohn.]

Figure 4–27
Parting lineation in fine-grained sandstone, Haymond Formation (Pennsylvanian), Pecos County, Texas. Fluvial current flowed parallel to hammer handle. [Pettijohn and Potter, 1964. Photo courtesy F. J. Pettijohn.]

Table 4–4
Checklist of Features of Detrital Rocks to Be Described in the Field

General characteristics

1. Extent of outcrop: height and width
2. Relative abundance of conglomerate, sandstone, and mudrock
3. Degree of intermixing: Is lower third mudrock, middle third conglomerate, and upper third sandstone? Or are rock types intimately intermixed?
4. If intermixed, is there a characteristic sequence, such as conglomerate units normally followed by sandstones?

Sedimentary structures

1. Thickness and lateral extent of beds
2. Nature of contacts with beds above and below: sharp or gradational
3. Character of bedding surfaces:
   a. Planar, wavy, or irregular; relation to lithology
   b. Ripple marks: symmetry, orientation, height, distance between crests
   c. Parting lineation: orientation
   d. Load casts
   e. Flute casts: scale and orientation
   f. Groove casts: description and orientation
   g. Organic tracks and trails: description
   h. Mud cracks, raindrop impressions, or other features
   i. Cut-and-fill structures
4. Internal character of beds:
   a. Laminations: cause (e.g., organic matter, grain-size change), thickness, continuity; fissility
   b. Organic burrows: abundance, type, relation to lithology
   c. Cross bedding: scale, orientation, angle of dip
   d. Graded bedding: thickness of graded unit, frequency, grain-size variation
   e. Convolute bedding, intraformational breccias
   f. Orientations of fossils (e.g., preferred orientation of elongate shells; brachiopod shells mainly concave up or down) or pebbles
   g. Imbrication: orientation

Comment on Field Observations

Many textural and most structural features of conglomerate and sandstone sequences are best seen in the field. For this reason, they should be described while you are in the field at the outcrop. After you have returned to the laboratory or office, it is difficult to recall such things as whether bedding contacts were sharp or gradational, whether the graded beds were abundant or exceptional, or whether imbrication was present at a particular locality. Each of these observations can be invaluable as a clue in paleogeographic interpretation.

Table 4–4 (*continued*)

Sedimentary textures
1. Mean grain size and sorting
2. Clay content
3. Shape of gravel particles: rod, disc, sphere
4. Rounding of grains: relation to grain size
5. Color: relation to rock type and grain size
Mineral composition
1. Percentages of quartz, feldspar, and rock fragments; variation with grain size
2. Degree of weathering: e.g., K-feldspar mostly altered to white kaolinite, but twinned plagioclase is fresh
Diagenetic features
1. Degree of lithification
2. Type of cement: quartz, calcite, hematite, or other
3. Porosity: Does water sink into a hand specimen?

Table 4–4 is a checklist of features that might be observed and described in the conglomerate or sandstone in the field setting. Many of them do not occur at any particular outcrop, but most are present somewhere in most stratigraphic sections.

LABORATORY STUDIES

Although some textural and most structural features of conglomerates and sandstones can be determined at the outcrop, an adequate evaluation typically requires laboratory study. This is particularly true of sedimentary textures because of their smaller scale. Details of texture not readily seen with a hand lens can reveal, for example, whether the transporting medium or environment of deposition was a desert or a periglacial setting.

Textures

Texture in rocks can be considered both as an aggregate property and as a property possessed by individual grains. For example, grain-size distribution, sorting, and permeability are aggregate properties; grain orientation and shape are properties of individual particles.

Size distribution and sorting are estimated in the field and determined more accurately in the laboratory with either sieves, pipettes, or settling tubes. The rock specimen is disaggregated with a 10% solution of cold or hot hydrochloric acid to dissolve calcite, dolomite, and hematite cements. Silica cement cannot be dissolved without also dissolving the detrital grains, so that the sizes of sand grains in silica-cemented rocks must be determined by measurements of grain lengths as seen in thin section. (Procedures are described in Chapter 15.)

Bimodality

One important aspect of a grain-size distribution seen more clearly with laboratory analysis than in the field is *bimodality* in a sandstone. The presence of more than a single mode in a size distribution can be a strong indicator of depositional environment, as shown by Folk (1968) for quartz-cemented quartz sandstones. Sandstones composed entirely of quartz and consisting of two modes—one in the coarse sand range, the other in the fine sand range (see Figure 4–28)—are quite common in early Paleozoic rocks of the United States and elsewhere. Folk explains them as residual deposits formed in desert environments. As the wind moves over an unsorted sandy sediment, the grains most easily moved are those in the fine and medium sand sizes and these are heaped into the familiar dunes downwind. The remaining detritus is thus made bimodal. Subsequently, marine transgression may occur, so that the bimodal sand finally comes to rest in a shallow marine environment, but the size distribution indicates that the sand is only briefly removed from the aeolian realm in which it was born.

Bimodality can also be produced by mechanisms that do not indicate a unique environment of deposition. For example, if a fine-grained sediment is deposited immediately above a coarser one, the finer grains may infiltrate the coarser ones by settling downward through the large pore spaces among the coarser grains. For this process to be effective, the diameter of the smaller grains must be less than about one-tenth the diameter of the larger; for example, a very fine sand infiltrating a very coarse sand. This could occur in any environment in which a wide range of grain sizes is available for deposition.

Burrowing organisms can create bimodality immediately below the water–sediment interface by mixing two well-sorted laminae. Probably this is more common in the shallow marine environment because of the greater abundance of bottom-feeding organisms there than in a fluvial setting.

Figure 4–28
(**A**) Highly idealized diagram showing transformation of unsorted, sandy detritus into two groups of sediment: well-sorted fine- to medium-grained sand in dunes; and bimodal, residual sand in upwind areas. (**B**) Photomicrograph showing resulting bimodality as seen in Lander Formation (Ordovician), Wyoming. Each mode is very well sorted, and the coarser sand is well-rounded, as is characteristic of coarse sand in environments of high kinetic energy. [Folk, 1968. Photo courtesy R. L. Folk.]

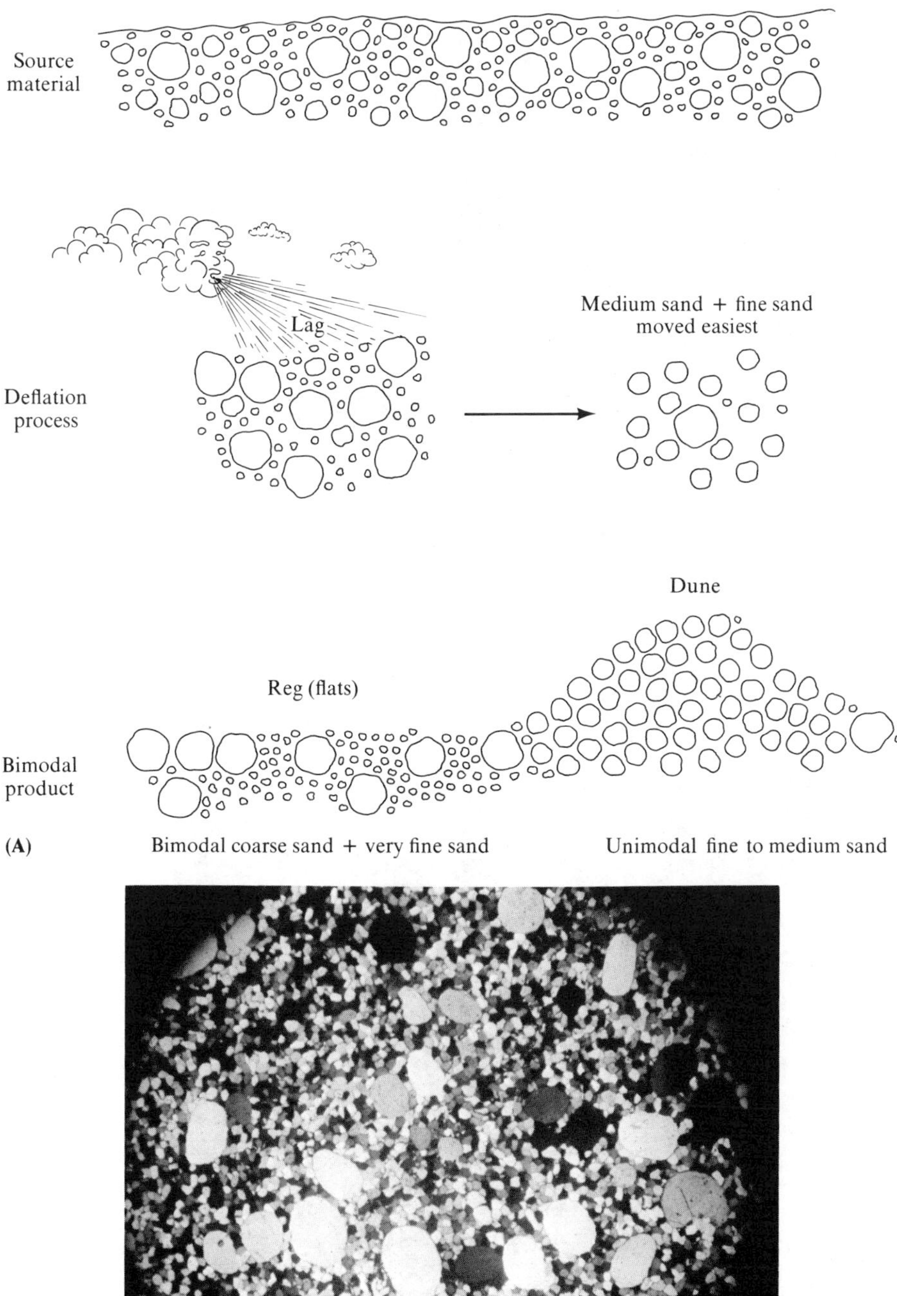
Unsorted fluvial detritus
Source material
Lag
Medium sand + fine sand moved easiest
Deflation process
Dune
Reg (flats)
Bimodal product
(A)
Bimodal coarse sand + very fine sand
Unimodal fine to medium sand
(B)

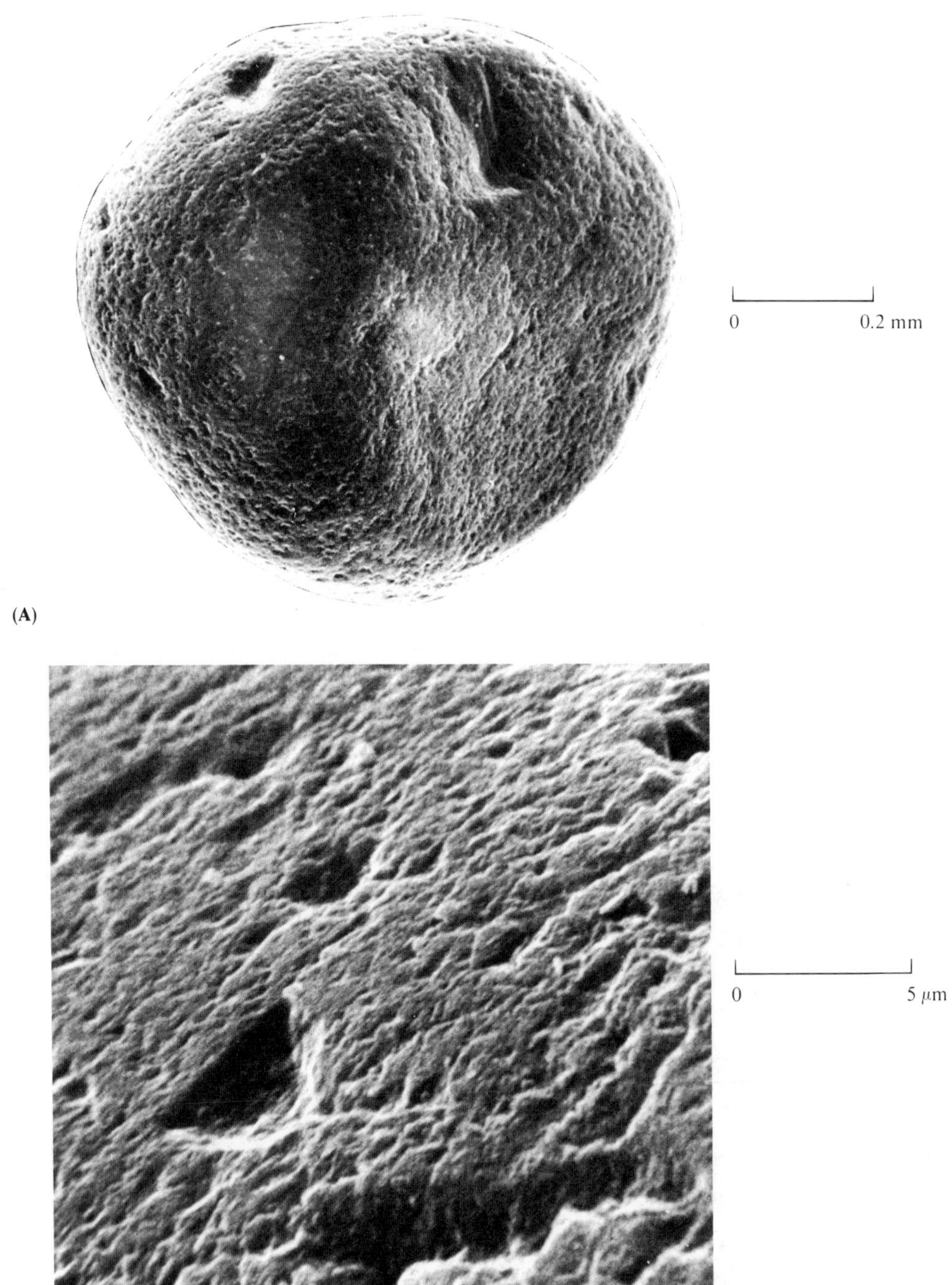
0
0.2 mm
(A)
0
5 μm
(B)

Surface Texture of Quartz Grains

When examined with an electron microscope, the surfaces of detrital quartz grains reveal a variety of markings, such as striations, V-shaped pits, and microsheeting. Field and laboratory studies during the past 20 years, largely by Krinsley and his co-workers and students, have demonstrated that the surface markings can reflect the last environment of deposition of the grains. On some grains, cross-cutting relationships can reveal previous depositional environments as well. Figure 4–29 illustrates two of the more common markings and their interpretation; more detailed treatment is provided by Krinsley and Doornkamp (1973). The principal difficulty in the application of this technique is the effect of diagenesis. After burial, percolating underground waters attack the surfaces of grains and remove the surface markings produced in the environment of deposition and replace them with V-shaped pits—etch pits characteristic of the trigonal symmetry of quartz. Hence, the surface texture of detrital quartz is most useful for Neogene sediments and sometimes can be used successfully for Mesozoic rocks; in Paleozoic rocks the quartz grains usually show only diagenetic effects on their surfaces.

Fabric

The term *fabric* refers to the spatial arrangement (*packing*) and orientation of the grains or crystals in a rock. Spatial arrangement is determined by measurement in thin section of two parameters: packing density and packing proximity. *Packing density* is the ratio of the sum of the lengths of grain intercepts to the total length of the traverse across the thin section (see Figure 4–30). It is a way to determine the percentage of void space in the detrital rock after compaction, the space now filled or partially filled with chemically precipitated cement. *Packing proximity* is the ratio of the number of grain-to-grain contacts (encountered in a traverse across the thin section) to the total number of contacts of all kinds; for example, grain to grain, grain to cement, grain to pore void, grain to matrix mud). In the cup of Jell-O with fruit purchased in a cafeteria, both the packing density and the packing proximity of the pieces of fruit typically approach zero.

Both packing density and packing proximity have been inadequately studied by sedimentary petrologists. They depend not only on burial depth (degree of compaction) but on the depth at which cement may have been precipitated and on the mineral and lithic-fragment composition of the rock. For example, if the rock con-

---

Figure 4–29
Surface textural features diagnostic of depositional environment of quartz grains as seen with scanning electron microscope. (**A**) Aeolian grain, central Libyan sand dune. Grain is well-rounded, and series of parallel plates of quartz about 1 μm in size is visible all over surface. Large indentation resulted from high-velocity grain-to-grain impacts. (**B**) Beach grain showing effects of littoral abrasion. V-shaped indentations of various sizes dominate surface. [Photos courtesy D. H. Krinsley.]

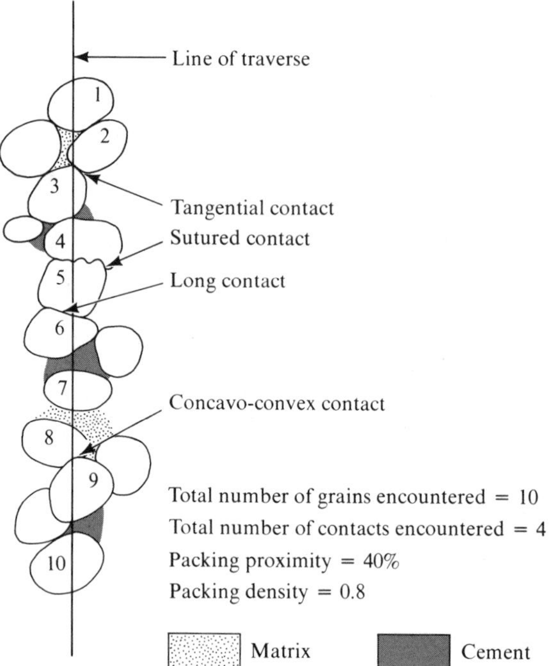

Figure 4–30
Schematic view of thin section, showing definitions of packing density and packing proximity and four types of grain-to-grain contacts commonly noted in detrital rocks. [Blatt et al., 1980. Copyright © 1980. Reprinted by permission of Prentice-Hall, Inc., Englewood Cliffs, New Jersey.]

tains a sizable proportion of ductile, foliated fragments such as schist, both packing density and packing proximity are very high. If the rock consists entirely of detrital quartz, the values of both packing parameters are low. If the grains have been forced apart by the force of crystallization of cement during early burial, the values are even lower. As yet, the number of studies of packing is too few to establish the interrelationships among the controlling variables.

Grain orientations in sandstones mimic those in conglomerates, although the orientations are not so well developed because of the smaller grain sizes in sandstones. To evaluate the presence and strength of sand-grain orientation, two thin sections must be cut. The first is cut parallel to stratification or lamination, so that the orientations of the long dimension of detrital grains can be seen. Normally, only the orientations of visibly elongate grains are determined, those with length/width ratios of at least 2/1. These grains show most clearly any orientation that is present. The second thin section is then cut perpendicular to the stratification but parallel to the direction of grain elongation. Imbrication of the grains can be seen in this plane. Grain orientations are only infrequently determined in sandstones because of the tediousness of the procedure. The same results may be obtained easily in the field with observations of sedimentary structures such as cross bedding and flute casts (Parkash and Middleton, 1970).

Porosity and Permeability

Porosity is defined as void/total-rock volume; *permeability,* the ability to transmit a fluid. The initial (depositional) porosity and permeability of a sediment are related directly to textural maturity. The presence of clay decreases porosity because the clay minerals are only 1/100–1/1,000 the size of sand grains and, as a result, lodge in the spaces between the sand. The clays also decrease permeability because of the very large surface/volume ratio of platy clay flakes. Attractive forces exist between the clay-mineral surfaces and the moving fluids (water, petroleum, or natural gas); and the more surface area the fluids must pass over, the slower their movement.

Sorting of the sand grains also affects initial porosity and permeability. Although the smaller grains of sand are closer to the size of their fellow travelers than are the clay particles, they tend to fill in the empty spaces between the larger grains. Poorer sorting (lower textural maturity) implies lesser porosity. Permeability is lowered as well by poorer sorting, although not so much as when clay minerals are present.

The roundness of sand grains has only a small effect on porosity and permeability. Increased roundness allows the grains to be packed closer together and therefore decreases porosity and permeability.

Both porosity and permeability are decreased greatly by diagenetic processes, with the effect of compaction related to the thickness of overburden and the types of detrital fragments in the sandstone. As noted in the discussion of packing, foliated lithic grains and micas deform under relatively low overburden pressures to increase packing density and decrease porosity. If the percentage of ductile fragments is sufficiently high, the porosity can be decreased to zero by compaction alone. Permeability is affected by compaction even more rapidly than is porosity. A decrease of 50% in porosity can cause a decrease in permeability of two orders of magnitude or more, depending on the nature of the grain surfaces over which the fluid must flow and on the nature of the fluid (water, petroleum, or natural gas).

During diagenesis, new substances are typically precipitated within the original pore spaces (or whatever is left of them after compaction). The familiar calcite, quartz, hematite, and authigenic clay cements are examples of these pore fillings. In extreme cases of cementation, both porosity and permeability are eliminated, and the resulting sediment becomes "hard as a rock."

Interconnected pores form a very irregular and tortuous three-dimensional pattern within the rock (see Figure 4–31) that cannot be detected by thin-section studies alone. Most pores in sandstones have diameters smaller than the 30 μm thickness of a standard thin section and are not seen during examination of the slide. Narrow pores that intersect the plane of the slide can be made more easily visible by vacuum-impregnating the rock with a colored epoxy resin before thin sectioning. The resulting

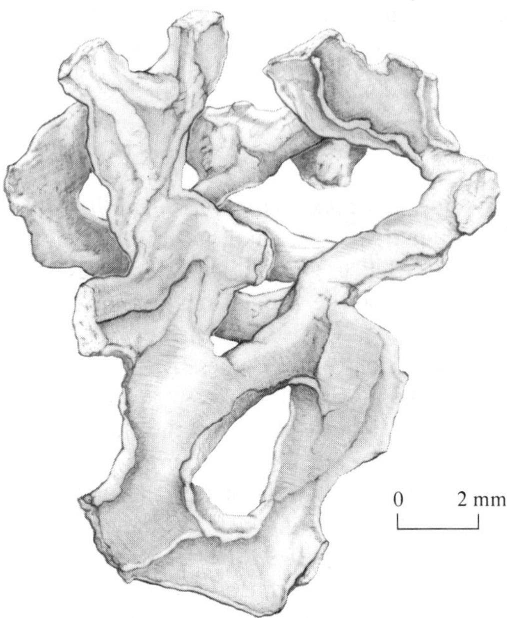

Figure 4–31
Metallic cast produced by impregnating pores of loose sand pack with metal that has a low melting temperature and then removing the sand. [Blatt et al., 1980. Copyright © 1980. Reprinted by permission of Prentice-Hall, Inc., Englewood Cliffs, New Jersey.]

colored streaks between the grains are clearly visible, but the true three-dimensional pattern is still invisible without the use of special techniques. Blatt et al. (1980) provide an extended discussion of porosity and permeability in detrital rocks.

TEXTURES, STRUCTURES, AND PLATE TECTONICS

Movement of the major plates that form the Earth's lithosphere causes the topography above sea level and the bathymetry below it. Thus, plate movements are responsible for creating both the sources of sediment and the sinks in which sands accumulate. In addition, the type of interplate contact, convergent or divergent, determines the slope of the seafloor immediately seaward of the beach. For example, a depth of 100 m is about 150 km offshore at Atlantic City, New Jersey (a divergent margin), but is only 15 km off Los Angeles, California (a convergent margin). The epicontinental seas that were so extensive throughout Paleozoic time formed along divergent margins that are believed to have had seafloor gradients more than an order of magnitude lower than the continental shelf off Atlantic City.

As we have seen, the textural maturity of a sand at the time of deposition is determined by the kinetic energy at the site of deposition. The zone of active movement of sand by waves extends much farther off the Atlantic coastline than off the Pacific coastline. Thus, plate movements determine the areas covered by each subsea environment and the relative volumes of texturally immature, submature, mature, and supermature sands that can be produced by sedimentary processes. Convergent plate margins are dominated by immature and submature sands; divergent margins, by mature and supermature sands. The texturally and mineralogically immature deep-water turbidite sands called *graywackes* (see Chapter 5) dominate the eugeosynclinal terrane of a convergent margin; sands containing at least 75% quartz and chert dominate divergent margins.

Plate movements also exert a strong control on the variety of sedimentary structures developed in sandstones. Sands deposited in shallow-water areas are normally dominated by cross bedding and ripple marks; sands deposited in deeper waters and on steeper slopes typically contain flutes and grooves, with cross bedding relatively uncommon. Thus, sands deposited in epicontinental seas rarely contain flutes and grooves; ancient sands deposited around the rim of the Pacific Ocean or along ancient convergent continental margins (e.g., the Ouachita trough in Oklahoma and Arkansas) contain abundant flutes and grooves.

SUMMARY

Many of the important textural and structural features of sandstone beds are best determined in the field at the outcrop, although adequate evaluation of some textural features requires laboratory study. The most important textural features of a conglomerate or sandstone are grain size and sorting. These can be used to determine distance of transport from a source area and the kinetic energy level of the depositional environment.

The most obvious but least understood characteristic of coarse-grained detrital rocks is bedding. Important observations of bedding include thickness, grading, bioturbation, and sediment transport directions indicated by the direction of cross bedding and by sole markings. Other important features of detrital sequences seen in the field are the ratios of beds of different types, such as conglomerates, sandstones, and mudrocks.

FURTHER READING

Atkinson, D. J. 1962. Tectonic control of sedimentation and the interpretation of sediment alternation in the Tertiary of Prince Charles Foreland, Spitsbergen. *Geol. Soc. Amer. Bull., 73,* 343–364.

Blatt, H., G. V. Middleton, and R. C. Murray. 1980. *Origin of Sedimentary Rocks,* 2nd ed. Englewood Cliffs, NJ: Prentice-Hall, 782 pp.

Dobkins, J. E., Jr., and R. L. Folk. 1970. Shape development on Tahiti-Nui. *Jour. Sed. Petrology, 40,* 1167–1203.

Ehrlich, R., J. J. Orzeck, and B. Weinberg. 1974, 1976. An exact method for characterization of grain shape. *Jour. Sed. Petrology, 44,* 145–150, and *46,* 226–233.

Folk, R. L. 1951. Stages of textural maturity in sedimentary rocks. *Jour. Sed. Petrology, 21,* 127–130.

Folk, R. L. 1968. Bimodal supermature sandstones: product of the desert floor. *Proc. 23rd Internat. Geol. Cong.,* Sec. 8, pp. 9–32.

Frey, R. W. (ed.). 1975. *The Study of Trace Fossils: A Synthesis of Principles, Problems, and Procedures in Ichnology.* New York: Springer-Verlag, 562 pp.

Krinsley, D. H., and J. C. Doornkamp. 1973. *Atlas of Quartz Sand Surface Textures.* Cambridge, England: Cambridge University Press, 91 pp.

Mills, H. H. 1979. Downstream rounding of pebbles—a quantitative review. *Jour. Sed. Petrology, 49,* 295–302.

Parkash, B., and G. V. Middleton. 1970. Downcurrent textural changes in Ordovician turbidite greywackes. *Sedimentology, 14,* 259–293.

Pelletier, B. R. 1958. Pocono paleocurrents in Pennsylvania and Maryland. *Geol. Soc. Amer. Bull., 69,* 1033–1064.

Pettijohn, F. J., and P. E. Potter. 1964. *Atlas and Glossary of Primary Sedimentary Structures.* New York: Springer-Verlag, 370 pp.

Potter, P. E., and F. J. Pettijohn. 1977. *Paleocurrents and Basin Analysis,* 2nd ed. New York: Springer-Verlag, 460 pp.

Potter, P. E., and W. A. Pryor. 1961. Dispersal centers of Paleozoic and later clastics of the Upper Mississippi Valley and adjacent areas. *Geol. Soc. Amer. Bull., 72,* 1195–1250.

Sneed, E. D., and R. L. Folk. 1958. Pebbles in the lower Colorado River, Texas: a study in particle morphogenesis. *Jour. Geol., 66,* 114–150.

Towe, K. M. 1963. Paleogeographic significance of quartz clasticity measurements. *Jour. Geol., 71,* 790–793.

Walker, R. G. 1967. Turbidite sedimentary structures and their relationship to proximal and distal depositional environments. *Jour. Sed. Petrology, 37,* 25–43.

Walker, R. G. (ed.). 1979. *Facies Models.* Geol. Assoc. Canada Reprint Series No. 1, 211 pp.

Yeakel, L. S., Jr. 1962. Tuscarora, Juniata, and Bald Eagle paleocurrents and paleogeography in the central Appalachians. *Geol. Soc. Amer. Bull., 73,* 1515–1540.

5

Conglomerates and Sandstones: Composition

Some circumstantial evidence is very strong,
as when you find a trout in the milk.

HENRY DAVID THOREAU

The majority of the detrital fragments in sandstones and conglomerates are coarser than 0.06 mm in diameter and are seen easily with either the unaided eye or a hand lens. Accurate identification of the mineralogy of the fragments, however, is another matter. Quartz is easy to spot, as is pink potassium feldspar; but chert can look like rhyolite, phyllite looks like shale, and those little black things you see through the hand lens may be basalt, magnetite, or black chert. Considerable experience is needed in order to achieve a reasonable degree of accuracy in the identification of sand-size particles in hand specimens. Normally, most of these mineralogic identifications are postponed until thin sections and a polarizing microscope are available. Areal changes in mineral and rock-fragment composition are very important in paleogeographic work, however, and as much as possible should be done in the field setting. The question of where to go next for a meaningful rock sample can commonly be answered by mineralogic trends pieced together from individual observations made at scattered outcrops. Time in the field is too valuable to be wasted wandering around wondering where to go next.

CONGLOMERATES

Field Observations

The lithologic identification of boulders, cobbles, and pebbles is made most easily in the field unless you are a weight lifter, your vehicle has a large trunk, and the outcrops are very near the road. The sampling technique is to select at random (every grain has an equal chance of being selected) 100–300 gravel-size grains, crack them open with a geology pick, and identify them with a hand lens. The fundamental categories for classification are (1) extrusive igneous, (2) plutonic igneous, (3) metamorphic, and (4) sedimentary. Each of these four groups may be subdivided further to the extent that it seems useful and is within your capabilities. For example, quartz pebbles must be identified, but the distinction between those that were derived from a quartz vein and those that were derived from a metaquartzite is probably not possible in most conglomerates.

A conglomerate is termed *oligomictic* if it is composed almost entirely of a single type of fragment (e.g., granite, rhyolite, or quartz); *polymictic*, if it is composed of a variety of fragment types. The composition of a gravel or conglomerate depends on several factors:

1. Lithology of the source area.
2. Initial size of the fragments.
3. Transport distance (see Figure 5–1).
4. Grain size of the gravel particles in the conglomerate.

The significance of the source area is obvious. If a rock type is not exposed upcurrent, it cannot occur in a downcurrent deposit. Normally, the object of a petrologic study of a conglomerate is to determine the nature of the source area.

The initial size of fragments released from the upcurrent sources differs for rocks of different lithology because of the varying boulder-forming capabilities of different rock types. For example, metaquartzite fragments initially are quite large because of their resistance to chemical weathering. Their initial size is determined by the thickness of bedding and the spacing of joints. Granite, in contrast, commonly shows spheroidal weathering on the outcrop and begins its sedimentary life at a size smaller than is the case for metaquartzites. Shale fragments typically begin sedimentary transport in the finer pebble sizes because of their fissile nature and closely spaced joint pattern in outcrop. Further, fragments have varying resistances to size reduction during transport: chert and rhyolite fragments are quite durable during transport; limestone and schist are not.

Because of these factors, the interpretation of the petrology of conglomerates is not so straightforward as it seems initially. Nevertheless, conglomerates are particularly valuable deposits for the determination of provenance (e.g., Green et al., 1968). Unlike most sandstones, they are typically rich in fragments of the undisaggregated source rock rather than being composed largely of individual mineral grains such as orthoclase or single crystals of quartz that might have originated in a wide variety of rock types. In addition, most accumulations of gravel-size fragments are located much nearer the source terrane than are sand deposits. As a result, they have undergone less modification, on the average, since being released from their parent rocks, and their interpretation in terms of paleogeology is more reliable.

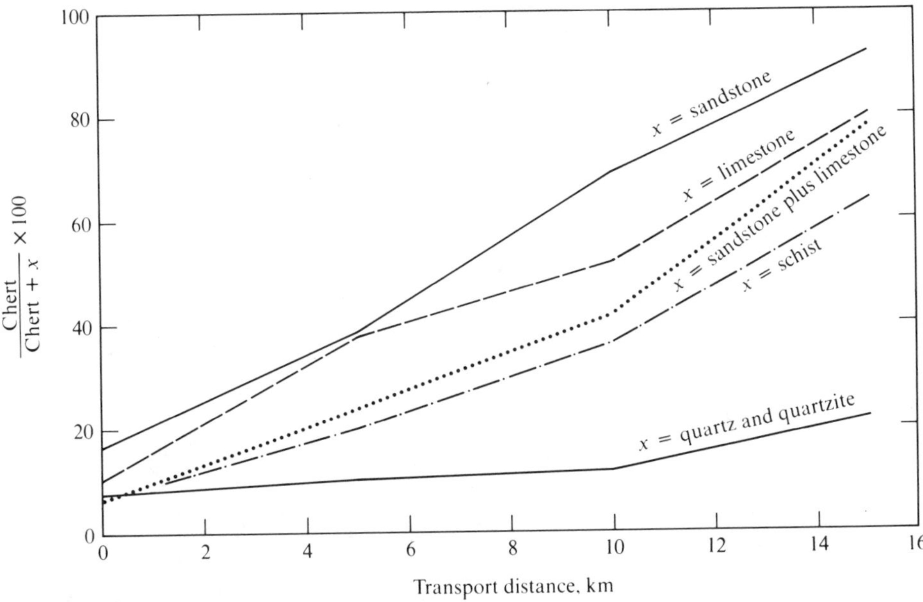

Figure 5–1
Ratio of chert to chert plus another constituent in gravel fraction of stream sediment in three streams draining eastward from Black Hills in South Dakota. Sandstone is weakly cemented with calcite and disintegrates very rapidly when transported. Limestone, though soft, must be abraded to sand size and thus lasts longer in gravel fraction. Schist splits along planes of foliation. Quartz and quartzite are only slightly less durable than chert, so that ratio of chert to chert plus quartz and quartzite changes very slowly during transport. [Data from W. J. Plumley, 1948, *Jour. Geol., 56.*]

The detritus from areas with high relief and rapid rates of erosion often contains abundant gravel of low resistance to weathering and transportation. Examples of such detritus include schists, granites, and limestones. Such polymictic conglomerates are characteristic of orogenic areas such as the Alps during the Tertiary Period, the Rocky Mountains during the Pennsylvanian Period (Hubert, 1960), and the Canadian continental margin during Archean time (Turner and Walker, 1973). The conglomerates can be either marine, emplaced in the deep oceanic basin by turbidity currents; or nonmarine alluvial fan deposits. Other occurrences are as beach deposits and as gravel bars in high-gradient streams.

Post Oak Conglomerate (Permian), Southwestern Oklahoma

The Post Oak Conglomerate has an outcrop pattern that rings the Wichita Mountain igneous complex (see Figure 5–2) and extends in the subsurface into the Anadarko Basin to the north and the Hollis Basin to the south. The conglomerate is as much as 5,000 m thick in the Anadarko Basin and 1,300 m thick in the Hollis Basin. Its lithologic composition clearly reflects its source rocks: limestones, rhyolite, granite, and gabbro.

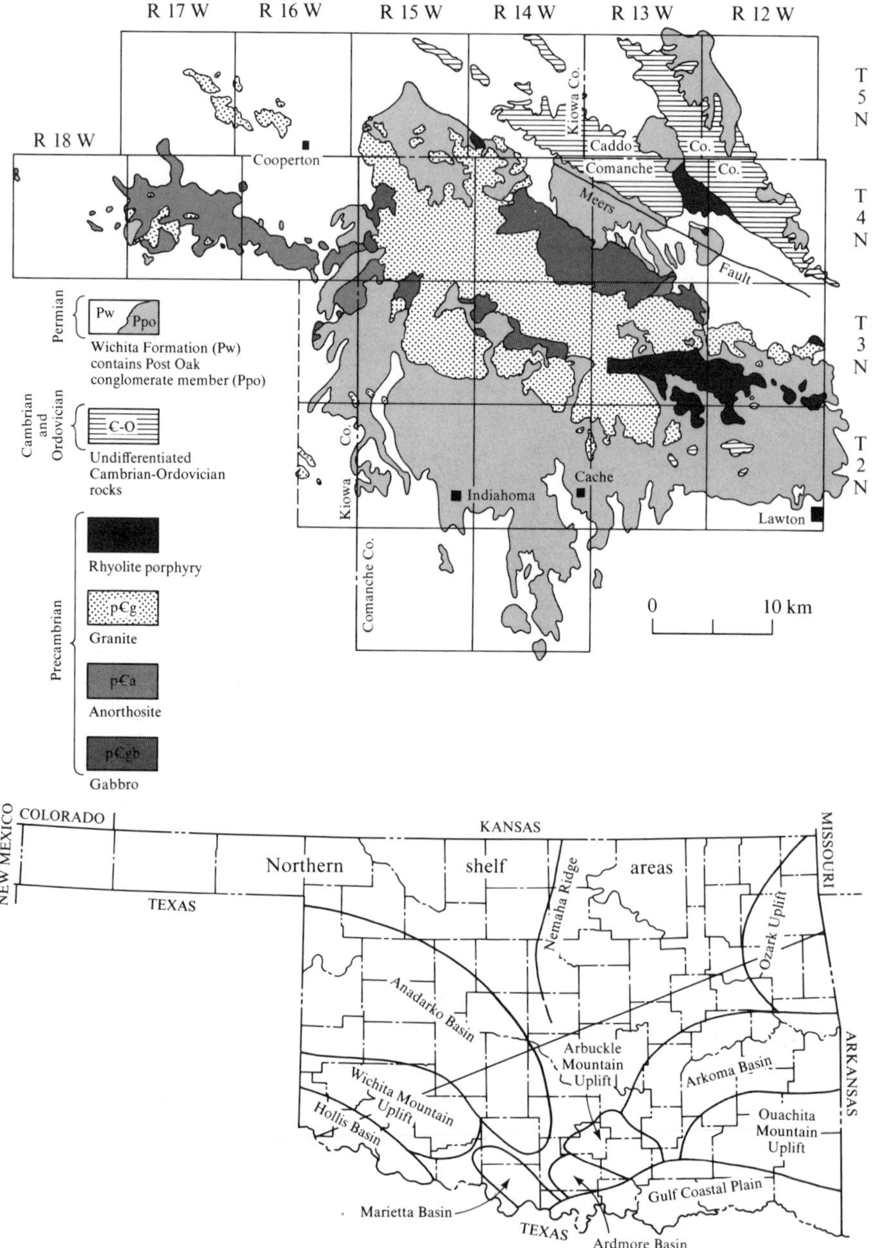

Figure 5-2

Geologic map of Post Oak Conglomerate (Permian) and surrounding rocks in vicinity of Wichita Mountains of southwestern Oklahoma. [Modified from G. W. Chase, 1954, *Amer. Assoc. Petroleum Geol. Bull., 38.*]

The Post Oak originated as a result of block faulting, the Wichita block behaving as a horst structure during Pennsylvanian tectonism in southwestern Oklahoma. Pennsylvanian movements followed older lines of structural weakness produced by structural deformation during the evolution of the southern Oklahoma Precambrian aulacogen. The Wichita block consists of up to 9,000 m of layered gabbro, granite, and rhyolite of Cambrian age that was overlain by 2,000 m of pre-Permian sedimentary rocks, mostly shallow-water carbonates. During Pennsylvanian time, the Wichita block was uplifted 8,000–10,000 m, shedding first limestone conglomerates, then rhyolite conglomerates, granite conglomerates, and, finally, gabbro conglomerates. Sandstones of similar composition are interbedded with the wedge-shaped conglomeratic units. Maximum thickness of the sediments at the mountain front is 600 m, decreasing to 400 m at a distance 40 km from the mountain front. Based on texture, structure, three-dimensional shape of the sequence, and absence of fossils, the conglomerates are interpreted as alluvial fan deposits (*fanglomerates*) deposited in a semiarid climate, much the same as the modern climate in this region of Oklahoma. Figure 5–3 shows the inferred paleogeography during the time of deposition of the Post Oak conglomerates and sandstones.

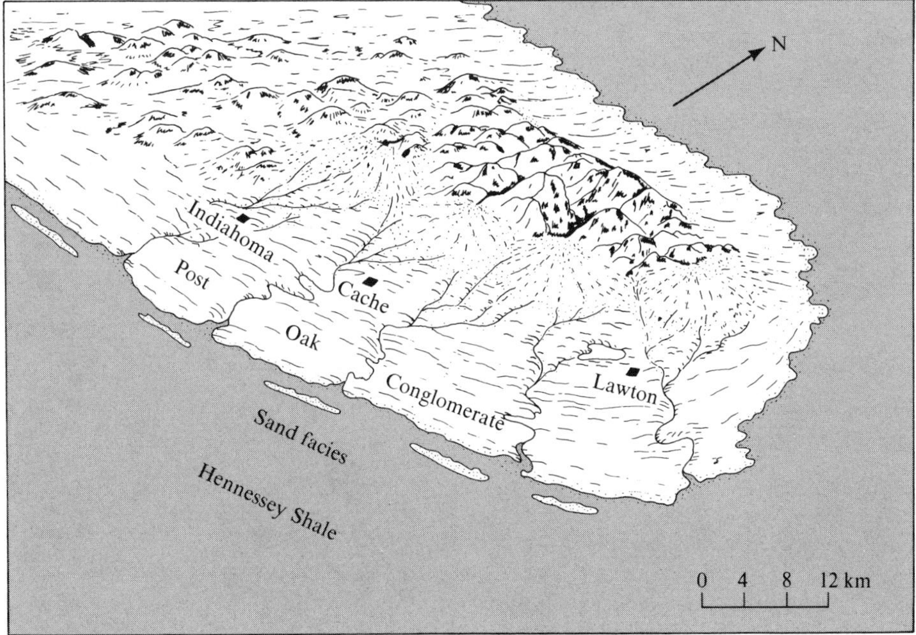

Figure 5–3
Paleogeography in area of Wichita Mountains during time of deposition of Post Oak conglomerates and sandstones, as seen by Permian dragonfly at 1,000 m heading northwest toward Denver. Hennessey Shale is marine and surrounds Wichita Mountain land area. Permian paleolatitude was 12°S. [Stone, 1977.]

The variation in lithology within the coarse fraction of the Post Oak was investigated by Stone (1977), who chose to study the coarse sand fraction of the conglomerate in detail in thin section rather than study gravel in outcrop (see Figure 5–4). Reconnaissance examination of the gravel sizes indicated that a study of either size fraction would produce very similar results. The type of result Stone obtained near the southeastern end of the Wichita Mountains is illustrated in Figures 5–5 and 5–6. Figure 5–5 shows the distribution of rock fragments with distance from the mountain front. Based on this and other, similar maps, Stone inferred the positions of the main stream channels on the alluvial fans that overlapped to form the Post Oak sediment accumulation.

Cardium Formation (Cretaceous), Alberta, Canada

Two interesting examples of oligomictic conglomerates occur in western Canada; the Cadomin Conglomerate (Lower Cretaceous), a deposit formed by a series of adjacent and overlapping alluvial fans (Schultheis and Mountjoy, 1978); and the Cardium Conglomerate (Upper Cretaceous), a shallow, marine gravel shoal (see Figure 5–7). Both units are composed of approximately the same materials. As described by Friedman and Sanders (1978), the single mineral constituent in the Cardium is quartz, but

Figure 5–4
Massive boulder conglomerate in Post Oak Conglomerate, southwest part of T3N, R14W. Hammer in lower center shows scale. Boulders are almost exclusively granite, reflecting nearest exposed crystalline rocks (see Figure 5–1). [Photo courtesy W. B. Stone, Jr.]

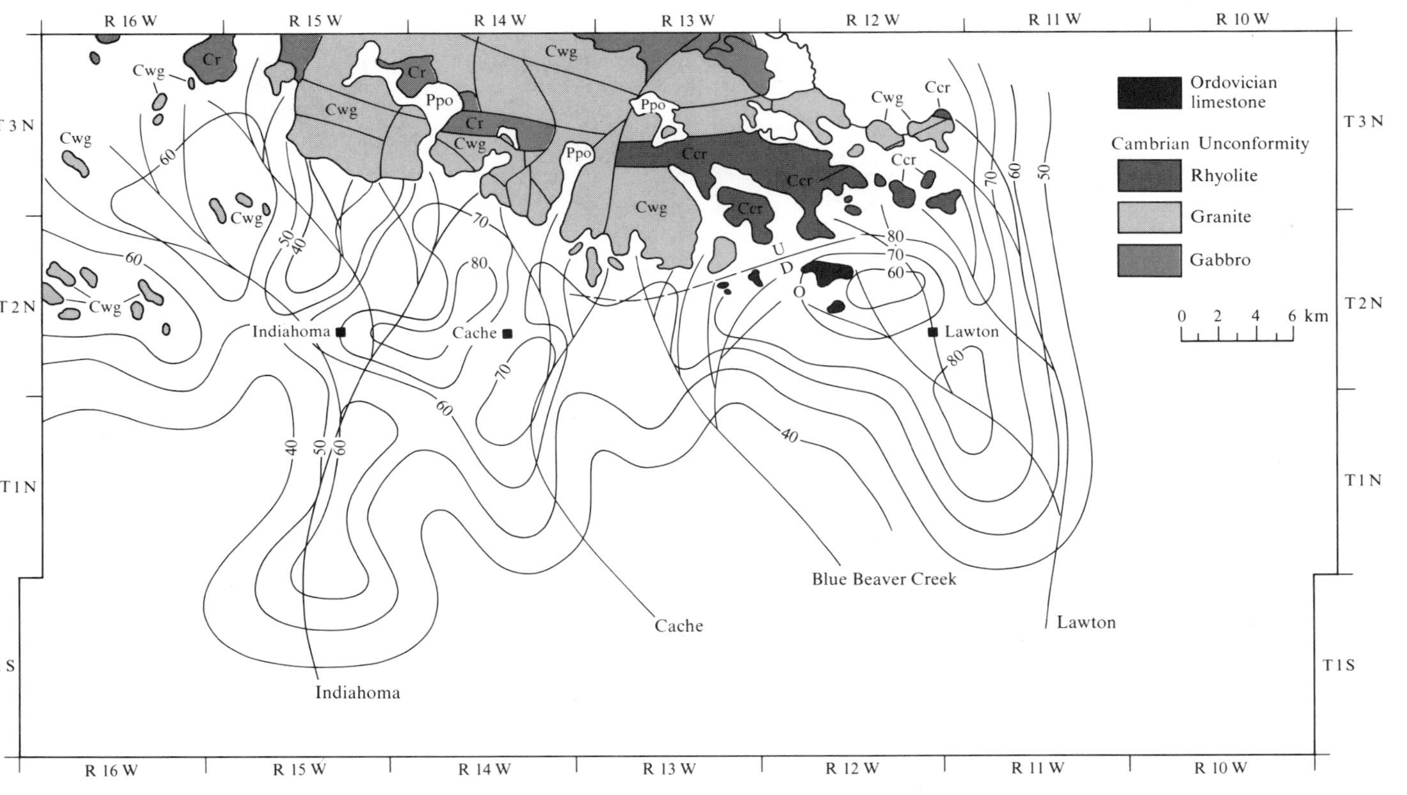

Figure 5–5
Variation in percentage of rock fragments of all types in coarse sand fraction as function of distance from Wichita Mountain front, showing inferred stream patterns. Four alluvial fans are inferred to have been present in this part of Wichita Mountains. [Stone, 1977.]

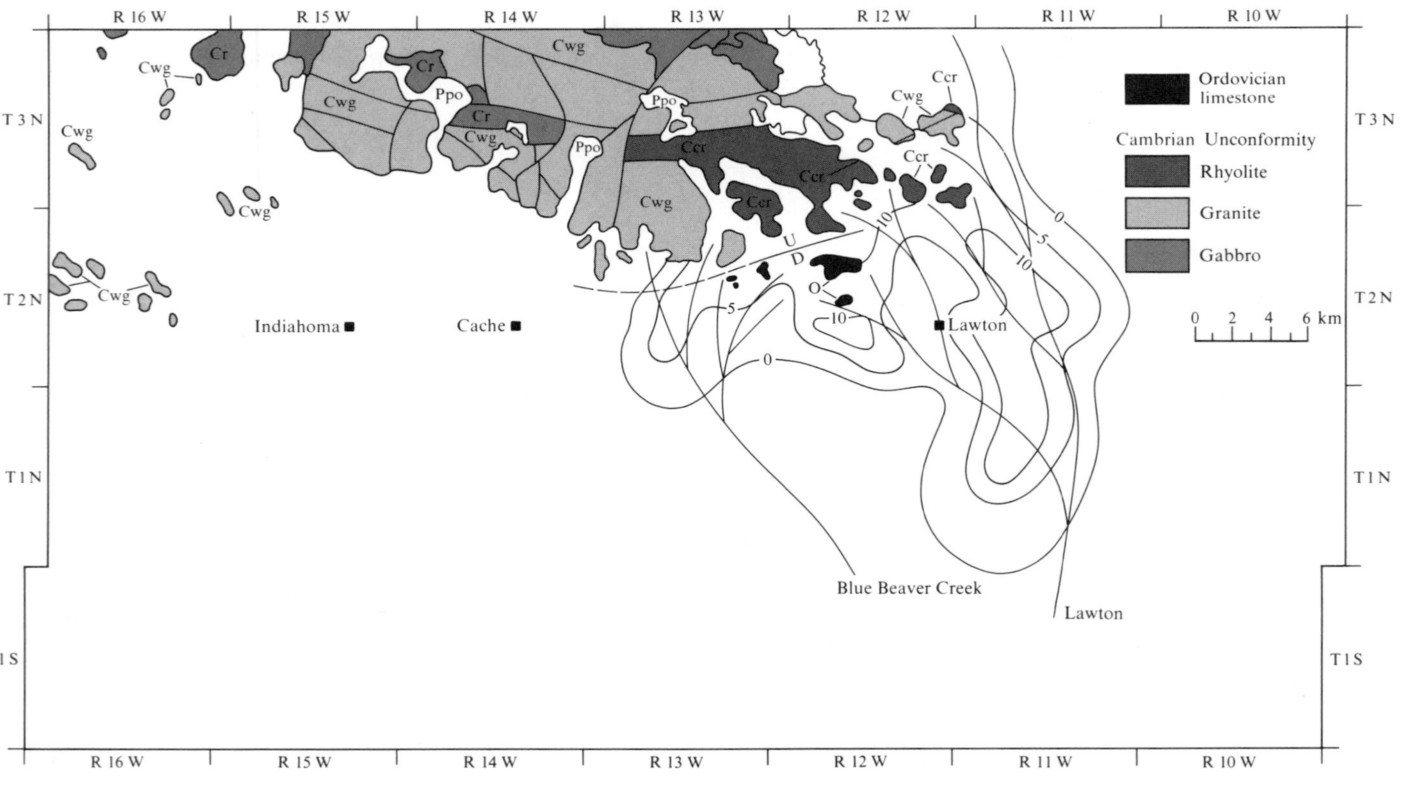

Figure 5–6
Variation in percentage of volcanic (rhyolitic) rock fragments in coarse sand fraction as function of distance from Wichita Mountain front, showing inferred stream patterns. Restriction of rhyolite to southeastern corner of Wichita Mountains is reflected clearly by distribution of rhyolite fragments in Post Oak detritus. [Stone, 1977.]

Figure 5–7
Core from borehole in Cardium Formation (Upper Cretaceous), Alberta, Canada, showing well-rounded chert pebbles of various colors. [Photo courtesy G. M. Friedman.]

the mineral occurs in a variety of forms, including metaquartzite, chert, siliceous shale, and radiolarian chert; some quartzite pebbles have been brecciated and healed with vein quartz. In keeping with the very mature mineralogic composition, the pebbles show evidence of severe abrasion. They are well-rounded with smooth and polished surfaces. The colors of chert pebbles include white, gray, green, blue, and black. Pebbles form 65–85% of the conglomerate. Ellipsoidal pebbles are oriented with their long axes parallel to the direction of current flow. The finer-grained particles between the pebbles consist of quartz and chert of sand and silt size.

Laboratory Studies

As indicated, the mineralogy or lithology of the gravel fraction of a conglomerate is best determined by boulder-cobble-pebble counts in the field at the outcrop. The size of gravel grains may be given as either the long or the intermediate dimension. Both lithology and size should be indicated in field notes, so that correlations between the two variables can be discovered.

Most conglomerates contain abundant finer sediment between the gravel particles. Commonly the sand is equal in abundance to the gravel; but whether or not this occurs, the matrix material should be sampled for later examination in thin section. Frequently the mineral composition of the sandy or muddy matrix is different from that of the gravel. Perhaps the difference is due simply to disaggregation: the gravel is composed of granite pebbles, and the sand is a mixture of feldspar, quartz, and mica. On the other hand, it may be that the gravel was locally derived, but the sand originated at some distance from the depositional site. The mud fraction in a conglomerate may have been derived from the same place as the gravel and was transported and deposited with the gravel, as in a turbidity-current deposit on a deep-sea fan. However, it may be that the mud entered the gravel through infiltration downward into an originally permeable gravel. For example, a mud layer deposited on the surface of a stream gravel filters downward. Examination of the matrix may resolve these uncertainties.

SANDSTONES

The mineral composition of a sandstone is studied to determine two things about the history of the rock: (1) the character of the source rocks from which the detrital grains were derived (provenance), and (2) the diagenetic events that have affected the sandstone. We discuss detrital mineral composition and provenance in this chapter. Consideration of diagenesis is deferred to the next chapter.

Quartz

About two-thirds of the detrital fraction of the average sandstone is quartz and, because of this, a large literature exists concerning its physical and chemical character in detrital rocks. Quartz is the most abundant mineral in most sandstones for several reasons:

1. It is abundant in the most common crystalline rocks: granitoid igneous rocks, gneisses, and schists.
2. It is mechanically very durable, with a hardness of 7 on the Mohs scale and a poor cleavage.
3. It is very resistant to chemical attack because the bonds between silicon and the shared oxygen ions are strong. Also, quartz contains no metallic cations that could be replaced by hydrogen ions during weathering, in contrast to, for example, feldspars.

But great stability in the sedimentary environment is a two-edged sword as far as provenance studies are concerned; sedimentary processes leave few readable marks on a detrital quartz grain. Trying to obtain information from a quartz grain is almost as difficult as trying to get blood from a turnip.

Provenance

In thin-section studies, detrital quartz grains (see Figure 5–8) are usually classified as either *monocrystalline* (composed of a single quartz crystal) or *polycrystalline* (composed of two or more quartz crystals). Thus, a polycrystalline quartz grain is a lithic fragment. With the exception of chert, nearly all silicate mineral grains in sandstones are derived ultimately from plutonic igneous rocks, gneisses, and schists. These

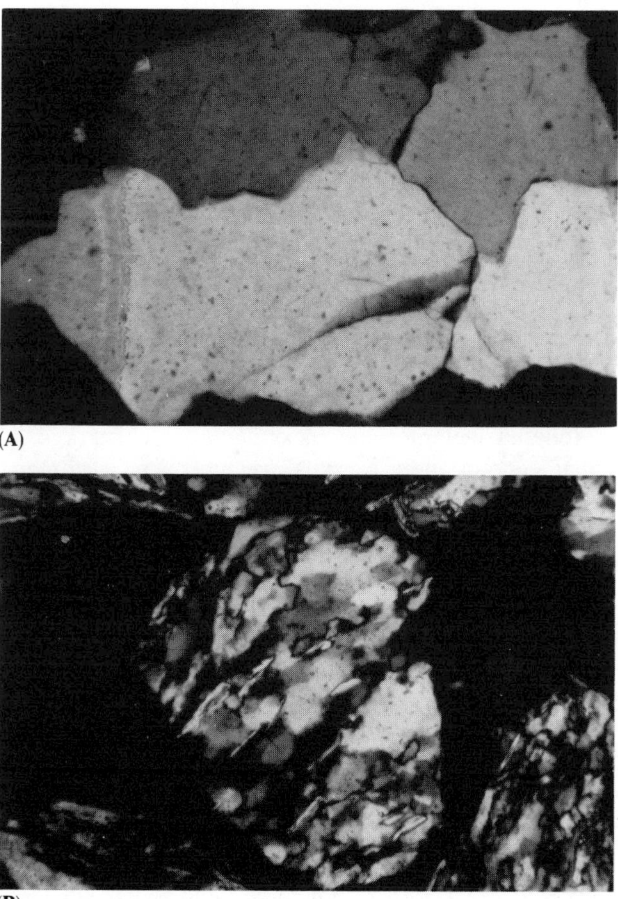

(A)

(B)

Figure 5–8
Photomicrographs of coarse sand-size quartz grains (crossed nicols) from **(A)** igneous rock and **(B)** metamorphic rock. Granitic quartz is composed of fewer crystals of more equant shape and has little or no intercrystalline suturing. Metamorphic grain consists of perhaps 10 times as many crystals, which are stretched and elongate and have intensely sutured contacts. These grains are extreme examples of polycrystalline grains from the two classes of crystalline rocks.

crystalline rocks reach the surface by means of tectonic uplift, and it is reasonable to expect that minerals in them will be plastically deformed, twinned, or fractured during the uplift process. Quartz can respond to stress by any of these mechanisms, but fracturing of detrital quartz is rarely seen in thin sections of sandstones and twinning cannot be detected. Plastic deformation, however, is common in nearly all quartz grains and is reflected in thin section as *undulatory extinction.* The crystal does not extinguish as a single unit on the slightest rotation of the microscope stage but instead extinguishes in sectors through a rotation of several degrees. The exact amount of stage rotation required for the entire crystal to pass from the grayish-white maximum birefringence color of quartz to the extinction position depends both on the degree to which the crystal has been plastically deformed and on the angular relationship between the *c* axis (axes) of the crystal and the plane of the thin section. The amount of stage rotation required for extinction is not an effective criterion to use in distinguishing between igneous and metamorphic origins for a quartz crystal. Undulatory extinction can also be produced in a sedimentary rock during folding and faulting.

Polycrystalline quartz grains can have several types of internal structures (see Figure 5–8):

1. If the individual quartz crystals within the grain are elongate, they have been deformed in a nonhydrostatic stress field. Such stretched quartz crystals are commonly found in foliated metamorphic rocks such as schists and gneisses. Hence, detrital grains that contain these crystals are mostly of metamorphic derivation. Stretched quartz crystals are also found along and adjacent to fault surfaces, where extreme stretching can be accompanied by granulation and recrystallization to produce the rocks mylonite and phyllonite.

2. Granitoid igneous rocks are coarse-grained, so that the quartz crystals in them typically exceed 0.5 mm; sizes of several millimeters are common. Quartz crystals in many metamorphic rocks, however, are fine-grained, for example, phyllites, most schists, and some gneisses. Therefore, the more quartz crystals in a detrital polycrystalline grain of a given size, the more likely the grain is to be of metamorphic derivation. A sand-size quartz grain composed of more than five separate crystals is probably of metamorphic derivation.

3. Metamorphic rocks are, by definition, recrystallized rocks. Recrystallization begins at points of stress concentration within the rock rather than simultaneously at all locations. Thus, when recrystallization ends, some quartz crystals are in a different stage of the process than others. This is reflected by a great variation in crystal size within a detrital quartz fragment, commonly a bimodal distribution of crystal sizes illustrating a recrystallization "caught in the act." The smaller crystals are the newly developing ones that have not yet grown to equilibrium size.

4. Intercrystalline suturing among quartz crystals is common in both igneous and metamorphic rocks, although it may be more intense in polycrystalline grains of metamorphic origin. It is, at best, an unreliable criterion to use for provenance.

Granitoid plutonic igneous rocks are, on the average, coarser-grained than metamorphic rocks. Therefore, monocrystalline grains of detrital quartz of medium sand size and coarser in sandstones are likely to have been derived from granites. Fine-grained monocrystalline grains in a sediment are not necessarily derived from metamorphic rocks, however. They are produced mostly by breakage and chipping of larger quartz grains of any provenance. Some fine grains of quartz are released in that size directly from phyllites and fine-grained schists, but the proportion is probably less than the amount produced by the size reduction of larger grains.

Quartz grains in silicic volcanic rocks and tuffs typically have euhedral crystal outlines of beta (high-temperature) quartz and have *nonundulatory extinction.* Both of these characteristics are quite uncommon in other types of crystalline rocks. A large proportion of granites crystallize in the beta-quartz stability field; but because of mutual interference during growth of the crystals in a granite, euhedral crystal outlines on quartz cannot be developed. The reason the quartz in flow rocks normally has nonundulatory extinction is that the magma is extruded at the Earth's surface to crystallize and typically is not subsequently deformed. Unfortunately, the volume of phenocrysts in rhyolitic flow rocks and tuffs is not large, and quartz crystals that have one or more crystal faces are uncommon as detrital grains in most sandstones. Such grains are swamped under by quartz derived from the more abundant igneous and metamorphic rocks.

Quartz veins are common in most areas of crystalline rocks and, as is the case for metaquartzites, typically release coarse gravel-size grains into the sedimentary environment. Veins form from solutions that are analogous to aqueous subsurface brines rather than to silicate magmas. As a result, vein quartz can contain unusually large volumes of water-filled vacuoles, which give the quartz a milky color. Not all quartz veins are milky, however, so that much vein quartz in sediments no doubt is unrecognized. Also, like quartz from rhyolites, vein quartz tends to be swamped under by the more abundant grains from granites, gneisses, and schists.

A high percentage of quartz crystals in igneous and metamorphic rocks contains inclusions. By far the most common type of inclusion is water bubbles. Typical mineral inclusions include rutile, mica, and apatite. In sedimentary rocks, quartz grains that contain mineral inclusions are rare because of the lesser mechanical stability during transport of these grains.

Differential Stabilities

The most stable variety of quartz in the sedimentary environment is nonundulatory monocrystalline quartz that contains no inclusions. Polycrystalline grains are weaker because of their internal discontinuity surfaces (crystal boundaries). Grains with undulatory extinction are weaker because they have been plastically deformed. Grains with inclusions are weak because they are composed of two distinct phases: either two solid phases or a solid and a liquid (or gaseous) phase. As with polycrystalline grains, discontinuity surfaces are present within the grain. Because of these dif-

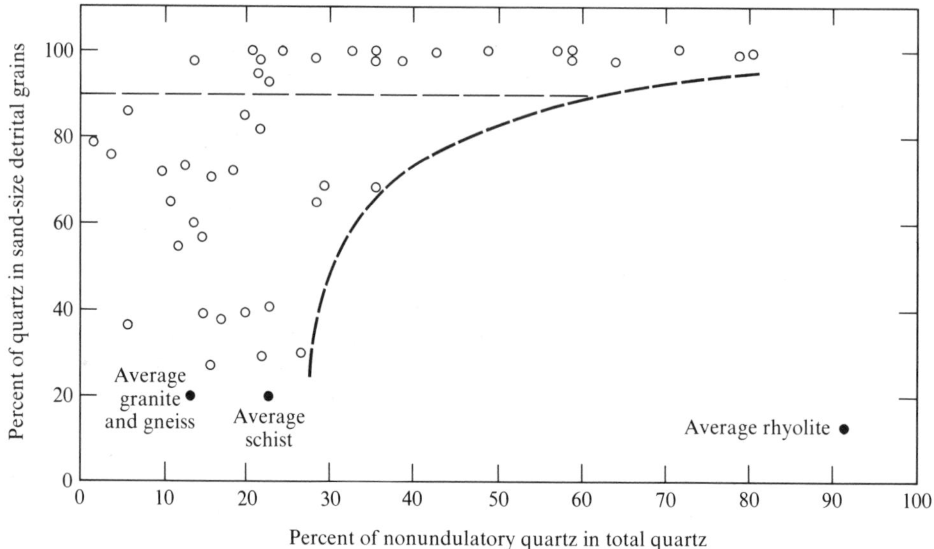

Figure 5–9
Relationship between percentage of quartz in crystalline rocks and sandstones and percentage of that quartz that has nonundulatory extinction. The dashed, curved line includes all sandstone sample points. Undulatory extinction is harder to see in smaller crystals, so that schists, which typically contain quartz crystals of fine sand size, appear to have higher proportion of nonundulatory quartz grains. [Blatt and Christie, 1963.]

ferences in stability, the evolutionary principle of survival of the fittest can be applied to quartz in the sedimentary environment. Assemblages of quartz grains that have spent more time in the sedimentary environment should be relatively enriched in nonundulatory monocrystalline grains and depleted in undulatory polycrystalline grains. Examination of thin sections of sandstones reveals that this is indeed the case (see Figure 5–9). The average crystalline rock contains less than 20% nonundulatory quartz crystals in its quartz population, as does relatively unabraded sand-size detritus. Sandstones composed mostly of feldspar and lithic fragments and less than 40% quartz still contain about 20% nonundulatory grains in the quartz population. Sandstones that contain more than 90% quartz average 40–45% nonundulatory grains. These nearly pure quartz sandstones contain less than 1% polycrystalline quartz. Quartz grains that contain mineral inclusions are also rare among grains in pure quartz sandstones.

Feldspar

The feldspars are the most abundant group of minerals in crystalline rocks, forming 60% of the mineral grains in igneous rocks and probably a similar percentage in metamorphic rocks. Feldspars are unstable in the sedimentary environment relative to quartz. Hence, although the feldspar/quartz ratio in crystalline source rocks of sandstones is 3/1, in the sandstones themselves the ratio is about 1/5. Feldspars form only 10–15% of the detrital fraction of the average sandstone.

The complex nature of the feldspar minerals has resulted in their being subdivided into many categories on the basis of their chemical, physical, and structural characteristics. In routine thin-section work, however, many of these characteristics are not determined. The categories of feldspar normally used by sedimentary petrologists are (see Figure 5–10):

1. Potassium feldspars: orthoclase, microcline, and sanidine.
2. Plagioclases: albite through anorthite.
3. Perthite: an intergrowth of sodium feldspar and potassium feldspar.

The ways by which each of these types of feldspar is recognized in thin section are given in Table 5–1.

Table 5–1
Thin-Section Characteristics of Feldspars Most Easily Used for Their Identification. (Potassium- and calcium-sensitive stains are commonly used to distinguish orthoclase from untwinned plagioclase, which is common in metamorphic rocks and sediments derived from them.)

Feldspar	Characteristics
Potassium feldspars	*Orthoclase:* untwinned or with Carlsbad twinning; refractive indices below Lakeside or quartz; birefringence less than that of quartz; large 2V; optically negative; internal alteration to illite flakes, commonly along cleavage planes. *Microcline:* unique and distinctive grid twinning pattern. *Sanidine:* Same as orthoclase except that the 2V is very small, 5–10°.
Plagioclase feldspars	Typically polysynthetically twinned ("albite twinning"); internal alteration to montmorillonite flakes, commonly along cleavage planes; can be concentrically compositionally zoned, reflecting change in Ab/An ratio; two directions of twinning at a large angle to each other (albite and pericline twinning); Ab/An ratio determined by Michel–Lévy method. Albite has indices of refraction lower than quartz, oligoclase is about the same as quartz in indices and birefringence, and andesine through anorthite have indices higher than quartz.
Perthite	Parallel intergrowth of two phases with contrasting indices of refraction and birefringence.

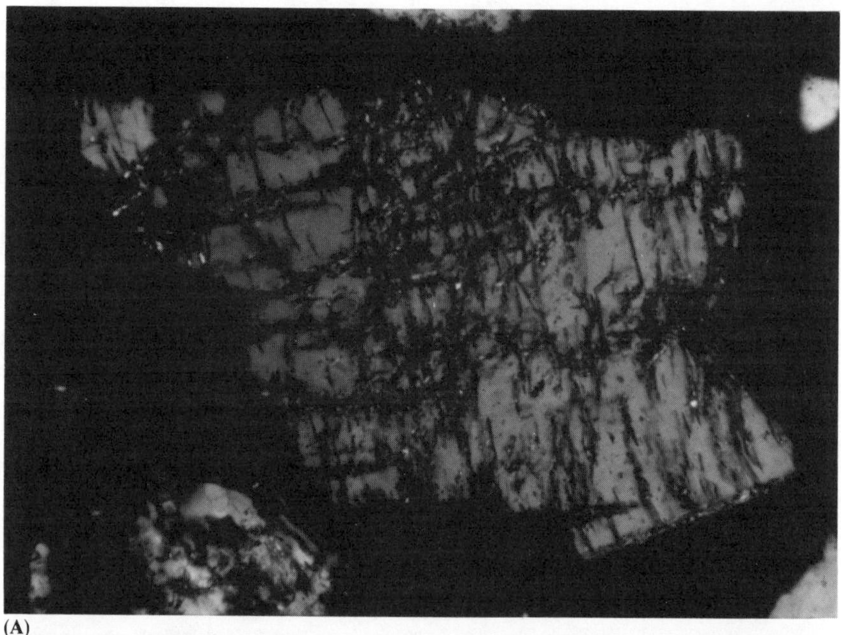

(A)

(B)

Figure 5–10

Photomicrographs of medium sand-size feldspar grains (crossed nicols). (**A**) Orthoclase, with alteration parallel to right-angle cleavages. (**B**) Plagioclase, showing polysynthetic albite twinning.

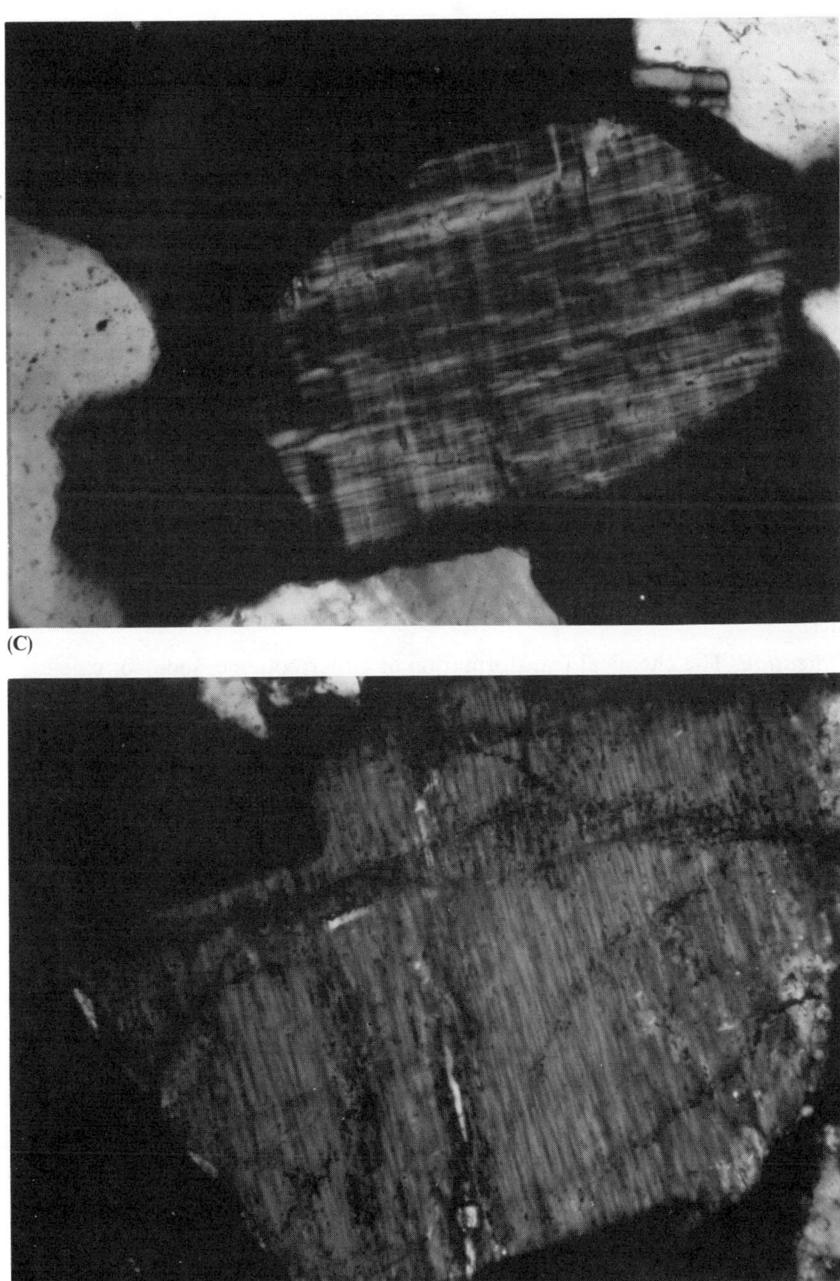

(C)

(D)

(C) Microcline, showing characteristic grid-twinning. (D) Microperthite, showing spindle-shaped, exsolution lamellae of sodic and potassic feldspar.

Types of Alterations

Several types of alterations of feldspars are commonly seen in thin sections of sandstones:

1. *Vacuolization.* Water percolates into the feldspar grain along cleavage planes, fractures, and twin composition surfaces, producing no visible chemical change in the grain. The grain appears turbid in transmitted light but has a whitish, cloudy appearance in reflected light. The whitish color can be seen by switching off the light under the stage of the microscope and shining a beam on the surface of the thin section.

2. *Sericitization.* In a chemical process that is not well understood, the tectosilicate crystal structure of a potassium-rich feldspar grain is converted in places to the phyllosilicate structure of sericite (illite), resulting in the appearance of illite flakes within the feldspar. Typically, the flakes appear first along surfaces where water can permeate, such as cleavages, fractures, and twin surfaces. The flakes are small, 5–10 μm in length, and have a straw-yellow birefringence.

3. *Montmorillonitization.* A tongue-twister type of alteration directly analogous to sericitization, except that it occurs in Na–Ca feldspars. The montmorillonite has the same birefringence as the sericite (illite) but lower indices of refraction than Lakeside and quartz.

4. *Kaolinization.* The chemical transformation of either potassic, sodic, or calcic feldspar to kaolinite reflects a more intense and/or prolonged process of alteration than either sericitization or montmorillonitization. Kaolinite has a very low birefringence (0.005) and is dark-gray in thin section. If the flakes are very small, kaolinite can look like chert.

Note that in describing these types of chemical change, we use the term "alteration" rather than the more specific term "weathering." The alteration of feldspars has been considered to reflect climatic conditions at the time and/or site of deposition of the feldspar grains: sericite and montmorillonite reflecting a temperate climate; kaolinite, a subtropical climate. Studies during the past decade, however, have emphasized that there is substantial alteration of feldspar grains after burial (see Chapter 6). Internal chemical alteration of feldspar grains can take place as easily and rapidly during diagenesis as during surface weathering. Therefore, no inferences concerning paleoclimate can be made based on the type or degree of chemical alteration of detrital feldspars in ancient rocks.

Provenance

Feldspar occurs in nearly all types of crystalline rocks, so that feldspar grains of sand size can be derived from granitoid igneous rocks, gneisses, and schists in large amounts. Lesser amounts are derived from gabbros and porphyritic, mafic volcanic rocks. Among the igneous rocks, orthoclase, oligoclase, microcline, and perthite are typical of granites; andesine, of granodiorites; and labradorite, of gabbros. Bytownite and anorthite are not common in igneous rocks and are very rare in sandstones because of their great instability in the sedimentary environment. Zoned plagioclase feldspars occur only in magmatic rocks and more commonly in volcanic rocks than in plutonic ones.

In metamorphic rocks the Na/Ca ratio of plagioclase is correlated with the grade of metamorphism. Greenschist facies rocks contain nearly pure albite; amphibolite facies rocks, oligoclase and andesine; granulite faces rocks, andesine and labradorite. The bulk of plagioclase in metamorphic rocks is untwinned. Orthoclase is stable only at the higher grades of metamorphism, in which dehydration causes muscovite to become unstable. Microcline occurs in migmatites as a crystallization product of the early stage of anatexis; that is, it occurs in the igneous part of the migmatite.

The percentage and type of feldspar in a sandstone depend on the rate and type of tectonic activity and on climate. In a tectonic setting characterized by block-faulted and uplifted crust such as occurred within the craton in Colorado during the Pennsylvanian Period, uplift, erosion, and burial are rapid and the resulting sands may contain 50% feldspar. What is meant by the expression "rapid uplift"? In the Pennsylvanian Period in Colorado there were at least 1,500 m of uplift in less than 40 million years (4 cm/yr), resulting in a wedge-shaped deposit of granitic conglomerate and feldspathic sandstone 1,500 m thick at the base of the Front Range, 60 km in width (east–west), and 350 km in length (north–south). The climate in the area was probably semiarid (see Chapter 6). The combination of high topographic relief and low intensity of chemical weathering is ideal for the accumulation of a thick sedimentary sequence.

In a quiescent cratonic setting such as during the Early Paleozoic Era in central North America, very little sediment can be produced from the low-lying granitoid crust. The small amount that is produced is reworked repeatedly by waves and currents in the beach–dune complexes that characterize this tectonic setting. Low topographic relief permits extensive transgressions and regressions of epicontinental seas, so that environments of high kinetic energy dominate the geographic setting and the resulting abrasion of sand grains removes nearly all the feldspar.

The variety of feldspar that is most abundant in a sandstone also depends to an important degree on the tectonic setting in which the sandstone forms. Plagioclase dominates when erosion and burial are rapid, and the granitoid rocks exposed are granodiorites and quartz diorites, as along convergent plate margins (e.g., Tertiary sandstones in California). Potassic feldspars (orthoclase and microcline) dominate the feldspar suite of sandstones formed in cratonic settings (e.g., Pennsylvanian sandstones in Colorado). In the California sandstones the plagioclase/potassic-feldspar ratio is typically 2/1, with abundant unstable calcic feldspar; in the Colorado sandstones there are only trace amounts of plagioclase of any composition. Quiescent cratonic settings also contain only very small amounts of plagioclase feldspar among the total feldspar grains. These relationships between the abundance and the composition of feldspars are summarized in Figure 5–11.

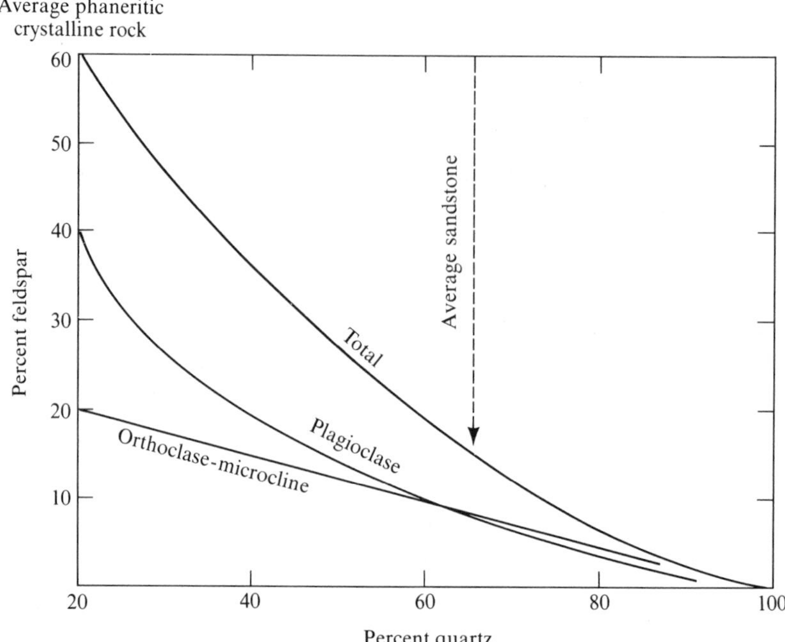

Figure 5-11
Generalized relationship between abundance and composition of feldspars in sandstones. Percentage of detrital grains not accounted for by quartz and feldspar is percentage of unstable lithic fragments plus clay minerals.

The proportion of total feldspar in a sandstone is also affected by the grain-size distribution in the rock. Detrital feldspars, because of their relative instability, are normally finer-grained than associated quartz grains. As a result, it is not uncommon to find, for example, that the medium sand fraction contains 90% quartz and 10% feldspar; in the fine sand fraction, 80% and 20%; and in the coarse silt fraction, 70% and 30%. The percentage of feldspar then decreases to about 5% in the medium silt fraction as the rapidly increasing surface/volume ratio of the smaller grains causes a rapid increase in the dissolution rate of the feldspars. The average mudrock contains about 5% feldspar (see Chapter 3). The clay-size fraction contains almost no feldspar.

Lithic Fragments

Pieces of polymineralic source rock (see Figure 5–12) form 15–20% of the average sandstone but can supply much more than 15% of the provenance information about the rock. Unlike quartz or feldspar grains, a piece of basalt or mica schist in sandstone is unequivocal evidence of the nature of the source rock. These fragments not only indicate whether the source rock was igneous or metamorphic, but also can reveal such things as the silica content of the magma, its rate of crystallization, or the character of the premetamorphic sedimentary rocks from which the metamorphic rock was formed.

Although any type of rock fragment can be found in a sandstone, some types are much more common than others. The factors that determine which types will occur are:

1. Areal abundance in the drainage basin.

2. Location in the drainage basin: whether in the highland or lowland areas.

3. Susceptibility of the rock fragments to chemical and mechanical destruction by sedimentary processes.

4. Size of the crystals within the fragments.

Obviously, the greater the areal extent of the source rock, the better the chance of finding pieces of it downstream, and we have already noted the greater rates of erosion from areas of high relief.

Factors 3 and 4 determine the *survival potential* of the fragment, and their importance can be illustrated by consideration of mudrock fragments. Mudrocks form two-thirds of the stratigraphic column, and probably most of the mudrock is shale. Therefore, we might anticipate that most lithic fragments in sandstone would be fragments of shale. But this is the opposite of what is found: shale fragments are quite uncommon in ancient rocks. The explanation is that shale fragments are practically untransportable because they are very soft and also split rapidly along fissility surfaces. They have a low survival potential. Extension of this principle leads to the expectation that fragments of gabbro will be poorly represented because of their chemical instability relative to granite; fragments of older sandstones will be rare because of the easy breakage during transport of the common cements calcite and hematite. Most fragments of older sandstone that survive will be either quartz-cemented quartz sandstone or chert.

The crystal size within the rock fragment determines the minimum size of fragment necessary for the fragment to exist. For example, a fragment of granite cannot occur in a fine-grained sandstone because fragments tend to break along crystal boundaries and the crystals in a granite are coarser than fine sand. Fragments such as rhyolite or chert, however, can occur with equal ease in sand of any size. Neglect of this factor can lead to erroneous paleogeologic inferences. It is clear that the interpretation of upstream paleogeology from sandstone petrology is not straightforward, even when the sandstone contains pieces of the source rock itself.

Plate Tectonics

Although many thick sequences of coarse-grained detrital rocks contain a diverse assortment of types of lithic fragments, many other such sequences are dominated by a single type. For example, Lockwood (1971) describes 29 known occurrences of sedi-

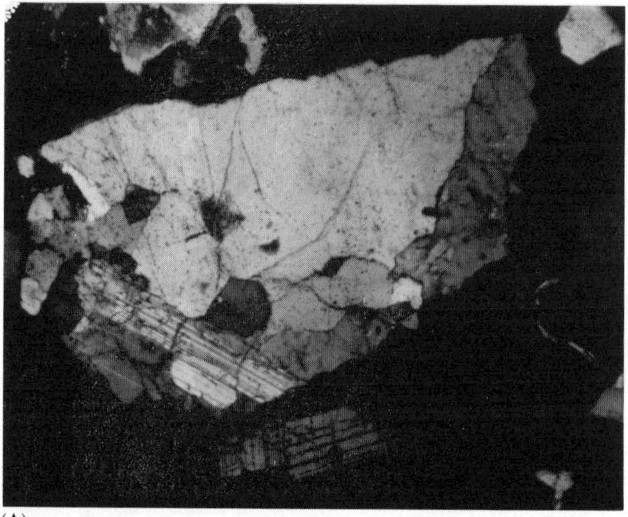

(A)

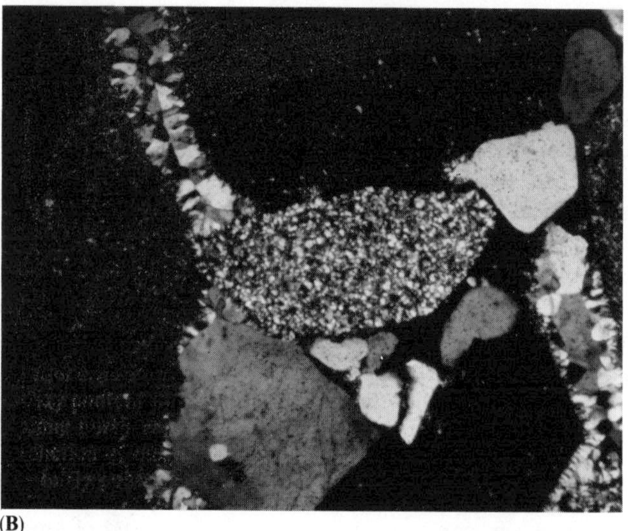

(B)

Figure 5–12
Photomicrographs of coarse sand-size lithic fragments (crossed nicols), showing characteristic appearances. (**A**) Granite fragment composed of large quartz crystal, orthoclase crystal (right, dark gray), and twinned plagioclase crystal (lower left). (**B**) Chert fragment (center) surrounded by three very finely microcrystalline chert fragments (black with gray speckles) and several quartz fragments (white and gray). Elongated crystals of quartz cement have grown from chert grain surfaces to bind the grains.

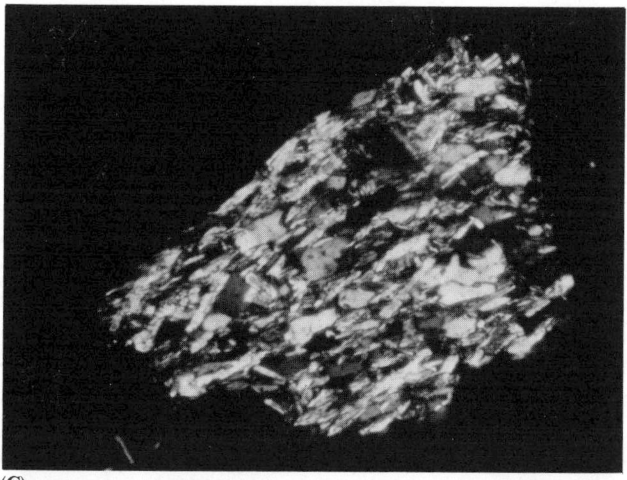

(C)

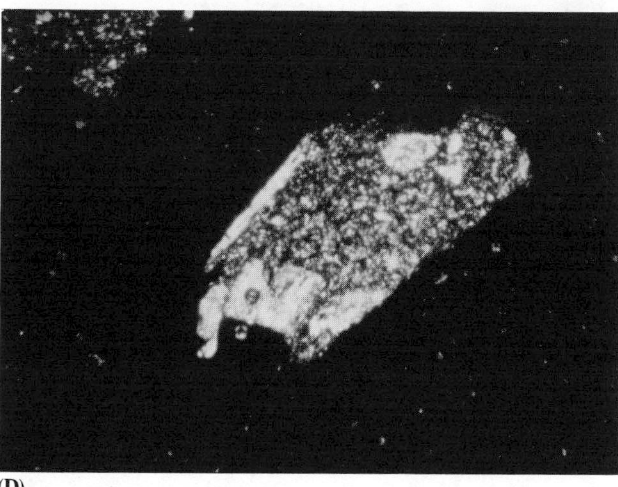

(D)

(**C**) Mica–quartz schist fragment containing a few opaque mineral crystals. (**D**) Volcanic rock fragment containing three large and altered K-feldspar crystals (one of which is euhedral) and large mica flake (upper left edge of grain) set in groundmass of felsitic crystals of low birefringence, probably quartz and K-feldspar.

mentary deposits composed of serpentine fragments; the deposits range in age from Early Paleozoic to Holocene and are associated with past and present plate convergences. The Devonian lithic sandstones of the Appalachian Mountain belt consist almost entirely of schist fragments produced by convergence of the North American plate with the African plate during the Paleozoic. Tertiary orogenic sandstones of the Alps tend to be dominated by lithic grains of sedimentary origin, particularly fragments of carbonate rocks. Sandstones resulting from convergence of the western Pacific plate with the eastern edge of the Asian plate during the past 200 million years are exceptionally rich in fragments of basalt. Is there a decipherable relationship between the type of lithic fragment in a sandstone and the plate-tectonic setting in which it occurs?

The clearest example of such a relationship is shown by the presence of mafic lithic fragments in the eugeosynclinal settings associated with island arcs and convergent plate margins. The rim of the Pacific Ocean basin provides many examples of marine volcanogenic sediments in this setting throughout Phanerozoic time. Examples include the Parry Group (Devonian-Mississippian) in New South Wales, Australia (Crook, 1960); Triassic rocks in Southland, New Zealand (Boles, 1974); Mesozoic rocks in central Oregon (Dickinson et al., 1979); and numerous Cenozoic basins in the northeastern Pacific Ocean basin adjacent to British Columbia, Canada, and the Aleutian Island chain south of Alaska (Galloway, 1974).

Examination of the sand fraction of modern large river systems is consistent with the data from ancient and modern marine environments. Figure 5–13 shows the relative amounts of the different types of lithic fragments in major rivers (Potter, 1978). It is clear that although these rivers contain a wide spectrum of types of fragments, two distinct fields are present. A group of rivers along the circum-Pacific mountains of the western hemisphere is very rich in volcanics. All of the other rivers that have significant amounts of lithic fragments define a continuous field extending between the plutonic and sedimentary poles, where volcanic fragments, although present, are definitely subordinate.

Another fairly clear relationship between the type of lithic fragment in a sandstone and its plate-tectonic setting is the association between potassium-rich granitic rock fragments and intracratonic block faulting. Examples of this relationship have been described by Krynine (1950) for Triassic rocks in Connecticut, by Hubert (1960) for Pennsylvanian rocks in Colorado, and by Al-Shaieb et al. (1980) for Permian rocks in southwestern Oklahoma. Each of these examples of nonmarine, lithic, granitic conglomerate and coarse sandstone on an alluvial fan was produced by tensional block faulting and the formation of horsts and grabens in cratonic basement rocks. The Triassic sequence in Connecticut (and along the entire east coast of North America) was formed several hundred kilometers inland from the continental margin that formed 40 million years later in Jurassic time (see Chapter 1). The intracontinental block faulting in Colorado during Pennsylvanian time was located 600–800 km to the east of the edge of the craton. The Permian feldspathic conglomerates and sandstones in southwestern Oklahoma accumulated on the craton 300 km to the northwest of the Permian cratonic margin.

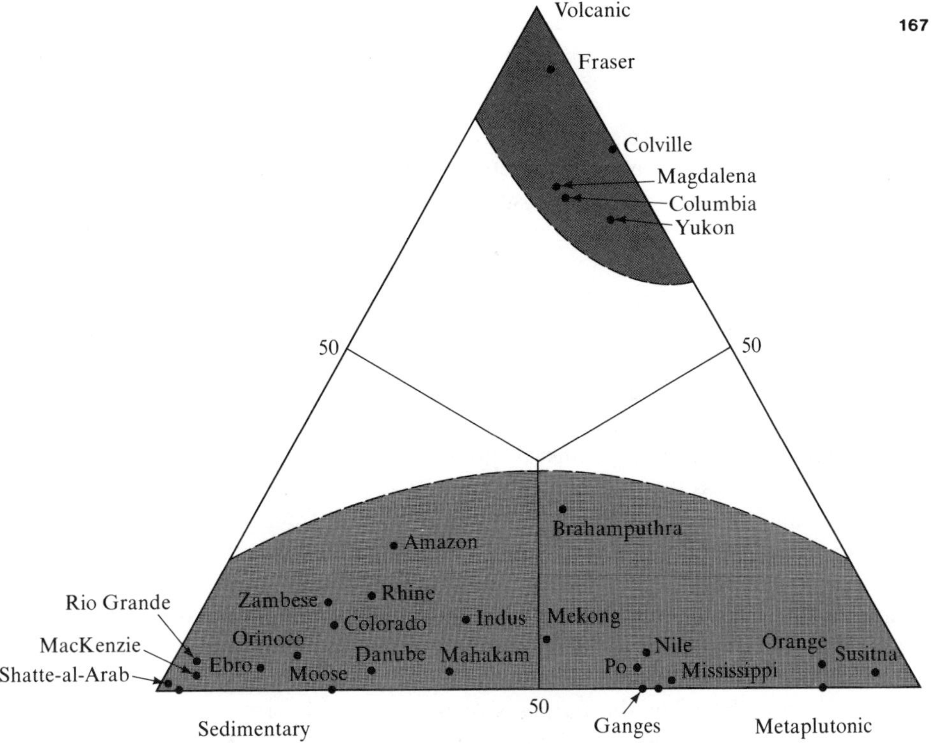

Figure 5–13
Proportions of three types of lithic fragments, mostly of fine sand size, in sandbars at downstream end of 29 large river systems. [Potter, 1978. Reprinted by permission of The University of Chicago Press. Copyright © 1978 by The University of Chicago.]

Convergent plate margins that lack volcanic island arcs can cause the formation of thick sequences of marine and nonmarine lithic detrital rocks that contain a wide range of sedimentary or schistose metamorphic rock fragments. Examples are provided by the Tertiary sandstones in many Alpine basins that are dominated by fragments of older limestones; by the Devonian sandstones in the central Appalachians (Krynine, 1940) that formed during the closing of the pre-Atlantic Iapetus Ocean and are composed almost entirely of schist fragments; and by the Siwalik Group of Late Tertiary age that formed during the collision of the India and China plates in the Himalayan region and are composed largely of shale and argillite fragments (Parkash et al., 1980).

Accessory Minerals

The accessory minerals in sandstones include all detrital minerals except quartz and feldspar, although micas are typically excluded from the accessory group because of their extremely platy shape and resulting anomalous behavior during transport. Any mineral that occurs in igneous and metamorphic rocks can occur in sandstones. The relative amounts of accessory minerals in a sandstone depends on the abundance of each mineral in the source rock, its survival potential during weathering, transport, and diagenesis, and its specific gravity. Because of the wide range in specific gravities of the common accessory minerals, there commonly is significant segregation among them during transport (placer deposits). The range in specific gravity among the common accessories is 3.0–5.2. In contrast, the range among quartz and feldspars is only 2.56–2.76.

Excepting micas, no common detrital minerals occur with specific gravities in the range 2.8–3.0, and the usual method for separating quartz plus feldspar from the accessory minerals is based on this fact. The loose sediment (or disaggregated sandstone) is dropped into a liquid with a specific gravity in the 2.8–3.0 range, with the result that the quartz and feldspar float while the accessories sink (see Chapter 9). For this reason, the accessory minerals are termed *heavy minerals*. The heavy minerals typically form less than 1% of a sandstone. The percentage is commonly related to the proportion of lithic fragments in the light-mineral fraction of the rock, particularly the proportion of metamorphic lithic fragments. A high percentage of metamorphic fragments suggests that a high percentage of heavy minerals may be present (perhaps 3%) because most species of heavy minerals in sandstones originate in metamorphic rocks. This is true because metamorphic rocks form in a much wider range of temperatures and pressures than do igneous rocks, permitting a larger number of species to crystallize. For the purposes of provenance determination, it is useful to group accessory minerals according to the type of crystalline rock in which they usually form (see Table 5–2). Unfortunately, many of the more common accessories in sandstones, such as zircon, tourmaline, and magnetite, form in abundance in both igneous and metamorphic rocks. Some minerals, such as tourmaline, occur in a variety of colors, and color variation may be related to provenance. For example, brown tourmaline is believed to be diagnostic of metamorphic rocks.

Most heavy minerals have a low survival potential because of both chemical and mechanical instability. One way to assess chemical stability in the sedimentary environment is to examine sandstones to determine which minerals can grow readily in sediments. Certainly, calcite, quartz, hematite, and a few other minerals grow as cements; occasionally, potassic and sodic feldspars precipitate from underground waters during diagenesis (see Chapter 6). The only accessory heavy minerals known to grow in sediments are zircon, tourmaline, and titanium oxides such as rutile and anatase. The rule of thumb is that the higher the temperature and pressure at which a heavy mineral forms in an igneous or metamorphic rock, the less stable it is in the sedimentary environment and the less likely it is to be found in a sedimentary rock. For example, in Holocene sediments of appropriate provenance, detrital augite, hypersthene, and olivine can be abundant and dominate the heavy-mineral suite. These minerals are rarely, if ever, found in ancient sedimentary rocks. Part of this disappearance results from chemical instability in the weathering zone; part, from chemical instability during diagenesis (see Chapter 6).

Heavy minerals are normally classified into two groups; *nonopaque* and *opaque*, based on their degree of transparency in thin section. Nearly all studies of heavy

Table 5–2
Common Accessory Minerals in Sandstones and Types of Crystalline Rocks in Which They Usually Originate

Igneous rocks	Metamorphic rocks	Indeterminate[a]
Aegerine	Actinolite	Enstatite
Augite	Andalusite	Hornblende
Chromite	Chloritoid	Hypersthene
Ilmenite	Cordierite	Magnetite
Olivine	Diopside	Sphene
Topaz	Epidote	Tourmaline
	Garnet	Zircon
	Glaucophane	
	Kyanite	
	Jadeite	
	Rutile	
	Sillimanite	
	Staurolite	
	Tremolite	
	Wollastonite	

[a]Common in both igneous and metamorphic rocks.

minerals in sedimentary rocks have been made of the nonopaque fraction, irrespective of whether it forms 10% or 90% of the heavy-mineral crop. In an average sandstone, perhaps one-half of the heavy minerals are opaque to transmitted light. Most of them are probably magnetite and ilmenite and intergrowths or alteration products of them.

One of the few published studies of opaque heavy minerals in a sandstone is by Hiscott (1979), who found 18% chromite in the fine sand-size heavy-mineral fraction of a Lower Ordovician sandstone in Quebec. Chromite in crystalline rocks is essentially restricted to ultramafic rocks such as peridotites, dunites, and serpentinites. Based on the abundance of chromite in the sandstone, Hiscott inferred the emplacement of a large sheet of ophiolite during Ordovician time in Quebec although very little ophiolite is now present in the area.

Mica

Detrital micas of sand size are a minor constituent of most sandstones. They are most abundant in fine-grained sandstones that contain abundant micaceous metamorphic lithic fragments. The relative abundances of biotite, chlorite, and muscovite in sandstones are unknown. Green or brown biotite can commonly be seen in thin section altering to pale-green chlorite with the development of "anomalous birefringence," a bluish color different from the blue color of normal birefringence. In a few sandstones

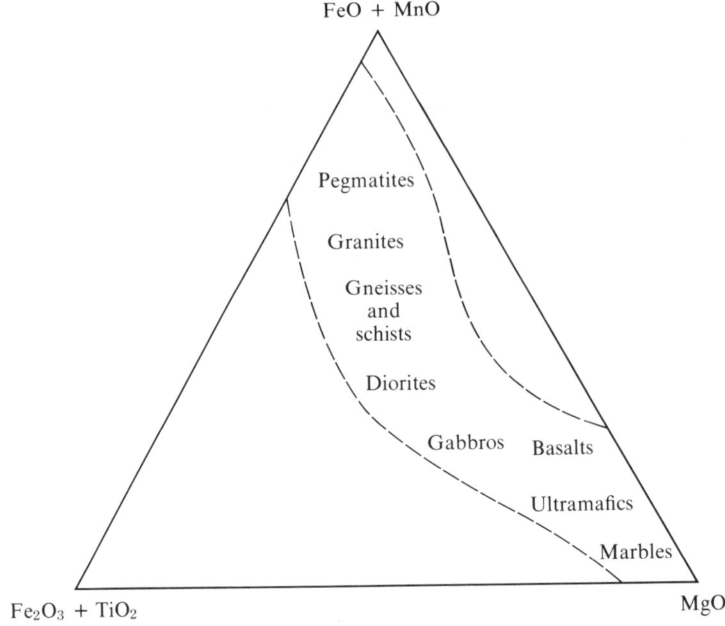

Figure 5–14
Relationship between contents of major metallic cations in biotites and phlogopites and type of crystalline rock in which crystals originate.
[A. E. J. Engel and C. Engel, 1960, *Geol. Soc. Amer. Bull., 71.*]

brown biotite occurs in euhedral hexagonal flakes—a perfection of crystal habit that reflects crystallization in a fluid (lava) and an absence of grain-to-grain impacts in the sedimentary environment. We infer that the biotite flakes were blown out of a volcano. Typically, such flakes occur in sandstones that contain other evidence of volcanic provenance, such as rhyolite fragments, sanidine grains, and quartz with beta outlines.

Muscovite is volumetrically uncommon in granitoid rocks because most of the potassium needed for the mineral to form goes instead into potassic feldspars during crystallization. Abundant muscovite in a sandstone suggests derivation from metamorphic rocks. Biotite, in contrast, occurs in a wide variety of crystalline rocks, and the proportions of iron and magnesium in the biotite reflect the composition of the rock in which the biotite formed (see Figure 5–14). Unfortunately, these chemical variations are not reflected clearly in properties of the mica that are seen in thin section (refractive indices, color, birefringence).

Glauconite

Detrital glauconite (McRae, 1972) occurs in sandstone as dark-green (brown when oxidized), ovoid pellets of approximately the same grain size as the associated quartz. It appears to form only in the marine environment. The glauconite generally is a structureless, crypto- to microcrystalline aggregate of potassium- and iron-rich phyllosilicate materials of varied origin. The bulk of glauconite pellets may be clay debris from the seafloor that has passed through the digestive tract of burrowing marine organisms, hence the ovoid shape.

Glauconite is not restricted to a particular type of sandstone, although the pellets are most common in essentially pure quartz sandstones of shallow marine origin. Typically, phosphatic debris is present as well, either as ovoid peloids and ooliths or as shell fragments of the brachiopod *Lingula*. Sandstones rich in glauconite (greensands) seem particularly common in rocks of Cambro-Ordovician and Cretaceous age, but the reason for this is unknown.

Grain Recycling

Our consideration of sand mineralogy thus far has concentrated on two questions: (1) What are the abundant minerals (and rock fragments) in sandstones? (2) Did these minerals originate in metamorphic or in igneous rocks? That is, we have been concerned with the ultimate sources of the grains. However, two-thirds of the continental surface is covered by sedimentary rocks, not by metamorphic and igneous rocks. If we are to construct an accurate paleogeographic map for the Devonian or Jurassic Period, we will have to determine which of the sand grains came directly from igneous or metamorphic sources and which were released from older, sedimentary rocks. We must distinguish between *ultimate sources* and *proximate sources*. Perhaps the quartz or garnet grain in a Jurassic sandstone emanated last from a Triassic sandstone, and before that resided in a Permian mudrock and an Ordovician conglomerate since being released from a Proterozoic gneiss.

Four approaches are currently used for distinguishing ultimate from proximate sources:

1. The percentage of quartz among the detrital grains. The principle involved is that repeated reworking over long periods of time is required to remove completely all the feldspars and lithic fragments from an assemblage of sand grains. Therefore, if a sandstone is composed entirely of quartz, the grains probably were derived from older sandstones rather than directly from an igneous or a metamorphic rock.

2. The percentage of superstable accessory minerals in the heavy-mineral assemblage. The principle is the same as for quartz. The most resistant nonopaque heavy minerals are zircon, tourmaline, and rutile, so that the *ZTR index* is a commonly used criterion of the importance of recycling.

3. The degree of rounding of the quartz grains. It requires repeated abrasion over long periods of time in a beach or dune environment to produce a well-rounded quartz grain from the angular grains released by crystalline rocks. Therefore, an assemblage of well-rounded grains indicates not only environments of deposition but recycling as well. Most pure quartz sands in the geologic column consist almost entirely of well-rounded grains.

Figure 5–15
Photomicrograph of medium sand-size detrital quartz grain (crossed nicols), showing multiple rounded overgrowths, Weber Sandstone (Pennsylvanian-Permian), Utah. Border of detrital part of grain is marked by inner oval ring of water-filled vacuoles. [Odom et al., 1976. Photo courtesy I. E. Odom.]

4. The presence of abraded secondary growths on quartz grains. It is common to find secondary quartz deposited from underground waters onto the surfaces of detrital quartz grains. Subsequently, the rock may be disaggregated and the enlarged quartz grains released and abraded. The abraded overgrowths (see Figure 5–15) can be seen in thin sections of the later sandstone deposit that includes the overgrown grains, and the overgrowths constitute excellent evidence of recycling. Unfortunately, this criterion of recycled grains is useful only with quartz grains and is uncommon even with quartz.

It is apparent that the key mineral for evaluating the abundance of older, sedimentary rocks in a drainage basin is quartz. Its percentage and grain shape are determined easily by petrographic studies; these variables serve as cornerstones of *drainage-basin* analysis. Several indices of recycling have been used to make quantitative evaluations of the proportion of older sediments that existed upstream; for

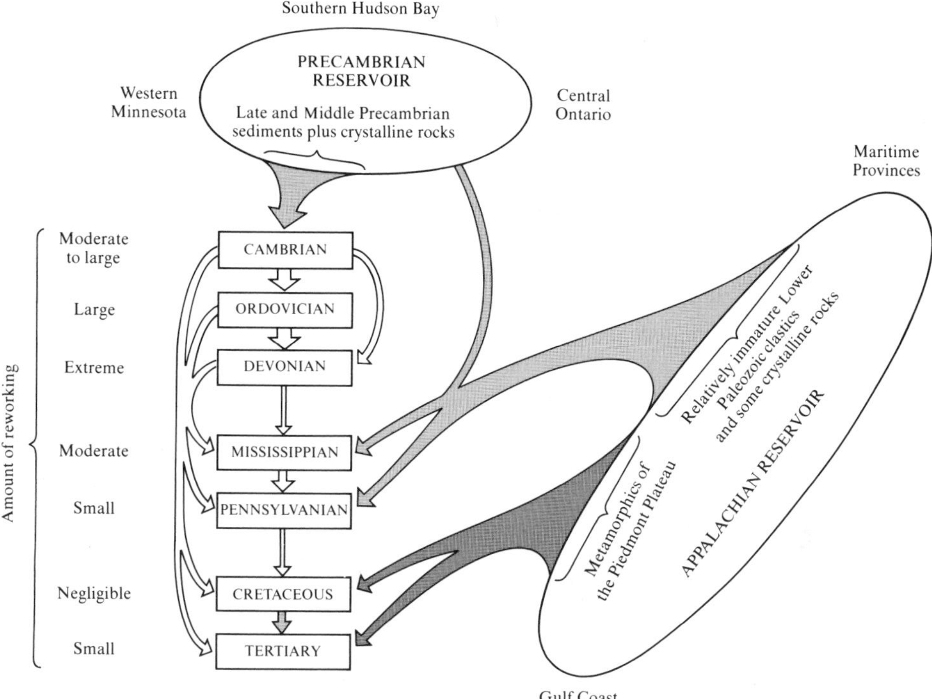

Figure 5–16
Dispersal centers, source relations, and recycling sequence of Phanerozoic clastic rocks in upper Mississippi Valley. [Potter and Pryor, 1961.]

example, the ratio of quartz plus chert to feldspar plus rock fragments, or the ratio of monocrystalline quartz to the more easily destroyed polycrystalline quartz grains. Other useful indices can be devised for particular circumstances.

An example of the complexities in paleogeologic interpretation caused by recycling is provided by a regional study of Phanerozoic sandstones of the upper Mississippi Valley conducted by Potter and Pryor (1961). Their integrated interpretation of petrology, cross bedding, major unconformities, and facies maps led them to the provenance interpretation shown in Figure 5–16.

Composition and Depositional Environment

The main controls of sediment mineral composition are tectonism, provenance (which can be related to tectonism), and climate. In some stratigraphic sections, however, it can be shown that the environment of deposition has exercised an important control on mineral composition. For example, in the Muddy Sandstone (Lower Cretaceous) of Montana and Wyoming, the fluvial facies contains an average of 14% lithic fragments; the deltaic facies, 11%; and the marine barrier-bar complex only 3% (Davies and Ethridge, 1975). The inference that can be made from these data is that environments of high kinetic energy cause the destruction of less stable detrital grains, presumably by fracturing and abrasion to silt size. Folk (1960) reported meter-by-meter vertical alternations in mineralogy and roundness in the Tuscarora Sandstone (Silurian) in West Virginia. Beds characterized by angular quartz grains, poor sorting, and high content of foliated metamorphic lithic fragments are interbedded with units consisting of well-sorted and well-rounded quartz grains and almost no lithic fragments. The less mature beds are believed on other criteria to be estuarine; the more mature beds, beach sands. The fact that these textural and mineralogic changes are repeated many times within a stratigraphic interval of less than 100 m and a time span of only 20 million years suggests that the changes did not result from either tectonic or climatic changes in the source area.

Hubert (1960), in his study of the Lyons Formation (Pennsylvanian) of Colorado, reported apparent environmental control of feldspar percentage by depositional environment. Environments were initially defined by sedimentary structures; subsequently, it was determined that the fluvial facies of the formation averaged 28% feldspar, whereas the beach–dune facies averaged only 8% feldspar. As was the case with the lithic fragments in Folk's (1960) study, the variation in feldspar percentage occurred frequently within a short stratigraphic interval, illustrating the speed and effectiveness of high kinetic energy as a disintegrator of mechanically unstable grains.

Many other studies of sandstone sequences, however, fail to reveal the strong control of mineralogy by environment described by Davies and Ethridge, Folk, and Hubert. The reasons for the different results are unknown. Clearly, the advisable policy is to keep in mind the possibility of environmental control and to look for it during studies of detrital sequences.

Classification

The first objective of a petrologic project is a thorough description of the rocks, both their outcrop characteristics and their appearance in thin section. Such a description may require several weeks or months and result in many pages of typescript. The purpose of the detailed work is to discover relationships among the various things being described, in the hope that they will provide insight into the origin of the rocks. Many researchers have noticed, however, that few scientists are willing to wade through rock descriptions of great length written by someone other than themselves. It is necessary to summarize results in order to communicate them effectively to others. This is the purpose of classification.

The object of a summary is to convey the most important information as briefly as possible. If we are to summarize by means of a classification scheme, we are limited to consideration of only a few variables—those that we believe give the most insight into the genesis of the rocks. For nearly all rocks (igneous, metamorphic, and sedimentary), texture and mineral composition are the most important variables. A large number of suggestions has been published concerning the best way to combine texture and mineral composition in a classification scheme. Practical and philosophical considerations that seem important in choosing among the suggested classifications are discussed by Blatt et al. (1980) and need not be detailed here. In this section we use a scheme that can be applied in field studies and will be only refined rather than completely changed by subsequent laboratory studies.

The easiest mineralogic separation to make in the field is among quartz, feldspar, and lithic fragments. These can form the poles of a classification triangle, and the inside of the triangle can be subdivided in any convenient manner. The triangle shown in Figure 5–17 is one of many in common use among sandstone petrologists. When classifying sandstones, the researcher must specify the poles and internal subdivisions of the triangle he or she is using. The variations among the triangles currently in use are so great that results will be unintelligible to readers unless this is done.

The texture of the sandstone is indicated by the use of the concept of textural maturity, discussed in Chapter 4. It is also useful to indicate in the rock name the type of cement that holds the detrital grains together. The presence in the sandstone

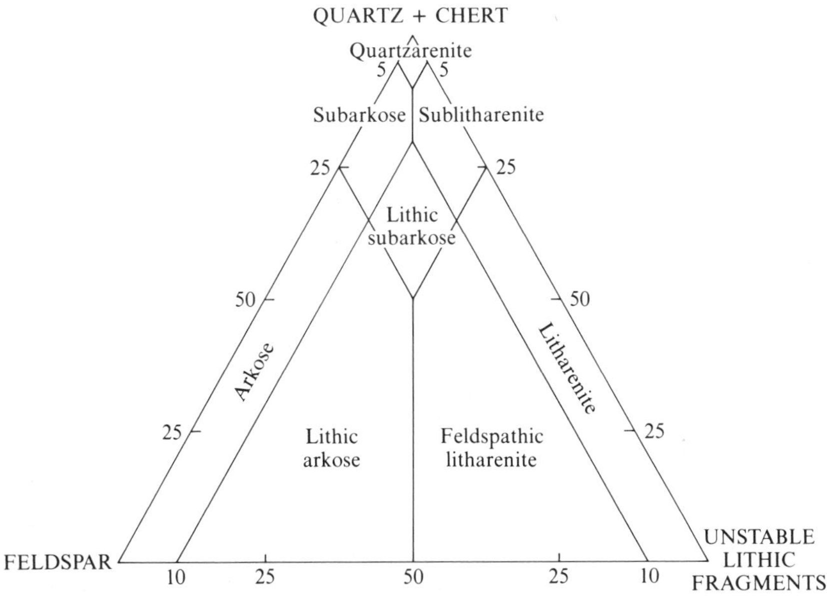

Figure 5–17
One of many mineralogic classifications of sandstones in common use.
[E. F. McBride, 1963, *Jour. Sed. Petrology, 33.*]

Table 5–3
Descriptions of Thin Sections of Six Detrital Rocks and the Summary Rock Name That Describes Them, Using the Textural Maturity Concept of R. L. Folk and the Mineralogic Triangle of E. F. McBride (see Figure 5–17)

Mineral composition, %	Texture	Diagenetic effects	Complete rock name
1) 95 monoxline quartz 4 orthoclase 1 mudstone fragments	no clay matrix grains well sorted grains well rounded mean grain size 0.7 mm	quartz cement	supermature medium-grained, quartz-cemented quartzarenite
2) 55 quartz, mono- and polycrystalline 5 chert 20 orthoclase 5 microcline 5 plagioclase 5 granite rock fragments 5 biotite	10% mud (kaolinite + illite + quartz silt) grains poorly sorted grains angular mean size of sand 0.2 mm	formation of hematite bending of mica kaolinization of orthoclase chloritization of biotite	immature, fine-grained, hematite- and clay-cemented arkose
3) 50 garnet schist fragments 15 amphibolite fragments 15 hornblende gneiss fragments 10 polycrystalline quartz 5 untwinned feldspar 5 metasiltstone fragments	no clay matrix grains poorly sorted grains well rounded mean grain size 5 mm	calcite cement	submature, pebbly, calcite-cemented lithic conglomerate
4) 40 quartz, mostly polycrystalline 25 twinned plagioclase 10 orthoclase 25 granodiorite fragments	no clay matrix grains well sorted grains subrounded mean grain size 1.2 mm	calcite cement	mature, very coarse-grained, calcite-cemented lithic arkose
5) 10 quartz, some with beta-habit 20 feldspar, mostly sanidine 70 rhyolite fragments	no clay matrix grains well sorted grains subangular mean grain size 0.6 mm	calcite cement	mature, coarse-grained, calcite-cemented, volcanic feldspathic litharenite
6) 60 monoxline quartz 25 chert 10 mudstone fragments 5 microcline	10% mud matrix grains well sorted grains rounded mean grain size 0.2 mm	squashed mudstone fragments	immature, fine-grained, mud-cemented, chert-bearing sublitharenite

of unusual constituents may also be worth noting if present in amounts greater than 5%. Examples of nomenclature that result from the use of the triangle in Figure 5–17 and the Folk scheme of textural maturity are given in Table 5–3.

COMMENT ON THE PETROLOGY OF SANDSTONES AND CONGLOMERATES

Many petrologic characteristics of conglomerate and sandstone sequences are best seen in the field. For this reason, they should be described while the investigator is standing at the outcrop. Back in the laboratory it is difficult to remember whether the conglomerates (which were too coarse-grained to sample adequately) contained a variety of types of rock fragments or only quartz pebbles or whether the conglomerates were more poorly cemented than the interbedded sandstones (which were sampled). Perhaps pebble compositions varied geographically or stratigraphically or with pebble size. Is there a relationship between the mean size of the gravel bed and that of the underlying or overlying sandstone bed? Each of these observations has the potential to contribute to the petrologic interpretation of the rock sequence.

Sandstone petrology is best studied in the laboratory. The most generally useful technique is thin-section analysis, but scanning electron microscopy and various types of microchemical analyses can be important tools in some investigations. The central control of sandstone mineralogy is plate tectonics, and this should be the framework into which mineralogic analyses are cast. A rock composed of 90% serpentine and basalt fragments clearly implies a different tectonic setting from that of a rock composed of 90% monocrystalline quartz or one composed of 90% granite fragments and potassic feldspar. Most sandstones, unfortunately, have less extreme compositions. Considerable experience and artistry are required to place them in their proper setting.

ANCIENT SANDSTONES

Cratonic Sandstones

As an example of a comprehensive study of an ancient cratonic sandstone, we can examine the study by Courdin and Hubert (1969) of the Fort Union Formation (Paleocene) in Wyoming. The Fort Union is a series of conglomerates, sandstones, and mudrocks exposed in the Wind River Basin of west-central Wyoming (see Figure 5–18). The formation crops out in narrow, discontinuous belts that trend parallel to the axis of the basin. Along the western and southern margins its thickness varies from 60 m to 300 m, depending largely on the extent of erosion prior to deposition of the overlying sediments (Lower Eocene). The Fort Union thickens northward and eastward toward the axis of the basin and locally exceeds 2,400 m in the subsurface (see Figure 5–19).

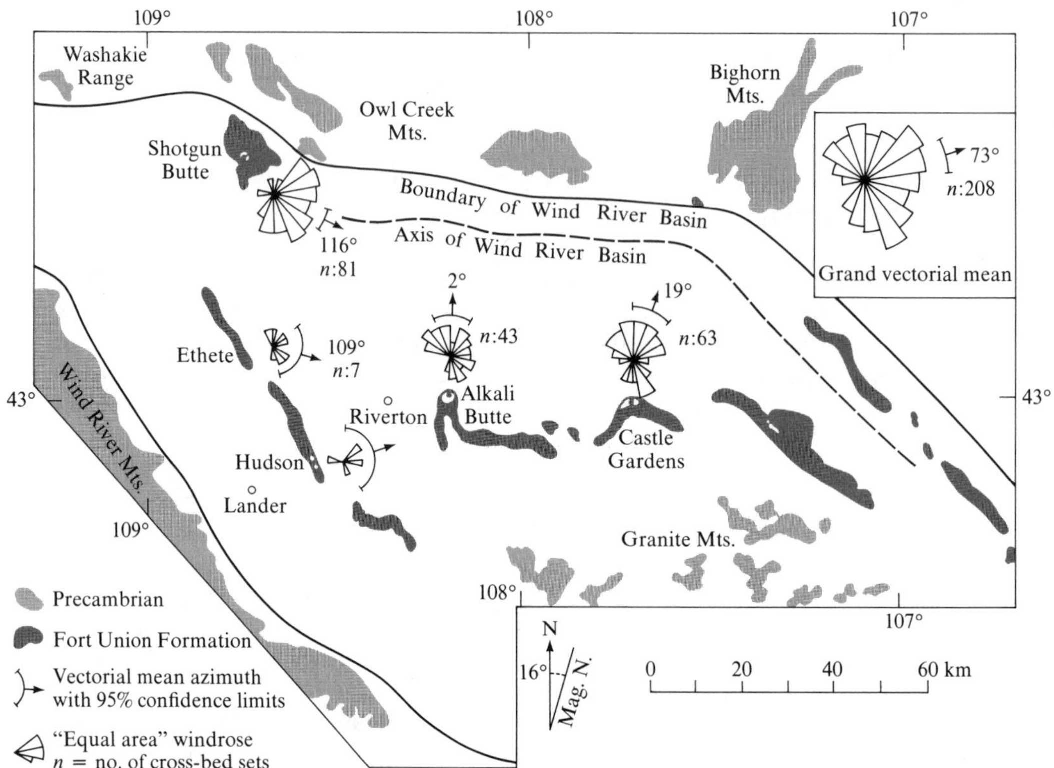

Figure 5-18
Outcrop map of Fort Union Formation (Paleocene) in west-central Wyoming, showing results of measurements of directions of cross bedding at each outcrop. [Courdin and Hubert, 1969.]

Stratigraphy

The lower part of the formation is characterized by white to gray, fine- to coarse-grained, massive to cross-bedded sandstone, interbedded with dark-gray to black shale, claystone, siltstone, and brown carbonaceous shale. Abundant thin, brown-weathering ironstone beds are present in most places, and lenticular coal beds occur locally. In places, much coarse-grained sandstone and conglomerate are present, and some of these units have concave-upward bases that cut across the bedding of the underlying deposits. Terrestrial plant fossils have been found in some of the finer-grained sandstones and mudrocks.

Overlying the lower Fort Union Formation is the Waltman Shale Member of the unit, 200 m thick at its type locality in the eastern part of the Wind River Basin but absent in most areas because of the erosion that preceded the deposition of the overlying, unconformable Wind River Formation. In the subsurface the Waltman is up to

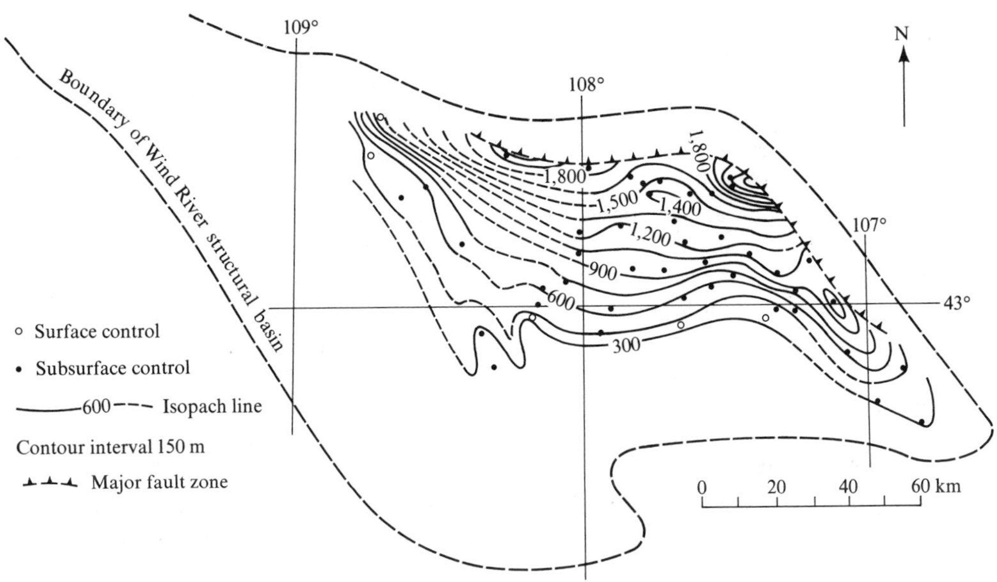

Figure 5–19
Isopach (thickness) map of Fort Union Formation. [Courdin and Hubert, 1969.]

1,000 m thick. The shale unit is a homogeneous dark-brown to black, silty, micaceous shale; small muscovite mica flakes are easily seen with a hand lens. In some areas the member contains several sandstone units that are 8–9 m thick. Some of this sandstone is conglomeratic, containing abundant pebbles of black chert and scattered cobbles of white granite as much as 15 cm in diameter in a coarse-grained, arkosic sandstone matrix.

Petrology

In the field Courdin and Hubert noted the character of the contacts between beds (parallel or cross-cutting), measured a total of 1,960 m of stratigraphic column in the formation at five locations, measured the orientation of 200 cross beds at these sites, and determined the lithology of 941 pebbles coarser than 5 mm in diameter. The results of the cross-bedding measurements are shown in Figure 5–18. The average direction of sediment transport was to the ENE, but a great deal of scatter is present in the measurements; this is interpreted to reflect either meandering streams or streams issuing in radial patterns from mountain areas onto overlapping alluvial fans. The paleoslope is clear, however, as almost no cross beds dip toward the WSW. The channel sands are fine-grained, with standard deviations of 0.4–0.7 ϕ—normal values for fine-grained fluvial sands. The mudrocks are silty and were interpreted to be overbank and floodplain deposits.

The pebble count revealed that 87% are sedimentary types, all from pre-Paleocene formations that still outcrop in the Wind River Basin. Chert is the most common type of pebble (76%). Varieties of chert present can be traced to specific sources and include black phosphatic chert from the Park City Formation (Permian), gray-banded chert from the Madison Limestone (Mississippian), and bird's-eye agate chert from the Morrison Formation (Jurassic). Most of the remaining 11% pebbles of sedimentary origin are composed of quartzarenite fragments from the Tensleep Formation (Pennsylvanian); a few (3%) are siliceous shale pebbles of the Mowry Formation (Cretaceous). About 13% of all pebbles are fragments of the Precambrian granite exposed at the border of the basin.

In the laboratory, thin sections and petrographic analyses were made of 49 randomly selected sandstones from the Fort Union Formation. The thin sections were cut perpendicular to lamination to avoid the possibility of slicing along a lamina with an unrepresentative mineral composition. In each slide, 300 detrital grains coarser than 0.03 mm were identified along linear traverses on the slide. Then 23 sandstones were disaggregated by dissolving their hematite and calcite cements in dilute, cold HCl; the freed grains were recovered, and a heavy liquid was used to separate the quartz and feldspar from the accessory minerals. The heavy minerals were then mounted in epoxy on a glass slide, and 200 nonopaque mineral grains were identified. During counting, features of the grains such as degree of rounding, euhedral character, and color were noted.

The light-mineral fraction consists of 77.5% quartz, 11.6% chert, 6.0% feldspar, and 4.9% rock fragments. Nearly all of the quartz (94%) is angular to subrounded single crystals that are undiagnostic of either ultimate or proximate source. Most were probably derived from older sandstone units that fringe the basin, but there is no way to be certain. A few of the monocrystalline quartz grains are well-rounded and, therefore, certainly were not derived directly from the granitic rocks in the area. The chert grains are of the same types found in the pebble count and occasionally contain "ghosts" of replaced fossils and other carbonate particles that reveal the chert to have formed by replacement of older limestone in the pre-Paleocene source rocks. The black chert sand grains contain phosphatic inclusions, reflecting their origin in the chert member of the Permian Park City phosphate formation.

About 95% of the feldspar grains in thin section are alkali feldspar. The frequency of occurrence of feldspars in the Fort Union is

$$\text{perthite} > \text{orthoclase} > \text{microcline} > \text{plagioclase}$$

The relatively high abundance of perthite, which is very easily decomposed and disintegrated, indicates that the feldspar was probably derived directly from nearby granitic rocks rather than being of recycled origin. Perthite has a low survival potential.

The rock fragments are 55% sedimentary (shale and siltstone), 35% metamorphic (schist), and 10% igneous (granite). The fact that rock fragments form only 4.9% of the detrital grains suggests that the dominant sources for the Fort Union Formation were older sediments rather than igneous and metamorphic rocks. The dominance of easily disaggregated particles such as shale and schist testify once again to the nearness of the source rocks of the formation.

The suite of accessory minerals is dominated by zircon (72%), garnet (8%), tourmaline (6%), and rutile (4%). Other minerals present include sphene, zoisite, apatite, staurolite, and anatase. The dominance of ZTR and garnet (which ranks just behind ZTR in stability in the sedimentary environment) supports the idea obtained from the light minerals that recycled sediments were the dominant source of the Fort Union.

Diagenesis

The sequence of precipitation of the major cements is iron oxides, silica, and carbonates. Limonite and hematite were deposited as coatings on the grains and as pore-filling cements. Secondary growths of quartz (overgrowths) were precipitated from migrating pore solutions on top of many iron-oxide-stained detrital quartz grains. Coarse, clear calcite is the dominant pore-filling cement in Fort Union sandstones. Occasionally, dolomite and siderite euhedra occur. The carbonate cements commonly corrode and replace quartz overgrowths, indicating that they formed later than the silica cement.

Other diagenetic growths present include feldspar overgrowths, tourmaline overgrowths, and euhedral crystals of anatase, kaolinite, and pyrite.

Comment on the Fort Union Formation

During Paleocene time, the granitic mountains that border the Wind River Basin were rising highlands that enclosed most of the subsiding basin. Fluvial, deltaic, lacustrine, and paludal (swamp) sediments of the Fort Union Formation accumulated in the basin. The cross-bedded sandstones were deposited as point bars on the inner slopes of meander bends of the shifting streams. The average grain sizes and standard deviations of these sands are typical of such deposits. Cross-bedding azimuths show that the rivers flowed basinward from the southern and western margins.

The initial detritus eroded from the Mesozoic and Paleozoic sedimentary rocks that overlie the crystalline basement consisted dominantly of quartz, chert, sandstone-siltstone fragments, and round ZTR grains. Further uplift and local unroofing of the granitic cores of the uplifts led to local influxes of granitic detritus, as revealed by changes in detrital petrology at Castle Gardens, Alkali Butte, and Hudson, where a transition is observed from cherty litharenite and sublitharenite in the lower parts of the sections to subarkose, lithic arkose, and arkose in the upper parts. The Precambrian cores of the Washakie, Owl Creek, and northern end of the Wind River Mountains were exposed only locally during the Paleocene.

Sandstones at Convergent Plate Margins

Sandstones deposited on the craton are typically characterized by very high percentages of quartz and chert—85–90% in the example of the Fort Union Formation. The mineral composition of these sandstones is dominated by resistant detrital minerals that have survived the repeated and vigorous destructive processes that typify a stable tectonic setting.

The opposite extreme of tectonic activity and detrital mineral composition occurs in the vicinity of the convergence of an oceanic plate and a continental plate, such as has existed around the rim of the Pacific Ocean for most of the past 200 million years. Sandstones deposited in forearc basins, between the oceanic trench and the volcanic island arc, have a mineral composition characterized by an abundance of basaltic volcanic rock fragments, calcic plagioclase grains, and noticeable amounts of mafic accessory minerals such as olivine and augite. These sediments have usually not suffered much transport, and the bulk of them has been deposited in marine waters below wave base. In addition, the area of convergent plate margins is of higher than average heat flow. The combination of high heat flow and an abundance of detrital particles that are chemically very unstable at low temperatures results in pervasive diagenetic alteration of the grains at relatively shallow depths of burial. The diagenetic production of clay matrix from the detrital grains and the crystallization of zeolite minerals are common phenomena in plate-margin sandstones.

As an example of plate-margin sandstones, we can consider the Mesozoic lithic sandstones in central Oregon studied by Dickinson et al. (1979). The sandstones occur in inliers surrounded mostly by Cenozoic plateau basalts of the Columbia River Sequence (see Figure 5–20A), and are similar in aspect to the sediments in the eugeosynclinal Fraser Belt exposed to the north in British Columbia and to the south in northern California. The most extensive outcrop of Mesozoic sandstones in central Oregon is the John Day inlier, approximately 2,500 km^2 in extent. Based on previous studies, the position of the inlier within the eugeosynclinal belt is shown in Figure 5–20B.

Stratigraphy

The Mesozoic sequence of the John Day inlier includes about 15,000 m of dominantly clastic strata that were deposited between mid-Triassic and mid-Cretaceous times. Most sandstones are turbidites, but facies changes are intricate in detail, and shelf deposits occur locally as lateral equivalents of nearby turbidite successions. At least six unconformities break the stratigraphic sequence, which has been subdivided into about 25 formally named stratigraphic units. Many of these are quite thin and have limited lateral extent.

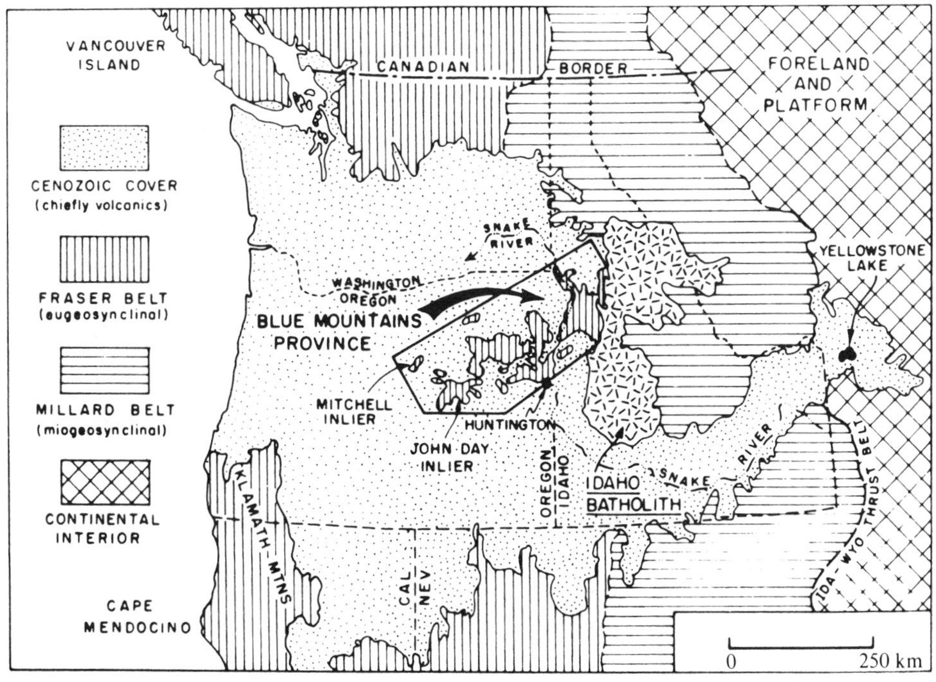

(A)

(B)

Figure 5-20
(**A**) Location of John Day and Mitchell inliers in relation to regional orogenic terranes. (**B**) Schematic restored profile across pre-Tertiary terrane of central Oregon, showing inferred position of John Day inlier within mid-Mesozoic arc-trench gap. [Dickinson et al., 1979.]

Petrology

As is typical of eugeosynclinal rocks, both depositional texture and mineral composition are obscured in many samples by intense alteration of framework grains, growth of diagenetic matrix, or replacement by secondary carbonates. Dickinson et al. concentrated their petrographic studies on sandstones whose texture and limited degree of alteration permitted positive identification of the framework grains. In each of the 34 thin sections examined, they identified 400 grains, noting not only whether the grain was quartz, feldspar, or a rock fragment, but also the type of feldspar (identified by selective staining for K-feldspar and plagioclase) and the internal structure of the quartz and lithic fragments.

Based on the results of the point counts, it was possible to group the 34 thin sections into three provenance categories (see Table 5–4):

a. 11 samples whose grains were derived entirely from the Upper Paleozoic to lowermost Mesozoic mélange terrane.

b. 9 samples of volcaniclastic turbidites derived mainly from contemporaneous Jurassic volcanic eruptions nearby.

c. 14 samples of lithic sandstones derived from mixed provenances exposing varied rock types not limited to volcaniclastic or mélange sources. Even in this provenance group, however, the most abundant grain types are volcanic rock fragments.

The source rocks of the mélange-derived sandstones were mainly chert, greenstone and keratophyric felsite, and slate or phyllite within the mélange terrane. Feldspar grains were evidently derived chiefly from the phenocrysts in the volcanic components of the mélange, for the feldspar content is highest in the samples that are richest in volcanic rock fragments. The very low percentages of quartz and K-feldspar indicate that no direct plutonic or basement source of these minerals was present (e.g., granite). Mélange-derived sandstones are most common in the lower part of the John Day inlier.

The volcaniclastic sandstone group has the clearest provenance of the three groups. About 60% of the framework grains are volcanic rock fragments, mainly microlithic, and 30% are volcanic feldspar grains, almost entirely plagioclase of phenocrystic origin. As we would expect from a provenance of mafic volcanic rocks, quartz and chert are nearly absent and clinopyroxene grains are uncommonly abundant, forming 3% of the framework of the sandstones. The rocks with volcaniclastic provenance are most prominent in the middle section of the John Day inlier.

The sandstones of mixed provenance have compositions intermediate between those of the other two groups, plus additions from a sedimentary or metasedimentary terrane. Volcanic rock fragments, which total 27% in the mélange-derived sandstones and 59% in the volcaniclastic sandstones, are 43% of the sandstones of mixed provenance. Chert plus sedimentary rock fragments total 48% in the mélange-derived sandstones, 3% in those of volcaniclastic provenance, and an intermediate 18% in

Table 5–4
Average Modal Compositions of John Day Inlier Sandstones Derived from Each of the Three Distinct Provenance Groups[a]

	Mélange-derived sandstones	Volcaniclastic sandstones	Mixed-provenance sandstones
Interstitial matrix and cement	7	7	9
Monocrystalline mineral grains			
Quartz grains	5	1	13
Plagioclase grains	12	30	21
K-feldspar grains	1	1	4
Polycrystalline quartz fragments			
Chert	38	1	5
Polycrystalline quartz	6	0	1
(Meta-) Volcanic rock fragments			
Microxenolithic grains	18	56	32
Felsite grains	6	3	10
Microgranular hypabyssal	3	0	1
(Meta-) Sedimentary rock fragments			
Argillite grains	1	2	7
Shale/slate grains	5	0	6
Quartz–mica tectonite	4	0	0
Clinopyroxene grains	0	3	0
Mica flakes	0	0	1
Miscellaneous framework grains	1	2	0

[a]Compositions total 100% excluding matrix and cement.
Source: Data from Dickinson et al., 1979.

sandstones of mixed parentage. The sandstones of mixed parentage are dominant in the upper parts of the John Day inlier sequence, as would be expected. The provenance is interpreted to have been an eroded arc terrance in which some of the folded and intruded substratum beneath the volcanoes had begun to undergo erosion.

Diagenesis

The samples studied by Dickinson et al. were specifically chosen to be largely free of diagenetic effects, but it is clear that all grades of alteration are present. In the most affected sandstones, pervasive diagenesis has largely obliterated the detrital mineralogy, so that the identification of the detrital grains is uncertain. Probable volcanic lithic fragments appear as roundish aggregates of clay and fine mica whose boundaries can sometimes be inferred and usually are indistinct.

In the 34 samples that were examined, however, the amount of silt- and clay-size matrix is less than 5% in all three groups of sandstones, indicating very little conversion of the unstable mafic volcanic fragments into clay. The sandstones of volcaniclastic and mixed provenance do, however, contain 3–5% of clearly authigenic green chlorite as a cementing material. The chlorite rims the framework grains and is oriented normal to their surfaces, which would not be possible if the chlorite were detrital in origin.

Comment on the John Day Inlier

Lithic sandstones that were deposited within a Mesozoic forearc basin in central Oregon include mélange-derived, volcaniclastic, and mixed-provenance suites in which relative proportions of feldspar grains, chert grains, volcanic-metavolcanic rock fragments, and sedimentary-metasedimentary rock fragments are different and diagnostic. The stratigraphic variations in sandstone petrology provide evidence about the tectonic evolution of the uplifted mélange ridge and the igneous arc terrane that flanked the forearc basin on opposite sides during the mid-Mesozoic. Although petrologic relationships in eugeosynclinal terranes may be obscured in many samples by diagenetic alteration, it is generally possible to obtain samples whose depositional character is well preserved.

MODERN SANDS

Cratonic Sands

The chief reason petrologists study the composition of the detrital fraction of ancient sandstones is to decipher paleogeology and paleogeography. But before we can do this, we must accumulate a data base from modern stream studies. How rapidly do schist fragments disintegrate during stream transport? How rapidly is the percentage of basalt fragments lowered by the input of quartz sand released from friable sandstones surrounding the area of outcrop of the basalt? Until we know the answers to questions such as these, it will not be possible to adequately interpret the mineral composition of ancient sands. With the important exception of turbidity-current deposits, sands accumulate in environments of high kinetic energy, such as the high-velocity sections of stream channels or alluvial fans, marine beaches or sandbars, and desert dunes. Therefore, the goal in the remainder of this chapter is to describe and interpret variations in the mineral composition of sands in modern environments, both nonmarine and marine. The present is the key to the past.

Elk Creek, South Dakota

Elk Creek is located in the northern part of the Black Hills, a partially unroofed, asymmetrical dome approximately 150 km long and 80 km wide (see Figure 5–21). The domed area rises 800 m above the surrounding plain. The core of the uplift is composed predominantly of fine-grained Precambrian schist, minor amphibolite, and lesser amounts of granitic gneiss. In the northern Black Hills were Elk Creek is located, numerous shallow intrusives of Tertiary age cut the schist and its overlying Paleozoic sedimentary rocks. Precambrian granite is exposed in the southern part of the Black Hills, but none crops out as far north as the drainage basin of Elk Creek.

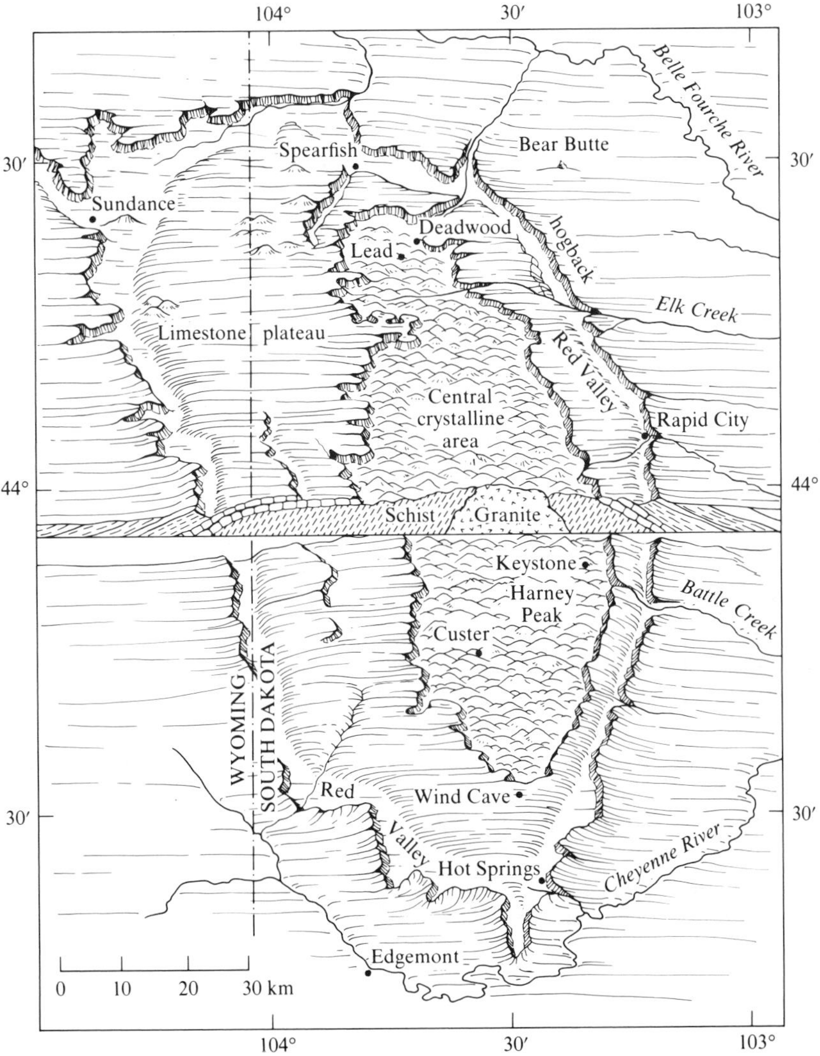

Figure 5–21
Bird's-eye view by Canada goose of physiography and general geology of Black Hills, as seen while flying north. [Cameron and Blatt, 1971.]

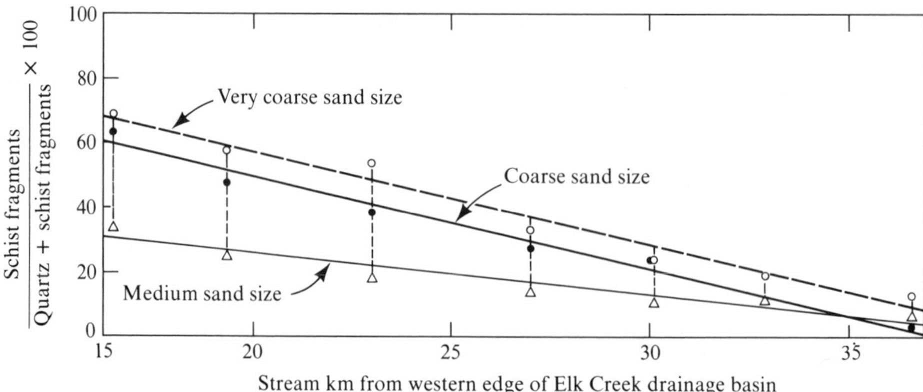

Figure 5–22
Decrease in abundance of schist fragments with distance of transport in Elk Creek, expressed as ratio of schist to quartz plus schist in the three coarsest sand fractions. [Cameron and Blatt, 1971.]

A study was conducted (Cameron and Blatt, 1971) to determine the rate of decrease of the schist fragments in Elk Creek sediment with increasing distance from their areas of outcrop. Samples were taken both from the residual soil developed in the source area and from the sand in the stream at arbitrarily selected distances downstream. Schist is exposed over 45% of the drainage area in the headwaters of the stream, and the sand fraction of the residual soil on the schist consists almost entirely of schist fragments. Nevertheless, after less than 0.5 km of stream transport, schist fragments form only 30% of the very coarse and coarse sand sizes and less than 20% of the two finer sand sizes. After 21 km of stream transport (13 airline km), schist fragments had decreased to less than 2% of the sediment (see Figure 5–22). It was concluded that schist fragments have a low survival potential because of mechanical instability. Apparently, soil-forming processes weaken the bonds between adjacent mica flakes sufficiently so that the schist fragments disaggregate almost immediately on entering the high-gradient part of the stream (20 m/km). Schist fragments are nearly untransportable in this semiarid area. This suggests that ancient sandstones containing large amounts of schist sand grains must be located either within a few kilometers of the source area or else adjacent to a source area with much greater topographic relief than is present in the drainage basin of Elk Creek.

Vogelsberg Area, West Germany

Subsequently, an attempt was made in West Germany to investigate the loss of volcanic fragments during stream transport (Blatt, 1978). In the state of Hessen there occurs a nearly circular mass of Tertiary basaltic rocks about 250 km^2 in area. The volcanic outcrop has a maximum relief of 400 m and a stream gradient in the upland area of 25–30 m/km; maximum relief is only half that in the Black Hills, but the stream gradient is slightly greater. Samples of stream sand were collected both within the area of volcanic outcrop and outward for several tens of kilometers, and the proportion of volcanic fragments determined. In the coarse-sand fraction the percentage of volcanic debris decreases to 5% or less within 10–20 km from the basalt outcrop; in the fine-sand fraction the decrease to 5% occurs within only 5 km (see Figure 5–23). The more rapid loss of volcanic debris from the fine-sand fraction was attributed to greater dilution by the fine-grained, friable quartz sandstone that surrounds the circular basalt mass.

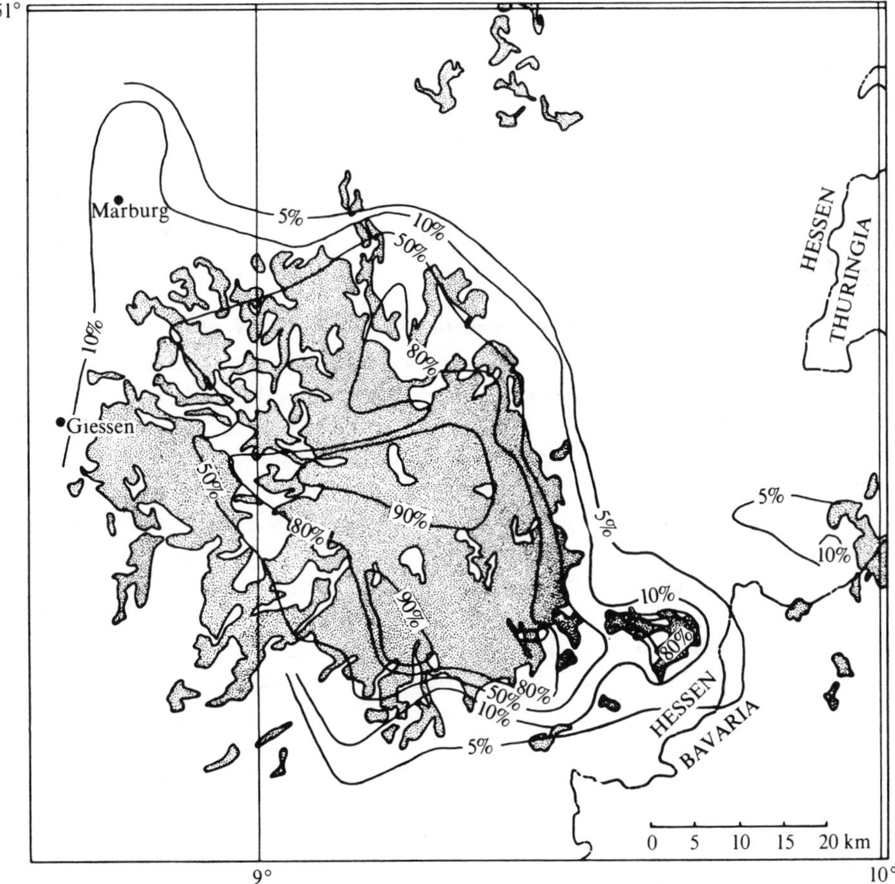

Figure 5–23
Map of outcrops of volcanic rocks in central Hessen. Isopleths are percentage of volcanic detritus in fine sand fraction of stream sediment. Note irregular contour interval. [Blatt, 1978.]

In general, then, the results of the Black Hills study and the German study were similar. In areas of crystalline rock outcrop a few hundred square kilometers in area and with topographic relief of a few hundred meters, debris from the outcrop disappears very rapidly. After a distance of stream transport of 20 km, less than 5% of the stream sediment will consist of crystalline rock debris. In the Black Hills the disappearance results from the mechanical instability of foliated fragments; in Hessen the disappearance results from intense dilution by sand released from surrounding sandstones. Clearly, areas of outcrop of crystalline rocks that are only a few hundred square kilometers in extent will be very difficult to detect in the drainage basin of an ancient sandstone.

Sands at Convergent Plate Margins

There have been numerous studies of volcaniclastic sands accumulating at convergent margins of an oceanic and a continental plate (Dickinson and Valloni, 1980). The bulk of these studies has been part of the JOIDES program of drilling and sampling Late Cenozoic sediments in the world oceans. As we would expect, the sands are dominated by volcanic lithic fragments and contain abundant pyroxene in the heavy-mineral fraction.

One of the more unusual studies of a tectonic setting associated with a convergent plate margin is the study of variations in composition of detrital feldspars in the Sea of Japan, a backarc sedimentary basin. Few sedimentary petrologists have attempted to characterize modern sediment distributions by variations in feldspar content or chemical composition because of the relatively time-consuming nature of feldspar studies. When dealing with lithic fragments and heavy minerals, petrologists can easily identify a large number of types. There exist many varieties of igneous, metamorphic, and sedimentary lithic fragments and as many species of heavy minerals. But there are few easily identifiable types of feldspars (orthoclase, microcline, plagioclase, and very rarely microperthite or sanidine), and determination of the compositions of individual plagioclase grains is usually not attempted.

A notable exception is a study of the provenance of turbidite sediments by Sibley and Pentony (1978). They obtained cores at two sites of a JOIDES sampling program from the sediment–water interface to a depth of about 500 m (see Figure 5–24). At site 299, 33 samples were taken from 8 cores; at site 301, 8 samples from 4 cores. The sediments are dominantly silts so that it was more convenient to examine the feldspars in the 74–34 μm size rather than in the fine sand size only. Results for the 41 samples from the two sites are summarized in Figure 5–25, and it is clear that two source areas for the sediments can be distinguished. The feldspars at site 299 contain significantly lesser amounts of orthoclase and greater amounts of plagioclase than at site 301 and, in addition, site 299 feldspars are more calcic than those at site

301. There is also a lower quartz/feldspar ratio in the site 299 sediment. These data led to the conclusion that site 299 receives most or all of its sediment from the Japanese volcanic arc via the north–south trending Toyama Trough (see Figure 5–25), whereas site 301 receives detritus from both the arc and the granitic Asian mainland to the west.

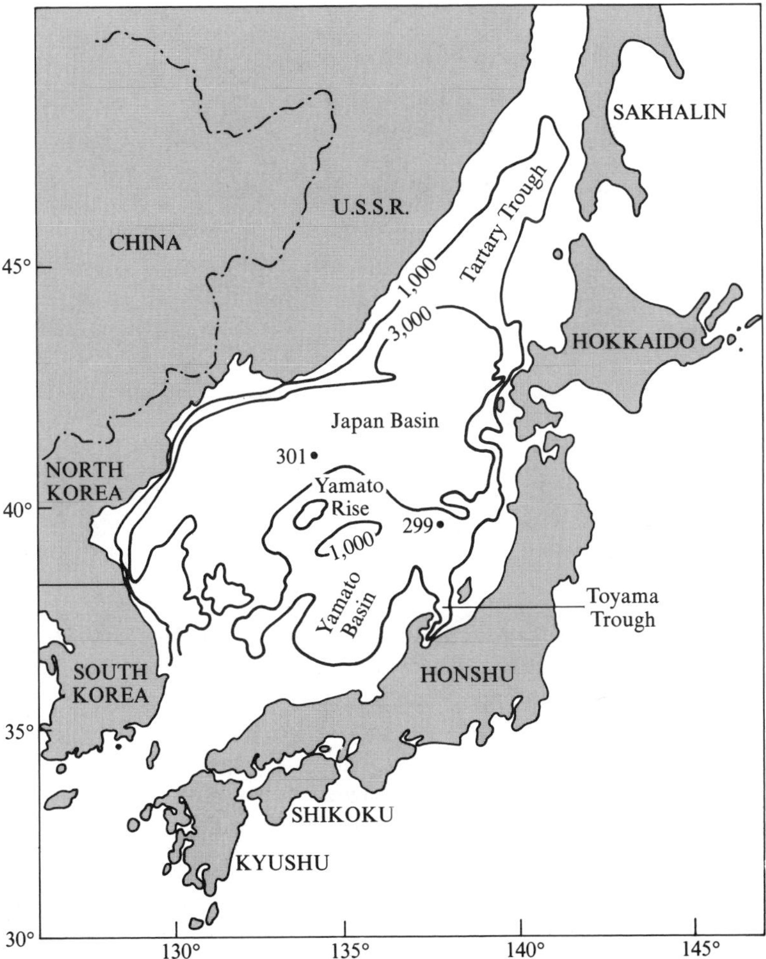

Figure 5–24
Location of sample sites 299 and 301 in relation to physiographic features in Sea of Japan. Bathymetric contours in meters. [J. C. Ingle, Jr., et al., 1975, *Initial Reports of the Deep Sea Drilling Project, 31.*]

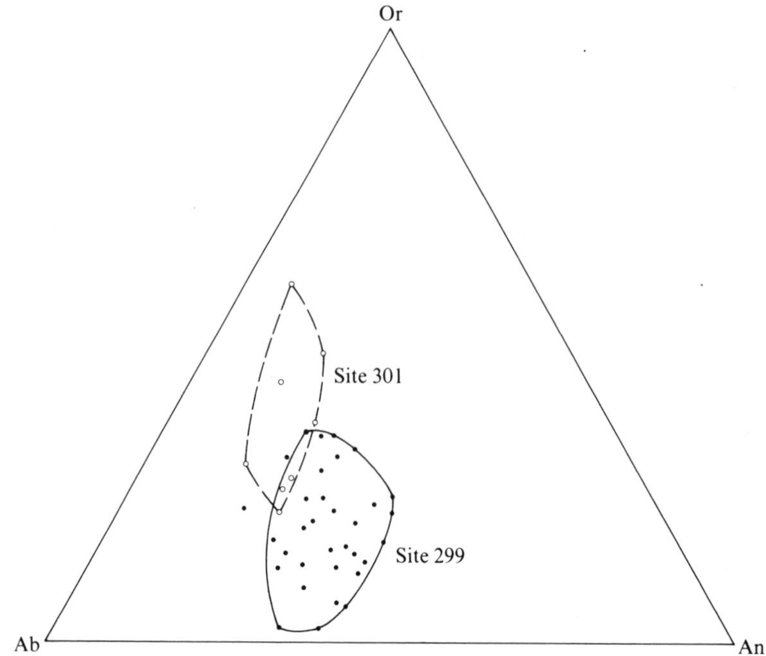

Figure 5–25
Compositions of unaltered feldspars in samples from sites 299 and 301 in Sea of Japan. Each data point indicates relative amounts of orthoclase and plagioclase as well as composition of plagioclase. [Sibley and Pentony, 1978.]

Samples from the two sites also show numerous stratigraphic changes in mineralogy through the 500 m length of the cores from each site (see Figure 5–26). It is clear that large changes in mineralogy occur over very short stratigraphic intervals, for example, the increase in calcic feldspar ($n > 1.54$) from 24% to 44% in the depth interval between 44.53 m and 44.98 m. So sharp a change within 0.45 m presumably reflects a sampling of different turbidity-current deposits, the more calcic feldspars reflecting a greater sediment contribution from the island arc.

SUMMARY

The essential features of conglomerates and sandstones that must be described and interpreted are texture, structure, and mineral composition. Textural features supply clues to distance of transport, mechanism of transport, and depositional environment. Sedimentary structures can indicate the fluid mechanics of the depositional process,

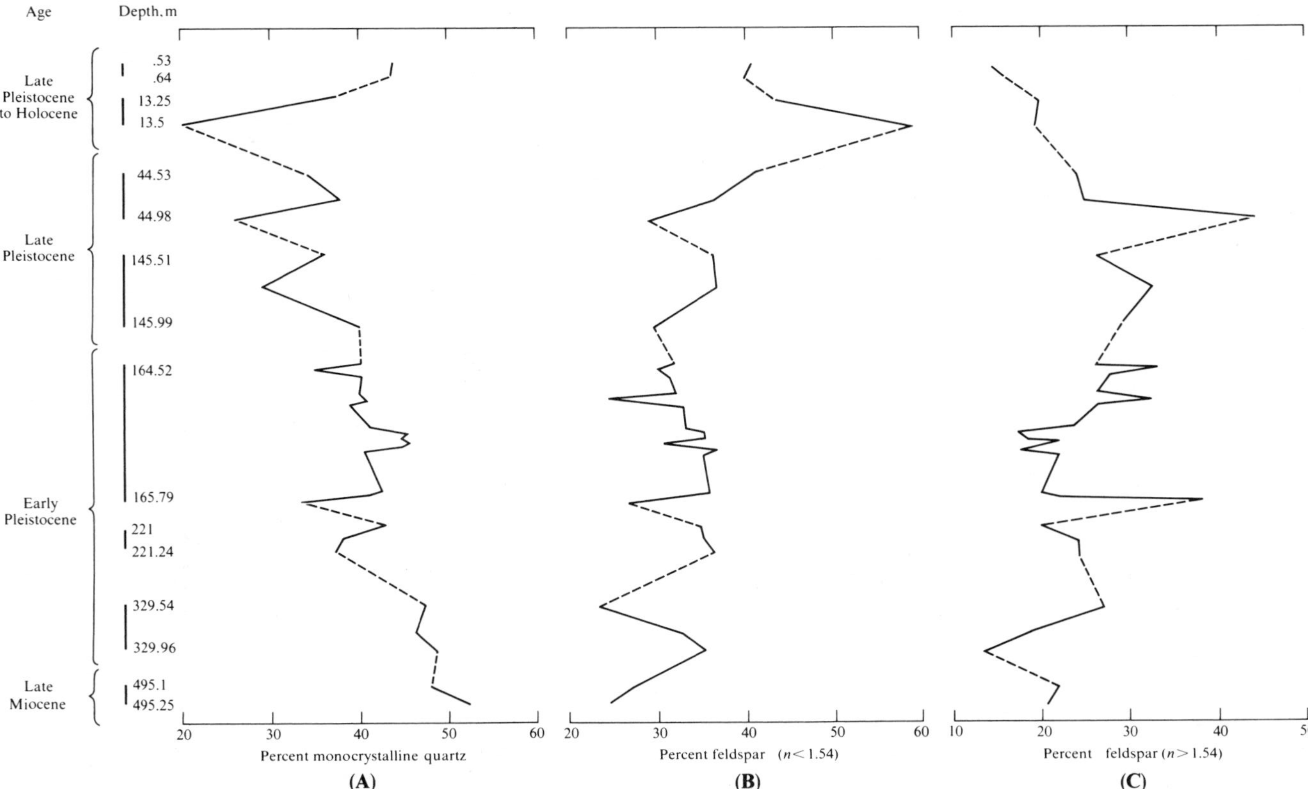

Figure 5–26
Stratigraphic changes in quartz and feldspars at site 299 in Sea of Japan based on petrographic distinctions. [Sibley and Pentony, 1978.]

the direction of fluid flow, and the water depth through changes in the type of trace-fossil (*ichnofossil*) assemblage (tracks, trails, burrows of bottom-dwelling organisms).

The mineral composition of the detrital rock is the best indication of the types of source rocks that supplied sediment to the rock being studied. Most provenance investigations require a sound understanding of igneous and metamorphic petrogenesis to be successful because mineralogic variations in sandstones are frequently quite subtle. Quartz grains may be plastically deformed or not, monocrystalline or polycrystalline; and polycrystalline grains may show a variety of types of internal structures characteristic of an igneous or a metamorphic parentage. Feldspar grains may be orthoclase, microcline, sanidine, a perthitic intergrowth, or plagioclase of variable composition. Different feldspar types and compositions are characteristic of different source terranes. Lithic fragments are the least equivocal grains with respect to their origin, but even with these grains, problems in interpretation exist. Minor accessory minerals are valuable sources of information, but many of the ones most abundant in sandstones occur in most types of crystalline source rocks.

The mineral composition of sandstones can be used to infer the tectonic setting in which the rocks formed (Dickinson and Suczek, 1979). Cratonic sands are rich in quartz and chert because of repeated reworking in shallow-water environments of high kinetic energy. Sandstones formed along divergent continental margins are somewhat less rich in quartz and chert and typically contain 20–30% of foliated sedimentary or metamorphic lithic fragments. In intracontinental rift zones such as the Triassic rocks in New England or the Pennsylvanian rocks of southern and southwestern Oklahoma, richly feldspathic sands are formed (dominantly potassic feldspars) as granitic rocks are exposed by tensional forces. Along convergent continental margins, volcanic lithic fragments tend to be abundant and are accompanied by a feldspar suite with a high proportion of fairly calcic (An content $> 30\%$) plagioclase grains. Within a convergent margin setting, many smaller tectonic units are present, and it may be possible to distinguish among them mineralogically.

FURTHER READING

Al-Shaieb, Z., R. E. Hanson, R. N. Donovan, and J. W. Shelton. 1980. Petrology and diagenesis of sandstones in the Post Oak Formation (Permian), southwestern Oklahoma. *Jour. Sed. Petrology, 50,* 43–50.

Blatt, H. 1967. Original characteristics of clastic quartz grains. *Jour. Sed. Petrology, 37,* 401–424.

Blatt, H. 1978. Sediment dispersal from Vogelsberg basalt, Hessen, West Germany. *Geologische Rundschau, 67,* 1009–1015.

Blatt, H., and J. M. Christie. 1963. Undulatory extinction in quartz of igneous and metamorphic rocks and its significance in provenance studies of sedimentary rocks. *Jour. Sed. Petrology, 33,* 559–579.

Blatt, H., G. V. Middleton, and R. C. Murray. 1980. *Origin of Sedimentary Rocks,* 2nd ed. Englewood Cliffs, NJ: Prentice-Hall, 782 pp.

Boles, J. R. 1974. Structure, stratigraphy, and petrology of mainly Triassic rocks, Hokonui Hills, Southland, New Zealand. *New Zealand Jour. Geol. Geophys., 17,* 337–374.

Cameron, K. L., and H. Blatt. 1971. Durabilities of sand size schist and "volcanic" rock fragments during fluvial transport, Elk Creek, Black Hills, South Dakota. *Jour. Sed. Petrology, 41,* 565–576.

Courdin, J. L., and J. F. Hubert. 1969. Sedimentology and mineralogical differentiation of sandstones in the Fort Union Formation (Paleocene), Wind River Basin, Wyoming. *Wyo. Geol. Assoc. Guidebook, 21st Field Conf.,* pp. 29–38.

Crook, K. A. W. 1960. Petrology of Parry Group, Upper Devonian–Lower Carboniferous, Tamworth–Nundle District, New South Wales. *Jour. Sed. Petrology, 30,* 538–552.

Davies, D. K., and F. G. Ethridge. 1975. Sandstone composition and depositional environment. *Amer. Assoc. Petroleum Geol. Bull., 59,* 239–264.

Dickinson, W. R. 1970. Interpreting detrital modes of graywacke and arkose. *Jour. Sed. Petrology, 40,* 695–707.

Dickinson, W. R., K. P. Helmold, and J.A. Stein. 1979. Mesozoic lithic sandstones in central Oregon. *Jour. Sed. Petrology, 49,* 501–516.

Dickinson, W. R., and C. A. Suczek. 1979. Plate tectonics and sandstone compositions. *Amer. Assoc. Petroleum Geol. Bull., 63,* 2164–2182.

Dickinson, W. R., and R. Valloni. 1980. Plate tectonics and provenance of sands in modern ocean basins. *Geology, 8,* 82–86.

Folk, R. L. 1960. Petrography and origin of the Tuscarora, Rose Hill, and Keefer Formations, Lower and Middle Silurian of eastern West Virginia. *Jour. Sed. Petrology, 30,* 1–58.

Friedman, G. M., and J. E. Sanders. 1978. *Principles of Sedimentology.* New York: Wiley, 792 pp.

Galloway, W. E. 1974. Deposition and diagenetic alteration of sandstone in northeast Pacific arc-related basins: implications for graywacke genesis. *Geol. Soc. Amer. Bull., 85,* 379–390.

Graham, S. A., et al. 1976. Common provenance for lithic grains of Carboniferous sandstones from Ouachita Mountains and Black Warrior Basin. *Jour. Sed. Petrology, 46,* 620–632.

Green, D. H., J. P. Lockwood, and E. Kiss. 1968. Eclogite and almandine–jadeite–quartz rock from the Guajira Peninsula, Colombia, South America. *Amer. Mineral., 53,* 1320–1335.

Hiscott, R. N. 1979. Provenance of Ordovician deep-water sandstones, Tourelle Formation, Quebec, and implications for initiation of Taconic orogeny. *Canad. Jour. Earth Sci., 15,* 1579–1597.

Hubert, J. F. 1960. *Petrology of the Fountain and Lyons Formations, Front Range, Colorado.* Colo. School of Mines Quart., *55,* 242 pp.

Ingersoll, R. V., and C. A. Suczek. 1979. Petrology and provenance of Neogene sand from Nicobar and Bengal Fans, DSDP sites 211 and 218. *Jour. Sed. Petrology, 49,* 1217–1228.

Krynine, P. D. 1940. *Petrology and Genesis of the Third Bradford Sand.* Pa. State College Bull., *29,* 132 pp.

Krynine, P. D. 1950. *Petrology, Stratigraphy and Origin of the Triassic Sedimentary Rocks of Connecticut.* Conn. State Geol. Nat. Hist. Surv. Bull., *73,* 247 pp.

Lockwood, J. P. 1971. Sedimentary and gravity-slide emplacement of serpentinite. *Geol. Soc. Amer. Bull., 82,* 919–936.

McRae, S. G. 1972. Glauconite. *Earth-Sci. Rev., 8,* 397–440.

Moore, G. F. 1979. Petrography of subduction zone sandstones from Nias Island, Indonesia. *Jour. Sed. Petrology, 49,* 71–84.

Odom, I. E., T. W. Doe, and R. H. Dott, Jr. 1976. Nature of feldspar–grain size relations in some quartz-rich sandstones. *Jour. Sed. Petrology, 46,* 862–870.

Parkash, B., R. P. Sharma, and A. K. Roy. 1980. The Siwalik Group (molasse)—sediments shed by collision of continental plates. *Sed. Geol., 25,* 127–159.

Pettijohn, F. J., P. E. Potter, and R. Siever. 1973. *Sand and Sandstone.* New York: Springer-Verlag, 618 pp.

Pittman, E. D. 1969. Destruction of plagioclase twins by stream transport. *Jour. Sed. Petrology, 39,* 1432–1437.

Potter, P. E. 1978. Petrology and chemistry of modern big river sands. *Jour. Geol., 86,* 423–449.

Potter, P. E., and W. A. Pryor. 1961. Dispersal centers of Paleozoic and later clastics of the upper Mississippi Valley and adjacent areas. *Geol. Soc. Amer. Bull., 72,* 1195–1250.

Russell, R. D. 1937. Mineral composition of Mississippi River sands. *Geol. Soc. Amer. Bull., 48,* 1307–1348.

Schultheis, N. H., and E. W. Mountjoy. 1978. Cadomin Conglomerate of western Alberta—a result of early Cretaceous uplift of the main ranges. *Bull. Canad. Petroleum Geol., 26,* 297–342.

Sibley, D. F., and K. J. Pentony. 1978. Provenance variation in turbidite sediments, Sea of Japan. *Jour. Sed. Petrology, 48,* 1241–1248. See also *DSDP Reports,* 1975, *31,* 507–514.

Stone, W. B., Jr. 1977. Mineralogic and textural dispersal patterns within the Permian Post Oak Formation of southwestern Oklahoma. Unpublished master's thesis, University of Oklahoma, 117 pp..

Turner, C. C., and R. G. Walker. 1973. Sedimentology, stratigraphy, and crustal evolution of the Archean greenstone belt near Sioux Lookout, Ontario. *Canad. Jour. Earth Sci., 10,* 817–845.

6

Conglomerates and Sandstones: Diagenesis

If you can look into the seeds of time,
And say which grain will grow and which will not,
Speak.

MACBETH

Diagenesis is defined as all the physical, chemical, and biological changes that a sediment is subjected to after the grains are deposited but before they are metamorphosed. Some of these changes occur at the water–sediment interface and are termed *halmyrolysis,* but the bulk of diagenetic activity takes place after burial. Examples of halmyrolytic activity include such things as the alteration of biotite flakes into the mineral glauconite on the seafloor and the destruction of clam shells by boring sponges as the shells lie on the seafloor. During deep burial the main diagenetic processes are *compaction* and *lithification.*

COMPACTION

Field Observations

Anyone who has ever stepped into a patch of solid-looking mud in a field or along a roadside is immediately aware that mud is a very compactible sediment. It is clear that freshly deposited mud contains a great deal of water that is easily squeezed out by a relatively small amount of pressure; indeed, the "mud" is mostly water rather than solid mineral matter. Measurements of modern muds reveal that a mixture of mud and water on the seafloor is typically at least 60% water, and values as high as 80% have been recorded—the "sediment" has a "porosity" of up to 80%! The water is squeezed out easily because the clay minerals in the mud are ductile (flexible) and platy and so can be compacted very tightly at relatively low pressures.

A marked contrast in compactibility is found by stepping on a beach composed of quartz sand. The crunching sound produced reflects the friction of grain against grain as the pressure causes an increase in the tightness of packing. Very little water spews out of the sand around one's feet. Quartz grains are subspherical to ellipsoidal in shape, relatively closely packed when deposited, and are not ductile at sedimentary pressures and temperatures. Hence, the thickness of the beach-sand accumulation is not decreased appreciably by sun-worshipers and joggers applying as much as several bars of pressure to its upper surface each day.

The differential compaction between sands and muds can be seen in ancient rocks as well, and a particularly graphic example on a small scale is shown by the crumpled, small sandstone dike in Figure 6–1. A sandstone dike is formed by an intrusion of loose sand under pressure into mud and is tabular in shape—the sedimentary analog of an igneous dike. Sandstone dikes typically occur in swarms and are produced by the cracking of an impermeable mud seal above an overpressured sand layer. The seal formed before much water was squeezed from the sand, with the result that the sand grains did not support the full overburden load: the load was

Figure 6–1
Part of subsurface core composed of thinly layered quartz sand (white) and organic-rich mud (dark), illustrating several contorted sandstone dikes, Dakota Sandstone (Cretaceous), North Dakota. [Shelton, 1962. Photo courtesy J. W. Shelton.]

partly supported by the water. When stresses cracked the seal, a sand–water mixture squeezed upward to form a sandstone dike, which subsequently was lithified. In the example shown, the dike formed before there was much compaction of the mud layers so that, when they did compact, the still unlithified tabular sand body was crumpled. When the crumpled structure was graphically stretched out to its original tabular shape, the original thickness of the mud layer was determined to be 2.0–3.4 times greater than at present; that is, the amount of compaction was 50–70%—in the range of what we would expect in a slightly silty mud.

The pressure exerted on a layer of sedimentary rock at depth is equal to ρgh; and if we assume that the sediment is permeable from its burial depth to the surface, ρgh is equal to the weight of a column of water of that height, approximately 10 bar/100 m. In some areas of rapid sedimentation, such as the Gulf Coast, many impermeable clay seals are formed in the stratigraphic column. In such cases the sediment pile is not permeable from certain layers of sand up to the surface, water in the sand supports some of the weight of the overburden (both rock and water), and the pressure of the water in the sand layer is greater than 10 bar/100 m. Fluid pressures may be as high as 25 bar/100 m—2.5 times the pressure exerted by a column of water at that depth. The upper limit is set by the density of the overlying column of rock.

Laboratory Studies

Examination of thin sections of sandstones reveals several effects of compaction. A freshly deposited quartz sand has a porosity of about 45% ± 5%, depending on sorting and grain angularity. A thin section of a well-lithified quartz sand, however, reveals that the percentage of secondary cement that completely fills the areas among detrital grains is perhaps 30% ± 5%, and there is no evidence that the detrital grains have been dissolved or otherwise altered since deposition. The difference between the original 45% porosity and the 30% paleoporosity that was available to be filled with cement must have been caused by compaction. The weight of overburden reduced the amount of empty space in the sand by about one-third the amount originally present.

Examination of thin sections of sandstones that contain ductile lithic fragments or clay supplies additional evidence of compaction. Sand-size pieces of mud and fragments of micaceous rocks, such as shale, slate, phyllite, and schist, are deformed. Their shapes are not the platy ones they had when deposited, and they bend around the more rigid quartz and feldspar grains and are squeezed into pore spaces (see Figure 6–2). In highly compacted lithic sandstones, porosity can be reduced from the original 45% to nearly zero simply by the squeezing of ductile grains into pores (see Figure 6–3). In such cases the original thickness of the sand must have been reduced by nearly 50%.

In summary, the amount of compaction of a freshly deposited sand depends on five factors: (1) clay content, (2) sorting, (3) percentage of ductile fragments, (4) angularity of the nonductile grains, and (5) burial depth or tectonic stresses.

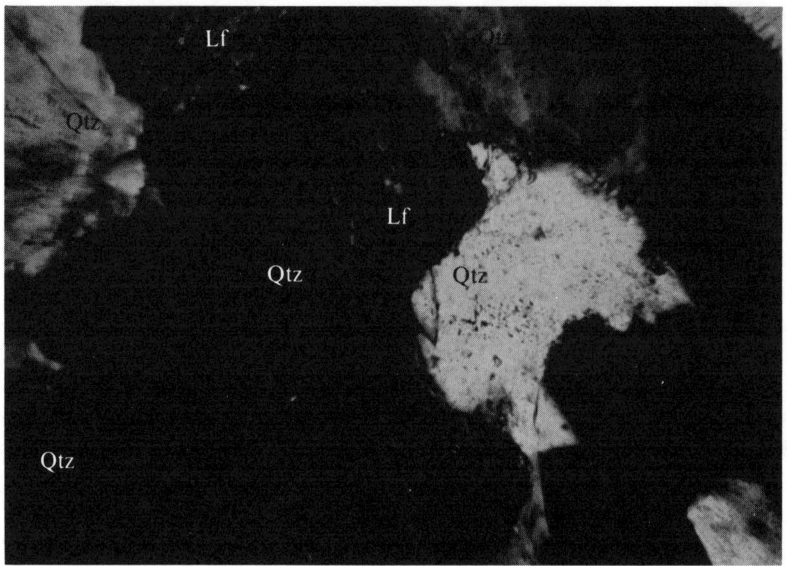

Figure 6–2
Photomicrograph (crossed nicols) of micaceous lithic fragment (Lf) plastically deformed by rigid quartz grains (Qtz) in compacted, coarse-grained lithic sandstone, Kingston, New York. [H. Blatt, 1966, *Wyo. Geol. Assoc. Guidebook.*]

CEMENTATION

How is an assemblage of loose detrital fragments converted into a solid mass that resists a sharp whack with a hammer? How is the process called *lithification* accomplished? Can compaction alone turn a sand into a sandstone? To answer such questions, we must go to the laboratory; fieldwork cannot supply the answers.

The question of whether compaction alone can produce lithification can be answered by laboratory experiments. We can take a cylindrical metal container open at one end, half-fill it with a sediment–water mixture, insert a piston into the open end of the cylinder, and apply pressure on the piston to compact the sediment. The base of the cylinder (or the piston) is microperforated to permit water to escape from the cylinder during the experiment.

Our first experiment is with a quartz–water mixture, and the results are negative. We apply pressures up to the limit of sedimentary conditions to no avail. When the piston is removed, the sand pours out as easily as it entered. We next repeat the experiment using 80% quartz and 20% schist fragments or mud and find that, although we have not produced a rock from the loose grains, we have produced some aggregates. Quartz grains have become indented into the mud or schist fragments to create a partial lithification—pieces of "rock" a few grain-diameters in size. Apparently, the presence of ductile fragments can lead to lithification by compaction. We decide to go all the way and run an experiment using 100% mud fragments and schist

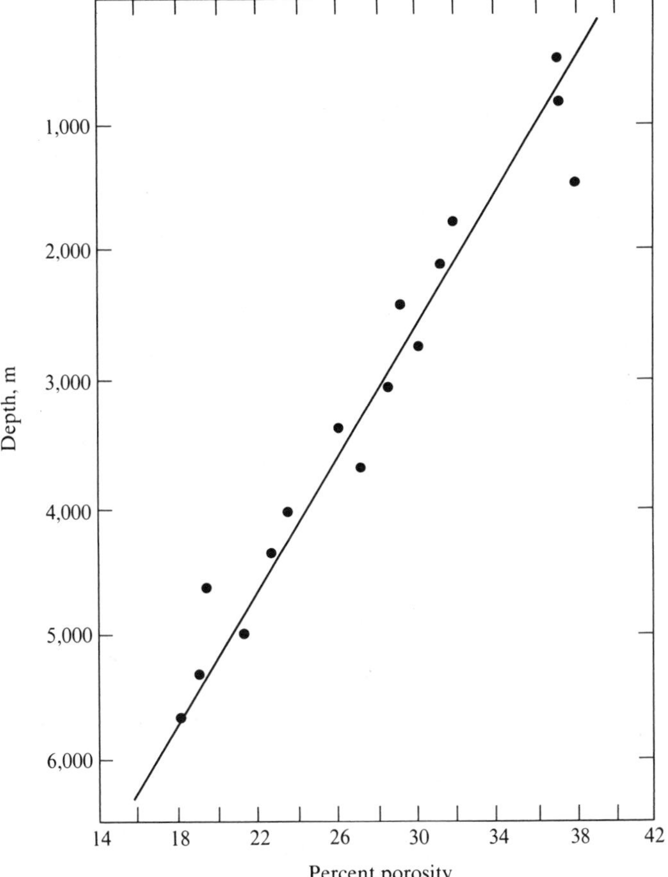

Figure 6–3
Effect of presence of ductile rock fragments on compaction of uncemented Tertiary sands in subsurface of southern Louisiana. Compaction is reflected by decrease in average porosity with depth; based on 17,367 cores averaged for each 1,000 feet. [H. Blatt, in Schluger, 1979.]

fragments plus water. The result confirms our inference: when the cylinder is inverted at the end of the experiment, a cylindrical piece of artificial rock falls out.

This result raises another question: Why does the presence of ductile fragments help to lithify a compacted sediment? To answer this, we make thin sections of the partially lithified fragments and fully lithified artificial rock to examine features too small to be visible otherwise. The polygranular fragments from the experiment with 80% quartz look exactly like the completely lithified "rock." It appears that indentation of grain against grain is sufficient to cause lithification; the reason the experiment using 100% quartz failed to produce lithification is that quartz is too rigid to permit adjacent grains to indent under the conditions of our experiment. The quartz grains are in contact only at points when deposited (poured into the cylinder), and compression alone cannot increase the area of grain-to-grain contact. When ductile fragments are in the sediment, however, the area of intergrain contact *can* be increased by pressure, leading to lithification. The success of lithification by compaction is correlated directly with the content of ductile material. Mud will be converted into mudrock by a relatively small thickness of overburden; lithic sandstones require a much greater thickness and perhaps even the intense squeezing caused by deformation and tectonic activity long after burial. It is probably not possible to lithify a pure quartz aggregate by compaction alone except under localized conditions such as immediately adjacent to a fault surface where stresses are unusually intense and the quartz grains are granulated.

The average sandstone contains only 15% lithic fragments, many of these fragments are not ductile (e.g., rhyolite, gneiss, and granite), and the amount of detrital clay in the average sandstone is probably about 5%. Therefore, intimate grain-to-grain contact in most sandstones must be achieved largely by the introduction of chemical precipitates—"cements." The growth of new mineral matter into the depositional pores creates the intimate surface–surface contact needed for lithification. The degree of lithification depends on the amount of cement-to-grain contact produced. If only a small amount of secondary mineral matter is precipitated in the pores, the rock can be disaggregated into individual grains by finger pressure; such rocks are termed *friable*. An increased amount of pore filling produces a rigid but still porous sandstone—the type of sandstone in which much of the world's petroleum, natural gas, and uranium is located. In the extreme, all porosity is lost and the sandstone is truly "hard as a rock."

Note that the exact nature of the pore-filling precipitate is of secondary importance in the lithification process. The intimate contact of quartz cement with a detrital quartz grain produces a stronger bond than the contact of calcite cement with the quartz grain, but both types of pore filling cause the rigidity that converts a loose pile of sand into a rock.

Quartz

Field Observations

Significant amounts of pore-filling quartz are, with few exceptions, restricted to sandstones whose detrital grains are nearly all quartz. Such sandstones occur mostly on cratons and divergent plate margins; they are rare near convergent margins. Pure quartz sandstones are most common in depositional environments of high kinetic energy and, therefore, quartz cement occurs most commonly in ancient beaches, marine bars, desert dunes, and some fluvial sandbars. It is rare in alluvial fans and turbidite sandstones.

Quartz-cemented quartz arenites are very resistant to weathering and typically stand out in relief at an outcrop. On a more regional scale, they are ridge formers, for example, in the central Appalachians, where they define clearly the pattern of plunging anticlines and synclines that form the structural pattern in the area (see Figure 6–4; Hack, 1980).

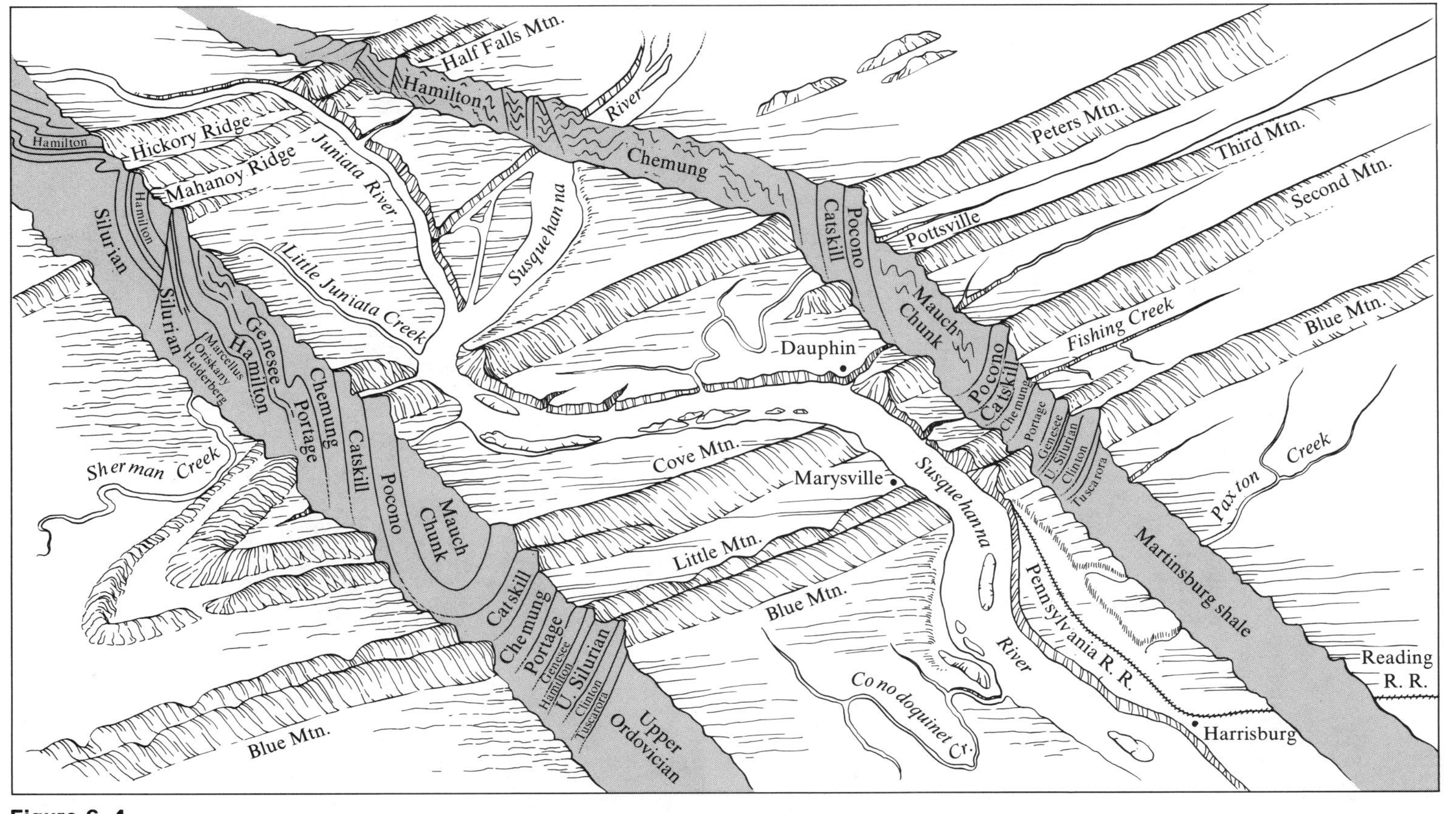

Figure 6–4
Idealized bird's-eye view looking north (parallel to Susquehanna River) near Harrisburg, Pennsylvania, with cutaway of Earth's surface showing close control of topography by lithology and differential resistance to erosion of Early Paleozoic bedrocks. Structures are steepened by vertical exaggeration. Width of diagram is about 20 km. Pocono Formation and Tuscarora Formation are quartz-cemented quartz sandstones, and the other ridges are underlain by well-lithified, hematite-cemented, coarse-grained sandstones and conglomerates. Topographically low areas are underlain by mudrocks. [G. W. Stose et al., 1933, *XVI Internat. Geol. Cong., Guidebook 10.*]

Laboratory Studies

Secondary quartz in detrital quartz sandstones grows as a coating on the detrital grains and in crystallographic continuity with them (see Figure 6–5). The growths may take root at several locations on each grain but never at a place where the boundary between zones of undulatory extinction (a surface of discontinuity) intersects the grain surface. The overgrowths nucleate on both sides of this line of intersection and join during growth directly above the line, propagating the discontinuity surface into the overgrowth.

The boundary between the detrital grain and the overgrowth may or may not be visible with standard petrographic techniques. When visible, the boundary is marked by substances different in petrographic character from quartz. Sometimes the substance is the red mineral hematite, sometimes dark-colored organic-rich material, sometimes clay minerals, and sometimes petrographically irresolvable material—"dirt" (see Figure 6–6). In any event, the substances that coat the detrital grain must be discontinuous; otherwise, secondary quartz will be unable to nucleate on the detrital grain. Secondary quartz in a quartz sandstone will not nucleate on the surface of a clay flake or bit of organic matter. It will nucleate on the exposed surface of the detrital grain, grow laterally, and perhaps cover the bits of extraneous material. Growths of secondary quartz are initiated as rhombohedra or prisms and grow to

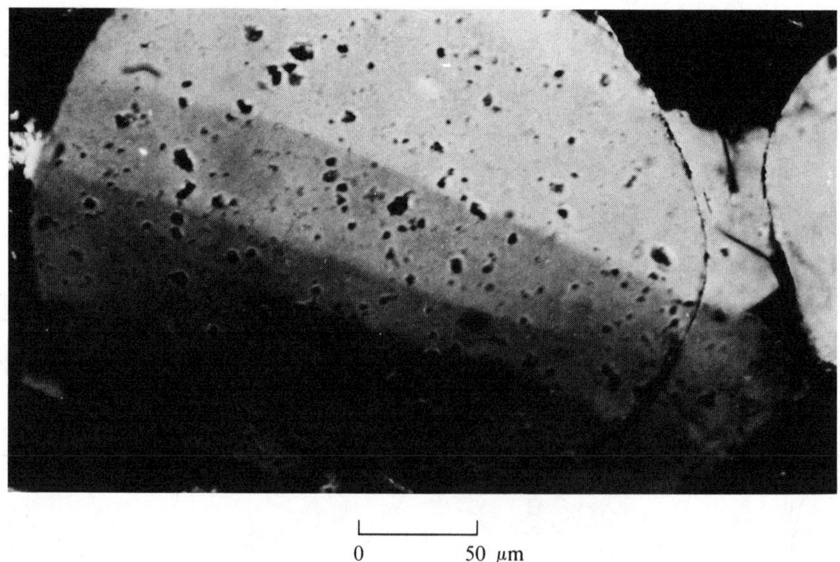

0 50 μm

Figure 6–5
Thin-section photomicrograph (crossed nicols) of detrital quartz grain with rhombohedral and prismatic overgrowths. Outline of detrital grain is marked by discontinuous coating of "dirt." Bands of undulatory extinction pass through both detrital grain and overgrowth. [Pittman, 1972. Photo courtesy E. D. Pittman.]

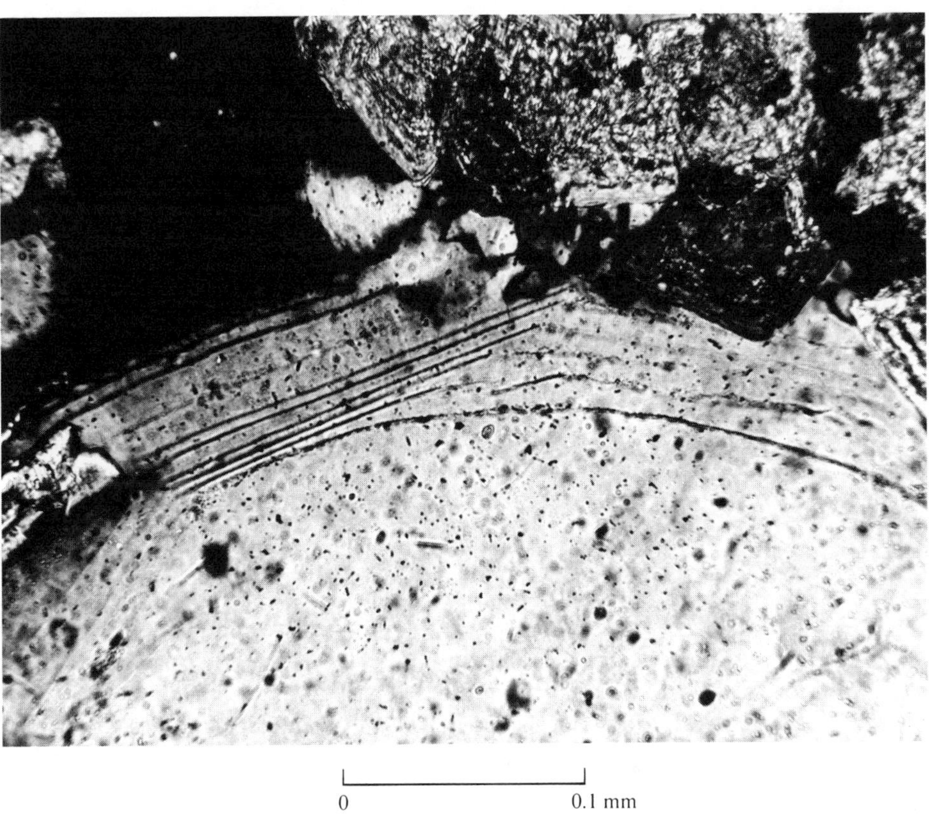

0 0.1 mm

Figure 6–6
Detrital quartz grain with nine distinct overgrowths, each bounded in part by well-defined crystal faces, Shakopee Formation (Ordovician), Minnesota. Dolomite rhombohedra occur in upper right. [G. S. Austin, 1974, *Jour. Sed. Petrology, 44*. Photo courtesy G. S. Austin.]

coalesce, resulting in large planar surfaces that abut somewhere in the pore space. When the surfaces abut, however, they generally lose their planar character and form crystallographically irrational compromise boundaries.

In many pure quartz sandstones the grains seem to interlock and adhere, but there appear to be no secondary growths. The perfectly contoured fit of grain against grain indicates that the texture could not have been produced by depositional processes. What are the alternative explanations? One possibility is that the surfaces of the detrital grains were scrubbed clean of extraneous substances by circulating pore waters before secondary quartz was deposited. That is, the rock has been lithified by secondary quartz that is indistinguishable from the host grains. A second possibility is that no secondary quartz is present, but that the detrital grains were welded together as part of a solution–compaction phenomenon. This latter process is termed *pressure solution.* The result of this process might appear identical to the rock in which the grains were scrubbed clean prior to the precipitation of secondary cement.

The distinction between the two alternatives of secondary infilling of silica and welding by pressure solution is important. If the first alternative is correct, we have to find a large, outside source of silica to lithify the quartz sandstone. If the second alternative is correct, we need to generate a pore solution capable of dissolving quartz and carrying the silica in solution to an area outside the area covered by the thin section. In the first alternative, the sandstone is a sink for silica in solution; in the second, the sandstone is a source of silica in solution.

The special technique that resolves this problem is *luminescence petrography* (Sippel, 1968). In this technique, a thin section is placed on a microscope stage and is bombarded by electrons from below, causing certain parts of minerals to luminesce. The parts that luminesce are those that contain either "activator elements" as trace impurities (commonly transition elements or rare-earth elements) or certain types or amounts of crystal defects. In the case of quartz, detrital grains nearly always luminesce, but secondary growths do not (see Figure 6–7). Presumably, this difference in response to electron bombardment can be related to the temperatures of formation of the quartz: above 300°C for the detrital grains but less than 150°C for the secondary growths.

Chemical Considerations

The solution chemistry of silica is relatively simple. Only two solid forms need to be considered for most purposes: amorphous silica (opal) and the crystalline equivalent, quartz. The solubility of amorphous silica is 100–140 ppm, depending on impurities; quartz has a constant composition and hence solubility, 6 ppm. The chief control of silica solubility in underground waters is temperature. The solubility of both the amorphous and the crystalline solid is increased significantly as temperature increases. Based on the graph in Figure 6–8, we can make several inferences:

1. Amorphous silica is much more soluble than quartz at all temperatures and, consequently, depths of burial.

2. The solubility of amorphous silica is increased greatly by even small increases in temperature, by only shallow burial. Thus, for practical considerations, it is impossible to precipitate opal as a cement at depth. There is no source of silica adequate to maintain supersaturation with respect to opal when values of several hundred parts per million are required simply to saturate the solution. (Normal surface waters—streams—average only 13 ppm.) Opal cement in a sandstone implies cementation within a few tens of meters of the surface.

3. Because both surface waters and shallow underground waters contain less than a few tens of parts per million of silica in solution, an extraordinary source of silica is required even near the surface to generate a solution containing more than 120 ppm. The source of this needed silica is revealed by examination of opal-cemented sandstones. In every case, the sandstone contains altered volcanic fragments such as basalt or glass shards (fragments). The alteration product is montmo-

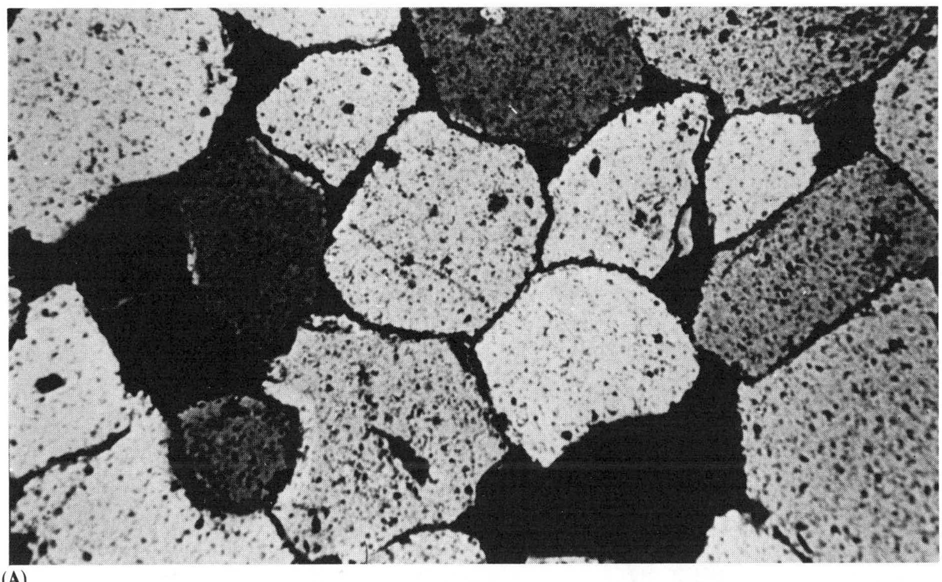

(A)

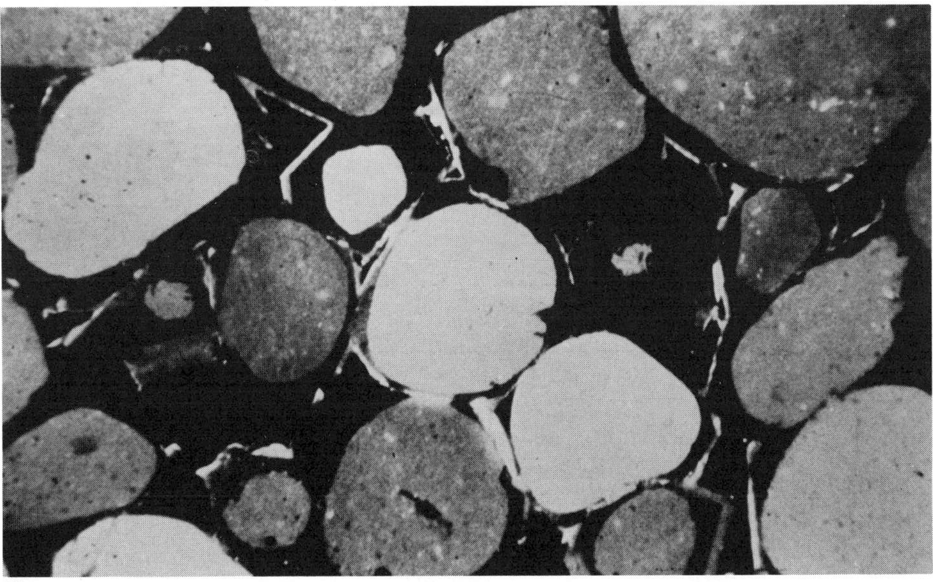

(B)

Figure 6–7
Photomicrographs of medium sand-size quartz grains in Hoing Sandstone (Devonian), Illinois. (**A**) Crossed nichols. (**B**) Cathodoluminescence. Apparent pressure-solution contacts are clearly seen in luminescence petrography to be formed of secondary quartz abutting in original pore space. [Scholle, 1979. Photos courtesy R. F. Sippel.]

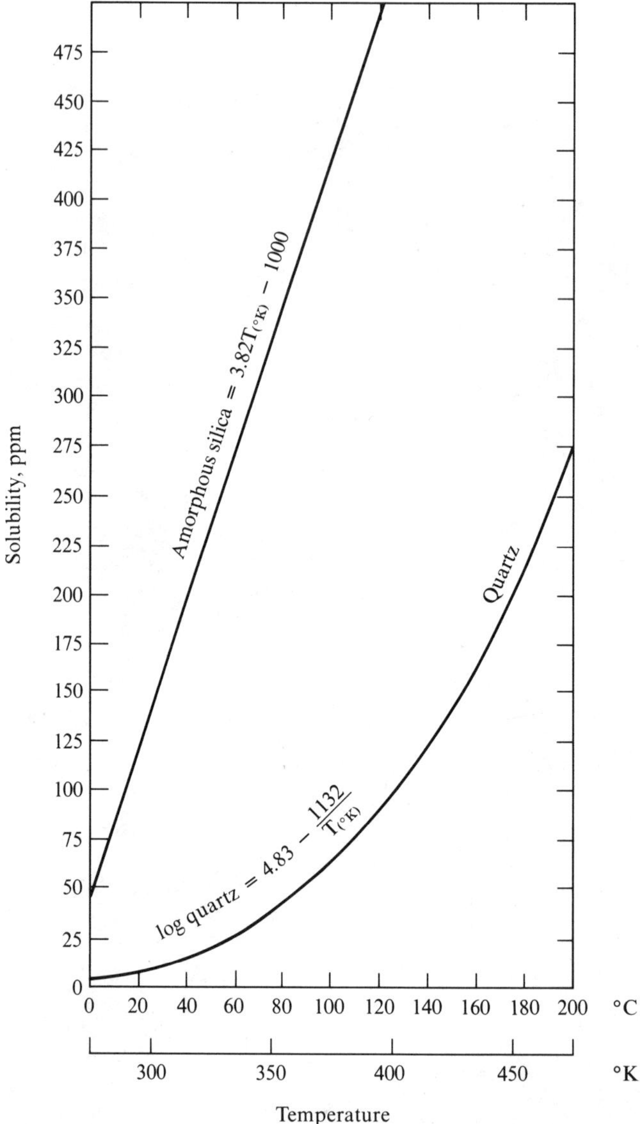

Figure 6–8
Solubilities of amorphous silica and quartz as function of temperature. [H. Blatt et al., 1980, *Origin of Sedimentary Rocks,* 2nd ed. Copyright © 1980. Reprinted by permission of Prentice-Hall, Inc., Englewood Cliffs, New Jersey.]

rillonite clay. Calculations reveal that the formation of the clay from the volcanic fragments does not require all the silica present in the fragments; some is released to the pore waters to increase the amount of silica in solution above the saturation level with respect to opal. The opal precipitates, filling the pore spaces and lithifying the sandstone.

4. The solubility of quartz rises much more slowly than that of amorphous silica, so that saturation of pore solutions with respect to quartz may be very common at temperatures less than 100–150°C. Many sedimentary petrologists believe that, on a world average, the content of silica in subsurface pore waters is determined largely by the depth (temperature) at which the water sample is taken. This is equivalent to saying that the solubility of quartz is the major buffer for the silica content of subsurface waters. However, data are not yet adequate to verify this hypothesis.

The solubility of silica is increased by increased hydrogen-ion concentration (pH). However, the effect is insignificant at pH values below about 9 for amorphous silica and 10 for quartz. It is unlikely that the precipitation of quartz cement is caused by a change in pH of a subsurface water.

Calcite

Field Observations

Coarsely crystalline calcite cement is common in sandstones deposited in all tectonic settings, sandstone compositions, and depositional environments. At the outcrop, it is easily identified by its immediate reaction with dilute hydrochloric acid. Carbonate-cemented sandstones may grade laterally into sandy limestones and pure limestones—an occurrence suggesting that original environmental conditions rather than diagenetic conditions are responsible for the cement.

The cementation of sandstones requires initial depositional permeability to permit the migration of saturated waters through the sand, and the cementation pattern at the outcrop can reflect this fact. Sometimes the scale of variation in original permeability is large, as where a well-sorted and clay-free fluvial sandbar is both overlain and underlain by muddy overbank sands. Sometimes the scale of variation is very small, on the order of a centimeter or less (see Figure 6–9), and the explanation is obscure. There is no apparent difference in the laminae other than the fact that one is cemented and the other is friable or unconsolidated.

Calcite cement can also form *concretions*, locally cemented areas within a more friable layer of sandstone. The concretions can occur in a wide range of shapes ranging from spherical to very irregular, probably in response to permeability variations within the sandy unit. The longest dimension of irregularly shaped concretions is nearly always in the plane of the bedding (see Figure 6–10), the surface of greatest permeability. The irregular shapes of concretions are evidence of the intricacies of permeability variation within sands and sandstones.

The relationship between sedimentary structures in the concretion and those in the enclosing sandstone can reveal the time of formation of the concretion. If laminations within the sandstone pass undeflected through the concretion, the concretion must have formed after the sand was compacted. If laminae curve around the concretion (see Figure 6–11), it must have formed very soon after deposition of the sand, probably before it was buried more than a few meters.

Figure 6–9
Outcrop face cross section of partially lithified modern desert sand dune near Ashkelon, Israel. Some cross bedding is defined by differential cementation to left of lower part of meter stick. Lithified laminae contain 40–55% $CaCO_3$; uncemented laminae, 4–10%. [D. H. Yaalon, 1967, *Jour. Sed. Petrology, 37*.]

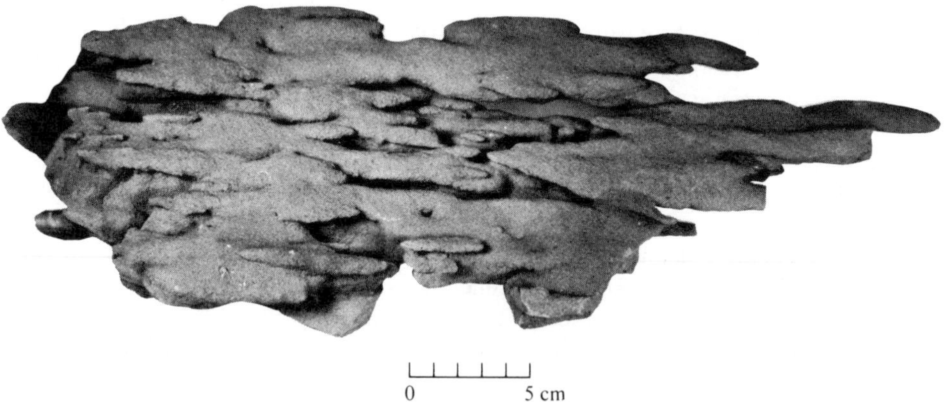

Figure 6–10
Concretion of hematite from Garber Sandstone (Permian), central Oklahoma, showing control of shape by permeability parallel to bedding in the sandstone.

Figure 6–11
Concretions of dolomite in Monterey Formation (Miocene), California. Bedding passes without deflection through middle of concretion, but enclosing beds bulge around it, indicating that concretion formed while enclosing sediments were still unlithified. [D. L. Durham, U.S. Geol. Surv. Prof. Paper No. 819, 1974. Photo courtesy D. L. Durham.]

Calcite is much more soluble than quartz, hematite, or clay minerals, so that its distribution as a cement in sandstone beds tends to be patchy. The sandstone may grade within a few meters laterally from well-lithified to friable for no apparent reason. Possibly, the water from which the calcite precipitated simply became depleted in calcium and/or carbonate ions as it moved through the sand layer. It is also possible, however, that the calcite cement is being dissolved by acidic rain and soil waters at the present outcrop. Such an occurrence was documented in the Bethel Sandstone in southern Indiana and northwestern Kentucky.

The Bethel is a sinuous, Mississippian, shallow-water channel sand (see Figure 6–12) composed of well-sorted, fine-grained quartzarenite. The channel is 225 km long, 800–1,300 m wide, and as much as 80 m thick. In the subsurface, cementation is well developed and calcite forms nearly 30% of the cement (see Figure 6–13). Other cements are quartz, illite, kaolinite, and pyrite. At the surface the Bethel contains no calcite and only minor amounts of other types of cement. Porosities and permeabilities increase upward as the degree of lithification decreases. Pyrite, present

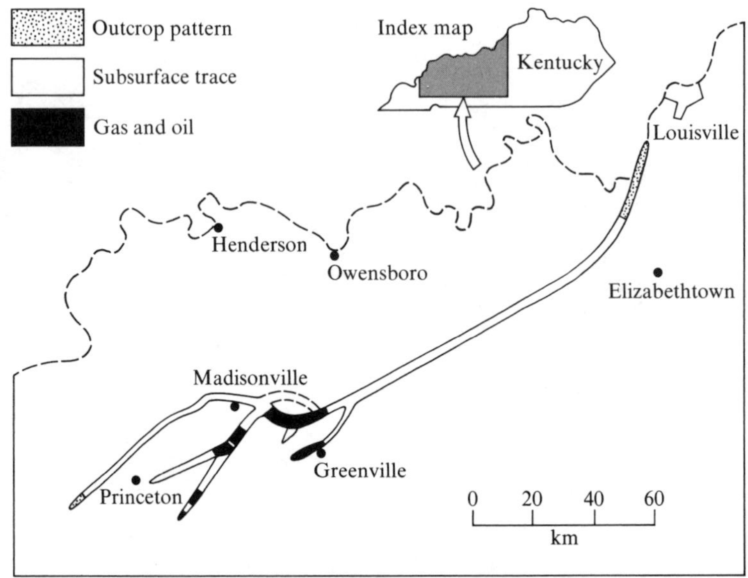

Figure 6–12
Bethel Sandstone distributary system in western Kentucky, showing locations of oil and gas accumulations. Sediment was transported from northeast to southwest. Dashed lines indicate probable channel. [D. W. Reynolds and J. K. Vincent, 1967, Kentucky Geol. Surv. Series X, Spec. Pub. No. 14.]

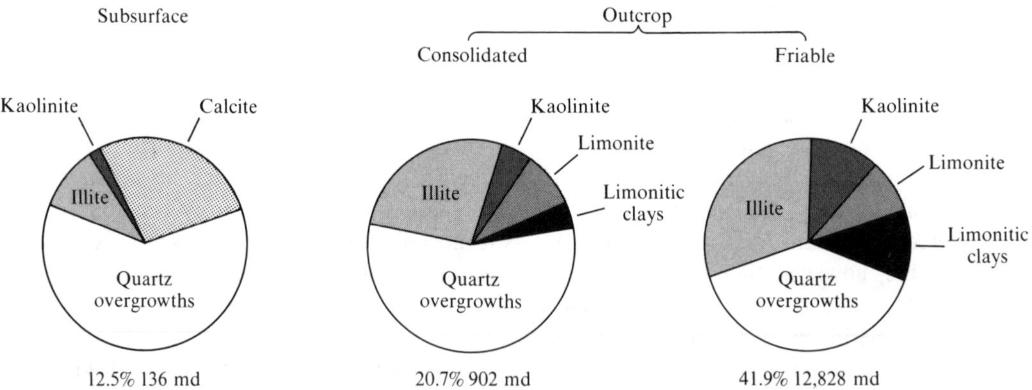

Figure 6–13
Pie diagrams showing relative abundances of cementing agents in Bethel Sandstone in subsurface and in outcrop. Below each pie are corresponding average porosities and permeabilities, which reflect decreasing amounts of total cement toward surface. [Sedimentation Seminar, 1969, Kentucky Geol. Surv. Report of Investigations No. 11.]

as a minor cement in the subsurface, is absent because it was converted to limonite at the surface. Laboratory studies reveal that the illite/kaolinite ratio decreases toward the surface because of the leaching of illite and the growth of kaolinite, and the percentage of feldspar decreases from 2% to 0.2% because of near-surface leaching.

Laboratory Studies

Calcite cement can form and be dissolved more rapidly than quartz cement and, as a result, euhedral growths of calcite are rarely seen in sandstones. The typical morphology in thin section is an anhedral mosaic of cement crystals 10 μm or larger with compromise crystallographic boundaries like those of quartz cement.

In some sandstones it appears that the calcite cement grows into the pore space spasmodically; periods of growth are interrupted by periods of nondeposition. Sometimes the composition of the pore waters is changed somewhat during these hiatuses, and this can be seen by luminescence petrography. As noted earlier, transition elements are activators of luminescence. Iron is a relatively abundant transition element; and if the change in composition of the pore fluid includes an increase in the amount of iron, the iron can be included in trace amounts in the growing calcite cement crystals, causing a change in the luminescence characteristics. The durations of the hiatuses are unknown, as are the causes of the changes in the composition of the pore waters. Perhaps they resulted from structural or topographic changes in the area where water entered the formation; perhaps they were caused by the mixing of different formation waters as a result of faulting in the depositional basin; perhaps they resulted from a change in the content of dissolved oxygen in the pore waters that caused a chemical reduction of ferric iron in an adjacent hematite-cemented unit.

In most sandstones the volume of carbonate cement does not exceed about 30%, the postcompaction porosity of a well-sorted sandstone. In some sandstones, however, the percentage of carbonate can be 50–60%, much greater than any postcompaction porosity in a rock composed only of quartz, feldspar, and lithic fragments. In rocks that contain an apparent "excess" of carbonate cement, several explanations are possible (Dapples, 1971):

1. The rock originally contained carbonate fossils as part of the grain-supported framework, so that the postcompaction porosity was significantly greater than would otherwise have been possible. Nonspherical shells such as clams, brachiopods, or ostracods can provide framework support with very few grains per cubic meter because of the "bridging effect" they provide. Following deposition, the fossils recrystallized without loss of volume, resulting in a sandstone that is now a limestone.

2. Calcite cement was precipitated in the sandstone at very shallow depths, and the pressure exerted by the growing crystals of calcite forced apart the framework grains of quartz. The resulting rock is now a sandy limestone rather than a calcite-cemented sandstone.

3. The calcite that precipitated in the pore spaces after burial has severely etched and dissolved parts of the detrital grains, reducing the percentage of silicate detritus in the rock. In these cases, the silicate grains are seen in thin section as irregularly shaped, fragile grains that could not survive transport and thus could not be detrital.

4. The "sandstone" is actually a limestone that contains a large percentage of wind-blown, fine-grained silicate detritus. The grains must be fine-grained, as wind velocities are normally not strong enough to carry coarser grains in suspension for any considerable distance.

Chemical Considerations

The chemistry of calcium carbonate in water involves a gas: carbon dioxide. Because of this, the solubility relationships are considerably different from those of silica. The relevance of CO_2 is that the solubility of calcium carbonate is severely affected by changes in pH, and the pH of pore waters is in part controlled by the partial pressure of carbon dioxide gas. The essential relationships among CO_2, H_2O, $CaCO_3$, and pH can be summarized by two chemical equations:

$$H_2O + CO_2 \rightleftharpoons H^+ + HCO_3^- \tag{1}$$

and

$$Ca^{2+} + 2HCO_3^- \rightleftharpoons CaCO_3 + H_2O + CO_2 \tag{2}$$

The first equation shows that carbon dioxide dissolved in water generates hydrogen ions and increases the acidity of the water. This causes any calcium carbonate present to dissolve or makes it more difficult to precipitate calcium carbonate if none of this solid is present. If CO_2 escapes from the system, perhaps because of an increase in temperature, the first reaction is forced to the left, hydrogen ions are eliminated, and the water becomes more basic. An increase of one pH unit—say, from 6.5 to 7.5—decreases the solubility of calcite in seawater from 500 ppm to 100 ppm. The second equation shows that the elimination of CO_2 causes more $CaCO_3$ to form because the reaction is forced continually to the right. Thus, both equations show that, other things being constant, it is easier to precipitate calcite cement at higher temperatures (increased depth) than at lower temperatures (shallower depth).

Interrelation of Quartz and Calcite

In thin-section studies of cements in sandstones, it is not uncommon to find shreds of calcite within the quartz cement or shreds of quartz within calcite cement. Clearly, the shreds are partly digested remnants of an earlier episode of cementation. What chemical changes in the pore waters could have caused these replacements? Based on the foregoing consideration of silica and calcite, the obvious possibilities are changes in pH and in temperature, and the interrelationships among the variables are shown in Figures 6–14 and 6–15. Several inferences of importance in considerations of diagenesis can be made from these graphs:

1. Figure 6–14 shows that the critical range of pH in which silica–calcite replacement reactions occur is 7–9—precisely the range of the vast majority of subsurface waters. At a diagenetic temperature of 100°C, a burial depth of 2,000–3,000 m, the point of calcite–quartz intersection is about 7.8.

2. Figure 6–15 shows that at CO_2 pressures likely to be present during diagenesis, the calcite–quartz intersection lies at 40–130°C, a burial depth of 1,500–3,000 m.

A depth of perhaps 2,000 m seems important in diagenesis. As noted in Figure 3–14, the depth range of 2,000–3,500 m is the zone in which most montmorillonitic sheets are converted to illitic sheets in a mixed-layer montmorillonite–illite.

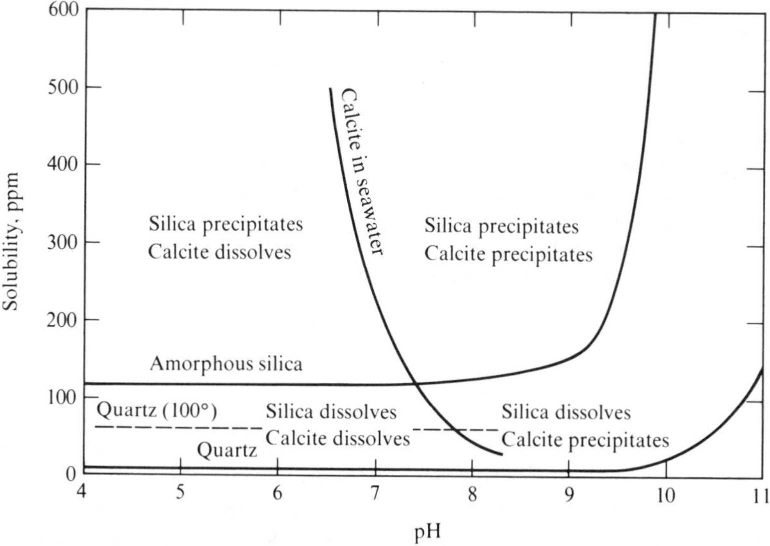

Figure 6–14
Relationships between pH and solubilities of amorphous silica, quartz, and calcite in seawater at 20°C. Intersections of curves divide graph into areas in which water is supersaturated or undersaturated with respect to each of the three solid phases. [H. Blatt et al., 1980, *Origin of Sedimentary Rocks,* 2nd ed. Copyright © 1980. Reprinted by permission of Prentice-Hall, Inc., Englewood Cliffs, New Jersey.]

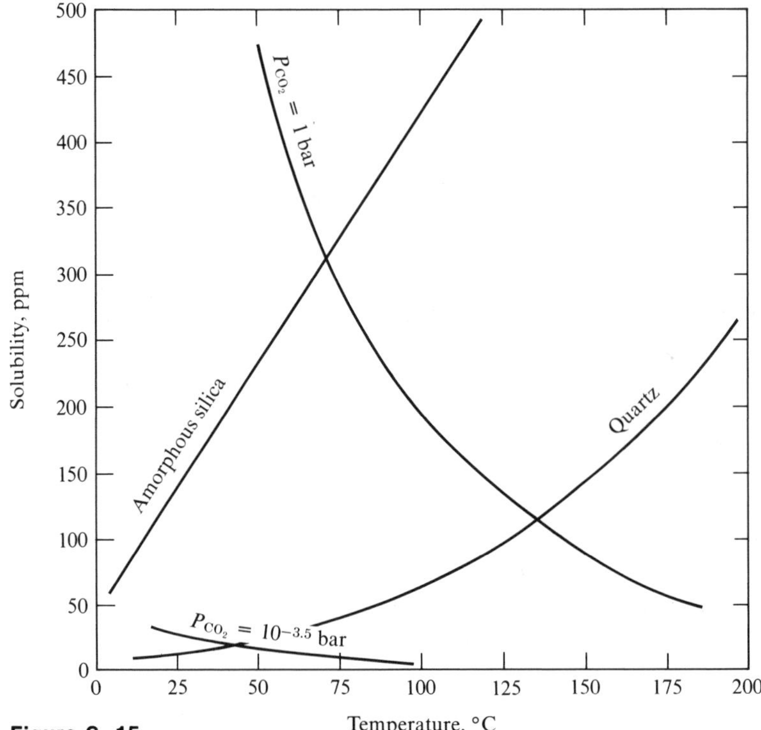

Figure 6–15
Relationships among solubilities of quartz, amorphous silica, and calcite in seawater solution at different partial pressures of carbon dioxide gas and temperature. Silica solubility is unaffected by carbon dioxide pressure. Areas of dissolution and precipitation of each of the three solid phases can be designated as in Figure 6–14, depending on assumed partial pressure of carbon dioxide in subsurface water. Lower carbon dioxide pressure is pressure in seawater in equilibrium with atmosphere. Note that curve intersections of quartz with calcite at likely CO_2 pressures occur in temperature range of normal diagenesis, 40–130°C.

Hematite

Field Observations

Hematite cement in sandstones is identified easily in the field by its red color, which is present with only 1% or less Fe_2O_3. Hematite is an intense pigmenting material. The iron atoms in hematite are derived mostly from weathering and leaching of the common iron-rich minerals such as magnetite, ilmenite, biotite, and hornblende, less commonly augite and olivine, and secondarily from iron-bearing clay minerals.

In some sedimentary sequences, different intensities of red coloration can be seen, reflecting the fact that most hematite forms after a sediment is deposited; some of these occurrences have been studied in detail by Walker (1967). One of them is located in the hot, arid Sonoran Desert of northeastern Baja California, Mexico (see Figure 6–16A). Mean annual rainfall is only 10 cm/yr at present, and the presence of evaporites and a flora of semiarid character in Pliocene sediments in the area indi-

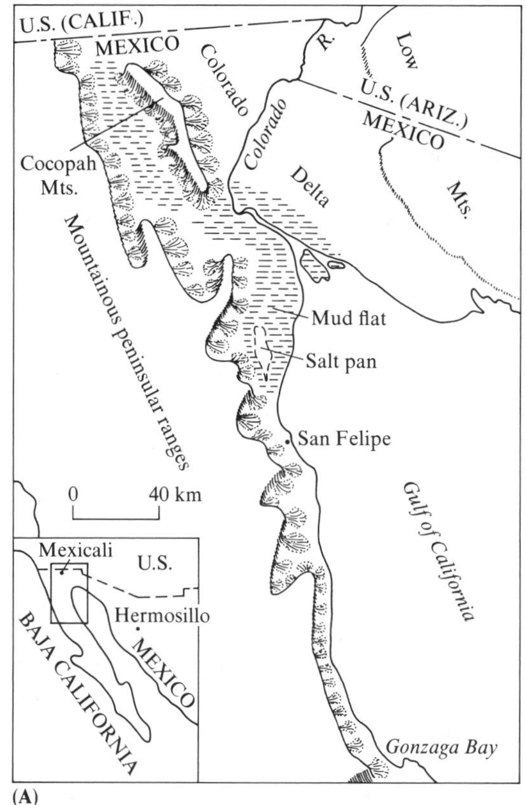

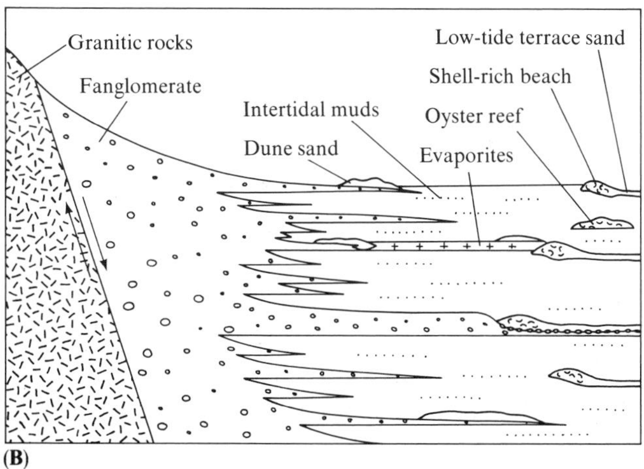

Figure 6–16
Northeastern Baja California, Mexico. (**A**) Map showing topographic and environmental setting in which hematite cement is forming. (**B**) Diagrammatic cross section showing facies relationships of Holocene, Pleistocene, and Pliocene sediments. [Walker, 1967.]

cates no major change for the past 5–10 million years. Throughout this period, alluvial fans derived from the crystalline highlands of the peninsular ranges have coalesced to form extensive bahadas, and these grade eastward into intertidal muds and salt flats (see Figure 6–16B) that border the Gulf of California. Normal faulting parallel to the mountain front has produced numerous alluvial fan cross sections, some more than 300 m in height. It is in these Holocene-Pliocene stratigraphic sections that the origin of the hematite can be seen.

The recent alluvium occurs as a veneer of sheetflood deposits on the surfaces of active fans and is rarely more than 1 m thick. It is composed of gray arkosic detritus, chemically unaltered fragmental granitic debris from the plutons that form the backbone of Baja California. This Holocene alluvium is conspicuously rich in chemically unstable ferromagnesian minerals, particularly hornblende and biotite. Underlying Pleistocene fan sediment has undergone mild chemical alterations that have produced distinctive characteristics: some degree of lithification and a reddish color.

Four major Pleistocene soils can be identified in the alluvium. The younger soils are yellowish but have a distinct reddish hue. Successively older soils become progressively more red. The intense red color of older soils is as red as some of the reddest ancient red beds. Given the evidence of constant climate during the time of deposition of the alluvial fans, and given the observation that the fan sediment changes in color from gray to red with increasing age, it is apparent that the red color has been produced after burial. Ferric oxide has formed during early diagenesis. We can even catch this process "in the act" by careful fieldwork (see Figure 6–17). Apparently, iron-bearing minerals are being leached by the infrequent rains that percolate through the sediment, with the result that the ferrous iron in the minerals is released, oxidized, and dispersed throughout the sediment.

It may be that the red coloration in many of the Upper Paleozoic redbed–evaporite sequences in the western United States has formed in this way (see Figure 6–18). A significant piece of data that supports this hypothesis is the fact that these ancient, intensely red alluvial fan deposits contain no hornblende, despite the fact that their source rocks, the Front Range granites, contain abundant hornblende, as do the modern stream sands that drain the granites.

Laboratory Studies

Hematite pore filling occurs as crystals less than 1 μm in size that cannot be resolved with a petrographic microscope. It appears in thin section as an irresolvable opaque area that is red in reflected light (yellow for limonite) and is typically thin or absent at grain contacts, reflecting its postdepositional origin.

Earlier we discussed the field evidence indicating that hematite cement in sandstones is diagenetic rather than depositional. We noted the reddish halo surrounding concentrations of accessory minerals in partially red sandstones. Detailed further documentation of the diagenetic origin of hematite can be made in the laboratory. Examination of iron-bearing accessory minerals in thin section reveals them to be intensely etched (see Figure 6–19) and reduced in size. Study of the sandstone with

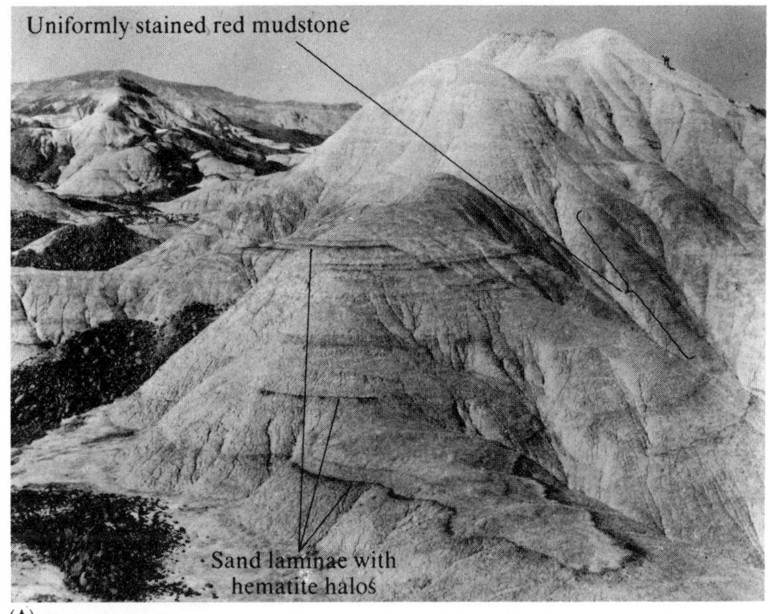

(A)

(B)

Figure 6–17
(**A**) Outcrop of gray mudrock and sandstone at crest of hill, underlain by uniformly red mudstone and sandstone that contains hematite development "caught in the act." (**B**) Close-up of hand specimen, showing red halo spreading downward from zone of concentration of ferromagnesian minerals. [Walker, 1967. Photos courtesy T. R. Walker.]

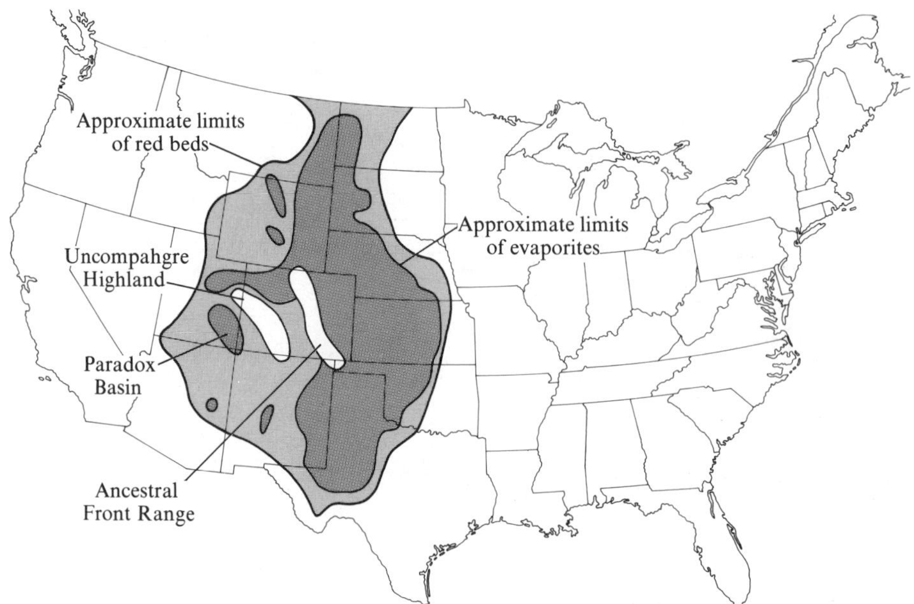

Figure 6–18
Areal distribution of redbeds and evaporites of Late Paleozoic age in western United States. [Walker, 1967.]

the scanning electron microscope reveals beautiful euhedral crystals of hematite growing in pore spaces (see Figure 6–20). In areas of the thin section where montmorillonite clay has formed in place from the altered hornblende grain, the clay is found by the electron microprobe to be iron-rich, with the amount of iron decreasing from more than 20% in the hornblende grain to less than 3% at the outer fringe of the clay as a result of leaching (see Figure 6–21). Hematite can form whenever iron-bearing minerals are present in a sandstone and oxidizing conditions exist.

Chemical Considerations

The chemistry of iron in sedimentary environments is complex because iron occurs at the Earth's surface in two oxidation states, the relative amounts of each determined by the balance between the amount of oxygen gas available to oxidize organic matter

---

Figure 6–19
Photomicrographs of intensely etched and partly dissolved hornblende grain in sandstone, Hayner Ranch redbeds (Miocene), north of Las Cruces, New Mexico. (**A**) Thin section (crossed nicols). (**B**) SEM. [T. R. Walker, 1976, in *The Continental Permian in Central, West, and South Europe* (Dordrecht, Holland: D. Reidel). Photos courtesy T. R. Walker.]

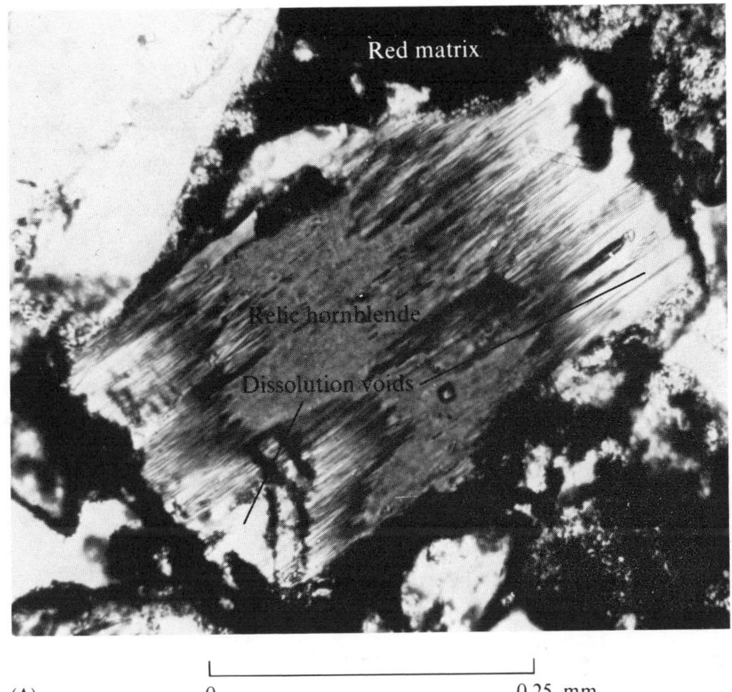
Red matrix
Relic hornblende
Dissolution voids
0
0.25 mm

(A)

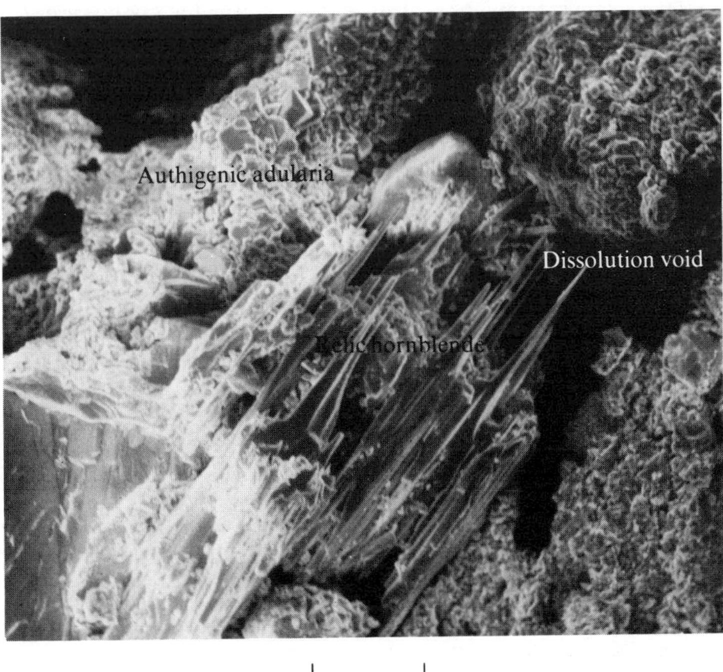
Authigenic adularia
Dissolution void
Relic hornblende
0
50 μm

(B)

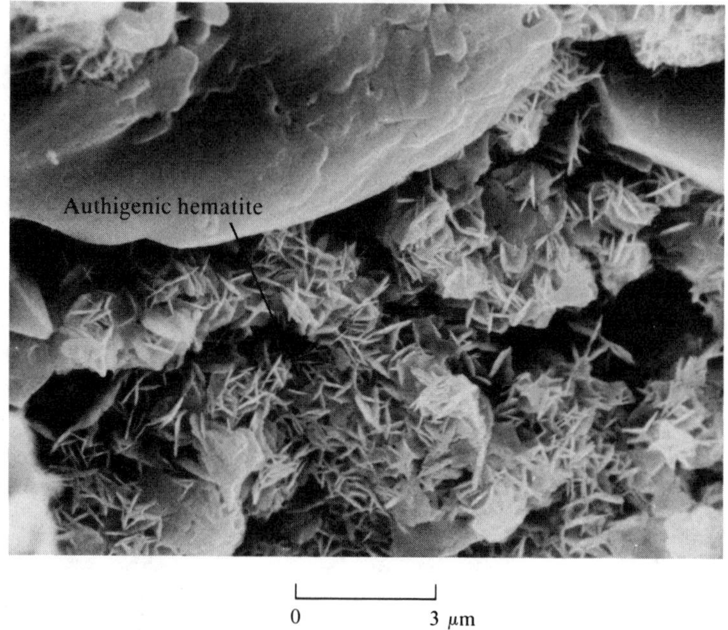

Figure 6–20
SEM photomicrograph of bladed, authigenic hematite crystals grown as clusters of rosettes, Moenkopi Formation (Triassic), Cameron, Arizona. Large grain at top is rounded detrital quartz grain with small overgrowths; larger, clearly euhedral overgrowths project from unseen quartz grains into the void. Rosettes are grown on surfaces of detrital and secondary quartz. [T. R. Walker, 1976, in *The Continental Permian in Central, West, and South Europe* (Dordrecht, Holland: D. Reidel). Photo courtesy T. R. Walker.]

and the amount of organic matter present to be oxidized. If there is a surplus of oxygen, the organic matter will be destroyed completely and the remaining oxygen will keep the environment in an oxidizing condition. A relative deficiency of oxygen gas results in partly decomposed and degraded organic compounds that accumulate in the reducing environment. For materials that are immersed in water, at and near the Earth's surface, water that contains dissolved oxygen must circulate from the surface downward to maintain oxidizing conditions; otherwise, the supply of oxygen below the water–sediment interface will be quickly exhausted.

Figure 6–22 shows the relationship between the oxidizing and reducing tendencies of an environment and the stability fields of the simple ferrous and ferric compounds in the iron–water chemical system. The ferrous compounds are very soluble and are never found as precipitated solids. The ferric compounds are extremely insoluble and, once precipitated, are difficult to reduce to their ferrous equivalents, even within the stability field of the ferrous materials. The location of groundwaters in terms of Eh and pH includes parts of both the ferrous and the ferric area, and is determined for

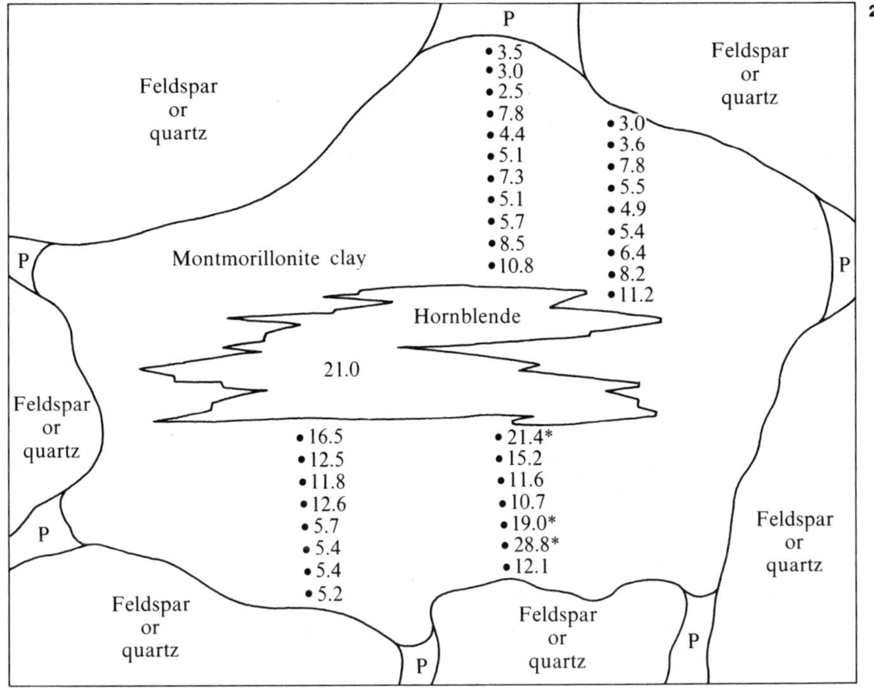

Figure 6–21
Results of electron-microprobe analysis of spots 4 μm in diameter in and around altered hornblende grain 0.4 mm in length in arkosic redbed in alluvial fan of Pliocene age, Sonoran Desert, northeastern Baja California, Mexico. Numbers are percentages of Fe_2O_3; high values within clay (asterisks) are local areas of hematite stain or microislands of hornblende. P = pore. [Walker et al., 1967.]

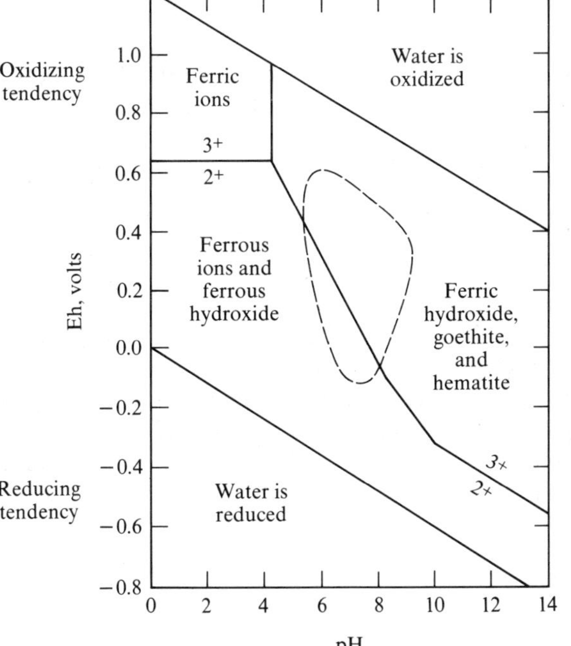

Figure 6–22
Stability fields of amorphous $Fe(OH)_3$, brown goethite, and red hematite in relation to Eh and pH at 25°C and activity of iron in solution that is typical of surface waters. Oval area in graph shows Eh–pH range of shallow groundwaters.

any specific groundwater by the balance between oxygen gas and organic matter discussed previously. Most near-surface environments are oxidizing; most organic matter in nature is destroyed, and the red color caused by the precipitation of hematite is very common in the sedimentary environment.

Clay Minerals

Field Observations

Individual clay flakes are, of course, too small to be seen with a hand lens, but sometimes it is possible to detect the presence of clay in a hand specimen. If present in amounts of at least 5–10%, clay minerals coat the sand grains to the extent that the boundaries of the grains appear fuzzy or indistinct. Even where clay is detected, however, it is not possible in the field to determine whether the clay is detrital or secondary. In either instance, it can serve as a binding agent.

The best way to recognize the presence of clay in a hand specimen is by inferring its presence from the absence of other cementing agents in a lithified sandstone. Quartz cement is prominent only in pure quartz sandstones, and it is easy to see in such rocks with a hand lens. Calcite cement is identified by its reaction with acid. (Break off a cubic centimeter of rock and drop it into a small porcelain dish with the acid. The extent of disaggregation reveals the importance of calcite cement.)

Iron-oxide cement is identified by its red, brown, or yellow (limonite) color. Most hematite-cemented sandstones also contain some clay, but laboratory studies are required to be sure. The reason for the association between hematite and clay is twofold:

1. If the hematite forms during early diagenesis by alteration of hornblende, biotite, etc., montmorillonitic clay is commonly generated, as noted earlier.

2. If the hematite entered the depositional environment as a coating on other particles, such as organic matter or clay minerals, it entered as particles hydraulically equivalent to clay, so that no separation during deposition was possible between the hematite and its fellow travelers.

Laboratory Studies

Laboratory studies of pore-filling clay cements are typically also concerned with the problem of *matrix*. The matrix of a detrital sediment is defined as the finer-grained material in the rock. In a conglomerate the sand grains are the matrix; in a sandstone the matrix is dominantly clay minerals. Sandstones that contain matrix normally have two distinct peaks in their grain-size distribution: one in the sand size, the second in the clay size.

In sandstones whose detrital sand grains are all quartz, any clay minerals that grow during diagenesis can be recognized and identified easily either in thin section or with the scanning electron microscope because of their euhedral shape (see Figure 6–23). This is also true for most arkoses. In lithic sandstones, however, the origin of clay matrix is less clear, and in many rocks we cannot be certain whether the clay is detrital, secondary, or perhaps simply crushed micaceous lithic fragments. The problem is particularly severe in deeply buried sandstones deposited in basins at convergent plate margins, such as along the western coast of North America and in southern Europe. In thin section we find an irresolvable clay mush that not only fills pore space completely but seems to grade imperceptibly into the detrital grains themselves. Much of the clay appears to have formed from chemical reaction among detrital volcanic fragments, phyllosilicate fragments, and the pore waters; but some of the mush may be detrital. Such matrix-rich sandstones are characteristic of convergent plate margins, although not restricted to them.

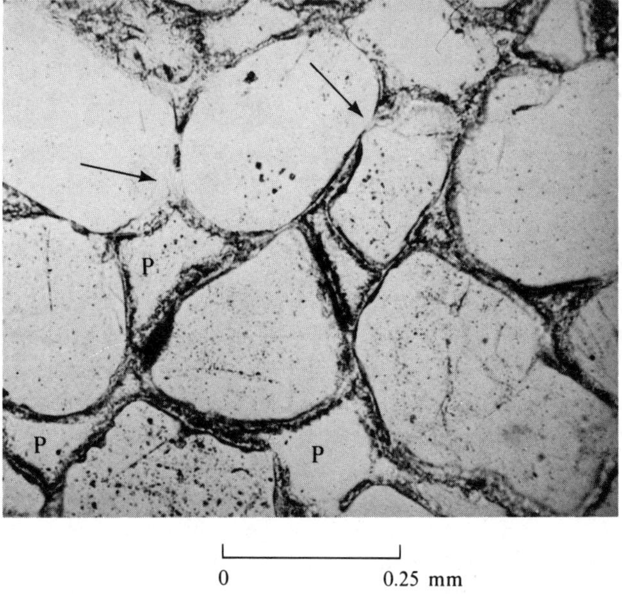

Figure 6–23
Thin-section photomicrograph of Spiro Sand (Pennsylvanian), Oklahoma, showing authigenic coatings of chlorite fibers or plates growing normal to surfaces of detrital quartz grains. Coatings are thin or absent at grain-to-grain contacts (arrows). Abundant porosity (P) remains in this gas-producing unit. [E. D. Pittman and D. N. Lumsden, 1968, *Jour. Sed. Petrology, 38.*]

Chemical Considerations

In most sandstones the volume of authigenic clay minerals is small, certainly less than 5% of the rock volume. The prominent exception is volcaniclastic sandstones (see below) in which secondary clays can form the bulk of the rock. Clay minerals can be potassic (illite and its polymorphs), sodic and calcic (montmorillonite and its polymorphs), or devoid of alkali and alkaline earth cations (kaolinite and its polymorphs). The chemical requirements for the growth of illite and/or montmorillonite from subsurface waters are at least 0.1 ppm dissolved alumina $[Al(OH)_4^-]$, at least 1 ppm of dissolved silica (H_4SiO_4), a value greater than 13 for the ratio log $(Na^+/H^+)^2$, and/or a value greater than 20 for the ratio log $(K^+/H^+)^2$. If the sodium is deficient, montmorillonite cannot form; if potassium is deficient, illite cannot form. If both sodium and potassium are deficient, kaolinite will form. If either alumina or silica is deficient, no clay mineral can form.

The sources of the required cations are varied. Suggested sources include dissolution of orthoclase and muscovite for the potassium, plagioclase for the sodium, and any available detrital aluminosilicate mineral for the silica and alumina. The detrital minerals that are destroyed may be within the bed in which the secondary growth occurs or in adjacent units. If the light-mineral detrital fraction of the rock consists only of quartz and chert, the accessory heavy minerals can provide the cations.

INTRASTRATAL SOLUTION

In our discussion of weathering (Chapter 2), we noted that the alteration of mineral grains in igneous and metamorphic rocks was not restricted to the air–rock interface. It also occurs below sea level at continental margins and on the seafloor. The only requirement for alteration to occur is the presence of unstable minerals and water. It comes as no surprise, therefore, to find that water–sediment interaction continues after burial and can result in the partial or complete destruction of certain mineral species. Most susceptible are those formed at very high temperatures and with low silicon/oxygen ratios, such as olivine, hypersthene, or augite. These minerals are extremely rare in pre-Tertiary sandstones. Preferential loss of some types of minerals results in the relative enrichment of others in the residue, such as zircon, tourmaline, and rutile. Paleozoic sandstones characteristically have high proportions of ZTR in their nonopaque accessory-mineral suite.

TEMPERATURE VARIATION

Experiments reveal that increased temperature is generally more important than increased pressure as a factor in lithification and secondary mineral growth. There are several reasons for this:

1. The range of pressures associated with diagenesis (as contrasted with metamorphism) is insufficient to deform rigid materials such as quartz, feldspar, and most nonfoliated rock fragments at room temperature. Increased temperature decreases the rigidity of materials, so that they can deform plastically.

2. Most materials are more soluble at higher temperatures. For example, the solubility of opal is 120 ppm at 20°C and 500 ppm at 100°C.

3. Some chemical reactions will not occur until an energy barrier has been overcome—the activation energy. Increased temperature can supply the energy needed to overcome this barrier.

The increase in temperature with burial depth is a consequence of the transfer of thermal energy from the interior of the Earth to the surface, where it is dissipated. Different geothermal gradients are observed, depending on the overall thermal conductivity of the rocks, regional heat flow, and subsurface water movement. The world average gradient is approximately 25°C/km, but gradients as low as 5°C/km and as high as 90°C/km have been measured. The lower values occur on the stable cratons; the higher values, on the continental side of convergent oceanic plate–continental plate margins. For example, in the Central Tertiary Basin of Sumatra the gradient averages 68°C/km.

The boundary between diagenesis and metamorphism is conventionally taken as 150–200°C, but the changes that we associate with metamorphism, such as the development of foliation or the recrystallization of quartz, occur over a range of temperatures. The changes are gradational, as are most natural processes. No specific temperature is adequate as the point at which diagenesis becomes metamorphism. Based on the range in geothermal gradients that have been measured, "sedimentary rocks" may change into "metamorphic rocks" at depths of 3–40 km, with an average depth for the change at perhaps 15 km at divergent plate margins and 5 km at convergent margins.

VOLCANICLASTIC SEDIMENTS

As noted in Chapter 5, sandstones deposited in basins adjacent to the convergence of an oceanic plate and a continental plate are rich in volcanic rock fragments, particularly basalt and its glassy equivalents. Furthermore, these sands are deposited in a zone of exceptionally high heat flow and access to saline waters rich in alkaline earth elements and alkalis, particularly sodium. As a result of these circumstances, the sands suffer intense diagenesis at depths of only a few thousand meters.

Types of Alterations

As revealed by many studies of Tertiary sands in basins rimming the Pacific Ocean and in similar settings in older rock sequences, several types of alterations of the detrital fragments are particularly common: (1) crystallization of zeolites, (2) albitization of plagioclase. (3) replacement of plagioclase by calcite, and (4) formation of clay matrix.

Crystallization of Zeolites

Zeolites are hydrous minerals of low specific gravity and, because of this, are relatively sensitive to changes in temperature and pressure. In most Tertiary volcaniclastic rocks around the Pacific Ocean basin, the common zeolite minerals are analcite ($NaAlSi_2O_6 \cdot H_2O$), heulandite [$(Ca, Na_2)Al_2Si_7O_{18} \cdot 6H_2O$], and laumontite ($CaAl_2Si_4O_{12} \cdot 4H_2O$), reflecting the dominance of sodium in seawater and of calcium in the basaltic fragments from which the zeolites form. Perhaps 10 other zeolite minerals have been reported as products in these rocks; detailed descriptions of facies relationships among the various zeolite minerals are given by Coombs et al. (1959), and a thorough field study of zeolite facies alteration of volcaniclastic sandstones in some rocks in New Zealand has been made by Boles (1977).

The zeolite facies can be considered as either high-grade diagenesis or low-grade metamorphism. Its occurrence is confined to sediments rich in volcanic detritus and lacking carbonate rocks. Typical reactions of the facies include the reaction of plagioclase (the anorthite molecule) to heulandite,

$$\underset{\text{anorthite}}{CaAl_2Si_2O_8} + \underset{\text{quartz}}{5SiO_2} + 6H_2O \rightarrow \underset{\text{heulandite}}{CaAl_2Si_7O_{18} \cdot 6H_2O}$$

the reaction of heulandite to laumontite,

$$\underset{\text{heulandite}}{CaAl_2Si_7O_{18} \cdot 6H_2O} \rightarrow \underset{\text{lamontite}}{CaAl_2Si_4O_{12} \cdot 4H_2O} + \underset{\text{quartz}}{3SiO_2} + 2H_2O$$

and the direct conversion of plagioclase to laumontite,

$$\underset{\text{anorthite}}{CaAl_2Si_2O_8} + \underset{\text{quartz}}{2SiO_2} + 4H_2O \rightarrow \underset{\text{lamontite}}{CaAl_2Si_4O_{12} \cdot 4H_2O}$$

Albitization of Plagioclase

The typical feldspar in basalt is labradorite that contains about 12% CaO and 4% Na_2O by weight; and when the plagioclase reacts to produce zeolites, the sodium is released. Abundant sodium ion is also present in the marine pore waters involved in the reactions; seawater contains more than 10,000 ppm Na^+. The result of cooking calcic plagioclase in this sodium-rich broth can be albitization of the detrital feldspar. Typically, the resulting feldspar is 95–98% albite molecule. For example, the Tanner graywacke (Upper Devonian–Lower Mississippian) in the Harz Mountains of northern West Germany contains 30–40% feldspar, of which 85–90% has a composition Ab_{90-97}.

Replacement of Plagioclase by Calcite

Another common characteristic of the diagenesis of volcaniclastic sandstones is the replacement of plagioclase by calcite, with clay matrix produced as a byproduct:

$$\text{anorthite} + CO_2 + H_2O \rightarrow \text{calcite} + \text{clay matrix}$$

Formation of Clay Matrix

Ancient volcaniclastic sandstones nearly always contain abundant clay matrix, sometimes to the extent that the rock might properly be called a volcaniclastic mudrock rather than a sandstone. The matrix material consists of chlorite, sericite, and quartz in older rocks, zeolites and montmorillonite in younger ones. Boundaries between recognizable detrital grains and the irresolvable clay paste are typically fuzzy and indistinct—a textural transition indicating a mineralogical transition between the two materials. It is impossible to make either an accurate or a precise point count in thin sections of such rocks. Four categories of matrix have been defined by Dickinson (1970): (1) protomatrix, being trapped detrital clay; (2) orthomatrix, recrystallized material; (3) epimatrix, a diagenetic product of the alteration of sand-size grains; (4) pseudomatrix, deformed and squashed lithic fragments. Dickinson discusses the criteria for the recognition of each type, but the character of the variables limits the accuracy of these distinctions.

COMMENT ON DIAGENESIS IN SANDSTONES

The diagenetic processes in sandstones are compaction and the formation of new minerals, either as pore-filling cements or as replacements for the original detrital fragments. The effectiveness of compaction depends on the ductility of the detrital grains and on ductility produced as part of the chemical diagenesis of the rock. Originally ductile fragments include shales and foliated metamorphic grains; secondary ductility can be induced in originally rigid grains by the transformation of these grains into clay matrix.

Pure quartz sands are lithified either by precipitation of secondary quartz, calcite, or other minerals in the pores or by pressure solution. Commonly, special laboratory techniques are required to distinguish between quartz cement and the effects of pressure solution of detrital quartz. Arkoses are rich in feldspar, usually orthoclase and microcline, and alter during diagenesis to illite and kaolinite. Alternatively, the feldspar grains can be dissolved and leave no clay residue. Cementing agents in arkoses are calcite, hematite, and detrital plus authigenic clays.

Lithic sandstones have a complex diagenetic history because of the great variety of fragment types they contain. Because of the combined effects of compaction and chemical alteration, it is commonly not possible to distinguish between detrital grains and primary or secondary matrix in these rocks. The problem is particularly severe in volcaniclastic sandstones deposited near convergent plate margins because of the unstable nature of the fragments and the exceptionally high heat flow in such areas.

Adequate evaluation of diagenetic history in volcaniclastic sandstones depends in part on experience and in part on good guesses.

DIAGENETIC INVESTIGATIONS

Most studies of sandstone diagenesis are made as a part of more comprehensive investigations aimed at unraveling the history of the unit. Because of this, diagenetic studies vary greatly in scope and detail. This section describes two studies to illustrate approaches that have been used.

Burial Depth and Permeability

Ventura Oil Field, California

Permeability of reservoir beds in oil fields generally decreases in progressively deeper producing zones. But the decrease may be due to several factors, such as finer grain size, poorer sorting, increased compaction, or increased precipitation of chemical cements. Each of these factors carries different implications with regard to the prospects for future oil production. For example, grain size and sorting are controlled largely by depositional environment, such as position on a delta or submarine fan. It is possible that completing a well a few hundred meters to one side of an existing hole might result in improved sorting and better production. If the decrease in permeability is due to increased compaction of the sand because of increased squashing of ductile rock fragments, we can probably forget about the possibility of significant production at greater depth but might consider broadening the scope of our exploration at shallower depths. If the decrease in permeability results from precipitation of cements in pore spaces, we need to consider the type of cement that is causing the problem. Can we expect the amount of cement to increase continually with depth until all pores are filled, or is the cement likely to be redissolved as depth increases? In some areas the amount of carbon dioxide increases with depth (as a result of decarbonation reactions, etc.), resulting in the solution of earlier-formed calcite cement and the regeneration of pore space. These are the kinds of questions that can be solved by a well-planned study of diagenesis.

K. J. Hsü (1977), a geologist then employed by the Shell Oil Company, wanted to determine the reasons for the decrease of permeability (and oil-producing potential) with depth in the Ventura Basin oil field in southern California. The Ventura oil field is located on the crest of the Ventura anticline and produces from several horizons in a Pliocene section of sands, silts, and shales more than 3,000 m thick. Regional geology and the sedimentologic character of cores from producing horizons are consistent with the interpretation that the sands were deposited by turbidity currents as linear bodies trending east–northeast in a submarine fan complex, much like the setting off the coast of southern California today (see Figure 6–24). Current indicators within the sands reveal an east–west flow direction.

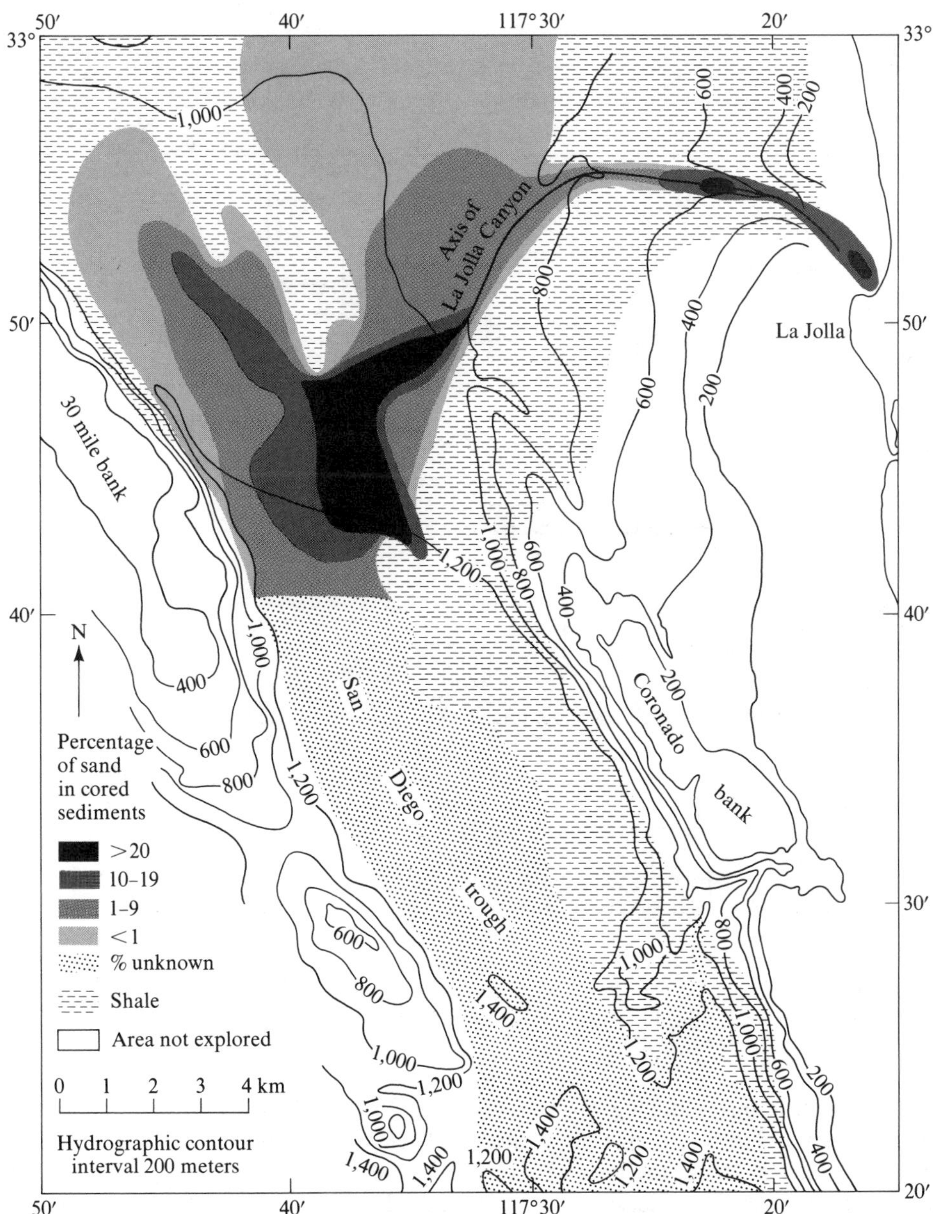

Figure 6–24
Lithologic and bathymetric map of seafloor off La Jolla, southern California, illustrating formation of modern submarine fan by sediment funneling down La Jolla Canyon. [Hsü, 1977.]

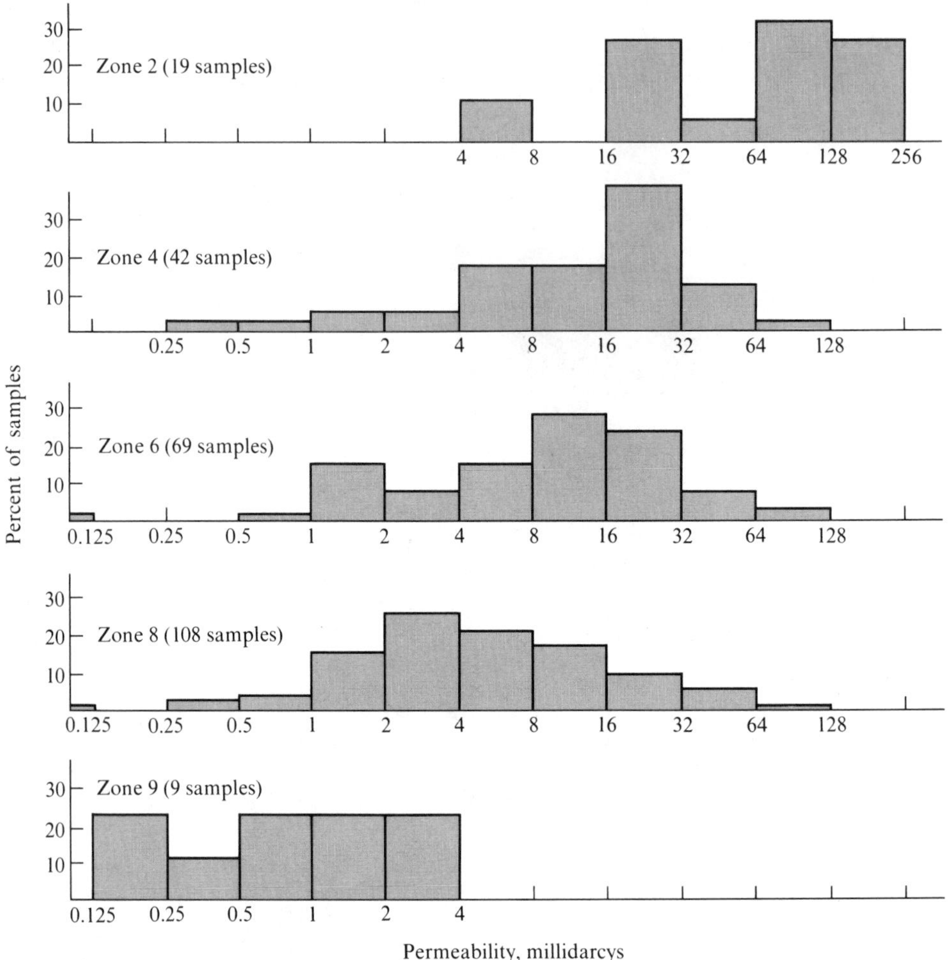

Figure 6–25
Range of permeabilities in sandstones in five of nine oil-production depth zones, Ventura field, California. [Hsü, 1977.]

The Pliocene oil-producing formations of the Ventura field have been subdivided on the basis of depth into nine producing zones of varying thickness. The shallowest, zone 1, is 800 m thick; the deepest, zone 9, is 300 m thick. Figure 6–25 shows the range of permeabilities measured on many samples in cores from five of the zones, and it is clear that a sequential decrease exists with increasing depth. To determine the reason for the decrease, Hsü made large numbers of sieve analyses of disaggregated samples from each zone and examined many thin sections of core samples. The results of these studies revealed that, although there were large differences in grain size and sorting among individual sandstone units, there was no difference among the nine zones. Further, there were no large or systematic differences among the zones in mineral composition, detrital grain shape, or grain angularity. Only one factor differed among the nine zones: the degree of closeness (packing) of the grains as seen in thin section (see Figure 6–26).

Based on studies of modern sediment accumulations, the closeness of the packing of sand grains in these sands at the time of deposition is known. By using this infor-

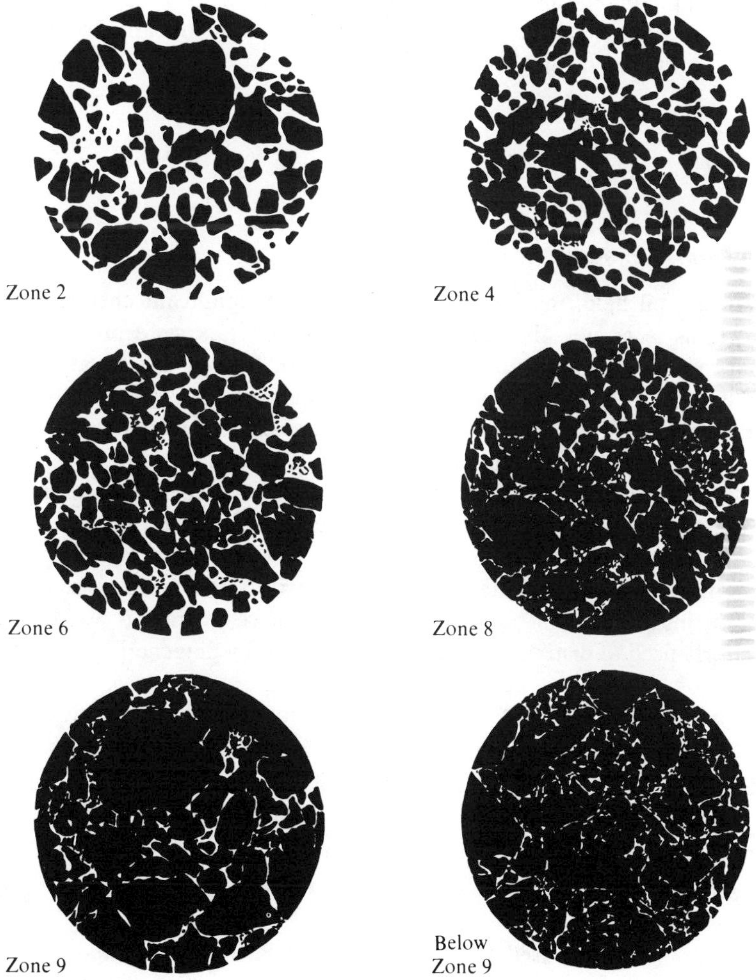

Figure 6–26
Tracings of outlines of framework grains as seen in typical thin sections of sandstones from Ventura oil field, California. Field diameters are 1 mm. Effects of increased depth on porosity and permeability are clearly shown. [Hsü, 1977.]

mation, Hsü was able to quantify the amount of compaction with depth. As grains get closer together, the sizes of the pores decrease, so that it is possible to correlate the measured permeabilities with the degree of compaction. The correlation is excellent in the producing zones of the Ventura oil field. Intense compaction is present in all zones and even in zone 2 (1,000–1,400 m depth), the permeability averages only 4.5% of depositional permeability (which was 1,300 md). Values decrease to 1.2% in zone 4; 0.5% in zone 6; 0.23% in zone 8; and 0.09% in zone 9. On the basis of this decrease, Hsü predicted that permeabilities 200 m below the base of zone 9 would be only 0.02% of depositional permeability, too low for economically useful oil production; i.e., he proposed drilling be stopped.

Chemical Diagenesis

Morrison Formation, New Mexico

In the Ventura oil field, there was little evidence of significant chemical diagenesis. Detrital grains were not dissolved, and no new minerals were precipitated by the fluids that migrated through the sands after burial. In part, the apparent absence of chemical diagenesis is due to the young age of the sandstones, 5–10 million years. In part, it probably results from a lack of significant change in the composition of the migrating pore fluids during this period. In part, it results from the fact that many of the sandstones were impregnated with oil very early in their diagenetic history. If water were not in contact with the grains, there would be no precipitation of new mineral matter in the pores. In older sandstones, however, extensive chemical interaction is common between detrital grains and the waters that have passed through the sediment during the tens or hundreds of millions of years since it was buried.

The criteria used to distinguish materials produced by diagenetic processes include:

1. Delicate external and internal morphologies that preclude sedimentary transport.
2. Spatial relationships of detrital or diagenetic components indicating an origin that postdates deposition or an earlier diagenetic stage.
3. Compositions that differ radically from similar materials of detrital origin.
4. Textures that are unlikely to have been produced by depositional processes.

The primary criterion used to determine the relative timing of diagenetic alterations is the assumption that pores fill inward. If a sequence of materials lines a pore, that material closest to the detrital grains is assumed to be the oldest and the material in the center of the pore is the youngest. It is also assumed that a cement tends to fill or line all types of available pores (with the exception of cements requiring a nucleus of similar structure or composition upon which they can develop). Thus, if cavities in partially dissolved grains are not lined or filled with cementing agents present in the rock, these cavities postdate the cements.

A good example of the complexity of these chemical interactions is provided by Flesch and Wilson (1974) in their investigation of outcrops of the Morrison Formation (Jurassic) in northwestern New Mexico. The study was a comprehensive one, concerned with environments of deposition and provenance of the detrital grains as well as diagenetic aspects, but our focus here is principally on diagenesis.

The Morrison Formation in this area consists of about 55% fine- to coarse-grained white to yellowish sandstone and 45% montmorillonitic claystone of various colors. Interpretation of depositional environments was made using the geometry of the sandstone units, sedimentary structures, and lithologies. Based on these field data, four members were recognized: the first (lowest) and third members are braided stream sands composed of more than 95% sandstone; the second and upper members are dominantly claystone and were interpreted as meandering stream deposits. The detrital mineral composition of the Morrison Formation sandstone is 55–80% quartz, 12–36% feldspar (subequal amounts of K-feldspar and plagioclase), 5–15% lithic fragments (almost all igneous), and 1–2% chert. These framework grains form only 67–77% of the whole rock, the rest being 9–31% clay plus chemical cement and 1–20% pore space.

Many chemical changes have affected the Morrison Formation in the area studied. The most common involve cementation by grain coatings of montmorillonite, interlayered montmorillonite–illite, chlorite, and chalcedony; overgrowths on quartz and feldspar grains; pore fillings of calcite and kaolinite; and dissolution of detrital feldspars. Less common alterations include the diagenetic formation of gibbsite, pyrite, and possibly anatase and the replacement of iron-rich materials by hematite and limonite. Some of the diagenetic features are illustrated in Figure 6–27.

It is clear from these changes that there have been many different types of pore waters passing through the formation during the 160 million years or so since it was deposited by Jurassic streams. For example, secondary growths of feldspar, calcite, montmorillonite, and illite require basic solutions rich in dissolved cations, such as calcium and potassium. The formation of kaolinite and gibbsite and the dissolution of detrital feldspars requires acidic, cation-poor solutions. Replacement of iron-rich minerals by hematite requires oxidizing waters—waters with a high content of dissolved oxygen. Precipitation of pyrite, on the other hand, requires the nearly complete absence of dissolved oxygen in pore waters. The chalcedony cement may have crystallized from an opaline predecessor, and the precipitation of opal from a pore solution requires an unusually high concentration of dissolved silica in pore waters, about 120 ppm. Normal near-surface waters average only 13 ppm. Quartz cement, on the other hand, forms from dilute solutions that contain only 6–10 ppm dissolved silica and crystallizes directly from solution with no opal or chalcedony precursor.

Any attempt at a complete analysis of the origin of all the cements, pore fillings, replacements, and dissolutions would take us deeply into physical chemistry and the general topic of sedimentary geochemistry. We will not pursue that type of analysis.

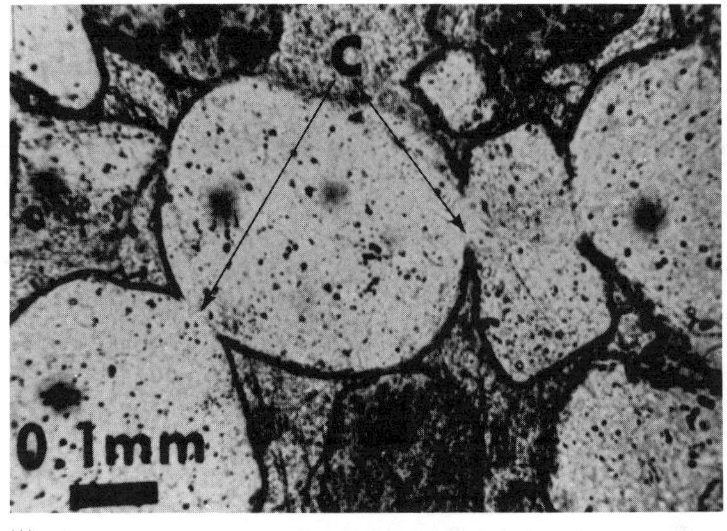

(A)

(B)

0 100 μm

Figure 6–27
Photomicrographs (crossed nicols) of some diagenetic features seen in sandstones of Morrison Formation (Jurassic), Sandoval County, New Mexico. (**A**) Thin section showing clay growths into pore spaces from surfaces of detrital grains. Note absence of clay growth at locations where grains are in contact (C); flow of pore waters was insufficient to form visible precipitate at these sites. (**B**) Coatings of clay (montmorillonite) on grains in part A as seen with scanning electron microscope. Clays are again seen to have grown normal to grain surfaces; honeycomb appearance is produced.

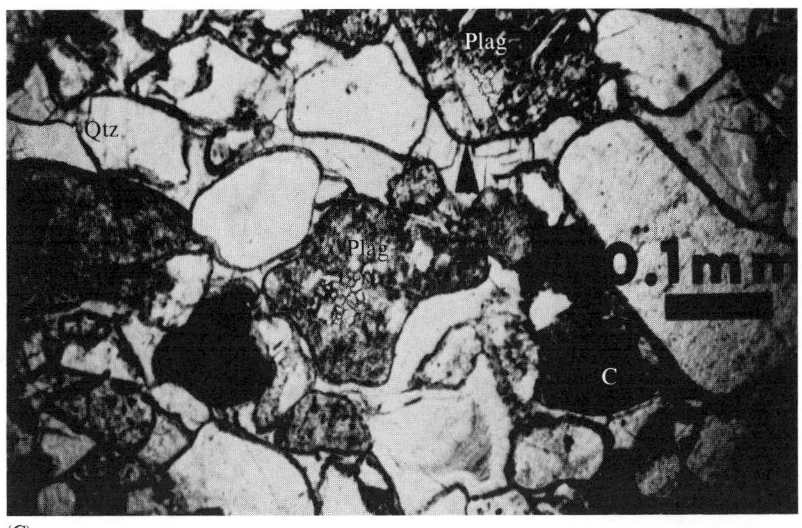

(C)

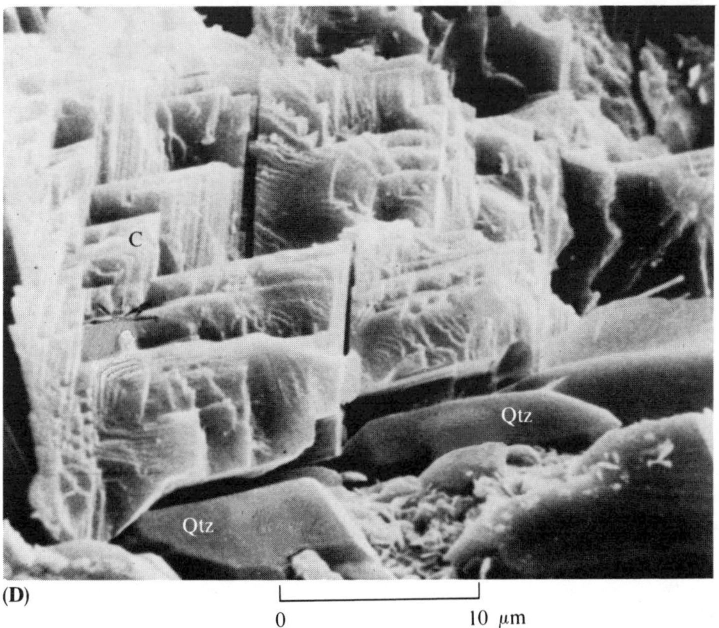

(D)

(**C**) Thin section showing subarkosic sandstone. Qtz = chalcedony cement (light color) fibers grown normal to grain surfaces. Plag = partly altered plagioclase grains. C = a chert grain. (**D**) Scanning electron micrograph showing rhombs of calcite cement (C) covering quartz overgrowths (Qtz), which, in turn, partially coat euhedral flakes of authigenic chlorite (between Qtz and Qtz at lower right). [Flesch and Wilson, 1974. Photos courtesy C. T. Siemers.]

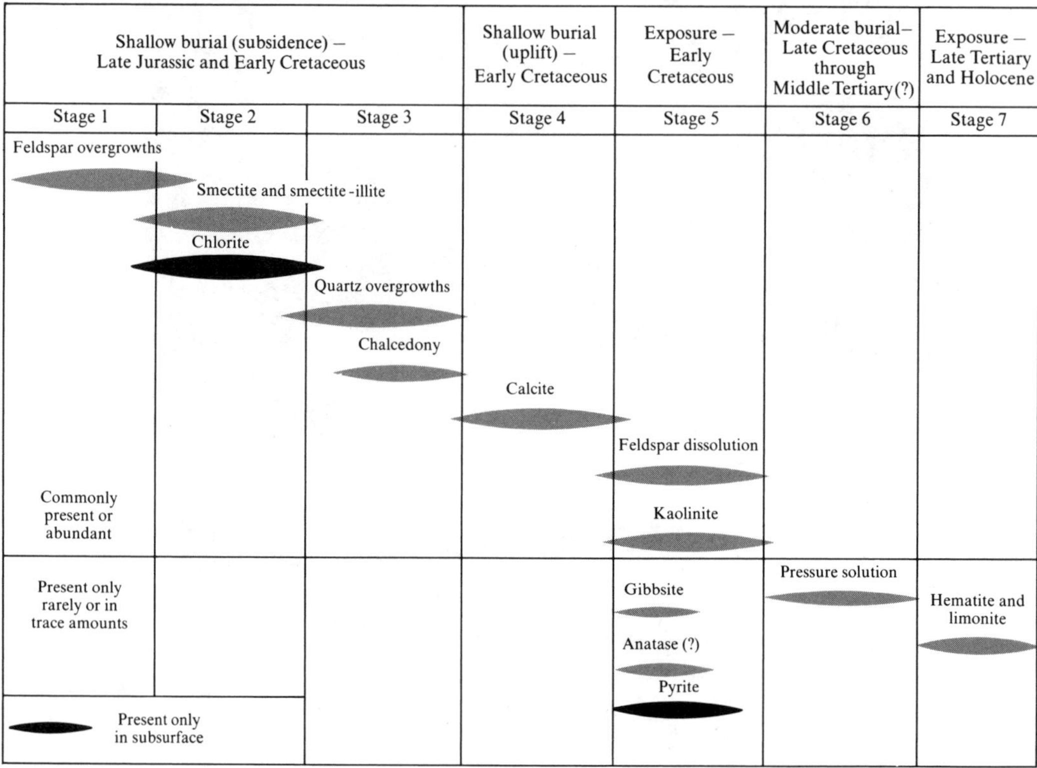

Figure 6–28
Types of chemical alterations and their timing in Morrison Formation sandstones, northwestern New Mexico, based on thin-section and SEM studies. [Flesch and Wilson, 1974.]

We can, however, construct a chronologic sequence of the chemical changes seen in the Morrison, based on textural evidence in thin sections and SEM observations. The result is shown in Figure 6–28, with the approximate timing of events based on structural events known from earlier studies of regional geology.

SUMMARY

Diagenetic processes in detrital rocks include some of the most complex phenomena in geology. They involve both physical stresses and extreme variations in chemical composition of natural waters, and the way in which these variables interact with the mineral and rock particles in a sediment is further compounded by the length of geologic time. Diagenetic studies, however, are rewarding in proportion to their complexity. The effort and background knowledge required to resolve most diagenetic complexities are great, but so are the rewards in the currency of better understanding of the history of the rock.

Compaction is a more effective process in sandstones that contain abundant ductile lithic fragments such as schist, phyllite, slate, shale, mica flakes, or fragments of floodplain and seafloor mud. With increased depth of burial, these fragments are squeezed into adjacent depositional pore spaces, thinning the stratigraphic section and diminishing the porosity and permeability of the sandstone.

The growth of secondary minerals in a sandstone requires that the pore fluid be oversaturated with respect to the mineral being precipitated. It is also necessary for very large amounts of pore fluid to pass through the rock over long periods of geologic time. Quartz and calcite probably form the bulk of authigenic mineral growths in sandstones, but others, such as hematite and clay minerals, can dominate in some sandstones. Most hematite seems to be produced very rapidly in modern semiarid areas by the leaching and oxidation of ferrous iron from ferromagnesian minerals. However, continued destruction of ferromagnesian minerals much later in the geologic history of the rock can cause hematite to be formed throughout the life of a sandstone.

FURTHER READING

Boles, J. R. 1977. Zeolite facies alteration of sandstones in the Southland Syncline, New Zealand. *Amer. Jour. Sci., 277,* 982–1012.

Coombs, D. S., A. J. Ellis, W. F. Fyfe, and A. M. Taylor. 1959. The zeolite facies, with comments on the interpretation of hydrothermal syntheses. *Geochimica et Cosmochimica Acta, 17,* 53–107.

Dapples, E. C. 1971. Physical classification of carbonate cement in quartose sandstones. *Jour. Sed. Petrology, 41,* 196–204.

Dapples, E. C. 1979a. Diagenesis of sandstones. In G. Larsen and G. V. Chilingar (eds.), *Diagenesis in Sediments and Sedimentary Rocks.* New York: Elsevier, pp. 31–97.

Dapples, E. C. 1979b. Silica as an agent in diagenesis. In G. Larsen and G. V. Chilingar (eds.), *Diagenesis in Sediments and Sedimentary Rocks.* New York: Elsevier, pp. 99–141.

Dickinson, W. R. 1970. Interpreting detrital modes of graywacke and arkose. *Jour. Sed. Petrology, 40,* 695–707.

Flesch, G. A., and M. D. Wilson. 1974. Petrography of the Morrison Formation (Jurassic) sandstone of the Ojito Spring Quadrangle, Sandoval County, New Mexico. *N. Mex. Geol. Soc. Guidebook, 25th Field Conf.,* pp. 197–210.

Hack, J. T. 1980. *Rock Control and Tectonism—Their Importance in Shaping the Appalachian Highlands.* U.S. Geol. Surv. Prof. Paper No. 1126-B, 17 pp.

Hsü, K. J. 1977. Studies of the Ventura Field, California, II: lithology, compaction, and permeability of sands. *Amer. Assoc. Petroleum Geol. Bull., 61,* 169–191.

Jour. Geol. Soc. London. 1978, *135,* Part 1, 1–156. An issue composed of 14 articles plus discussions of the state of the art in studies of sandstone diagenesis.

McBride, E. F. 1979. *Diagenesis of Sandstone: Cement–Porosity Relationships*. Soc. Econ. Paleontol. Mineral. Reprint Series No. 9, 233 pp.

Pittman, E. D. 1972. Diagenesis of quartz in sandstones as revealed by scanning electron microscopy. *Jour. Sed. Petrology, 42,* 507–519.

Schluger, P. R. (ed.). *Diagenesis as It Affects Clastic Reservoirs*. Soc. Econ. Paleontol. Mineral. Spec. Pub. No. 26, 443 pp.

Scholle, P. A. 1979. *A Color-Illustrated Guide to Constituents, Textures, Cements, and Porosities of Sandstones and Related Rocks*. Amer. Assoc. Petroleum Geol. Mem. No. 28, 201 pp.

Sedimentation Seminar. 1969. Bethel Sandstone (Mississippian) of western Kentucky and south-central Indiana, a submarine-channel fill. *Ky. Geol. Surv. Report of Investigations* No. 11, pp. 7–24.

Shelton, J. W. 1962. Shale compaction in a section of Cretaceous Dakota Sandstone, northwestern North Dakota. *Jour. Sed. Petrology, 32,* 873–877.

Siever, R. 1979. Plate-tectonic controls on diagenesis. *Jour. Geol., 87,* 127–155.

Sippel, R. F. 1968. Sandstone petrology, evidence from luminescence petrography. *Jour. Sed. Petrology, 38,* 530–554.

Walker, T. R. 1967. Formation of red beds in modern and ancient deserts. *Geol. Soc. Amer. Bull., 78,* 353–368.

Walker, T. R., P. H. Ribbe, and R. M. Honea. 1967. Geochemistry of hornblende alteration in Pliocene red beds, Baja California, Mexico. *Geol. Soc. Amer. Bull., 78,* 1055–1060.

Wilson, M. D., and E. D. Pittman. 1977. Authigenic clays in sandstones: recognition and influence on reservoir properties and paleoenvironmental analysis. *Jour. Sed. Petrology, 47,* 3–31.

7

Limestones

There is an island in the silent sea,
Whose marge the wistful waves lap listlessly—
An isle of rest for those who used to be.

THOMAS SAMUEL JONES, JR.

Carbonate rocks (limestones and dolomites) total 10–15% of the sedimentary column and are nearly always quite pure. Impurities total less than 5% of the rock, and are typically confined to clay minerals, fine sand- and coarse silt-size quartz grains, and very finely granular to powdery quartz of uncertain origin. Because of the essentially monomineralic nature of limestones and dolomites, their study is largely a study of textures and structures, supplemented by geochemistry. Because the minerals calcite and dolomite are very soluble at near-surface conditions, carbonate rocks recrystallize easily and frequently, obliterating many of the diagnostic depositional textures and structures needed to interpret the origin of the rocks. As a result, the study of carbonate rocks is, at least initially, more difficult than the study of sandstones. Experience is even more necessary for proper analysis of limestones and dolomites than it is for sandstones. In this chapter we consider limestones; in the next, dolomites.

FIELD OBSERVATIONS

Limestone is recognized in outcrop by the bubbly evolution of carbon dioxide gas when a few drops of cold, dilute hydrochloric acid are dropped on it,

$$CaCO_3 + 2HCl \rightarrow Ca^{2+} + 2Cl^- + CO_2 + H_2O$$

by its softness, typically white color, and the interlocking texture of its crystals. Limestone is distinguished from dolomite by the fact that dolomite does not react visibly to dilute hydrochloric acid unless powdered. Also, dolomite commonly weathers with a dull brownish-yellow cast (buff color) because it usually contains some ferrous iron as a substituent for magnesium in the crystal structure. The iron is released from the carbonate during weathering and oxidizes, causing the color.

The relative proportions of calcite, dolomite, and quartz (or other silicate minerals) in a limestone can be estimated by etching the rock surface at the outcrop. Dilute hydrochloric acid is dripped onto a clean, flat surface of a hand specimen, such as a bedding plane, until the calcite is dissolved in sufficient amounts so that the less soluble materials stand out in positive relief. Quartz is identified by its translucency and glassy appearance, chert by its opacity and hardness (if the grains are large enough to be scratched and examined with a hand lens), and dolomite by its white color, softness, and rhombohedral crystal outlines. Calcite only rarely occurs as scalenohedra or rhombohedra, but dolomite nearly always occurs as scattered rhombs in limestones. The etching technique also reveals the distribution in the rock of these relatively insoluble constituents. For example, chert may be located along bedding planes; fossils may have been converted into chert, but the remainder of the rock is still limestone; the limestone–dolomite contact in the outcrop may be at an angle to the bedding of the rocks, proving a replacement origin for the dolomite.

Textures

The textures of limestones are extremely variable because of the complex origins of carbonate rocks. Limestones can have textures that are identical to those of detrital rocks (grain rounding, sorting, etc.; see Figure 7–1) or to those of chemical precipitates (equigranular, interlocking crystals, "porphyritic," etc.). Many carbonate rocks display both types of textures (see Figure 7–2). In addition, limestones commonly have biologically produced textures characteristic of the growth habits of living organisms, such as algae (see Figure 7–3) or corals. Some of these biological textures are so intricate that they defy adequate written description (see Figure 7–4). In these cases, a labeled photograph is worth 10^3 words.

The texture of most limestones can be described adequately in the field by determining the types of gravel- and sand-size particles, the presence or absence of calcium carbonate mud matrix, and the presence or absence of coarsely crystalline calcite cement (visible with a hand lens).

Allochemical Particles

Allochems are the gravel-, and sand-, and coarse silt-size carbonate particles that occur and typically form the framework in mechanically deposited limestones. They are the equivalent of quartz, feldspar, and lithic fragments in sandstones. Four types of allochems are common: (1) fossils, (2) peloids, (3) ooliths, and (4) limeclasts.

Figure 7–1
Well-rounded, poorly sorted pebbles of algal micritic limestone in matrix of black (organic matter) microcrystalline limestone, Cool Creek Limestone (Ordovician), Arbuckle Mountains, Oklahoma. Algal nature of pebbles is visible only in thin section.

Figure 7–2
Laminated limestone composed of interlocking microcrystalline calcite crystals a few micrometers in diameter, Kindblade Formation (Ordovician), Arbuckle Mountains, Oklahoma. Laminations indicate original clastic nature of crystals that has been obliterated during recrystallization from aragonite to calcite.

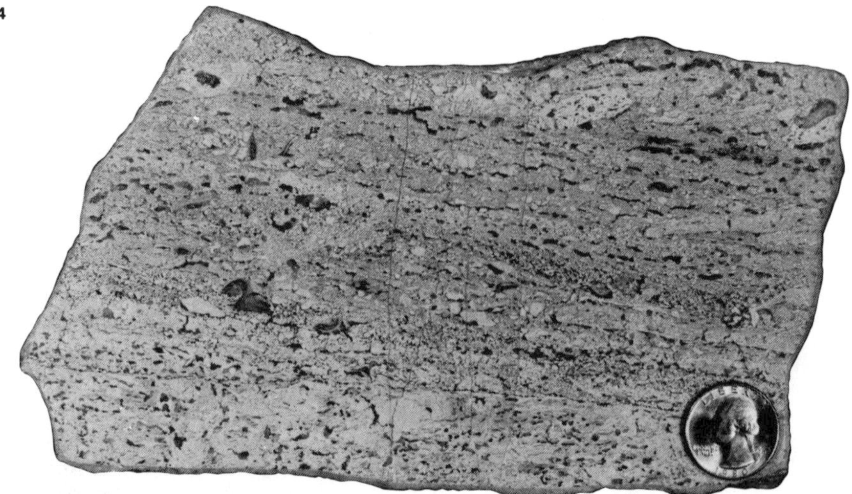

Figure 7–3
Limestone composed of fragments of calcified algal mat, showing platy limestone fragments characteristic of such algal mats, McLish Limestone (Ordovician), Arbuckle Mountains, Oklahoma. Also visible are original void spaces now filled with coarser, translucent calcite (sparry calcite), dark-gray areas roughly parallel to bedding.

Figure 7–4
Polished slab of reef rock, Capitan reef (Permian), New Mexico, composed of coelenterate (?) *Tubiphytes* (white), sponges and bryozoa (light gray), and alga *Archaeolithoporella* (laminated white crusts). Cements are shades of gray bands. [J. A. Babcock, 1977, SEPM Permian Basin Section, *Field Trip Guidebook to Guadalupe Mountains*. Photo courtesy J. A. Babcock.]

Fossils

The clearly distinguishable fossils in limestones in outcrop are those that have inequant biologically determined shapes, such as the concavo-convex outline of pelecypods, brachiopods, and ostracods; the segmented, tubular shape of a crinoid stem; or the leafy pattern of a bryozoan (see Figure 7–5). Because they are identified by their shapes, however, they are more difficult to recognize if they were broken into small pieces before burial, and this is a common phenomenon in the shallow marine environment in which most limestones are deposited.

Both high current velocities and predators can cause the fragmentation. The minimal size required for recognition depends on the microstructure of the organism. For example, a fragment of a coral 10 mm in diameter in a hand specimen might be indistinguishable from a bryozoan or an algal fragment; a similar size piece of crinoid column would be easily identifiable. An echinoid fragment of this size might not be

Figure 7–5
Large, unbroken fossils in matrix of microcrystalline calcite, about actual size. Easily visible are brachiopods and crinoid stems. Probable gastropod at upper right; dark objects may be fish plates. Note extremely poor sorting, indicating quiet-water deposition, as does micritic nature of the matrix. [F. J. Pettijohn, 1975, *Sedimentary Rocks,* 3rd ed. (New York: Harper & Row). Photo courtesy F. J. Pettijohn.]

recognized as a fossil fragment at all because echinoderm skeletons disaggregate into sand-size crystals of calcite and might look like cement crystals in limestone hand specimens.

Peloids

Peloids are aggregates of microcrystalline (aphanitic) calcium carbonate that lack internal structure. Most peloids have an ellipsoidal to roughly spherical shape (see Figure 7–6), and many are believed to be fecal pellets; they contain organic matter and assorted detritus normally ingested by organisms during feeding. In many limestones the peloids are of rather uniform coarse silt to fine sand size, presumably reflecting the anal dimensions of the organisms that produce them. Studies of the feeding habits of marine worms reveal that they burrow through carbonate sediments on the shallow seafloor, swallowing anything that is small enough to ingest that con-

Figure 7–6
Peloids (structureless micritic intraclasts) at left center, Arbuckle Formation (Ordovician), southern Oklahoma. Also visible are stylolite seam just above matchstick and several large fragments of clastic limestone, the one at top center being highly peloidal.

tains nourishing organic matter. The ingested sediment passes through the alimentary tract of the organism and finally is excreted as the microcrystalline aggregates we call fecal pellets.

Peloids may also be produced by other mechanisms. In many modern carbonate environments, carbonate sand and silt grains of various kinds are micritized by endolithic (boring) algae, with the destruction of the original fabric of the grain. The grains then appear as particles of structureless microcrystalline calcite. *Peloid* is the general term used when the origin of the grain is unknown.

Peloidal limestones seem to be underrepresented among ancient limestones in comparison to the abundance of peloids in modern marine areas. Probably this results from the initially soft state of fecal pellets, which causes them to merge when compacted and to be indistinguishable from pure micrite in ancient limestone. Indeed, micrite is the most abundant type of carbonate in ancient limestones—a fact that may be closely related to the abundance of fecal pellets in modern carbonate areas.

Ooliths

Ooliths are nearly spherical, polycrystalline carbonate particles of sand size that have a concentric or radial internal structure (see Figure 7–7). Unbroken ooliths in hand specimen can be mistaken for peloids. Ooliths, however, always contain a nucleus of some sort, such as a quartz grain or fossil fragment, around which the oolitic coating

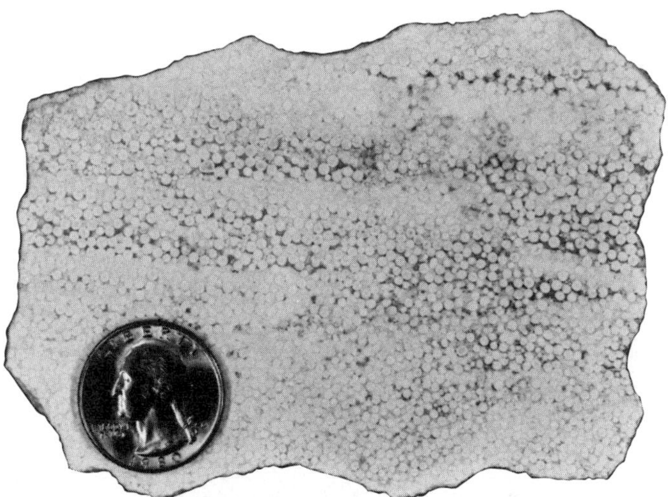

Figure 7–7
Ooliths cemented by translucent, coarse calcite crystals, Chimney Hill Limestone (Silurian), Arbuckle Mountains, Oklahoma. Upper part of slab shows horizontal bedding; at lower right oolitic layers are cross-bedded, dipping to left.

has formed. These nuclei can be seen with a hand lens. Ooliths are the only abundant type of carbonate particle that is essentially inorganic in origin. The coating on the nucleus is chemically precipitated from agitated water and, therefore, the presence of ooliths is evidence that the particle has been transported by strong currents. Oolitic limestones commonly are cross-bedded, a reflection of the high kinetic energies in their environment of deposition.

Particles similar in appearance to ooliths include pisoliths and oncoliths. *Pisoliths* are concentrically laminated bodies of inorganic origin that are larger than 2 mm in diameter (see Figure 7–8). Although the distinction between ooliths and pisoliths depends only on size, pisoliths are much more irregularly laminated and occasionally adjacent pisoliths fit together like a puzzle of polyhedral pieces. The origin of pisoliths is controversial. Some workers believe they are formed as part of caliche crusts in the vadose zone of a soil (Dunham, 1969); others believe they are formed by inorganic precipitation in hypersaline brines (Pray and Esteban, 1977). However, pisoliths have not been seen in modern carbonate environments, either in caliche or in hypersaline bodies of water.

Oncoliths are essentially identical to pisoliths in hand specimen. In thin section, however, the oncoliths reveal a filamentous structure characteristic of algal encrustations. That is, oncoliths have an organic origin and form in marine waters of normal salinity. They are found frequently in modern carbonate environments.

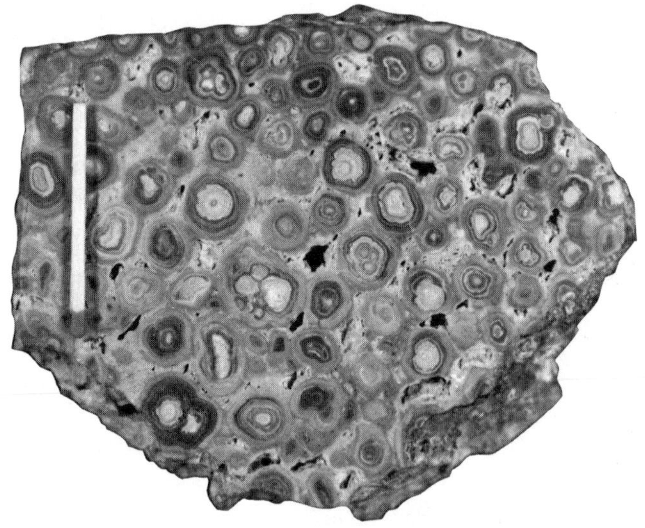

Figure 7–8
Pisolith, Tansill Formation (Permian), southeastern New Mexico.

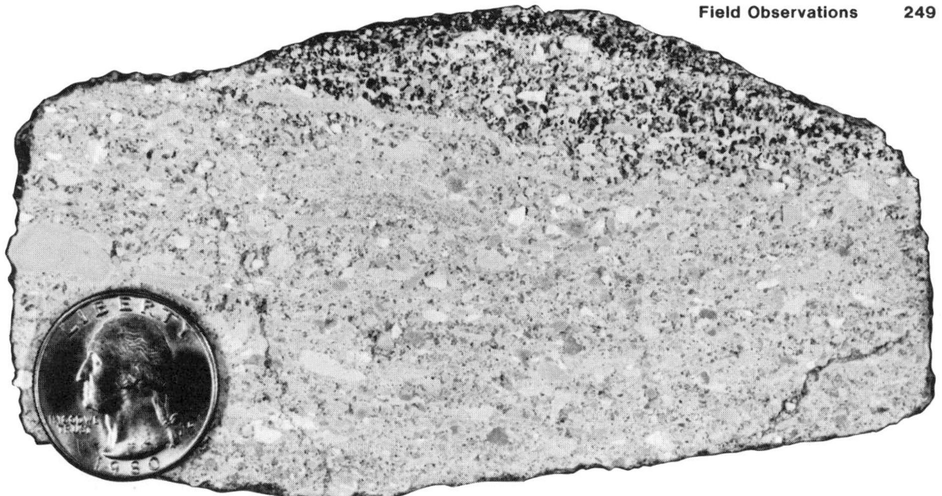

Figure 7–9
Microcrystalline intraclasts cemented by sparry calcite and oriented with long dimensions parallel to bedding, West Spring Creek Formation (Ordovician), Arbuckle Mountains, Oklahoma. Dark-colored layer (hematitic) of finer-grained intraclastic debris overlies coarser layer.

Limeclasts

Limeclasts are fragments of earlier-formed limestone. They may originate in a number of ways. Most are *intraclasts,* pieces of penecontemporaneous lithifield carbonate rock from within the basin of deposition (see Figure 7–9). Perhaps they are pieces of semiconsolidated carbonate mud torn from the seafloor by a winter storm on January 3, 100 million years B.C. Perhaps they are aggregates of peloids (*grapestone*) that stuck together because of mucilaginous organic coatings or were cemented together. Perhaps they are fragments produced by drying and cracking of intertidal mud from carbonate sediment on the margin of the basin. Any of these origins qualifies the particles as intraclasts.

There also exist limeclasts carried into the basin of deposition from the surrounding area, e.g., a piece of Mississippian fossiliferous limestone in a Cretaceous limestone. Most limestones do not contain such fragments, and those that do typically also contain other evidence of externally derived (*terrigenous*) detritus, such as noticeable percentages of detrital quartz, feldspar, or silicate lithic fragments. Commonly, the limestone that contains such fragments was deposited in a tectonically active area such as the Alpine region during the closing of the Tethys seaway or in the Marathon region of western Texas during its Pennsylvanian orogenic episodes.

Orthochemical Particles

Orthochems are the calcium carbonate matrix and clearly secondary cement that bind the allochems to lithify the sediment. This interallochem material is of two types: microcrystalline calcite or aragonite (both termed *micrite* as an abbreviation of microcrystalline carbonate) and coarsely crystalline carbonate cement. The difference in crystal size between micrite and the coarsely crystalline cement is adequate to permit easy distinction in hand specimens of most limestones.

Microcrystalline Carbonate

Micrite crystals are predominantly 1–5 μm in diameter (see Figure 7–10) and seem to be considerably more abundant than coarse cement in limestones. In hand specimen, micrite appears dull, opaque, and aphanitic, like a piece of rhyolite or chert. It can vary in color from white to black depending on impurities, particularly the amount of organic matter. In may be present in small amounts between the allochems or form most of the rock, with the allochems dispersed through the rock like plums in a pudding. Many limestones seem to lack allochems entirely. In hand specimen it is not possible to determine the origin of the micrite matrix any more than it is possible to determine the origin of the clay flakes in a mudrock.

Nearly all microcrystalline carbonate sediment originates as disarticulated algal material (Stockman et al., 1967). From the viewpoint of the carbonate geologist, the algal world can be divided into two groups: those that have hard parts (calcareous algae) and those that do not (seaweed and others). The calcareous algae can be usefully subdivided into encrusting types that coat and bind fauna such as corals or

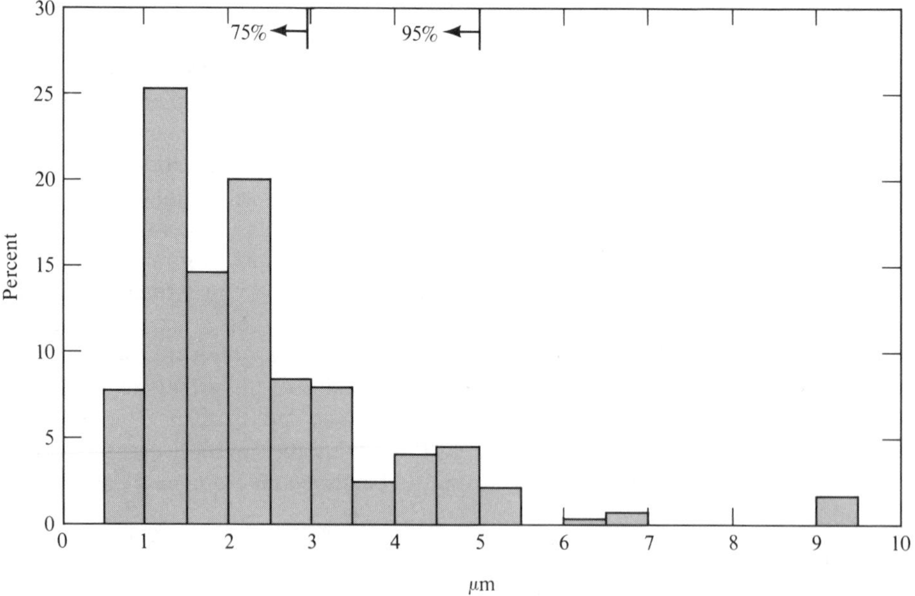

Figure 7–10
Size frequency distribution of microcrystalline calcite crystals in six limestones as determined with scanning electron microscope. [Data from Bathurst, 1975.]

Figure 7–11
Miniforest of *Penicillus* plants (shaped like asparagus tips) surrounded by aragonite needles produced by decay of their ancestors. Assorted organic debris litters "forest floor." Depth of water is a few meters. [Photo courtesy R. N. Ginsburg.]

bryozoa, and solitary types such as *Penicillus* or *Halimeda.* Some of the modern solitary calcareous algae have an internal structure formed of needles of aragonite a few micrometers in length connected by organic tissue (see Figure 7–11). When the organism dies, the organic tissue decomposes, releasing the needles onto the intertidal or shallow seafloor. It is these needles, subsequently recrystallized to equant blocks of calcite 1–5 μm in size, that we see in the fine-grained limestone in the stratigraphic column.

Coarsely Crystalline Cement

This material, termed *sparry calcite,* is transparent or translucent in hand specimen, like a crystal of clear quartz. This appearance is in sharp contrast to aphanitic micrite, so that the two types of calcite are distinguishable in most limestones. Most sparry calcite among allochem particles is coarser than 15 μm in diameter; micrite, smaller than 5 μm. Occasionally, crystals 5–15 μm in size occur in limestones (termed *microspar*) and typically will be incorrectly identified in hand specimen as micrite.

Sparry calcite among allochems has the same origin as the calcite cement in a sandstone. The cement is precipitated from a supersaturated pore solution, commonly very early in the history of the carbonate unit, within a million years after the formation of the sediment. The waters in which carbonate sediment forms can be assumed to be supersaturated, so that calcium and carbonate ions are readily available to be released from solution as a suitable hydraulic gradient arises to move large volumes of fluid through the sediment. Typically, sparry calcite cement and micrite matrix do not occur together because the presence of micrite causes a sharp decrease in the permeability of the sediment. And, of course, the presence of the micrite decreases the amount of pore space available for the sparry calcite to fill.

Grain Size, Sorting, and Rounding

The interpretation of these textural features in limestones is more difficult than in sandstones, largely because of the biologic character of fossils and peloids. For example, the fossils in a limestone can be whole ostracods of a particular species that were buried in the carbonate mud in which they lived. The fact that these allochems are of a certain size and are "very well sorted" is not directly related to current strengths in the depositional environment. Both the mean size and the sorting are biologic in origin rather than hydrodynamic.

We noted earlier the biologically determined, excellent "sorting" of fecal pellets. Rounding can be similarly biologically determined. Crinoid columns and fecal pellets are always round, irrespective of whether the depositional environment is of high or low kinetic energy. The variety of shapes and sizes of biologic particles makes "hydraulic parity" very difficult to determine.

The energy level of the environment of carbonate deposition is evaluated mostly from the presence or absence of calcium carbonate mud. It is assumed that aragonite needle producers are ubiquitous, so that microcrystalline ooze is always available in carbonate environments. Therefore, if a limestone lacks these aphanitic particles, it means that current strengths were high enough to remove them. If the limestone is rich in microcrystalline carbonate, we interpret the depositional environment as having been of low kinetic energy. It must be remembered, however, that the "source area" of most carbonate particles is very close to the site of final deposition; allochems may be produced rapidly and in large numbers, thus forming a rock with many allochems and little mud in an environment of low kinetic energy.

Other factors used to evaluate the kinetic energy level of the environment are:

1. Evidence of mechanical abrasion during transport.
2. Presence of ooliths.
3. Current structures such as cross bedding.

Noncarbonate Mineralogy

The average limestone contains only 5% noncarbonate material, of which nearly all is quartz, clay minerals, and chert. The quartz and clay minerals are silicates derived from outside the basin of deposition (as are a small percentage of carbonate lime-

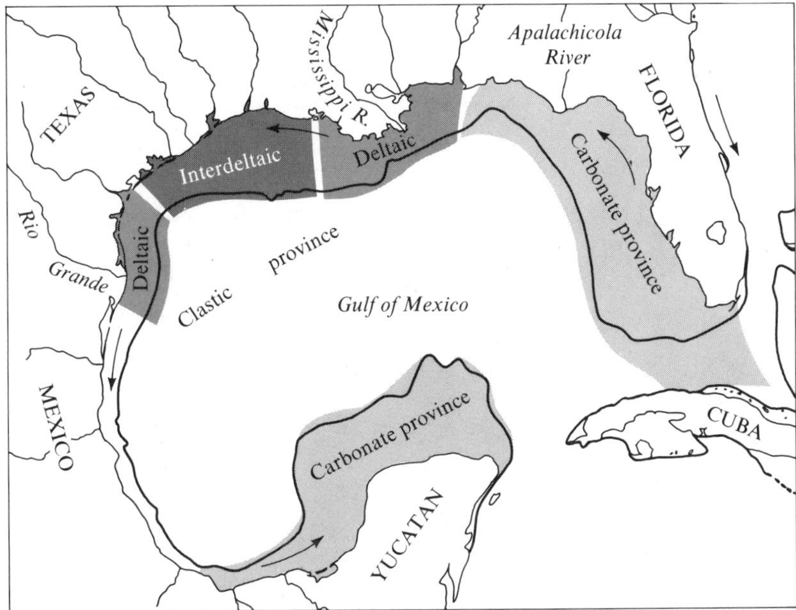

Figure 7–12
Modern sediment types in nearshore area of Gulf of Mexico. Modern carbonate deposition occurs along western coast of Florida and along northern side of Yucatán peninsula. Both Florida and Yucatán are underlain by Tertiary carbonates. Eastern coast of Florida receives quartz and clay from southern Appalachians; central Gulf Coast, from mid-continental United States; eastern coast of Mexico, from Sierra Madre Oriental. Arrows indicate direction of nearshore current flow and sediment transport. Heavy line is shelf boundary. [C. E. B. Conybeare, 1979, *Lithostratigraphic Analysis of Sedimentary Basins* (New York: Academic).]

clasts) and are considered terrigenous detritus. They reflect the presence of source areas somewhere within the regional drainage, with the grain size of the terrigenous sediment reflecting the velocities of currents as they enter the depositional basin. The reason such detritus is so uncommon in limestone-depositing basins is the biologic origin of nearly all carbonate particles.

An influx of silicate detritus implies an influx of fresh water. Most carbonate-secreting organisms live in very shallow water and cannot tolerate a significant change in salinity. In addition, their reproductive rate is inadequate to keep up with the rate of influx of mud characteristic of most streams, so that they are killed off rather quickly. For example, limestones are not forming today near the Mississippi Delta or for hundreds of kilometers to the west where the Mississippi detritus is spread by the westward-flowing nearshore surface currents in the Gulf of Mexico. Similarly, carbonates are absent along the northeastern coast of Mexico because of sand and mud originating in the Sierra Madre Oriental. Limestone reefs are abundant, however, along the western coast of Florida beyond the reach of Mississippi detritus and along the southeastern coast of Mexico and the Yucatán peninsula, which are unaffected by drainage from mountainous areas (see Figure 7–12).

Figure 7–13
Chert nodules and lenses developed along surfaces parallel to bedding, Onondaga Limestone (Middle Devonian), Albany, New York. Length of ruler at lower left is 15 cm. [Photo courtesy R. C. Lindholm.]

Chert in carbonate rocks is of two origins. A minor amount is extrabasinal, with the same source as other terrigenous particles. Nearly all chert in carbonates is intrabasinal and forms by crystallization of the amorphous silica from shells of siliceous organisms. The bulk of the amorphous silica is secreted by siliceous sponges (Cambrian–present), radiolarians (Ordovician–present), and diatoms (Jurassic–present), with the marine diatoms doing about 80% of the secreting at present. These organisms remove dissolved silica from seawater (diatoms also live in fresh water) and precipitate it as an amorphous, solid support for their soft tissues. When they die, the tissues decompose, increasing the specific gravity of the microcarcasses, and they sink to the seafloor. Some of the silica shells are buried in the carbonate sediment, dissolved, and later crystallized as microcrystalline quartz (chert). The shells of diatoms and radiolarians are mostly of silt size and, therefore, tend to accumulate in quiet-water areas with texturally immature sediment.

The intrabasinal chert in limestone occurs either as very finely crystalline to powdery quartz or as nodules centimeters to meters in length (see Figure 7–13). When in the nodular form, it is concentrated along visible bedding planes, presumably reflecting the migration paths of the silica dissolved from the siliceous shells. The reason the dissolved silica migrates to the centers of crystallization we see as nodules is not known.

Classification

The classification of limestones is based almost exclusively on textural variations because of the lack of mineralogic variations. But despite the constant mineralogy of ancient carbonates, a useful classification scheme can be constructed by analogy with sandstones (Folk, 1959). The allochemical grains, calcium carbonate mud, and sparry calcite cement are the analogs of sand grains, clay matrix, and cement in sandstones. The closeness of the analogy between classification schemes for sandstones and limestones is shown in Figure 7–14.

In the construction of descriptive names for limestones, the key terms are modified so that fossil becomes *bio-*, peloid becomes *pel-*, oolith is shortened to *oo-*, intraclast is *intra-*, microcrystalline calcite is *mic-*, and sparry is *spar-*; thus, we have biosparites, pelmicrites, and oosparites. The main part of the name is based on the major allochem and orthochem; appropriate modifiers may precede the main part of the

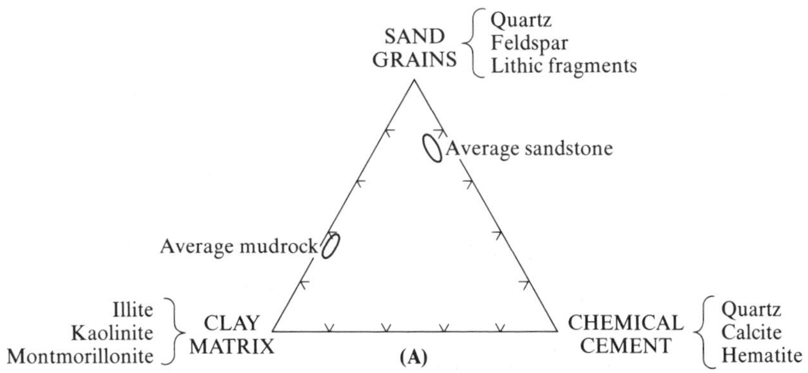

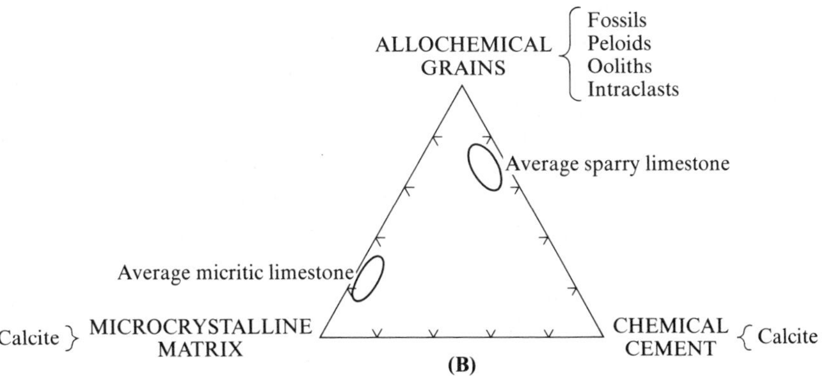

Figure 7–14
Triangles showing analogy between components of (**A**) sandstones and (**B**) limestones.

Table 7–1
Representative Compositions of Limestones and Their Best Descriptive Names

Composition of limestone		
Allochems	Orthochems	Appropriate name
70% Pelecypods 30% Ooliths	Sparry cement	Oolitic pelecypod biosparite
80% Ooliths 5% Fossil fragments 15% Glauconite	Sparry cement	Glauconitic oosparite[a]
60% Pellets 30% Fossil fragments 10% Intraclasts	Microcrystalline carbonate matrix	Fossiliferous pelmicrite
70% Intraclasts (gravel size) 25% Trilobites 5% Pellets	Sparry cement	Trilobite intrasparudite[b]
40% Crinoid fragments 40% Brachiopods 20% Clay minerals	Microcrystalline carbonate matrix	Clayey crinoid–brachiopod biomicrite

[a]If terrigenous detritus forms 10% or more of the sand- and gravel-size debris, it is added to the rock name as a modifier.

[b]If the allochems are of gravel size (> 2 mm), *rudite* is added to the rock name.

name. Table 7–1 illustrates typical limestone compositions and their descriptive names. As in all classifications, rocks occur in nature that are difficult to place in pigeonholes, but the concept behind the nomenclatural scheme is sound and leads to better communication.

An alternative and widely used system of limestone classification has been proposed by Dunham (1962). The key concept in Dunham's system is grain support (see Figure 7–15). When deposited, did the sediment consist of a self-supporting framework of allochems, or do the allochems float in a micrite matrix? Application of this system is not so simple as it first appears because of the exotic and extremely irregular shape of some allochems, particularly fossils. Solid spheres such as are approximated by quartz grains or ooliths form a self-supporting pack with about 60% of the volume being grains; that is, the porosity of packed spheres is about 40%. Arcuate shells, however, form a self-supporting framework with only 20–30% grains, and many limestones contain allochems of a variety of irregular shapes. In practice, one examines the hand specimen and thin section of the limestone and estimates whether or not an allochem framework exists. A point count to determine the allochem/orthochem ratio may not be conclusive. The concept of an allochem-supported fabric may be very useful because such fabrics commonly facilitate important diagenetic modifications because of their high porosities and permeabilities compared to non-grain-supported, micrite-rich limestones.

<table>
<tr><td colspan="4">Original components not bound together during deposition</td><td rowspan="4">Original components were bound together during deposition . . . as shown by intergrown skeletal matter, lamination contrary to gravity, or sediment-floored cavities that are roofed over by organic or questionably organic matter and are too large to be interstices.</td></tr>
<tr><td colspan="3">Contains mud
(particles of clay and fine silt size)</td><td>Lacks mud</td></tr>
<tr><td colspan="2">Mud-supported</td><td colspan="2">Grain-supported</td></tr>
<tr><td>Less than
10% grains</td><td>More than
10% grains</td><td></td><td></td></tr>
<tr><td>Mudstone</td><td>Wackestone</td><td>Packstone</td><td>Grainstone</td><td>Boundstone</td></tr>
</table>

Figure 7–15
Classification of limestones according to R. J. Dunham (1962).

Structures

A large variety of sedimentary structures occurs in limestones. Some of the structures reflect biologic origin (e.g., reefs); some, current origin (e.g., cross bedding); and others, diagenetic origin (e.g., stylolites).

Bedding and Lamination

These structures are found in carbonate rocks of any origin and any composition. As is true of sandstones, the origin of bedding and lamination in many carbonates is obscure. Consider bedding in micrites, for example. The micrite is essentially identical on both sides of the bedding plane (even in thin section), and no concentrations of organic matter, clay minerals, or other materials are evident. Perhaps there was incipient cementation or recrystallization of the carbonate ooze before the next lamina of ooze was deposited. But why? What characteristic of the sea-bottom water or exposure to air on a tidal flat was responsible for the very rapid lithification? Sometimes such questions cannot be answered.

Sometimes the origin of bedding and lamination is clear. This is true where there is a change in size or type of allochem or in the ratio of micrite to allochems or where an obvious surface of subaerial exposure is present.

Laminated carbonate sediment can be produced by blue-green algal mats that lack hard parts. The soft-bodied algae grow as filamentous mats in the intertidal zone. The filaments are mucilaginous and, as they are repeatedly swept over by wave-generated currents and tides, they trap and bind microcrystalline carbonate particles in the water, resulting in the formation of laminated layers consisting of a mixture of organic tissue and micrite (see Figure 7–16). These structures are called *stromatolites* and are common in rocks of all ages, from Precambrian to the present. In pre-Holocene stromatolites the organic tissue has decomposed and only laminated car-

(A)

(B)

Figure 7–16
(**A**) Subtidal stromatolites in water 2 m deep, Shark Bay, Western Australia. Elongation of mounds is in direction of wave movement. Mounds are built by colloform mat growth and are soft on top, increasingly lithified with depth into the mound. [P. E. Playford and A. E. Cockbain, in M. R. Walter (ed.), 1976, *Stromatolites* (Amsterdam: Elsevier). Photo courtesy P. E. Playford.] (**B**) Plan view of eroded surface of limestone, showing cross sections of algal stromatolites, Hoyt Limestone (Cambrian), Saratoga Springs, New York. White scale is 10 cm in length. Comparison with part A shows that, for stromatolites, the present is a good key to the past. [Wilson, 1975. Photo courtesy J. L. Wilson.]

bonate sediment remains. In many deposits the layers appear to have accumulated with a bulbous upper surface—a growth topography caused by erosive effects on the accumulating algal mat.

Nodular Structure

Nodular structure in limestones is of two distinct types. One type occurs in micritic, shallow-water carbonates (see Figure 7–17A) and is believed to be formed by burrows of the trace fossil *Thalassinoides* (Fürsich, 1973). This fossil is most common in intertidal and shallow subtidal sediments of Mesozoic age. The branching burrows disrupt the partially lithified but plastic micritic carbonate to produce the nodular appearance of some ancient limestones.

Another type of nodular structure seems to be formed in deep-water pelagic micrites such as are common in the Mesozoic and Tertiary limestones of the Alps (see Figure 7–17B; Jenkyns, 1974). These nodules look very much like those in the *chicken-wire structure* of gypsum (see Figure 9–3B) and anhydrite (see Figure 9–5). Limestone nodules ranging in size up to several centimeters occur in a matrix of micrite and clay (marl). According to Jenkyns (1974), these nodules form during early diagenesis in an area of slow sedimentation in the pelagic environment. Aragonite is dissolved and migrates to centers of crystallization where the carbonate precipitates as calcite in a marly matrix. Subsequent compaction may cause differential movement of the nodules and a still plastic marly matrix, giving rise to complex microtextures within the matrix and at the boundaries of the nodules.

Stromatactis Structure

This peculiar structure (see Figure 7–18) is common in micritic reef knolls and consists of layers or irregular masses of coarsely crystalline calcite within an otherwise homogeneous micrite. The features are 1–5 mm thick, can be 10 cm or more in length, and are oriented parallel to the bedding of the limestone. The base of the structure is flat and commonly is floored with laminated micritic sediment, but the bulk of the structure is sparry calcite. The origin of these structures is uncertain, but they seem to form during very early diagenesis. Cavities are partially filled by micritic sediment, and then the upper part of the cavity is filled by the coarse spar that typically forms the bulk of the stromatactis structure. They are useful for defining the bedding in otherwise uniform micritic mounds.

Bird's-Eye Structure

Bird's-eye structure consists of blebs, spots, tubes, or irregular patches of sparry calcite found in many limestones. It seems to form as sparry-calcite fillings in cavities resulting from localized organic or inorganic disturbances of the depositional limestone fabric. Suggested causes of the disruption include algal or burrowing activity,

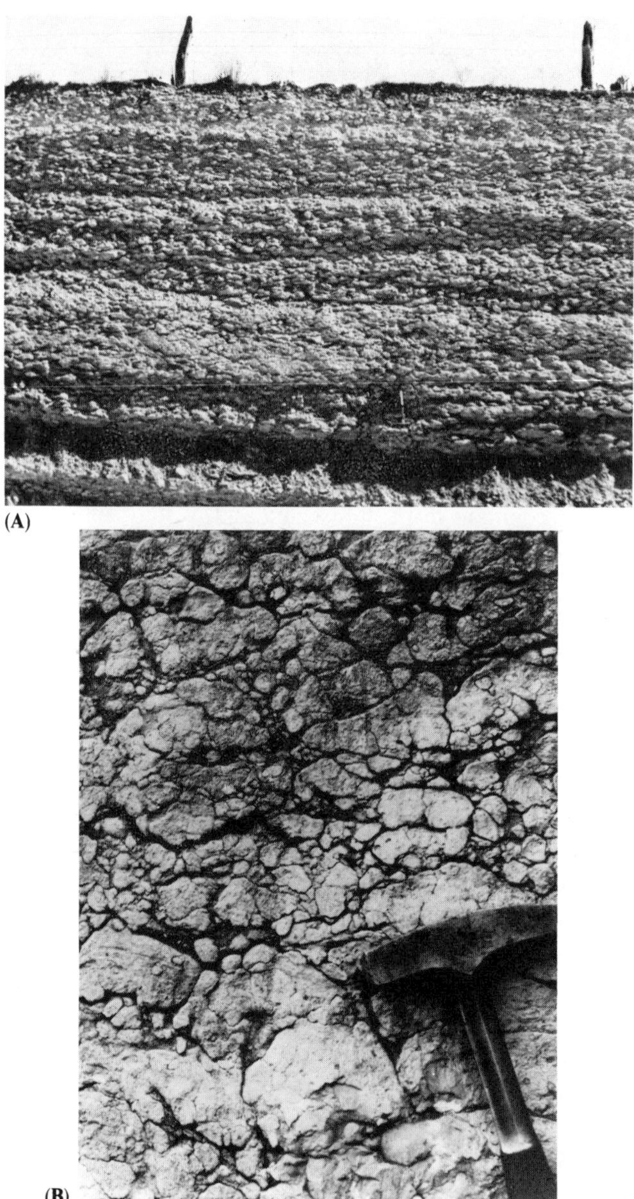

Figure 7–17
(**A**) Well-developed nodular structure in apparently homogeneous, shallow-water, micritic limestone in Cretaceous, central Texas. [F. J. Pettijohn and P. E. Potter, 1964, *Atlas and Glossary of Primary Sedimentary Structures* (New York: Springer-Verlag). Photo courtesy P. E. Potter.] (**B**) Nodular structure in Upper Jurassic limestone, Erbezzo, Italy. Nodules are composed of biomicrite. Reddish material among nodules is more clayey and contains fragments of pelagic crinoid. [E. L. Winterer and A. Bosellini, 1981, *Amer. Assoc. Petroleum Geol. Bull.*, *65*. Photo courtesy E. L. Winterer.]

Figure 7–18
Stromatactis structure in micritic mud mound, Ireland. Lighter-colored filling of structures is sparry calcite. Length of specimen is 15 cm. [H. Blatt et al., 1980, *Origin of Sedimentary Rocks,* 2nd ed. Copyright © 1980. Reprinted by permission of Prentice-Hall, Inc., Englewood Cliffs, New Jersey. Photo courtesy R. C. Murray.]

escaping gas bubbles, shrinkage cracking, and decomposition of plant roots. It has not been found in deep-water limestones and is, therefore, generally believed to be evidence of shallow-water deposition of the host rock.

Dune Forms, Ripples, and Cross Bedding

Allochemical limestones can display the same variety of current structures as sandstones, although they typically are not so evident at first glance as in sandstones. As with sandstones, the most common are dunes, ripples, and cross bedding. Many marine limestones are cross-bedded, with the thickness of cross-bedded sets ranging up to several meters. In general, the azimuths of the cross bedding in marine limestones show a bimodal distribution (see Figure 7–19), an expression of reversing tidal currents like that described earlier from intertidal sandstones. Two opposed directions of cross bedding are a more common feature of limestones than of sandstones because many preserved limestones were formed in the intertidal zone. Sandstones form in a much greater variety of environments.

Some carbonates are transported and deposited at great depth by turbidity currents and show graded bedding, flute casts, and many of the other sedimentary features characteristic of sandstone beds formed by this mechanism. Limeclasts of extrabasinal (terrigenous) origin are common in such limestones.

Figure 7–19
Bimodal cross bedding, reflecting two opposing directions of water movement (opposing tidal currents) in sparry limestones of Kansas City Group (Upper Pennsylvanian), eastern Kansas. Upper cross beds dip to right; those below dip to left. [W. K. Hamblin, 1969, *Kans. Geol. Surv. Bull., 194*. Photo courtesy W. K. Hamblin.]

Mounds and Reefs

Sandstones and mudrocks are generally deposited in topographically low areas. This is not so true for limestones. Because the carbonate sediment is mostly produced in place by living organisms, local sites may build thick accumulations that are topographically higher than their surroundings. These prominences can strongly influence surrounding sedimentation. The locally thick limestone section may consist of fossils that have grown attached to one another to form a wave-resistant structure called a *reef* (see Figure 7–20). Numerous modern examples of reefs occur along the coasts of Florida, Australia, and elsewhere, and around the edges of mid-oceanic volcanoes. The internal structure of these limestone accumulations is extremely complex, reflecting the biologic shapes and ecologies of the organisms involved. The deposit lacks the usual two-constituent fabric of framework and cement that characterizes allochemical limestones. Often, whole, undisturbed fossils are abundant and are interwoven with micrite and allochems that settled into crevices among the growing organisms. Ooliths and intraclasts are rare within the reef itself, but intraclasts are abundant in the talus deposits that fringe the reef. The reef core is typically massive and unbedded, although there are exceptions, such as laminations of coralline algae.

In Paleozoic times, framework-building organisms were less common than in modern seas, and many of the topographic constructional highs in carbonate areas were formed by accumulations of micrite. These highs are called lime–mud *mounds*. Many of these mounds may have originated as accumulations of sand-size pellets swept together by current activity (Lake, 1981), as occurs today in some areas of Florida Bay (see Figure 7–21). Perhaps there was also a biologic control in some cases.

Figure 7–20
Polished slab of reef rock from Sausbee Formation (Pennsylvanian), northeastern Oklahoma. Lower two-thirds of rock is complex growth structure of tabulate corals and laminated algae; upper one-third is composed of concentric laminae of algae *Osagia* and *Ottonosia*.

Figure 7–21
Aerial photograph of lime–mud mounds in Florida Bay. Mounds are several hundred meters long and composed of micrite pellets swept together by currents, partially lithified, and further held together by roots of mangrove trees that form bulk of vegetation on the islands. Mounds are only slightly above mean sea level. [Photo courtesy R. N. Ginsburg.]

Sometimes the structureless micritic core of a mud mound is capped by micrite that contains organisms capable of trapping fine carbonate sediment, such as sponges, algae, or bryozoa.

Stylolites

A stylolite is an irregular surface within a bed and is characterized by mutual interpenetration of the two sides—the columns, pits, and teethlike projections on one side fitting into their counterparts on the other (see Figure 7–22). The seam is made visible by a concentration of insoluble constituents, such as clay or organic matter.

Figure 7–22
Stylolites in Mississippian brachiopod–crinoid biosparite used as dividing panel between toilet stools in Oklahoma Memorial Union, Norman, Oklahoma. Stylolite seams are formed of material less soluble than limestone, a mixture of clay and carbonaceous material in this example.

Stylolites can be seen cutting across allochems such as fossils, with the upper or lower part of the fossil apparently dissolved away. Hence, the seams are a diagenetic feature. The orientation of the seams is nearly always approximately parallel to the bedding, implying that the dissolution was caused by the same type of interaction between overburden stress and pore waters that forms the pressure-solution surfaces in some quartzarenites. The only difference between the two is that the stylolites have greater lateral continuity (often several meters in length) and are marked by material of a different composition from the main rock mass (noncarbonate sediment). Stylolites are very common structures in limestones because of the relatively great solubility of carbonates, and the abundance of stylolites testifies to the huge thicknesses of carbonate sediment that can be removed by stylolite formation.

LABORATORY STUDIES

The prime tool used in laboratory studies of limestones is the standard polarizing microscope, but in recent years increasing use has been made of luminescence petrography, the scanning electron microscope, and the electron microprobe. These newer tools have provided data and insights into the evolution of limestones (and dolomites) that were not available previously.

Allochemical Particles

Fossils

The outer form and internal microstructure of the hard parts of organisms are extremely complex. An extensive description with photomicrographs of many types of microstructures is given by Bathurst (1975, pp. 1–76), and many other photos are given by Horowitz and Potter (1971), Majewske (1969), Milliman (1974), and Scholle (1978). Figure 7–23 shows a few examples of the appearance of fossils in thin section. Proper identification of some types of fragments in thin section poses no problem, as is true of the same fossils in hand specimen. Pelecypods, brachiopods, and ostracods all are concavo-convex, but in thin section it is apparent that their shells have markedly different internal structures. The pelecypods have a structure in ancient rocks composed of a mosaic of equidimensional calcite crystals; the brachiopods generally have a fibrous structure parallel to the outline of the shell; and the ostracods, like all arthropods, have a shell structure formed of rod-shaped calcite crystals oriented perpendicular to the outline of the shell. Comparison between modern representatives of these organisms and their ancient counterparts reveals that the pelecypod shells have been recrystallized, but the brachiopod and ostracod shells have not. This results from the fact that pelecypods build their shells largely of aragonite, which is unstable relative to calcite in diagenetic waters and recrystallizes soon after

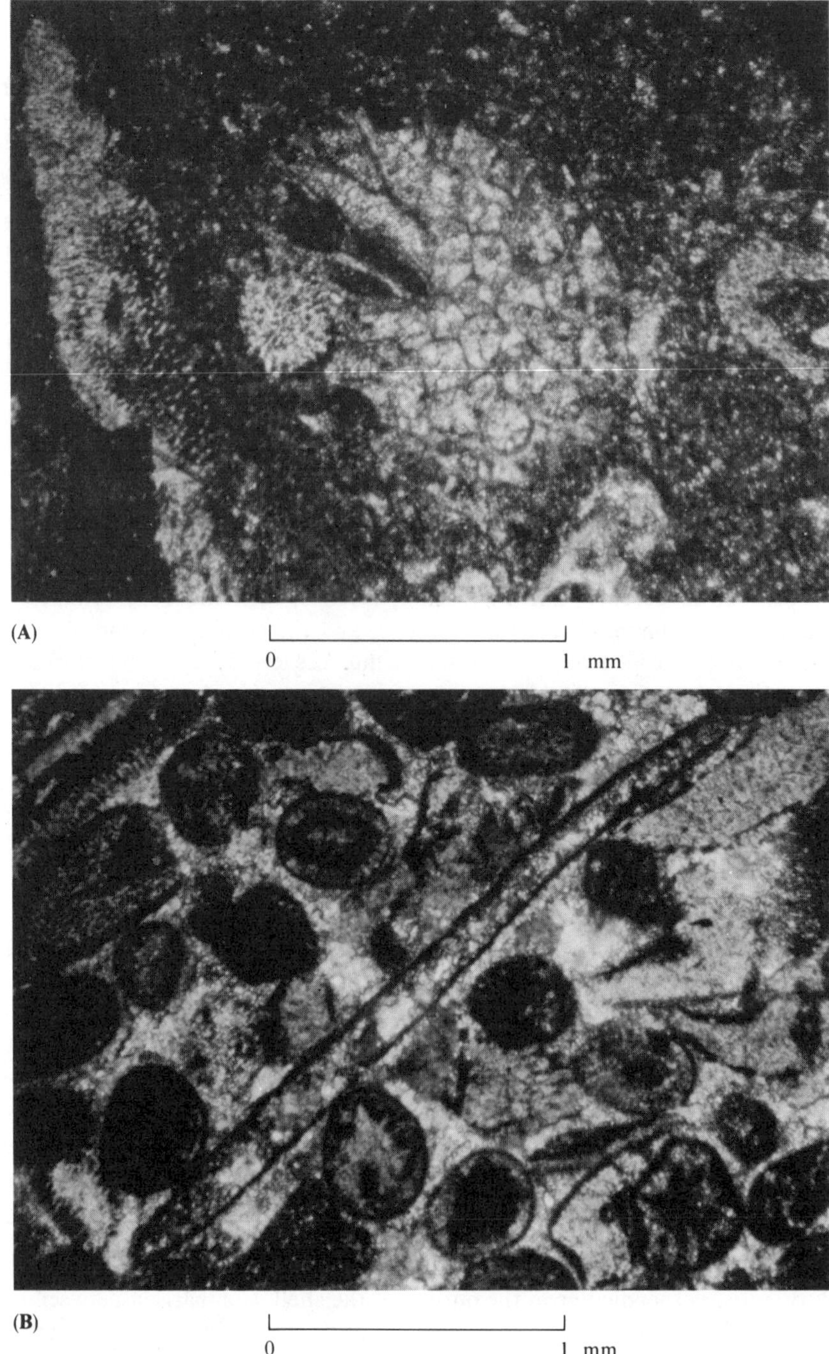

Figure 7–23
Characteristic appearances of carbonate-shelled fossils in thin section (crossed nicols). (**A**) Bryozoan frond (center) and assorted echinoderm fragments in micrite matrix. (**B**) Long pelecypod fragment with thin micritic coating formed by boring algae, surrounded by ooliths with nuclei of echinoderm fragments and cemented by sparry calcite. Pelecypod shell has been replaced by equigranular sparry crystal mosaic, as have nearly all ancient pelecypod shells.

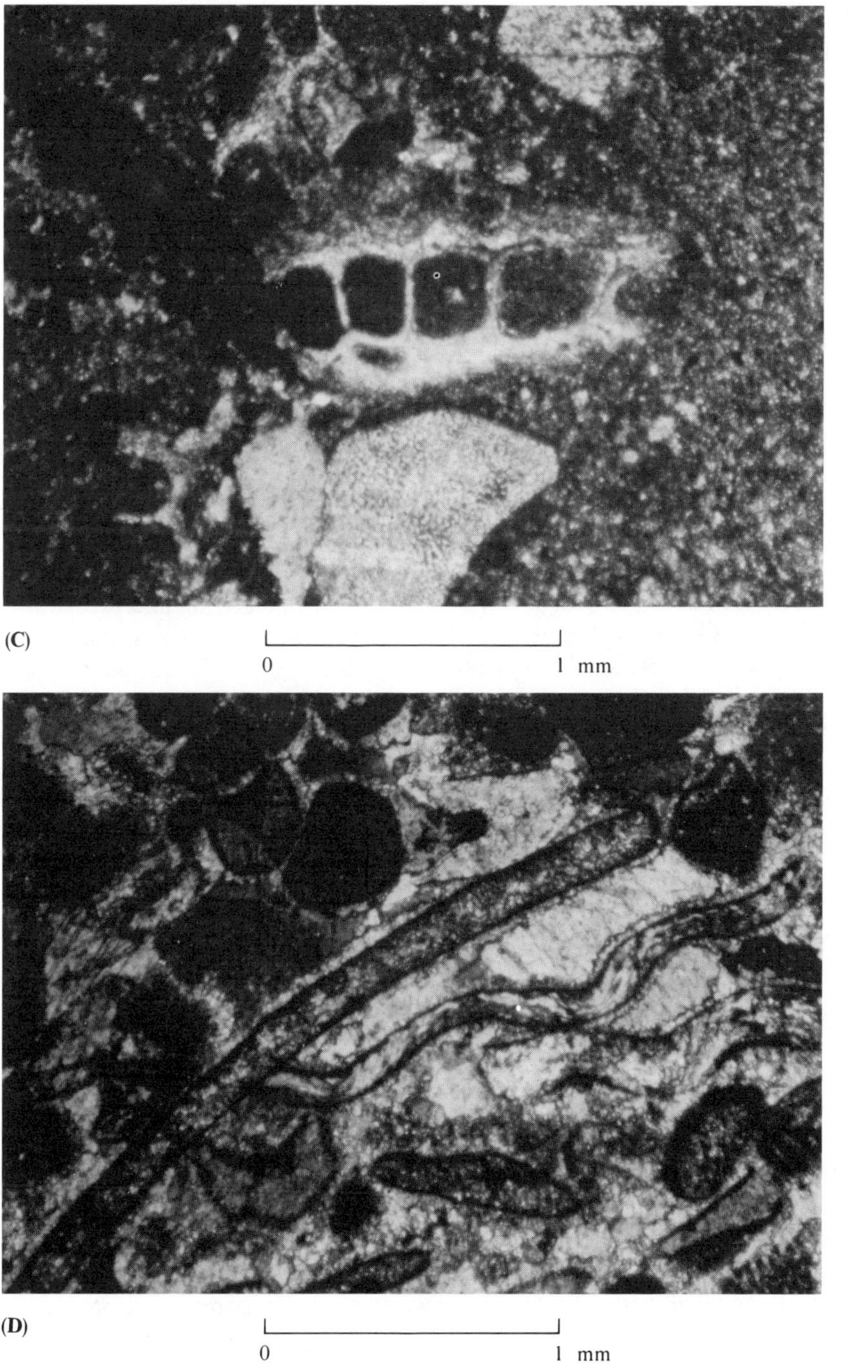

(C) Encrusting foraminifera with chambers filled with micrite surrounded by echinoderm fragments and set in matrix of fossil hash and micrite. (D) Sinusoidal brachiopod, showing its fibrous structure parallel to length of shell. Also present are long pelecypod fragment and numerous crinoid fragments coated with micrite generated by boring algae, set in sparry calcite cement.

the death of the organism. Brachiopods and ostracods, on the other hand, build shells of nearly pure calcite, and these are stable after the death of the organism. Analysis of variations such as these has led to the finding that shells composed of calcite containing less than 2–4% magnesium are very resistant to recrystallization; those formed of calcite that contains more than a few percent of magnesium commonly are recrystallized; those that are aragonitic during the life of the organism are nearly always recrystallized in pre-Holocene limestones. This has permitted petrographers to establish the mineral composition of shells of extinct organisms, such as trilobites. Trilobite hard parts show the same internal structure as do their arthropod relatives, the ostracods: rod-shaped calcite crystals oriented normal to the shell wall. Apparently, the trilobites built shells of calcite with a very low content of magnesium.

Many identifications of fossils in thin section are much more speculative than those of the organisms we have considered so far. Some corals can be difficult to distinguish from some bryozoans, and crinoid plates may look much like echinoid fragments. Calcareous algae show an astonishing variety of internal architectures. The only way to learn the appearances of the major types of fossils is to examine a variety of fossiliferous limestones, with one eye peering down a polarizing microscope and the other eye examining various illustrated books until the appropriate photomicrograph is discovered. For most shell fragments a suitable match will be found, but it is common to observe problematica in thin section for which one person's guess is almost as valid as another's, particularly when the fossil fragments are small. Proper identification of fossils is worth the effort, however, because of the ecologic information that can be obtained. For example, some organisms like a higher salinity than others, some prefer colder water than others, and a repeated assemblage of whole shells of several species can establish a useful biofacies for stratigraphic correlation.

Peloids

Peloids are more easily seen in thin section than in hand specimen, where they are usually difficult to distinguish from the micrite matrix. Even in thin section there exist all gradations from clear and sharply bounded peloids to those with diffuse and vague borders to micrites in which the previous existence of peloids can be suspected but not proven. An additional advantage of viewing peloids in thin section is that the origin of the peloid is more evident (see Figure 7–24). Some seem to be partly replaced fossils; others, micrograpestone lumps.

Ovoid peloids that appear to be fecal pellets seem to be more common in sparry limestones than in limestones with a micritic matrix. It makes no ecologic sense to conclude that an organism will scavenge for food in both agitated and quiet waters but move to agitated waters to deposit its waste products. A more reasonable explanation is that fecal pellets are composed of structureless micrite that can be distinguished from the micrite matrix only by the outlines of the pellets. When deposited, the pellets are ductile; and if they are compacted before they are rigidified, they will be squashed and become indistinguishable from the micrite matrix. Hence, pelsparites are seen more commonly in ancient rocks than are pelmicrites.

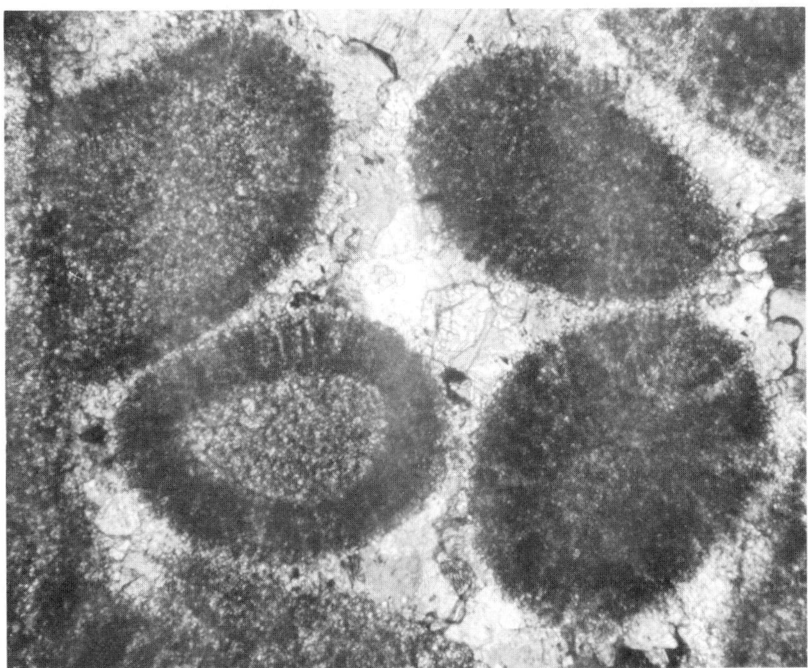

Figure 7–24
Peloidal oosparite, showing stages of peloidization of ooliths, West Spring Creek Formation (Ordovician), Arbuckle Mountains, Oklahoma. Oolith in lower left has clearly discernible oolithic coating around crinoid plate. In oolith at upper left thin coating is less distinct but still clearly visible. Oolith at lower right has been largely pelletized, although vague radial texture is still visible. Grain at upper right is a pellet and has no internal fabric; it has been completely micritized. Diameter of pellets is 2.7 mm.

Ooliths

Ooliths are nearly always of marine origin, although a few nonmarine examples have been found. When examined in thin section (see Figure 7–25), the nature of the clastic nucleus can be seen (assuming the thin section happened to slice deeply enough into the oolith to reveal the nucleus), as can the number of layers and the character of the oolithic coating. Some particles barely qualify as ooliths, with only one or two thin layers of calcium carbonate surrounding the nucleus; others have thick coatings. Presumably, the difference reflects the length of time during which the oolith formed.

In cross section, the crystals in the carbonate coating can appear either concentric or radial to the nucleus. In some ooliths the regular concentric structure is interrupted, apparently by breakage or erosion of the original body followed by regeneration or renewed growth. The result is a microunconformity between the outer concentric shells and those of the internal core. Modern ooliths are composed of aragonite

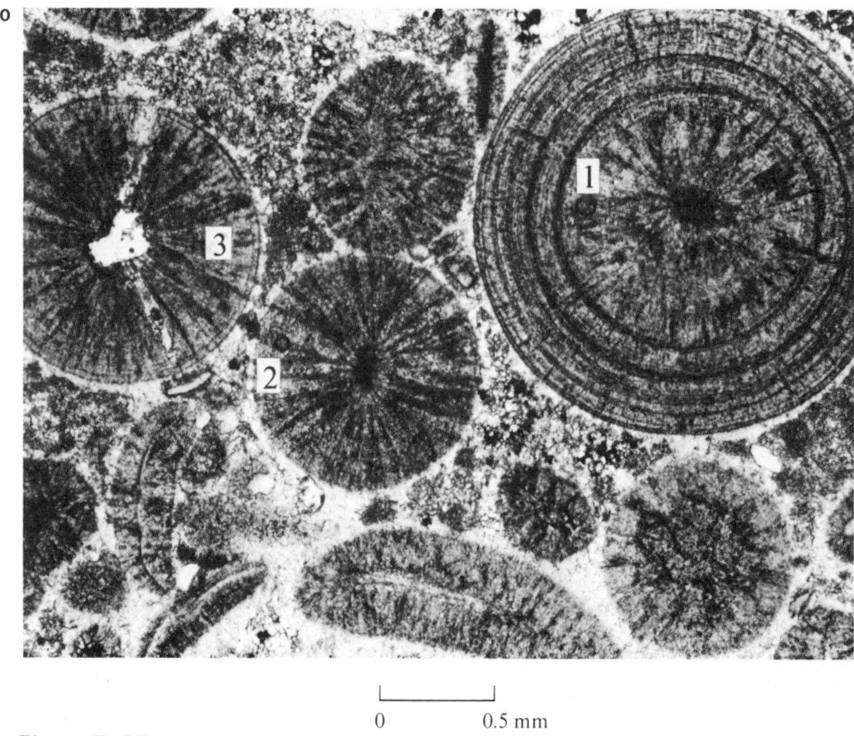

Figure 7–25
Oosparite showing internal structure of ooliths in thin section, Warrior Formation (Cambrian), Pennsylvania. (1) Radial-concentric oolith with peloidal core and purely radial center; diameter is 1.7 mm. (2) Purely radial oolith. (3) Radial oolith with two concentric coatings. Note that diameters of inner, radial parts of ooliths are nearly identical. [P. L. Heller et al., 1980, *Jour. Sed. Petrology, 50.* Photo courtesy D. R. Pevear.]

needles that, in most cases, are oriented tangentially to the oolith surface. In ancient limestones the ooliths are composed of calcite (unless replaced by another mineral such as hematite or chert).

Diagenesis, in addition to converting the original aragonite to calcite and the original concentric structure to a radial one, can also completely obliterate the internal oolithic structure. Recrystallization can result in a granoblastic internal microstructure that may contain faint inclusions marking the original concentric structure. In other cases the original internal structure is converted to dense, structureless, microcrystalline calcite. Such ooliths may be confused with peloids and micritic intraclasts. In some oolithic limestones the ooliths have been selectively leached to create spherical voids called *oomoldic porosity.*

Limeclasts

Limeclasts in thin section can show any internal structure that occurs in carbonate rocks. Thus, there are biomicrite intraclasts, pelsparite intraclasts (see Figure 7–26),

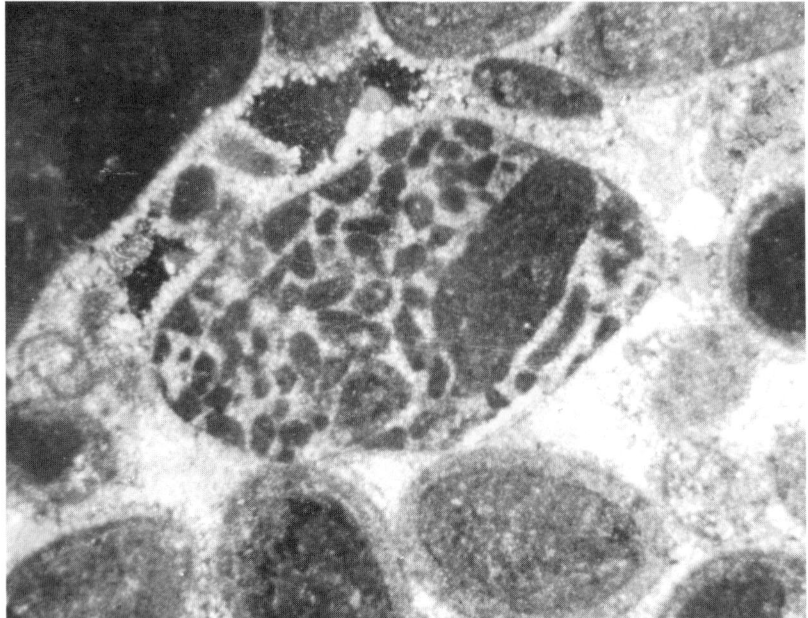

Figure 7–26
Pelsparite intraclast in oosparite, Kindblade Formation (Ordovician), Arbuckle Mountains, Oklahoma. Ooliths have peloidal cores and thin coatings. Note difference in appearance between aggregates of micrite crystals that form peloids and coarser and translucent sparry crystals that form cement between peloids. These are characteristic appearances of micrite and spar in carbonate rocks. Diameter of intraclast is 18 mm.

and intraclasts composed of dolomitized micrite (dolomicrite). It may be possible to distinguish in thin section between intraclasts and terrigenous carbonate fragments by determining the age of the fossils in fragments of bioclastic carbonate in a limestone. For example, a fragment of trilobite biosparite in a Mesozoic limestone is clearly extrabasinal.

Orthochemical Particles

Micrite

In thin sections of ancient limestones, micrite appears as subequant blocks of featureless, polyhedral calcite 1–5 μm in diameter. As noted earlier, microcrystalline carbonate in modern environments is composed largely of bioclastic aragonite, so that, if we assume that the present is an accurate key to the past for this material, we must conclude that ancient micrites are thoroughly recrystallized rocks.

Micrite is very fine-grained material, analogous in hydraulic behavior to clay minerals. Thus, it is normal to find, in insoluble residues of limestones, that micritic units contain a higher percentage of organic matter and clay minerals than do sparry units.

(**A**)

(**B**)

Sparry Calcite

In thin section, sparry calcite appears mostly as crystals of pore-filling cement, as it does in sandstones. In some limestones, however, the percentage of spar greatly exceeds any possible original porosity in the limestone, indicating that previously existing allochems or micrite has been converted to spar by recrystallization. In the Great Basin of the western United States, there exist thick limestone units of Paleozoic age composed largely of essentially pure sparry calcite whose original character is unknown. Stratigraphic relationships suggest that the limestones were normal shallow-water marine units, so that they must have had a variety of original textures.

Cementation of carbonate particles by sparry calcite can occur intermittently and over long periods of time (see Figure 7–27). At present, there is no known method of obtaining absolute dates for these episodes of cementation.

Microsparry Calcite

Micrite crystals that form the matrix in many limestones have diameters in the range of 1–5 μm. Sparry calcite pore filling has crystal diameters coarser than 15 μm. In some limestones that appear micritic in hand specimen, however, the presumed micrite crystals are seen in thin section to be clear crystals of microsparry calcite in the size range 5–15 μm (see Figure 7–28). Except for its crystal size, the textural relationships of the microspar are those of normal micrite. It can form the entire rock (obviously impossible if the microspar were a cement), it is most common in rocks that contain few allochems as contrasted to current-sorted limestones, and it occurs as fringes around allochem grains within a micrite matrix. Further, these silt-size spar crystals are not sorted in layers, as would be expected if they were detrital. Based on these observations, microspar is interpreted as resulting from the recrystallization of micrite, although the reason for the rather uniform crystal size is unknown.

Insoluble Residues

Although most limestones contain only a few percent of sediment that is insoluble in cold, dilute hydrochloric acid, these insoluble grains can be useful as indicators of the nature of the rocks surrounding the carbonate basin. They can also be used for correlating widely separated outcrops of limestone or subsurface sections (Stevenson

---

Figure 7–27
Crinoidal biosparite from Lake Valley Limestone (Mississippian), New Mexico. (**A**) Normal thin-section view (uncrossed nicols) shows crinoid plate (left center) surrounded by apparently uniform, one-stage growth of calcite cement. [Scholle, 1978. Photo courtesy P. A. Scholle.] (**B**) Approximately same area viewed under cathodoluminescence shows at least five generations of cement that can be correlated from sample to sample and related to a variety of tectonic and erosional events. [W. J. Myers, 1974, *Jour. Sed. Petrology, 44*. Photos courtesy W. J. Myers.]

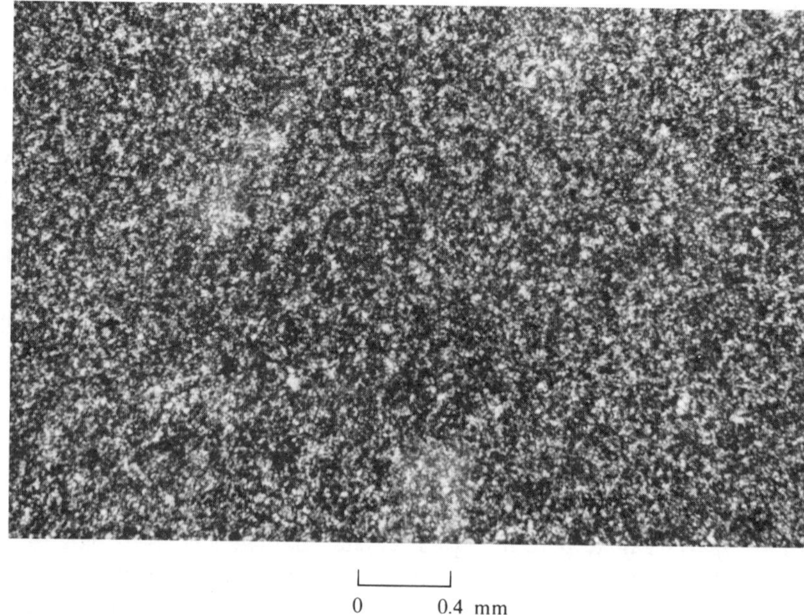

Figure 7–28
Photomicrograph (crossed nicols) of Solenhofen Limestone (Jurassic), West Germany. In hand specimen this rock is so finely crystalline that individual crystals cannot be resolved and it appears to be a homogeneous micrite. Thin section reveals that much microspar is present (clear, translucent areas) among micrite crystals (darker, opaque areas). Micrite seems vaguely peloidal in places. [Scholle, 1978. Photo courtesy P. A. Scholle.]

et al., 1975). In practice, the insoluble constituents are isolated and concentrated by digestion in HCl and examined with a binocular microscope at magnifications as high as 20×; only sand-size and coarse silt-size sediment can be identified with this technique. X-ray diffraction is used to identify the clay-mineral species.

Geopetal Fabric

Many fossils have curved outlines (pelecypods, brachiopods, ostracods, trilobites), and after death the hydrodynamically stable resting position for them is concave-downward. Thus, they form a bridge over the underlying sediment that subsequently can be partially or completely filled with internal sediment or sparry calcite. Those subbridge areas that are only partly spar-filled commonly serve as *geopetal structures* (see Figure 7–29)—structures that permit the up direction of the bed to be determined in highly deformed stratigraphic sections.

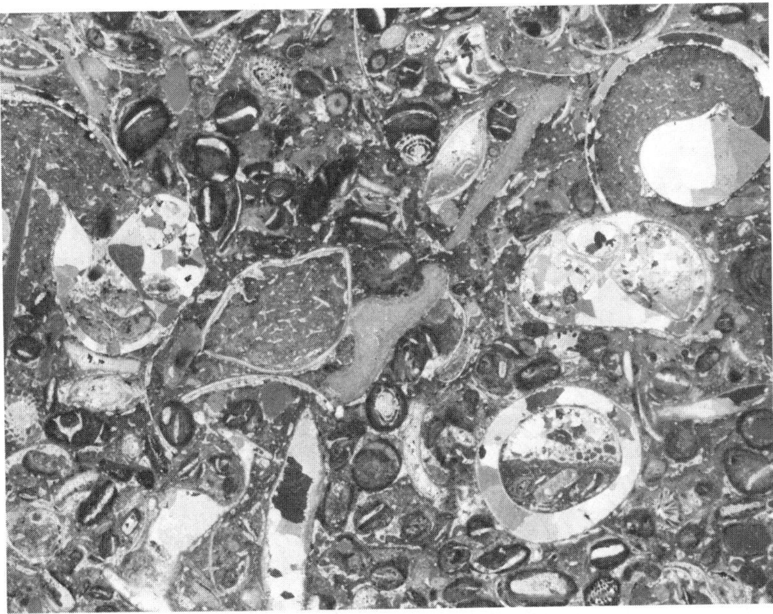

Figure 7–29
Geopetal structure inside gastropod shells, Winfield Limestone (Pennsylvanian), Kay County, Oklahoma. Oval shell in lower right is floored by micrite, with upper part of shell filled with later, coarse sparry calcite. Boundary between the two types of calcite is perpendicular to gravity. Diameter of shell is 3 mm. [Photo courtesy C. Gasteiger.]

Directional Fabric

In some allochemical limestones it is possible to determine from thin-section study the principal direction of current flow during deposition. An excellent example is provided by Stauffer (1962), who isolated a single bed within the Bar B Formation (Pennsylvanian) in the Caballo Mountains of southwestern New Mexico. Stauffer contoured the percentages of allochems of different types over an area of about 1 km^2 and the orientation of the long axis of the allochem grains. A typical result is shown in Figure 7–30, and a pronounced NE–SW orientation is evident. The unit is a nonreefal, intertidal limestone, as determined by the types of fossils present, and the NE–SW direction was interpreted to be the orientation of tidal currents, normal to the shoreline. Studies of the orientation of elongate allochems in two other limestones in the area—one Ordovician, the other Pennsylvanian—also indicated pronounced preferred orientations (see Figure 7–31). Such results are to be expected because there is no difference in concept between the effect of fluid movement on the orienta-

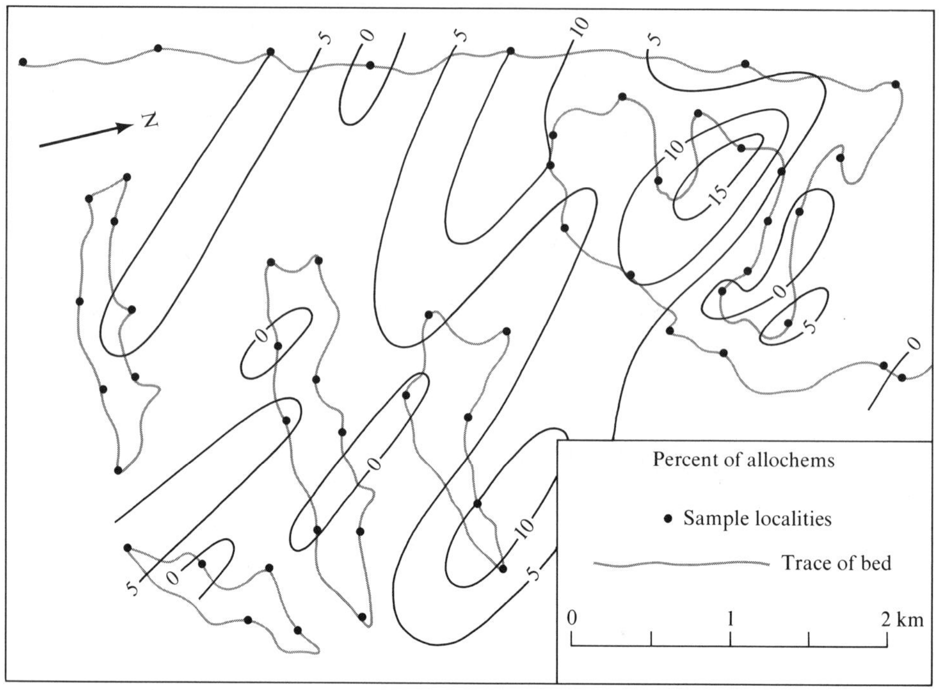

Figure 7–30
Contour map of percentage of intraclasts in single bed of Bar B Formation (Pennsylvanian), Caballo Mountains, New Mexico. [Stauffer, 1962.]

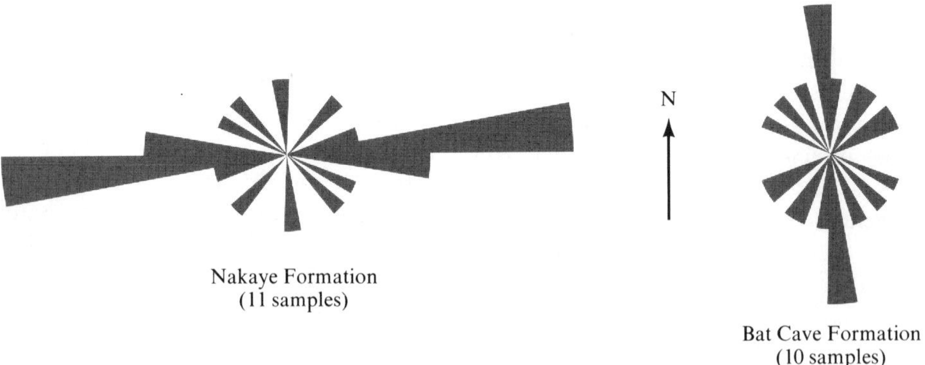

Figure 7–31
Rose diagram showing orientations of long axes of allochemical grains in Bat Cave Formation (Ordovician) and Nakaye Formation (Pennsylvanian) in Mud Springs Mountains, southwestern New Mexico. Rose petals are in 10° classes, each sample representing at least 200 grains. [Stauffer, 1962.]

tion of silicate grains and that of carbonate grains. Both will tend to assume a rest position that offers the least resistance to the moving current. (In Chapter 4 we considered the orientation of elongate grains of sand-size quartz.)

REEFS AND PALEOCLIMATE

Limestones are composed of calcite and are essentially organic remains; that is, nearly all limestones are macroscopic or microscopic coquinas. Biologists tell us that marine organisms reproduce more rapidly in warm waters. Chemists tell us that calcite is less soluble in warm water than in cold water. From this information, we would anticipate that modern carbonates and reefs should be located near the equator, and this is indeed the case (see Figure 7–32A). Nearly all reefs are located within 30° of the equator where the temperature of the surface water is at least 20°C. This suggests that the locations of ancient reefs might reflect paleolatitudes of ancient

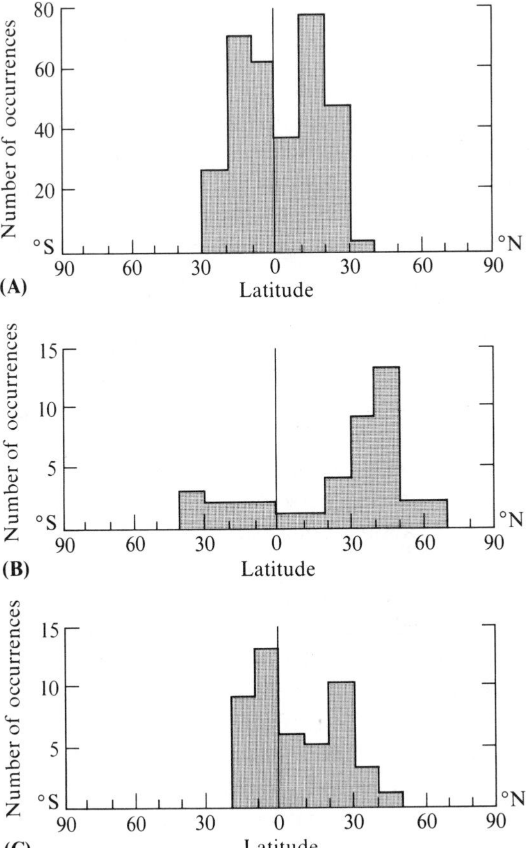

Figure 7–32
Latitudinal histograms for organic reefs. (**A**) Present latitude of modern reefs. (**B**) Present latitude of ancient reefs. (**C**) Paleolatitude of ancient reefs. [B. F. Windley, 1977, *The Evolving Continents* (New York: Wiley).]

landmasses. At present, only about 30% of the ancient reefs are located within 30° of the equator (see Figure 7–32B); but when the present latitudes are replaced by the paleolatitudes as determined from paleomagnetic measurements (see Figure 7–32C), the proportion rises to 90%.

CALCIUM CARBONATE DEPOSITION

As we have observed, carbonate sediments are, with the exception of ooliths, collections of the hard parts of organisms in various states of disaggregation or recombination. We have also noted that the formation of calcium carbonate is favored by shallow waters and warm temperatures. These facts determine the sites of accumulation of carbonate sediments and, therefore, the sites of origin of limestones (and dolomites). Three types of settings are generally recognized: (1) epeiric seas, (2) shelf margins, and (3) deep-sea basins.

Epeiric Seas

Epeiric seas can be defined as those arms of the ocean that spread over broad areas of the central parts of continents, such as the seas that existed on the North American craton during much of the Paleozoic Era. These seas covered 10^5–10^7 km^2; and when the continents were located in low latitudes, thin but extensive limestones were formed in them. Based on the ubiquitous occurrence and great abundance of fossils in these limestones and on the sedimentary structures in the rocks, it is probable that the maximum depth of water in the epeiric seas did not exceed 30 m. The slope of the seafloor must have been about 2 cm/km—an extraordinarily low gradient. For comparison, we might note that the present average slope of the world's continental shelves is 125 cm/km. Examples of limestones formed in such epeiric seas include the Ordovician and Mississippian rocks of mid-continental United States.

The shallowness of epeiric seas over wide areas tends to damp out lunar tidal effects that create turbulence and mixing of waters. These seas are also thought to have been too shallow to permit extensive wave action. Long-period swells propagated from a deep-sea basin would have been damped at the shallow margins of an epeiric sea. Within the epeiric sea, the mechanism for agitating the water would be mostly local winds. This scenario of relatively restricted water agitation and lack of appreciable influx of water from the deep oceanic basin results in a general shortage of nutrients, so that reef growth would be unlikely. Hence, epeiric seas are dominated by particulate carbonate sediment rather than by massive reef growth.

The carbonate sediments in the central parts of these seas tend to be micrites. Toward the margins of the basin, sand-size fossils become more abundant; and in the more saline, shallower water near the basin edge, peloidal muds occur. In the tidal flat and supratidal environment (see Figure 7–33), dolomite and gypsum typically are formed (see Chapter 8).

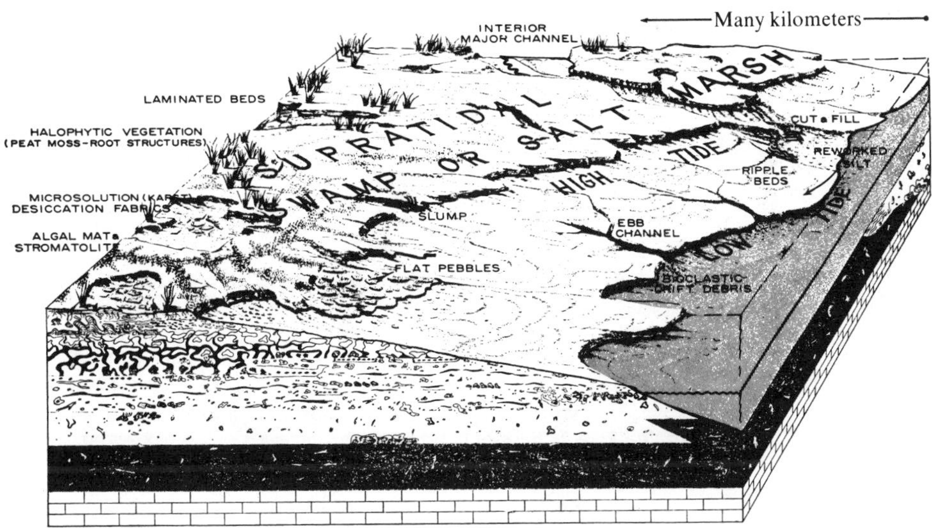

Figure 7–33
Idealized view of epeiric sea shoreline, showing appearance of tidal flat and supratidal environment of low-energy carbonate deposition. Vertical and horizontal scales are unequal. [P. O. Roehl, 1967, *Amer. Assoc. Petroleum Geol. Bull., 51.*]

Shelf Margins

A growing reef has an exceptionally large need for nutrients. It is also true that the formation of these carbonate structures is favored by agitation of the waters. (Agitation drives off carbon dioxide dissolved in the water and causes the pH to increase.) It follows, therefore, that the most favorable site for reef development is a location where (1) cool water is being warmed and agitated so that CO_2 is driven off, and (2) the cold water is rising from relatively deep areas of the ocean because deep waters are rich in nutrients. The place where such conditions are common is at a fairly sharp break in slope, such as at the edge of the continental shelf or around the edge of a mid-oceanic volcano. Modern examples of such sites include the reef tract along the southern tip of Florida, the Great Barrier Reef along the northeastern coast of Australia, and the circular reef growths that partially surround many volcanoes in the Pacific Ocean in the lower latitudes. Figure 7–34 shows an idealized cross section of a reef environment and its associated facies.

Deep-Sea Basins

Carbonate sediment is found in bathyal and abyssal depths in the modern oceanic basins, and limestones formed in such depths are known from rocks of Mesozoic and Cenozoic age. The Holocene deposits consist of the shells of planktonic organisms,

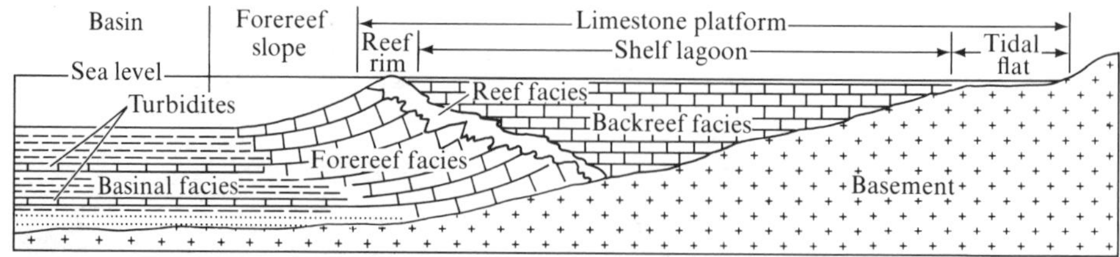

Figure 7–34
Schematic cross section illustrating principal environments of deposition near a reef. Vertical growth of reef is made possible by gradual rise in sea level. Probably only 10% of reef complex consists of *in situ* reef organisms. Most of complex is formed of debris broken from living reef (mostly on seaward side of reef), supplemented by remains of carbonate-shelled organisms that lived in the environment created by reef growth. [F. J. Pettijohn, 1975, *Sedimentary Rocks,* 3rd ed. (New York: Harper & Row). Based on a diagram by P. E. Playford, 1972, *Ann. Soc. Geol. Belgique, 95.*]

particularly of the foraminifera *Globigerina,* but also including hard parts of the algal family Coccolithophoridae and the planktonic, microscopic mollusks called pteropods. About 48% of the ocean floor is presently covered by sediment in which the remains of these organisms form at least one-third of the particles.

The calcium carbonate shells in these deep-sea deposits were formed in the upper few meters of the sea, as were the shells and rigid reef structures we considered previously. But the organisms whose hard parts accumulate in shallow waters are bottom-living forms (e.g., clams, crinoids, bryozoa); those in deep water are chambered, floating forms.

The areal extent of calcium carbonate accumulations in the deep sea is limited by two factors. The first is the same as is true for shallow-water accumulations: sea surface temperatures. Surface seawater in low latitudes is supersaturated with respect to calcite and aragonite, so that it is not difficult for marine organisms to remove calcium and carbonate ions from the water and to form it into these minerals as skeletal material. *Globigerina* and pteropods are more common in tropical waters than in higher latitudes.

The second limitation on the occurrence of deep-sea carbonates is depth. Seawater is colder at depth than at the surface, and colder waters contain more dissolved carbon dioxide than do warm waters. The increased carbon dioxide causes an increase in carbonic acid (H_2CO_3) in the water and results in the dissolution of the calcitic and aragonitic shells as they settle toward the seafloor after the death of the organism. Few shells survive below about 5,000 m. The increased hydrostatic pressure at depth also increases the solubility of calcium carbonate, but this effect is of less importance than the temperature–CO_2 effect. The depth below which no calcium carbonate accumulates is called the *carbonate-compensation depth,* about 5,000 m in equatorial regions but rising gradually toward the sea surface at higher latitudes because of the lower temperature of surface seawater in polar regions (see Figure 7–35).

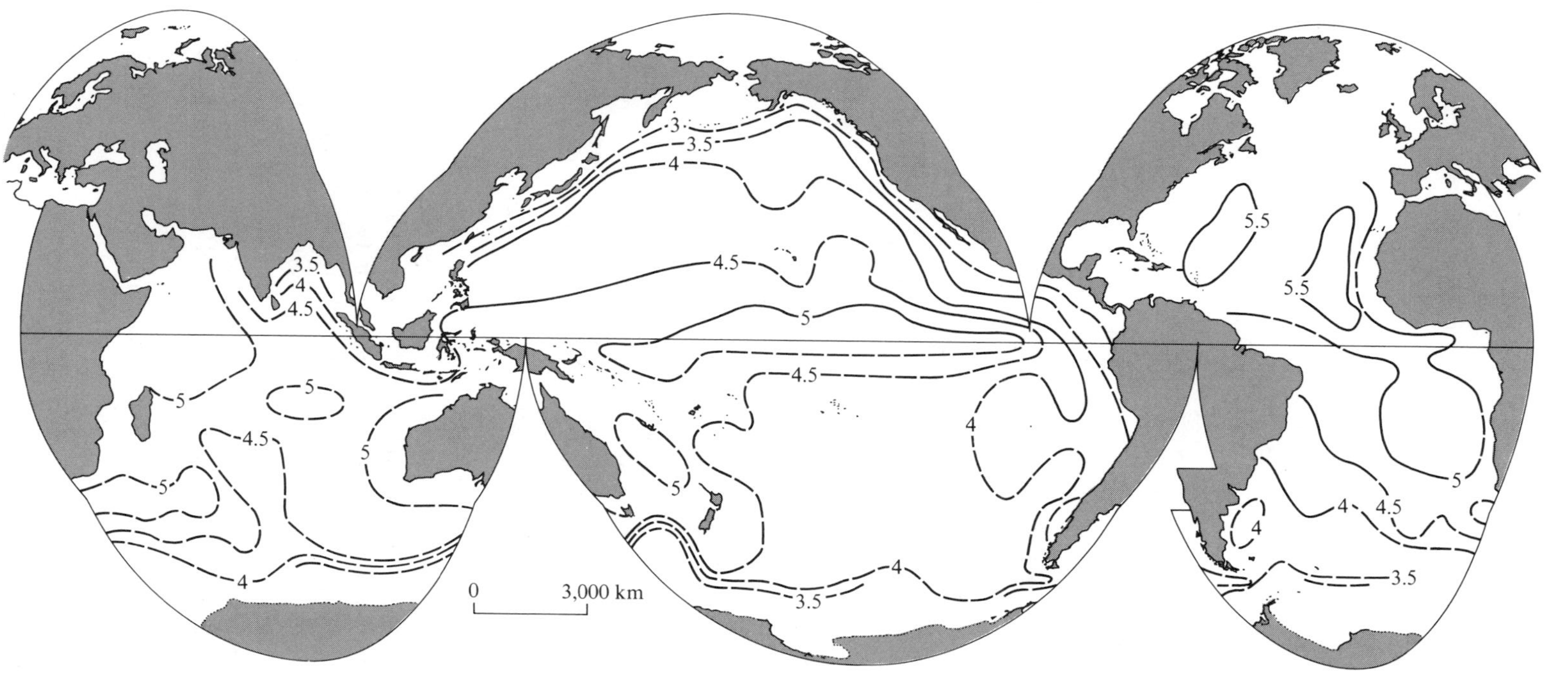

Figure 7–35
Bathymetry of calcium carbonate compensation surface, defined by interpolating local facies boundaries between calcareous sediments and sediments containing less than a few percent of carbonate. Dashed lines are based on fewer than 20 control samples. Depths are in km. [Berger and Winterer, 1974.]

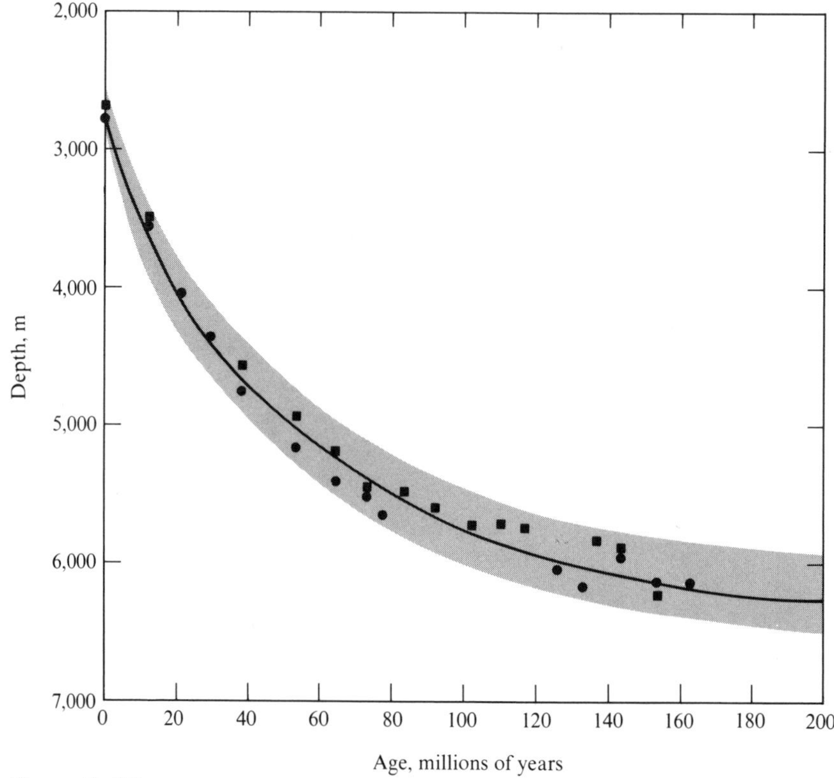

Figure 7–36
Plot of depth of ocean floor as function of its age (= distance from ridge crest) for Atlantic Ocean (squares) and North Pacific Ocean (circles). Band surrounding curve of best fit is uncertainty of depth data. [J. G. Sclater and C. Tapscott, "The History of the Atlantic." Copyright © 1979 by Scientific American, Inc. All rights reserved.]

Sediment composed of calcite and aragonite accumulates on the slopes of the thermally raised areas on the deep seafloor where divergent sections of oceanic plates occur. As the carbonate sediment moves laterally away from the crest of the rise, it must move into deeper water (see Figure 7–36) and is typically carried below the compensation depth. By this time the carbonate sediment has been covered by clay and siliceous ooze that, if close enough to shore, is covered in turn by terrigenous sediments carried by turbidity currents from the continental shelf. The deep-sea clay, siliceous sediment, and terrigenous sediment protect the calcium carbonate deposits from the seawater, which would dissolve them if it had contact with them. Thus, carbonate deposits can be preserved below the carbonate-compensation depth.

Deep-sea limestones in ancient rocks are identified in outcrop chiefly by the stratigraphic and tectonic settings in which they occur. Obviously, it requires rather severe tectonic activity to cause bathyal and abyssal sediments to be raised upward several thousand meters and incorporated into the continental mass. Deep-sea micrites are known from the Franciscan Formation (Jurassic) in California and from several units in the Alps—both locations where subduction of oceanic crust has occurred, apparently with the incorporation of some of the seafloor deposits (the carbonate sediment) into the continental plate. In these limestones, the most abundant fossils typically are the silt-size plates of planktonic algae called coccoliths, seen with an electron microscope (see Figure 7–37). As with the *Globigerina,* the plates settled to the deep seafloor following the death of the organism. Carbonate-secreting planktonic organisms did not evolve until the Jurassic Period so that there are no pre-Jurassic "microbiomicrites." Triassic and older deep-sea limestones consist of sand- and gravel-size limeclasts carried into deep oceanic waters by turbidity currents. The source of the detritus is an actively rising highland, such as the Alpine area created by the continental plate collisions that closed the Tethys seaway through Europe and Asia.

Figure 7–37
Transmission electron micrograph of Laytonville Limestone (Cretaceous), Franciscan Formation, Sonoma County, California, a coccolith coquina showing whole, segmented, oval plates and a hash of separate segments. Length of chipped plate in left center is 6 μm. [Photo courtesy R. E. Garrison.]

CARBONATE ROCKS AND PLATE TECTONICS

The biologic origin of carbonate sediments requires warm, shallow water for significant accumulations to form. In addition, almost no silicate detritus is permitted to enter a carbonate basin. Where might such settings exist, viewed in the context of divergent (extensional), strike-slip (extensional or compressional), and convergent (compressional) plate margins?

1. One important setting is near a low-lying landmass that apparently is sufficiently distant from the nearest plate margin so that the character of the margin is irrelevant. A modern example of such a site is the Great Barrier Reef off the northeastern coast of Australia. Ancient examples are the extensive carbonate deposits of Early and Middle Paleozoic age that formed along the rim of the emergent part of the North American craton. The Iapetus Ocean lay to the east in the location of the Atlantic Ocean, with a plate margin somewhere within it; the ancestral Pacific Ocean lay to the west. The marginal location of the carbonate accumulation with respect to the craton is reflected clearly by an isopach map (see Figure 7–38). The halo of

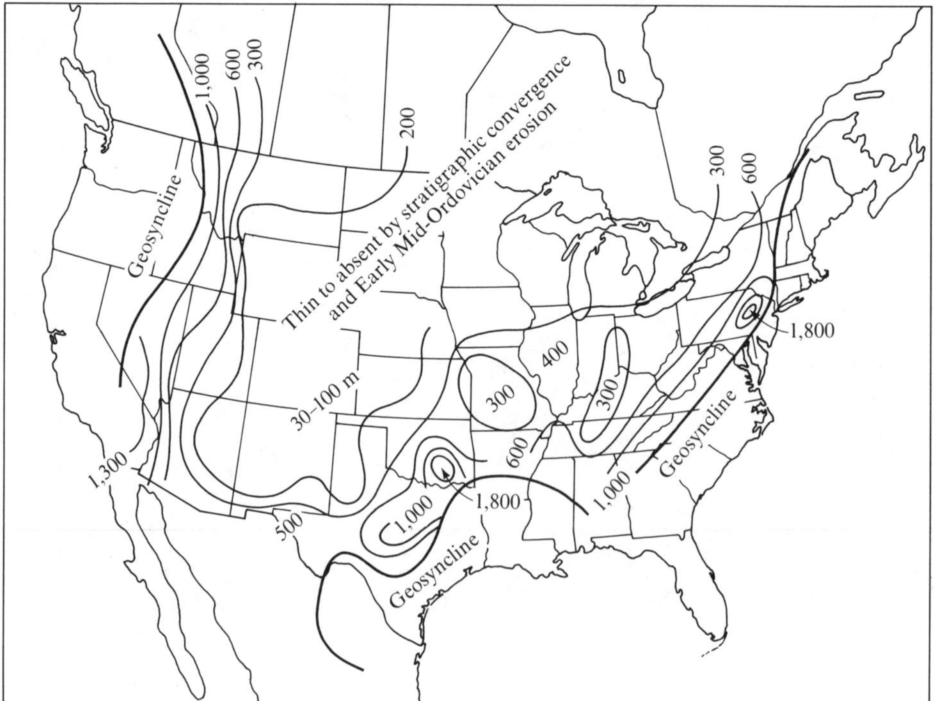

Figure 7–38
Thickness of Lower Ordovician carbonate sediments around North American craton (meters). [Wilson, 1975.]

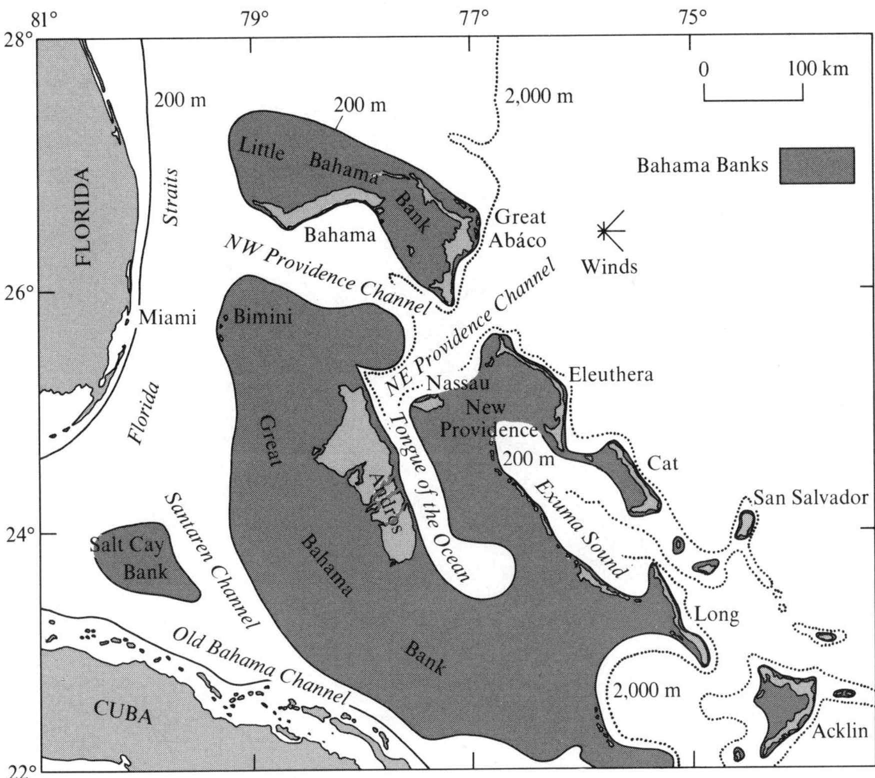

Figure 7–39
Great Bahama Bank and surrounding area, showing areal extent of carbonate sediments. [N. D. Newell, 1955, Geol. Soc. Amer. Spec. Paper No. 62.]

shallow marine and intertidal limestones and dolomites testifies to the general lack of topographic relief on the craton, as well as to the great distance from the site of carbonate accumulation to the nearest convergent plate margin and the topographic relief associated with it.

2. Strike-slip movements within a continent-dominated plate can result in the isolation of continental fragments, which find themselves rafted into waters of oceanic depths. A modern example of this type of occurrence and the associated accumulation of carbonate sediments is the Great Bahama Bank and surroundings (see Figure 7–39). An ancient example is the Permian Basin in western Texas and southeastern New Mexico (see Figure 7–40).

The carbonate accumulations on these isolated structural highs may originate at low stands of the sea and, once started, manage to maintain themselves and grow upward in spite of considerable subsidence of the continental fragment on which they are located. The Great Bahama Bank has been accumulating carbonate sediment

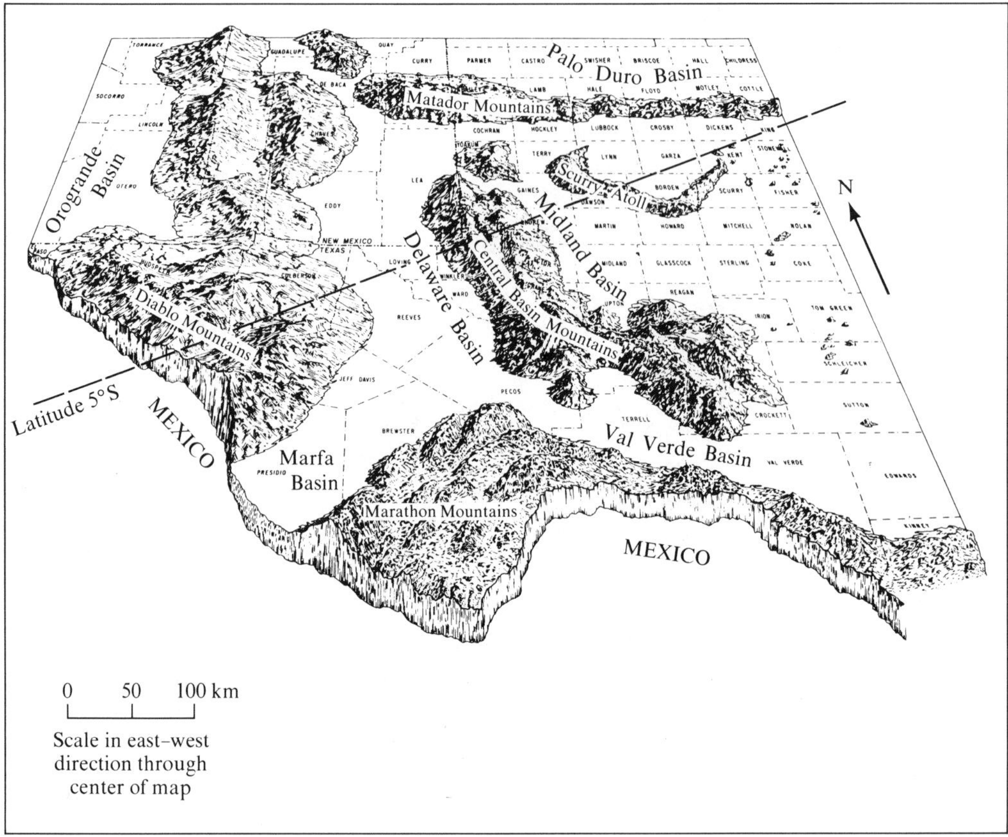

Figure 7–40
Paleogeographic perspective map of western Texas and southeastern New Mexico at close of Pennsylvanian time, showing numerous islands surrounded by basins hundreds of meters in depth. Also shown is approximate position of 5°S latitude, based on paleomagnetic data. Reefs grew prolifically around margins of land areas. [W. F. Wright, 1979, *Petroleum Geology of the Permian Basin* (Midland, Texas: West Texas Geol. Soc.).]

since early Cretaceous time, and the resulting limestone and dolomite now have a thickness of about 5,000 m—a rate of accumulation of 1 mm/25 yr. Around the Central Basin topographic feature in the Permian Basin of western Texas (see Figure 7–40), 800 m of limestone was formed in about 15 million years, a rate of 1 mm/20 yr.

3. Although the vast majority of accumulations of carbonate sediments forms in shallow marine waters, these sediments need not be many hundreds or thousands of kilometers from a major plate boundary. For example, the Jurassic-Cretaceous petroliferous limestones and dolomites of the Arabian Peninsula and southwestern Iran accumulated for 60 million years at a rate of 1 mm/40 yr on a shallow shelf adjacent to the boundary between the African and the Eurasian plate (Murris, 1980). The

boundary is marked by extensive ophiolites and deep-water radiolarian cherts of Late Cretaceous age that extend along the suture line NW–SE through southwestern Iran (see Figure 7–41). The boundary between the two plates became compressional about 105 million years ago. Deepening occurred in the marine carbonate basin to the west of the suture zone, and the coarsely fossiliferous, shallow-water sparry carbonates were succeeded in the deeper areas by argillaceous micritic limestones rich in globigerinid and other planktonic calcareous organisms. Carbonate deposition continued to be dominant in the Saudi Arabia–southwestern Iran area, however, until Miocene time, about 25 million years ago, when they were succeeded by evaporites and then by detrital silicate sediments from the Arabian craton to the west and the uplifted suture zone to the east.

4. There exist many examples at present of carbonate accumulations immediately adjacent to convergent plate boundaries, such as those along the western side of

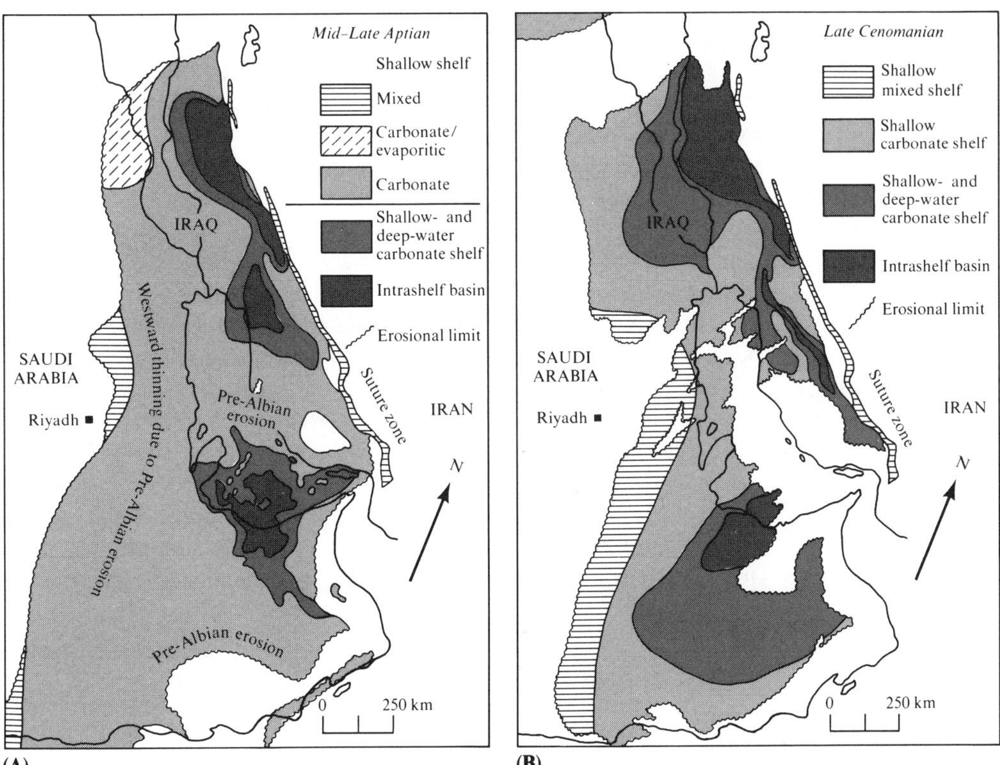

Figure 7–41
Depositional environments in the vicinity of Persian Gulf, showing change from shallow-water to deep-water carbonates that resulted from compression and suturing of African and Eurasian plates in southwestern Iran. (**A**) 108 million years ago. (**B**) 96 million years ago. [Murris, 1980.]

Sumatra and near the Philippine Islands. Because of their location adjacent to convergent plate margins, however, they are unlikely to be preserved in the geologic record. They may be killed quickly (in the geological sense) by inundation with volcanic detritus or by rapidly deepening water. In addition, a large part of the reef detritus may be carried directly into the adjacent deep-sea trench and subducted into the mantle.

CHEMICAL CONSIDERATIONS

Crystal Chemistry

Crystalline calcium carbonate occurs at the Earth's surface in two polymorphs—calcite (hexagonal) and aragonite (orthorhombic)—of which calcite is the more stable (less soluble). In surface seawater at 25°C in contact with atmospheric carbon dioxide, the solubility of calcite is about 15 ppm; aragonite, 16 ppm. Geochemists have calculated that normal, surface seawater is supersaturated with respect to calcite by 2.8×; with respect to aragonite by 1.9×. These data indicate that seawater should be perpetually clouded with a white precipitate of calcite and aragonite crystals but, as anyone who has been to the seashore is aware, no such precipitate occurs.

There are two reasons for the lack of inorganic precipitation of calcium carbonate from normal seawater. The first is that it is very demanding from the viewpoint of chemical energetics to form crystal nuclei in an aqueous solution. A very high degree of supersaturation is required—perhaps an order of magnitude or more. We noted previously (in Chapter 6) the absence of quartz precipitation in rivers despite the fact that average river water is supersaturated with respect to quartz by a factor of 2, and the nucleation of calcite and aragonite is probably as difficult as the nucleation of quartz. The second reason for the absence of inorganic nucleation of a calcium carbonate mineral is the large amount of magnesium ion in seawater, about 1,300 ppm. The crystallochemical properties of calcium ion and magnesium ion are sufficiently similar that a nucleating calcite crystal cannot accurately distinguish between them. More than 17,000 ppm (7 mol % $MgCO_3$) of magnesium ion is incorporated "by mistake" into the developing unit cells of the calcite in seawater, decreasing the stability of the mineral to the extent that it dissolves as soon as it forms. That is, a lower energy state is present when all the calcium, magnesium, and carbonate ions are dissolved in the seawater than when a calcite crystal that contains abundant magnesium is present. So calcite will not nucleate in normal, surface seawater—the place where nearly all allochems and micrite form.

Where, then, does this calcite and aragonite in marine limestones come from? The answer is that, with the exception of ooliths, it originates as calcareous fossil fragments. Living organisms can perform chemical feats not possible by purely inorganic mechanisms, and marine calcareous organisms have no difficulty in removing calcium and carbonate ions from supersaturated seawater to form their hard parts. In fact, some calcitic organisms include 25,000 ppm Mg^{2+} in their skeletons; bryozoa and echinoderms are examples. Other calcite secreters are able to discriminate very effectively against magnesium ions when building hard parts and contain almost no magnesium ion in their shells; brachiopods are an example. The reason for such differences in power of discrimination among different phylogenetic groups is unknown. Thus, a fresh accumulation of calcitic shells and hard parts of organisms can have a magnesium ion content between nearly zero and several tens of thousands of parts per million.

The aragonite crystal structure does not accommodate a significant amount of magnesium ion. It can, however, accommodate larger divalent ions such as strontium. The average aragonitic shell contains about 9,000 ppm (1 mol % $SrCO_3$) of Sr^{2+}. Some organisms build their shells of separate layers of calcite and aragonite and, therefore, can contain significant amounts of both magnesium and strontium; cephalopods are an example.

Pellets and ooliths are composed of aragonite in modern carbonate environments. Intraclasts, because of their varied origins, can consist of aragonite, calcite, dolomite, or combinations of these minerals. Many of the calcite crystals and all of the dolomite crystals in intraclasts are the product of recrystallization and replacement of preexisting aragonite.

In ancient rocks, it is extremely rare for aragonite to occur, although a few examples are known of aragonitic fossils of Paleozoic age. In all of these cases the shells have been entombed soon after burial either in volcanic ash or in organic matter, so that they have been protected from contact with pore waters during diagenesis. The key to understanding both the original formation and postburial changes in calcium carbonate crystals is the chemical composition of the waters in contact with the crystals and in interparticle pore spaces.

Mechanisms of Calcium Carbonate Equilibria

The chief control of the solubility of calcium carbonate is the hydrogen ion concentration (pH), which is controlled by the partial pressure of carbon dioxide according to the linked series of reactions below:

$$CO_2 + H_2O \rightleftarrows H_2CO_3 \tag{1}$$

$$H_2CO_3 \rightleftarrows H^+ + HCO_3^- \tag{2}$$

$$HCO_3^- \rightleftarrows H^+ + CO_3^{2-} \tag{3}$$

$$CO_3^{2-} + Ca^{2+} \rightleftarrows CaCO_3 \tag{4}$$

The net result of these interactions can be summarized as

$$CO_2 + H_2O + CaCO_3 \rightleftarrows Ca^{2+} + 2HCO_3^- \tag{5}$$

This summary equation makes it clear that carbon dioxide gas dissolved in water is responsible for the dissolution of calcite and aragonite (or the prevention of their formation). We noted previously that geochemical calculations reveal normal, surface seawater to be slightly supersaturated with respect to both minerals and that a higher degree of supersaturation might make inorganic precipitation possible. This is true whether we are interested in the formation of limestone at the Earth's surface, the dissolution of limestone after burial, or the precipitation of carbonate cement in the pores of limestones or sandstones.

What are the mechanisms that Mother (or Father) Nature uses to accomplish a decrease in the amount of carbon dioxide dissolved in water and a consequent precipitation of calcium carbonate? Five mechanisms seem important: (1) temperature increase and evaporation, (2) agitation of the water, (3) increased salinity, (4) organic activity, and (5) changes in P_{CO_2} in the soil vadose zone.

Temperature Increase and Evaporation

All gases are less soluble in warm water, and this is a reason carbonate rocks form only in warm surface waters in tropical and subtropical seas, rather than in the cooler latitudes nearer the polar region or at depth.

Agitation of the Water

As anyone who has shaken a bottle of soft drink or beer is aware, the carbon dioxide dissolved in the liquid at the factory or brewery is released when the container is shaken. The rapid evolution of the gas causes the liquid to overflow the container. The same process operates in the sea, where water is blown by the wind over shallow substrates such as cratonic margins and volcanic peaks. The calcareous organisms that form reefs benefit from the agitation, which occurs at depths less than one-half the wavelength of the moving surface water.

Increased Salinity

Carbon dioxide is less soluble in saline waters than in fresh waters, so that an increase in salinity causes CO_2 to be released. An increase in salinity normally results from evaporation, which increases the amount of calcium in the solution as well as the amount of carbonate ion.

Organic Activity

A reef is a symbiotic community of plants (algae) and animals (mostly corals in modern seas but bryozoa, sponges, or other animals in ancient seas). Plants and animals have contrasting metabolisms in that plants ingest CO_2 (during photosynthesis) and emit O_2, while animals (such as humans) ingest O_2 and emit CO_2. An additional difference is that plants have a peak in metabolic activity during the daylight hours, while animal metabolism is less affected by sunlight. Thus, during the daylight hours the net effect of the gas exchange involving the reef community is the ingestion of CO_2 from the seawater and the precipitation of $CaCO_3$ (formation of shell material); that is, the reef grows. During the night, plant metabolism is depressed and the effect of the animal community dominates; the CO_2 content of the water increases and the reef dissolves. As a result, the pH of seawater in solution basins within the reef is about 0.3 pH unit higher in the afternoon than 12 hours later because changes in the CO_2 content of the water are reflected in its hydrogen ion concentration (Equations 1–3 above).

Changes in P_{CO_2} in the Soil Vadose Zone

Rainwater contains an amount of carbon dioxide tied to the amount in the air mass through which the water falls. The rainwater must then pass through the soil zone, in which the partial pressure of carbon dioxide is much greater than in the atmosphere. As a result, soil water is markedly enriched in carbon dioxide relative to water in the air. If the soil water then enters a cave, in which the P_{CO_2} resembles that of normal air, carbon dioxide will be released from the water, resulting in features such as stalactites and stalagmites.

MODERN CARBONATES AND ANCIENT ANALOGS

Reefs are abundant in modern seas, and their study by geologists and biologists during the past 30 years has provided a good understanding of the ecologic and physicochemical factors that govern the formation of ancient reefal limestones. Examples of extensive modern reef development include the Great Barrier Reef along the northeastern coast of Australia, the Belize barrier reef and atoll complex, the reef tract along the southern tip of Florida, and the many atolls that surround volcanic peaks in the Pacific Ocean basin. These areas are all located where the sea surface temperature during the coldest winter months is at least 18°C—between about 25° north and south of the equator.

Southern Florida Reef Tract

The reef complex located at the southern tip of Florida is the only areally extensive carbonate depositional environment along the coast of the United States. It was one of the first modern carbonate environments to be studied in detail (Ginsburg, 1956), and the insights it provided still guide much present-day thinking about the biologic and environmental controls of limestone formation.

The southern Florida reef tract is located in a subtropical area at 25°N latitude, bordered on the northwest by the mangrove swamps of the Everglades and on the southeast by the warm Gulf Stream waters of the Atlantic Ocean (see Figure 7–42).

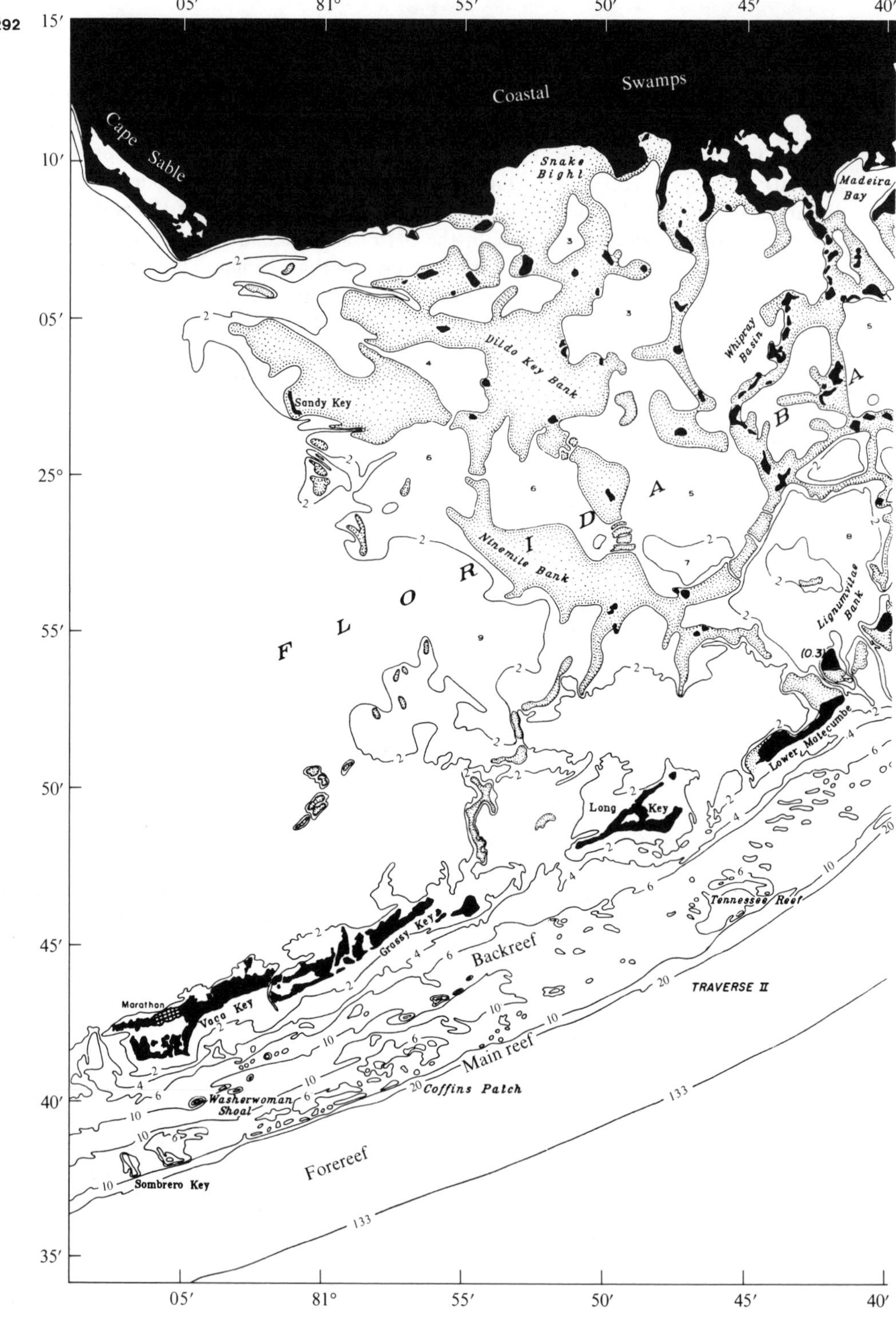

Figure 7–42
Bathymetric chart of southern Florida reef tract, showing locations

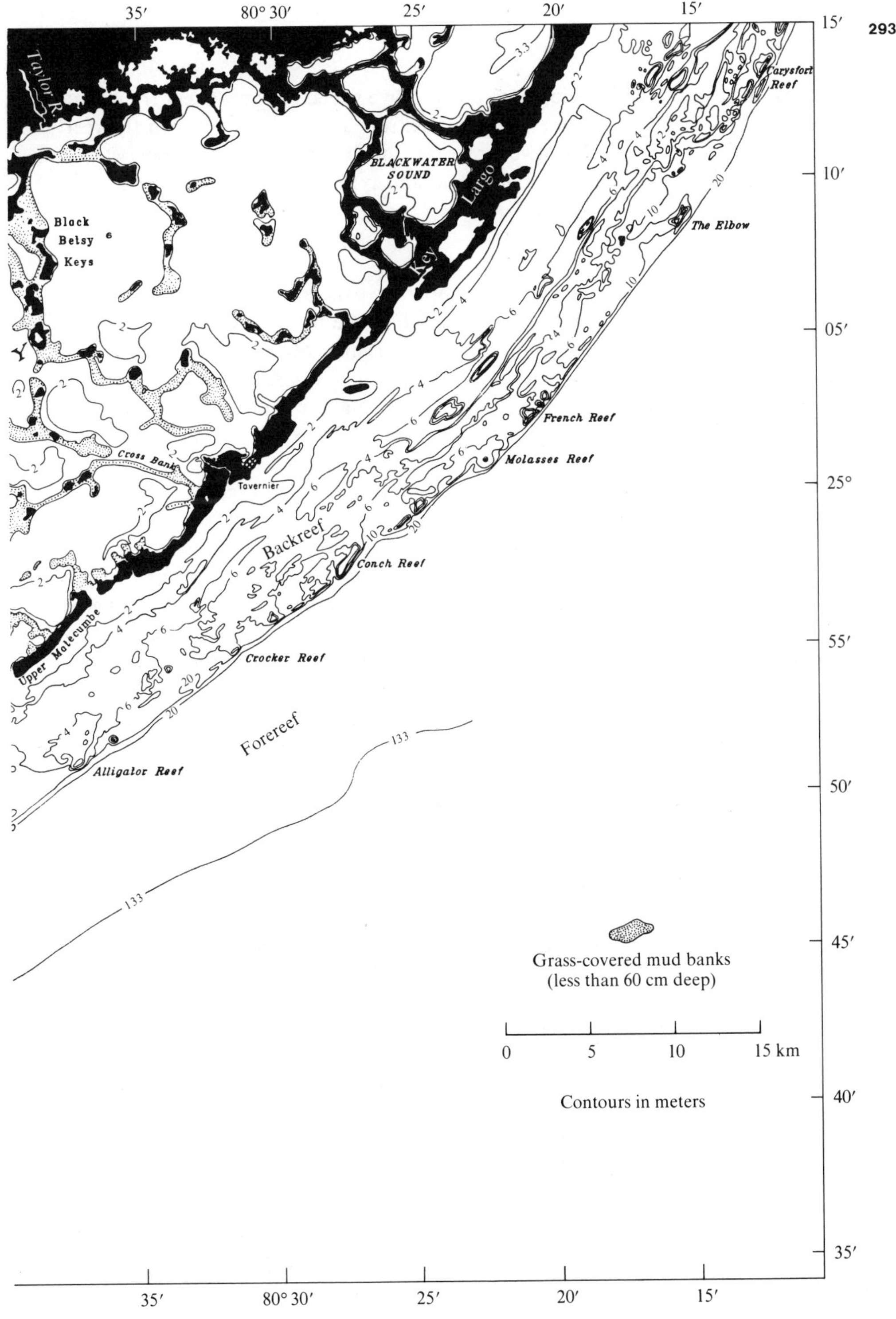

of forereef, reef, backreef, and Florida Bay. [Ginsburg, 1956.]

The area can be considered as being composed of two parts: the linear reef front that is washed by the warm Gulf waters, and the backreef area located between the reef front and the Florida Keys. The two areas are distinguished by different water-circulation patterns, salinities, flora and fauna, and types of carbonate particles.

The Reef

On the main reef tract the annual range in temperature of surface waters is 15–32°C; salinity, 35,000–37,000 ppm. This relatively small range in salinity is characteristic of marine areas that contain abundant living organisms. Most organisms are very sensitive to salinity and cannot survive large changes. The calcareous flora of the reef is dominated by two families of algae: the green Codiaceae and the red Corallinaceae. The fauna is extremely varied but is dominated by corals, mollusks, echinoderms, Foraminifera, worms, bryozoans, and crustaceans.

The reef is conveniently divided into three parts: forereef, main reef, and backreef, which grades westward into Florida Bay.

1. The *forereef* is the area in front of the growing reef below the depth of effective light penetration, 50–100 m. The sediment in this part of the reef setting is composed of poorly sorted gravel and sand derived from the disintegration of the growing reef—a result of the pounding by surface waves and the boring activities of marine predators. It is a submarine talus deposit. In its lower parts, there may be some submarine cementation by fibrous calcium carbonate; at the upper part, there can be cementation by algal growth where the forereef grades into the shallow, actively growing heart of the reef complex.

2. The *main reef* consists of a series of elongate, living reefs and rocky shoals separated by areas of deeper waters that are usually floored with ripple-marked carbonate sand and gravel. As shown in Figure 7–42, the reefs with distinct topographic relief rise from water shallower than 20 m, and more commonly less than 15 m, reflecting the depth below which more than half the visible light has been absorbed by the seawater (see Figure 1–4). The primary structural element of the main-reef mass is the moosehorn coral, *Acropora palmata* (see Figure 7–43), which grows at a rate of about 2 cm/yr. Sand- and gravel-size detritus broken from the reef accumulates among the coral branches, and the entire mass is stabilized by a covering of the hydrozoan *Millepora alcicornis* and encrusting algae. Behind the wall-like masses that take the full force of the waves (which generate the forereef talus), many other genera of corals grow and their broken fragments accumulate. In summary, the actively growing part of the reef is an intricately complex mixture of *in situ* coral–algal growth (termed *biolithite* in ancient rocks) and its debris.

3. The *backreef* area is floored either by sediment with or without a thick cover of marine grass or by patch reefs. The marine grass is the species *Thalassia testudinum* ("turtle grass"; see Figure 7–44). The bafflelike carpet of grass provides a protected habitat that is inhabited by a variety of calcareous algal and coral genera.

Figure 7-43
Branching colonies of *Acropora palmata* in upper half, wavy bladed hydrozoan *Millepora complanata* in foreground, and two varieties of soft-bodied Gorgonian corals (lower right and left center), Key Largo, Florida. Depth is 3 m; width of photo is 2 m. [H. G. Multer, 1969, *Carbonate Rock Environments: Florida Keys and Western Bahamas* (Published by the author, Fairleigh Dickinson University, Madison, NJ). Photo courtesy R. C. Murray.]

The prominent algae are *Halimeda, Penicillus,* and *Goniolithon; Porites* is the abundant coral genus. The grass baffles also trap microcrystalline carbonate sediment suspended in the water that is continuously washing over the area.

The Bay

Surface water temperatures in Florida Bay behind the reef have a range similar to that over the reef, but salinities are much more variable because of seasonal freshwater runoff from mainland Florida. As a result, the flora and fauna are more limited both in diversity and in abundance. Most of the bay is floored by micrite trapped by

Figure 7–44
Sharp blades of *Thalassia* grass, solitary *Penicillus* (top center; also two clusters in upper third of photo), and hairy filaments of *Batophora,* Berry Islands, Bahamas. Depth is 1.5 m. Blades of *Thalassia* are about 15 cm long. [Bathurst, 1975. Photo courtesy R. G. C. Bathurst.]

the *Thalassia* grass. Carbonate mud mounds rise to the sea surface in many areas where carbonate mud has locally accumulated (see Figure 7–21). The relatively small amount of sand-size sediment in the bay is composed of 80% molluscan fragments, with almost all of the remainder being foraminiferal debris. Algal pieces form only 1% of the sand. The striking difference between the sediment in the reef tract and that in Florida Bay is shown in Figure 7–45.

El Abra Reef (Cretaceous), Mexico and Texas

The southern tip of Florida illustrates the physiography and depositional facies along the current southern shoreline of the North American plate. Very similar physiographies and carbonate facies are found in ancient limestone units along the paleoshorelines of the plate, one of which is the El Abra reef complex (Lower Cretaceous), which crops out west of Tampico, Mexico, and in the Edwards Formation in the subsurface of southeastern Texas (see Figure 7–46). Analysis of the El Abra reef indi-

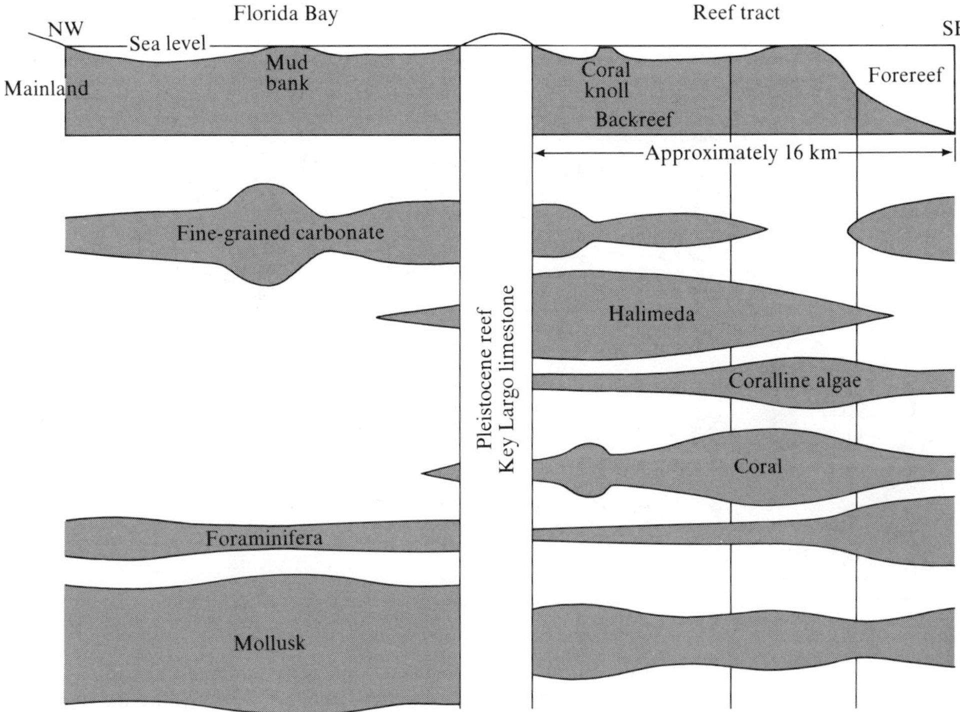

Figure 7–45
Generalized variation in relative abundance of grain types in sediment of southern Florida reef tract. [Ginsburg, 1956.]

cates that there are four major facies present. Each facies is composed of several types of particles (as is each of the Florida facies) but can be conveniently referred to by its dominant constituent. The stratigraphic arrangement of the facies is shown in Figure 7–47, which is a cross section through quarry outcrops near the southern end of the reef complex. Because of the very large scale of the cross section, the El Abra reef seems more complex than the modern Florida tract, but in fact it is not. Our discussion of the modern reef complex considered only the facies pattern existing at the present seafloor surface and ignored the complexities that would appear in a cross section. The complexities in the third dimension are caused by changes in the pattern of reef growth, current patterns, and shoreline position through time. The period of time represented by the 40–50 m thickness of limestone in the cross section is about 10 million years—plenty of time for many shifts in sediment distribution pattern to occur.

In the El Abra reef, the framework builder is an extinct and aberrant type of attached pelecypod called a rudistid (see Figure 7–48), which was a major reef

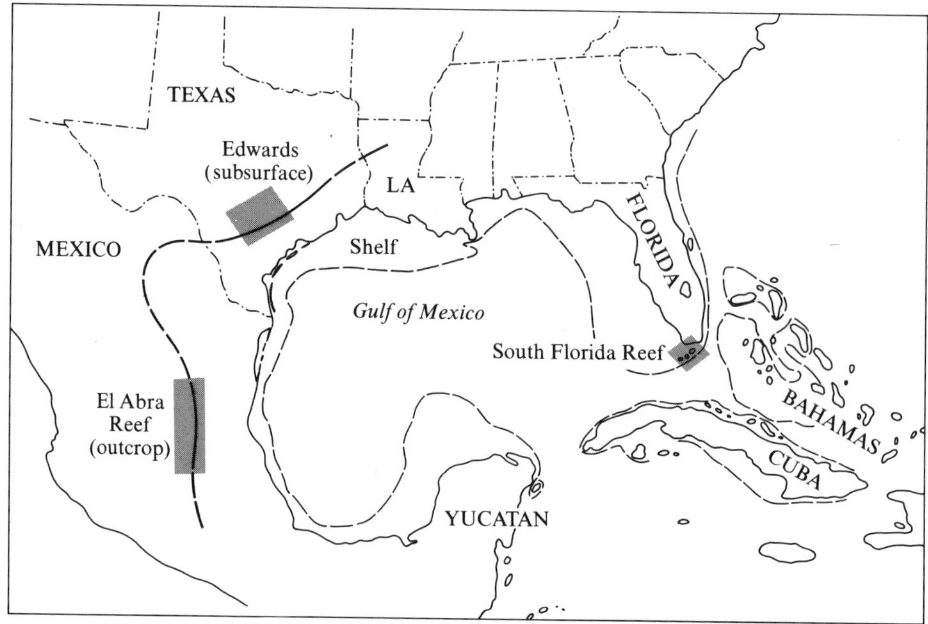

Figure 7–46
Map showing locations of El Abra reef and Edwards Formation reef (Cretaceous) and modern reef of southern Florida. Heavy dashed line indicates trend of reef. Light dashed lines indicate shelf boundary. [Griffith et al., 1969.]

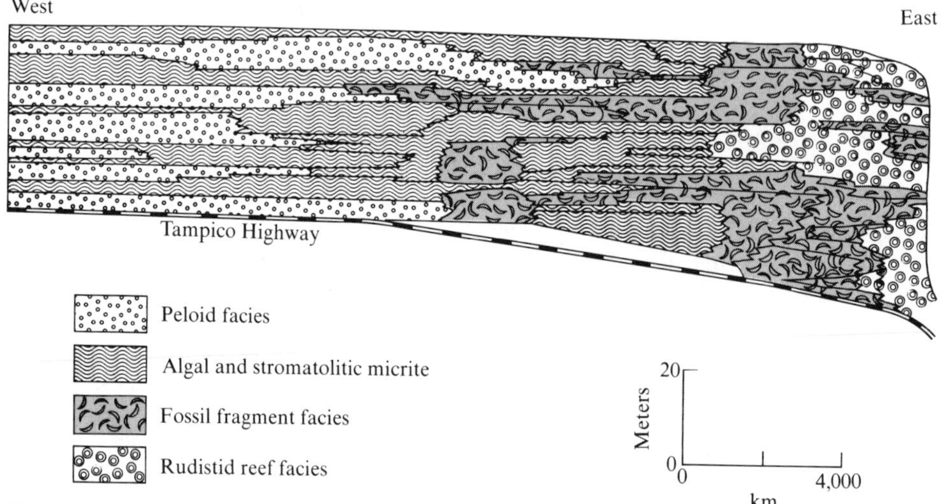

Figure 7–47
Cross section showing distribution of four major facies of El Abra Limestone. [Griffith et al., 1969.]

Figure 7–48
Colony of Upper Cretaceous rudistids (*Hippurties mexicana*) in limestone bed of Cuautla Formation near San Lucas, Mexico. [K. Segerstrom, 1962, *U.S. Geol. Surv. Bull. 1104C.* Photo courtesy K. Segerstrom, U.S. Geol. Surv.]

former during Cretaceous time. The skeletal sand facies to the west of the organic reef is massive, shows little bedding, and is composed of detritus derived from the physical and biological destruction of the rudist reef. To the west, the skeletal sand changes abruptly into stromatolites, algal micrite, and peloids—the same type of sediment seen in Florida Bay.

The reef facies in the Edwards Formation (Lower Cretaceous) of southeastern Texas were described from well cuttings and cores obtained during exploration for oil and gas. The distribution of facies reveals a depositional pattern identical to that observed in the El Abra outcrops, with the addition of a dolomite–anhydrite facies shoreward of the backreef area. This latter facies represents a supratidal zone in which calcium sulfate precipitated and dolomite then replaced original limestone (see Chapter 8).

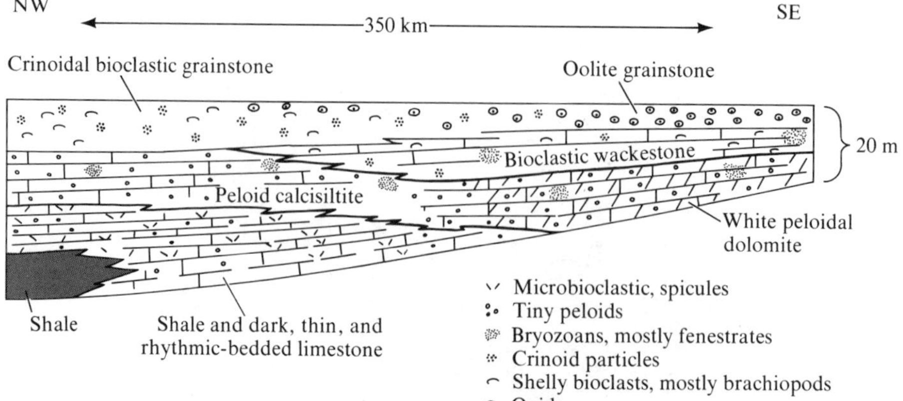

Figure 7–49
Idealized shallowing-upward subtidal (shale) and intertidal epeiric sea carbonate cycle, Lodgepole Formation (Mississippian), Montana. Terminology is that of Dunham (1962). [Wilson, 1975.]

Epeiric Sea Limestones

The only modern locations that qualify as epeiric seas are the Baltic Sea and Hudson Bay, both located above 50°N latitude and, therefore, unable to support the accumulation of carbonate sediments. The present is a poor key to the past for the nonreefal, tabular limestone units that characterize, for example, the Paleozoic rocks of mid-continental North America. Yet such biostromes are probably the most common limestone deposits. Their shallow-water origin is demonstrated by abundant mud cracks, stromatolites, and current-generated structures, as well as by the dominance of ooliths and shelled benthonic fossils such as crinoids among the allochems.

Carbonate sediments characteristically accumulate at rates much greater than the rate of subsidence of the shelf or platform on which they are deposited. This occurs because carbonate sediments are produced in the environment of deposition, in shallow water where conditions for the biochemical fixation of calcium carbonate are optimum. As a result, carbonate accumulations repeatedly build up to sea level and slightly above, resulting in a characteristic sequence of deposits in which each unit is deposited in progressively shallower water. Commonly, these sequences are capped by supratidal dolomites and evaporites as the final stage. This shallowing-upward sequence is commonly repeated many times in a succession of shallow-water deposits. An idealized shallowing-upward cycle is shown in Figure 7–49.

Manlius Formation (Devonian), New York

One of the better petrologic–stratigraphic studies of a shallowing-upward carbonate sequence is of the Manlius Formation (Lower Devonian) by Laporte (1967). He distinguished three facies, which he termed subtidal, intertidal, and supratidal, each with a characteristic group of sedimentologic and petrologic features.

1. The *subtidal facies* is a skeletal, peloidal micrite that commonly contains well-developed tabular masses of stromatoporoids (an extinct probable coelenterate; see Figure 7–50). The relatively diverse biota of this facies required continuous marine submergence, but the close association and juxtaposition with rocks that clearly indicate periodic subaerial emergence suggest that water depths were very shallow, just below mean sea level. The stromatoporoids grew either as encrusting masses within tidal creeks or channels or as individual heads at the subtidal–intertidal boundary. Rates and effects of burrowing organisms on the carbonate sediment are higher in the subtidal facies than in the intertidal zone, and there is a consistent lack of planar orientation of skeletal remains as a result.

2. The *intertidal facies* is composed of thin beds (2–10 cm) of sparsely fossiliferous, peloidal micrite and biosparite; individual beds commonly lie unconformably on those below (see Figure 7–51). Many primary sedimentary structures are present, especially scour and fill, cross bedding, and limestone pebble conglomerate. The pebbles in the conglomerate are fragments of the underlying subtidal micrite. The

Figure 7–50
Weathered surface of stromatoporoid-rich bed from subtidal facies. Individual stromatoporoids are irregularly globular and commonly intergrown with each other. Matrix is biomicrite. [Laporte, 1967.]

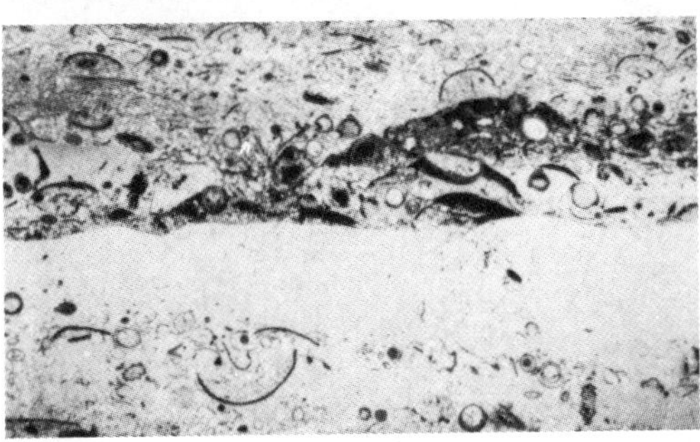

Figure 7–51
Negative print of thin section in intertidal facies, showing biosparite lying unconformably on micrite. Biosparite grades upward into micrite, which is truncated above by a second biosparite, above which lies a second micrite (not visible). Abundant fossils include gastropods, bryozoa, and ostracods. [Laporte, 1967.]

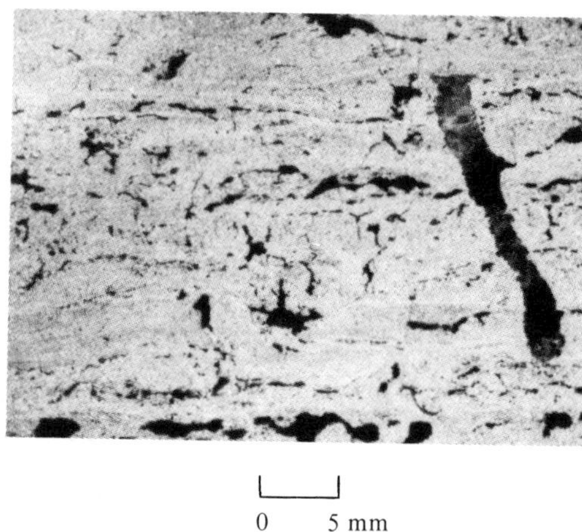

Figure 7–52
Negative print of thin section in supratidal facies, showing spar-filled, irregular vugs (bird's-eye structure), which are interpreted as internal desiccation cracks; larger, vertical vug may be gas trackway. Matrix is slightly dolomitic pelmicrite. [Laporte, 1967.]

biota in the intertidal facies contains few taxa but abundant individuals. Many of the elongate fossils show a strong preferred orientation, consistent with tidal influence during deposition. Algal stromatolites are common. Some dolomite rhombs and quartz silt occur, and commonly there are shaly partings between the thin limestone beds. A few mud cracks and minor erosional relief on micrite beds indicate intermittent subaerial exposure.

3. The *supratidal facies* is characterized by nearly unfossiliferous, laminated, mud-cracked, dolomitic, peloidal micrite. Disrupted mud-crack polygons form intraclasts. The mud cracks and spar-filled vugs (bird's-eye structure) that occur indicate frequent subaerial exposure; thin bituminous films separating individual carbonate laminae suggest the presence of algal mats.

The supratidal facies formed several centimeters to tens of centimeters above mean high-tide level. Occasional storms or unusually high tides inundated broad supratidal flats, depositing sediment and occasionally fine-grained fossil debris. As the waters receded, algal mats temporarily flourished on the wet substrate (see Figure 7–52). Subsequent desiccation of the flats caused the mud cracks to form. Dolomitization occurred penecontemporaneously with calcium carbonate accumulation, as occurs today in the Persian Gulf and elsewhere (see Chapter 8).

SUMMARY

Limestones are essentially monomineralic rocks of bioclastic origin. In nearly all limestones the only inorganic crystals of calcium carbonate are those in ooliths and in sparry cement. As a result, it is not possible to understand these rocks without considerable knowledge of the biochemical interactions between marine organisms and the waters in which they live.

The textures and structures in limestones are more complex than those in sandstones because of these biologic influences. In addition to the textural and structural features of sandstones, such as grain-size variation, cross bedding, and replacement of early cement by later cement, there occur in limestones algal boring of clastic particles to produce micritic envelopes around grains, voluminous amounts of fecal pellets, and lithification of loose calcareous debris by the growth of living organisms on grain surfaces. A good limestone petrologist must be part biologist, part biochemist, part petrographer, and part sedimentologist.

The major environments of carbonate deposition are epeiric seas, reefs, and the deep oceanic basins, probably in that order of abundance in the geologic record. The biostromes formed in epeiric seas usually form shallowing-upward sequences because of the rapid growth rates of organisms compared to rates of subsidence. A change from subtidal micrite to intertidal biosparite and oosparite to supratidal dolomitic micrite is characteristic, but extensive lateral variation occurs in this stratigraphic norm because of local shoreline and current variations.

Reefs and carbonate mounds create their own environments of deposition within a carbonate basin by forming a moundlike, wave-resistant structure that may reach hundreds of meters above the level of the adjacent basin floor. A reef is composed of forereef talus, the framework core, and the backreef lagoon or shallows. In many reefs the talus forms the bulk of the sediment, and the framework the smallest part, but much variation occurs. Some "reefs" are actually lithified mounds of mud formed by the accumulation of calcareous sediment by currents in the environment of deposition.

The carbonate sediment in deep-sea accumulations forms within a few meters of the sea surface and represents the fraction of the shell material that survived partial dissolution while settling thousands of meters to the ocean floor. The greater the depth, the smaller the percentage of surface production that is preserved. In addition, the bulk of the carbonate debris that reaches the ocean floor is probably carried into an oceanic trench and subducted into the mantle, so that deep-sea limestones are uncommon on the continents.

FURTHER READING

Bathurst, R. G. C. 1975. *Carbonate Sediments and Their Diagenesis,* 2nd ed. New York: Elsevier, 658 pp.

Berger, W. H., and E. L. Winterer. 1974. Plate stratigraphy and the fluctuating carbonate line. In K. J. Hsü and H. C. Jenkyns (eds.), *Pelagic Sediments: On Land and Under the Sea.* Internat. Assoc. Sedimentol. Spec. Pub. No. 1, pp. 11–48.

Cook, H. E., and P. Enos (eds.). 1977. *Deep-Water Carbonate Environments.* Soc. Econ. Paleontol. Mineral. Spec. Pub. No. 25, 336 pp.

Dunham, R. J. 1962. Classification of carbonate rocks according to depositional texture. In W. E. Ham (ed.), *Classification of Carbonate Rocks.* Amer. Assoc. Petroleum Geol. Mem. No. 1, pp. 108–121.

Fischer, A. G., S. Honjo, and R. E. Garrison. *Electron Micrographs of Limestones*. Princeton, NJ: Princeton University Press, 141 pp.

Folk, R. L. 1959. Practical petrographic classification of limestones. *Amer. Assoc. Petroleum Geol. Bull., 43,* 1–38.

Folk, R. L. 1962. Spectral subdivision of limestone types. In W. E. Ham (ed.), *Classification of Carbonate Rocks*. Amer. Assoc. Petroleum Geol. Mem. No. 1, pp. 62–84.

Friedman, G. M. (ed.). 1969. *Depositional Environments in Carbonate Rocks*. Soc. Econ. Paleontol. Mineral. Spec. Pub. No. 14, 209 pp.

Frost, S. H., M. P. Weiss, and J. B. Saunders (eds.). 1977. *Reefs and Related Carbonates—Ecology and Sedimentology*. Amer. Assoc. Petroleum Geol. Studies in Geol. No. 4, 421 pp.

Fürsich, F. T. 1973. *Thalassinoides* and the origin of nodular limestone in the Corallian Beds (Upper Triassic) of southern England. *Neues Jahrbuch für Geologie und Paläontologie, Monatshefte,* pp. 136–156.

Ginsburg, R. N. 1956. Environmental relationships of grain size and constituent particles in some south Florida carbonate sediments. *Amer. Assoc. Petroleum Geol. Bull., 40,* 2384–2427.

Griffith, L. S., M. G. Pitcher, and G. W. Rice. 1969. *Quantitative Environmental Analysis of a Lower Cretaceous Reef Complex*. Soc. Econ. Paleontol. Mineral. Spec. Pub. No. 14, pp. 120–138.

Horowitz, A. S., and P. E. Potter. 1971. *Introductory Petrography of Fossils*. New York: Springer-Verlag, 202 pp.

Hurst, J. M. 1980. Paleogeographic and stratigraphic differentiation of Silurian carbonate buildups and biostromes of North Greenland. *Amer. Assoc. Petroleum Geol. Bull., 64,* 527–548.

James, N. P. 1979a. Introduction to carbonate facies models. In R. G. Walker (ed.), *Facies Models*. Geol. Assoc. Canada Reprint Series No. 1, pp. 105–107.

James, N. P. 1979b. Reefs. In R. G. Walker (ed.), *Facies Models*. Geol. Assoc. Canada Reprint Series No. 1, pp. 121–132.

James, N. P. 1979c. Shallowing-upward sequences in carbonates. In R. G. Walker (ed.), *Facies Models*. Geol. Assoc. Canada Reprint Series No. 1, pp. 109–119.

Jenkyns, H. C. 1974. Origin of red nodular limestones (Ammonitico Rosso, Knollenkalke) in the Mediterranean Jurassic: a diagenetic model. In K. J. Hsü and H. C. Jenkyns (eds.), *Pelagic Sediments: On Land and Under the Sea*. Internat. Assoc. Sedimentol. Spec. Pub. No. 1, pp. 249–271.

Kelts, K., and K. J. Hsü. 1978. Freshwater carbonate sedimentation. In A. Lerman (ed.), *Lakes: Chemistry, Geology, Physics*. New York: Springer-Verlag, pp. 295–323.

Lake, J. H. 1981. Sedimentology and paleoecology of Upper Ordovician mounds of Anticosti Island, Quebec. *Canad. Jour. Earth Sci., 18,* 1562–1571.

Laporte, L. F. 1967. Carbonate deposition near mean sea-level and resultant facies mosaic: Manlius Formation (Lower Devonian) of New York State. *Amer. Assoc. Petroleum Geol. Bull., 51,* 73–101.

Laporte, L. F. (ed.). 1974. *Reefs in Time and Space*. Soc. Econ. Paleontol. Mineral. Spec. Pub. No. 18, 256 pp.

Longman, M. W. 1980. Carbonate diagenetic textures from nearsurface diagenetic environments. *Amer. Assoc. Petroleum Geol. Bull., 64,* 461–487.

McIlreath, I. A., and N. P. James. 1979. Carbonate slopes. In R. G. Walker (ed.), *Facies Models.* Geol. Assoc. Canada Reprint Series No. 1, pp. 133–143.

Majewske, O. P. 1969. *Recognition of Invertebrate Fossil Fragments in Rocks and Thin Sections.* Leiden: E. J. Brill, 101 pp.

Milliman, J. D. 1974. *Marine Carbonates.* New York: Springer-Verlag, 375 pp.

Murris, R. J. 1980. Middle East: stratigraphic evolution and oil habitat. *Amer. Assoc. Petroleum Geol. Bull., 64,* 597–618.

Park, W. C., and E. H. Schot. 1968. Stylolites: their nature and origin. *Jour. Sed. Petrology, 38,* 175–191.

Perkins, R. D., and C. I. Tsentas. 1976. Microbial infestation of carbonate substrates planted on the St. Croix shelf, West Indies. *Geol. Soc. Amer. Bull., 87,* 1615–1628.

Playford, P. E. 1980. Devonian "Great Barrier Reef" of Canning Basin, Western Australia. *Amer. Assoc. Petroleum Geol. Bull., 64,* 814–840.

Pray, L. C., and M. Estaban (eds.). 1977. Upper Guadelupean facies, Permian reef complex, Guadelupe Mountains, west Texas and New Mexico. Soc. Econ. Paleontol. Mineral. Pub. No. 77–16, pp. 79–118.

Scholle, P. A. 1978. *Carbonate Rock Constituents, Textures, Cements and Porosities.* Amer. Assoc. Petroleum Geol. Mem. No. 27, 241 pp.

Stauffer, K. W. 1962. Quantitative petrographic study of Paleozoic carbonate rocks, Caballo Mountains, New Mexico. *Jour. Sed. Petrology, 32,* 357–396.

Stevenson, D. L., T. L. Chamberlain, and T. C. Buschbach. 1975. *Insoluble Residues of the Sauk Sequence (Cambrian and Lower Ordovician) Rocks of the Fairfield Basin, Illinois: An Aid in Correlation and in Petroleum Exploration.* Ill. Geol. Surv., Ill. Petroleum No. 106, 12 pp.

Stockman, K. W., R. N. Ginsburg, and E. A. Shinn. 1967. The production of lime mud by algae in south Florida. *Jour. Sed. Petrology, 37,* 633–648.

Wanless, H. R. 1979. Limestone response to stress: pressure solution and dolomitization. *Jour. Sed. Petrology, 49,* 437–462.

Wantland, K. 1975. *Belize Shelf—Carbonate Sediments, Clastic Sediments, and Ecology.* Amer. Assoc. Petroleum Geol. Studies in Geol. No. 2, 599 pp.

Wilson, J. L. 1975. *Carbonate Facies in Geologic History.* New York: Springer-Verlag, 471 pp.

8

Dolomites

Observe always that everything is the result of change,
and get used to thinking that there is nothing
Nature loves so well as to change existing forms
and to make new ones like them.

MARCUS AURELIUS ANTONINUS

The term *dolomite* is used for both the mineral, $CaMg(CO_3)_2$, and the rock composed mostly of the mineral. Occasionally, *dolostone* is used to designate the rock, but the word has not caught on in the 35 years since it was proposed. It is a good example of a term whose need is dictated by logic but, nevertheless, has been successfully resisted by the petrologic community. In most cases, the context in which the term dolomite is used serves to reveal whether it is the mineral or the rock that is meant.

Dolomites occur in approximately the same tectonic and physiographic settings as limestones: on the shallow shelves of low-lying continents, most commonly far from the nearest convergent plate margin. Like limestones, dolomites are essentially monomineralic, so that their study emphasizes variations in textures and structures. Geochemical analyses are an even more important feature of dolomite investigations than of limestone studies because, as we shall see, nearly all dolomites form as replacements of limestones.

Extensive field studies during the past century have documented that dolomite occurs in rocks of all ages but is conspicuously more abundant in older rocks. It is rare in modern carbonate environments, forms less than 5% of the carbonate rocks of Tertiary age, about 10% of those of Mesozoic age, 35% of those of Paleozoic age, and more than two-thirds of Precambrian carbonates (Garrels and Mackenzie, 1971). Field studies have also revealed that carbonate beds tend to be composed of either all calcite or all dolomite (see Figure 8–1). Apparently, whatever the conditions are that produce limestones and dolomites, a marked tendency exists to form one or the other but not subequal mixtures of the two. Why is this so? Does it reveal something about

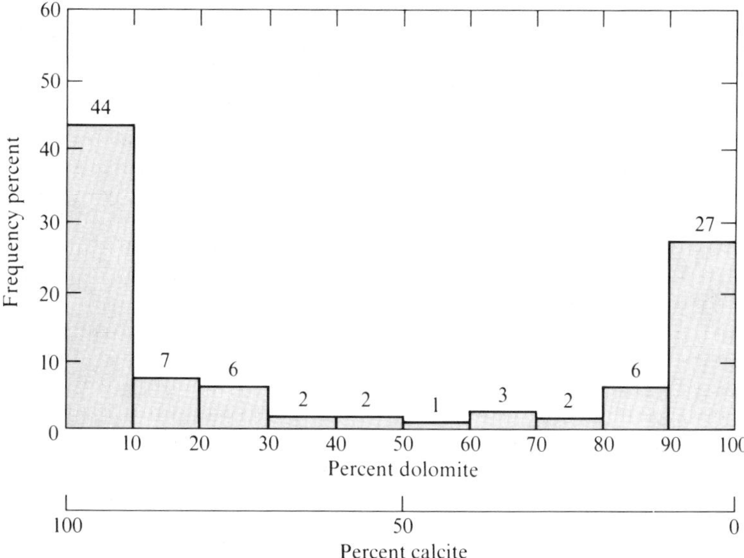

Figure 8–1
Computed percentages of calcite and dolomite for 1,148 modal analyses of North American carbonate rocks. [E. Steidtmann, 1917, *Geol. Soc. Amer. Bull.*, 28.]

the environments of deposition of the two types of rocks? Or does it tell us something about the diagenetic history of carbonate accumulations? As usual, the place to begin the search for answers to such questions is in the field.

FIELD OBSERVATIONS

As was true of limestones, the bulk of dolomites is marine, as indicated by stratigraphic relationships and the presence of fossils, although fossils are noticeably less common in dolomites. Most dolomite units occur as distinct beds or formations interlayered with other nondetrital sedimentary rocks of similar thickness. The contacts with limestone units above and below are usually sharp, but a dolomite bed may grade laterally into another rock type, typically limestone or evaporite. In most cases, the evaporite is gypsum or anhydrite. In some limestone–dolomite sequences the contact between the two types of rocks is irregular or cuts across bedding planes at a high angle (see Figure 8–2). This relationship clearly indicates that the dolomite has formed by replacement of preexisting limestone. Etching a hand specimen of the limestone associated with the dolomite may reveal that easily identifiable and originally calcitic fossils such as crinoid columns or brachiopods are composed of dolomite, although the matrix of the rock is not. Sometimes it is possible with a hand lens

(A)

(B)

Figure 8–2

(**A**) Contact between dolomite (light, right) and limestone (dark, left) at high angle to bedding (upper left to lower right), Lost Burro Formation (Devonian), California. The dolomite–limestone contact is sharp but irregular. [Chilingar et al., 1979. Photo courtesy D. H. Zenger.] (**B**) Contact between dolomite (light-colored, positive relief) and limestone as seen on bedding plane surface of Monte Cristo Limestone (Mississippian), Nevada. Some dolomite blebs appear isolated from main mass of dolomite. [Chilingar et al., 1979. Photo courtesy J. S. Shelton.]

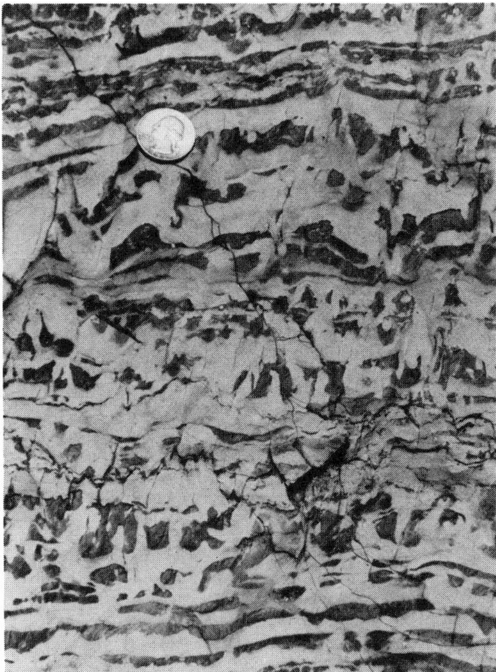

Figure 8–3
Carbonate rock composed of subequal percentages of calcite and dolomite, Rockdale Run Formation (Ordovician), Maryland. Light-gray micritic limestone is mottled very thoroughly by dark-gray dolomite along zones of paleopermeability during early diagenesis. Three stylolite seams can be seen in lower half. [F. J. Pettijohn and P. E. Potter, 1964, *Atlas and Glossary of Primary Sedimentary Structures* (New York: Springer-Verlag). Photo courtesy P. E. Potter.]

to see that rhombs of dolomite partly transect some of the fossils, although this is more easily seen in thin section in the laboratory. There are, however, many beds of pure dolomite that have perfectly conformable, sharp contacts with overlying and underlying limestones and, in addition, are microcrystalline (*dolomicrite*) and contain no visible allochems. In these rocks it is not possible to determine from field observations whether the dolomite bed was originally a limestone.

Mottling

Although most carbonate beds are nearly all calcite or all dolomite, some are not. It is not uncommon in these bimineralic rocks to find that the dolomite forms a mottled structure within the limestone, best seen on an etched or weathered surface. These mottles may have several origins (Osmond, 1956). Mottles approximately parallel to the bedding of the host limestone (see Figure 8–3) represent zones of relatively high

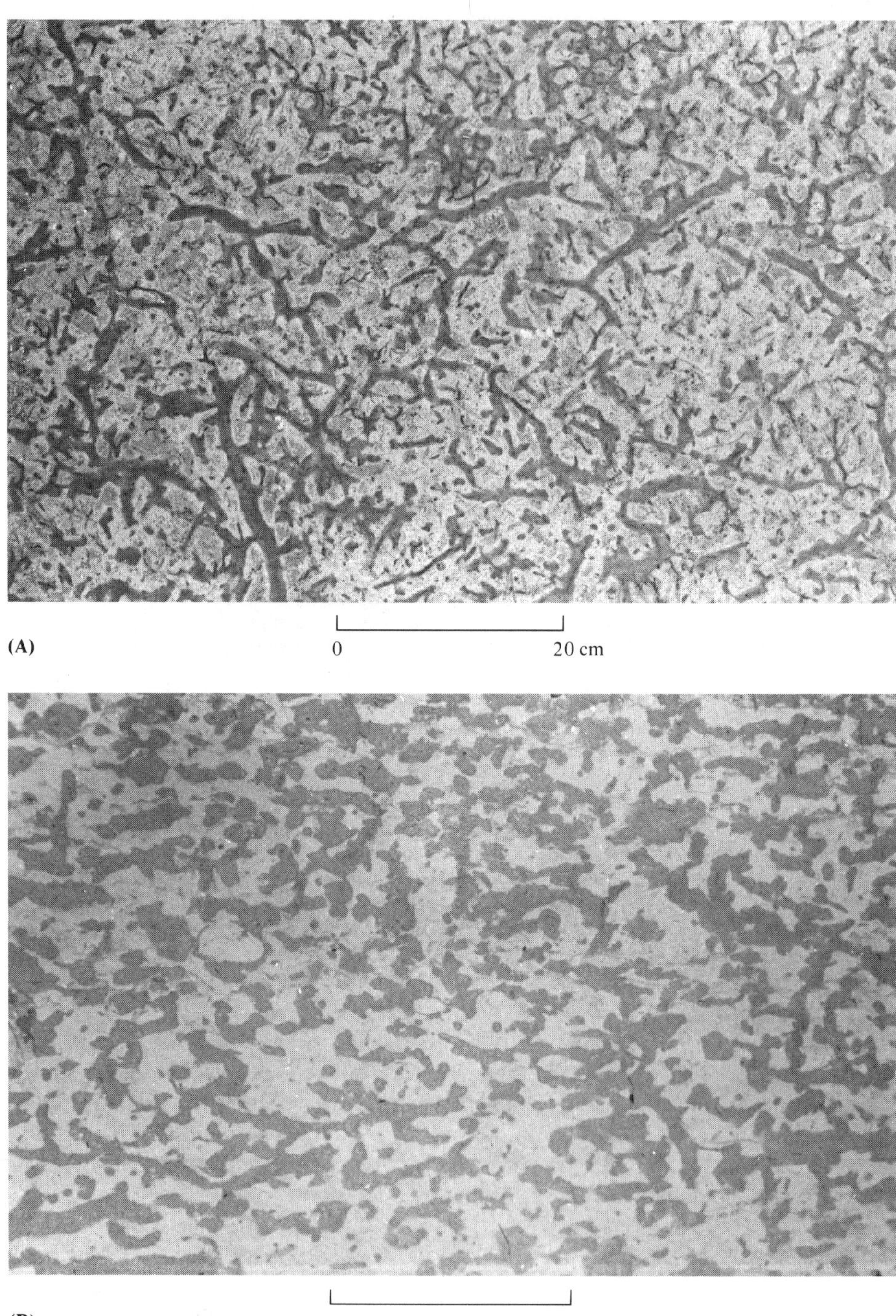

Figure 8–4
Dolomite mottles (dark gray) in limestone (light gray) resulting from filling of burrows, Selkirk Limestone (Ordovician), Manitoba, Canada. (**A**) Parallel to bedding. (**B**) Perpendicular to bedding. [Kendall, 1977. Photos courtesy A. C. Kendall.]

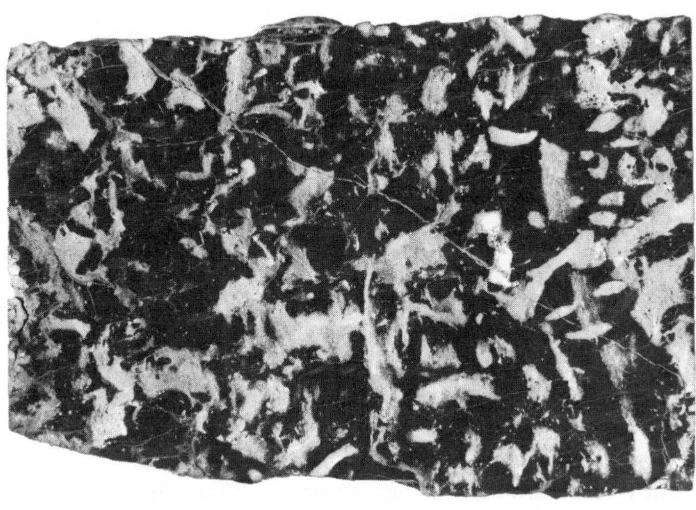

0 5 cm

Figure 8–5
Light-colored mottles of dolomite in black micritic limestone, Palliser Limestone (Devonian), Alberta, Canada. Irregular outlines and irregular distribution of mottles suggest that dolomite infilled vugs in host limestone during early diagenesis. [Photo by C. Weber. Courtesy F. J. Pettijohn.]

porosity and permeability during emplacement of the dolomite. The fact that in some places these mottles transect surfaces of stratification in the limestone indicates that the dolomite is secondary—a diagenetic phenomenon. In some occurrences the mottles are tubular (see Figure 8–4), suggesting that they formed by the infilling of borings by benthonic organisms (Kendall, 1977) in unlithified carbonate mud. In still other occurrences, the dolomitic mottles are more irregular (see Figure 8–5). Other origins of large-scale, macroscopically visible mottling are possible, and deciphering the most probable origin of these structures depends to a great extent on the investigator's understanding of structures, textures, and mechanism of formation of the host limestones.

Paleogeography

Although it is clear from field relationships that both limestones and dolomites form along the fringe of the low-lying craton, there is both local and regional evidence indicating that dolomite commonly forms nearer the shoreline than does limestone and on the shelf rather than within the basin. In ancient reef environments where the reef is growing at the edge of a continental mass, the seaward front of the reef is typically limestone. The landward, backreef part has commonly been replaced by dolomite. On a regional scale and over extensive stratigraphic intervals, the same ten-

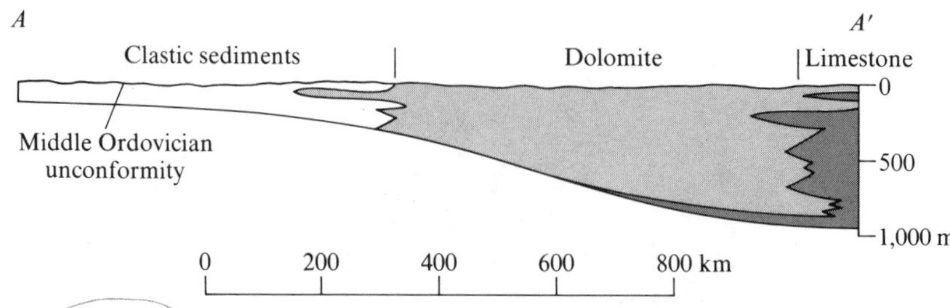

Figure 8–6
Generalized paleogeographic map and cross section of Late Cambrian and Early Ordovician time, showing regional distribution of facies of Knox Group and equivalent rocks on cratonic shelf. [Harris, 1973.]

dency is present (see Figure 8–6). When we consider that the limestone was formed only a few meters below high tide, the association between dolomite formation and extremely shallow water becomes even more striking.

Paleolatitude

In our discussion of limestones, we noted that they occurred in low latitudes, generally within 30° of the equator. If dolomites are formed by geologically instantaneous replacement of these limestones, that is, replacement within a million years after the limestone was formed, we would expect ancient dolomites to have the same latitudinal distribution as limestones. As shown in Figure 8–7, we are not disappointed. As was true of limestone reefs (see Figure 7–32), the present locations of ancient dolomites cluster in the high latitudes, but the latitudes in which they formed were almost all within 30° of the equator.

LABORATORY STUDIES

Distinguishing Dolomite from Calcite

The best general technique for distinguishing between dolomite and calcite in thin section is staining (see Chapter 15). The most commonly used stain, Alizarin Red S, turns calcite pink but leaves dolomite unaffected. In unstained thin-section slides, other criteria are useful: (1) euhedral crystal habit, (2) zoning, and (3) twinning.

Euhedral Crystal Habit

The mineral dolomite has a very strong tendency to form euhedral crystals, and these can be seen even in dolomicrites with careful examination. Calcite occurs in euhedral

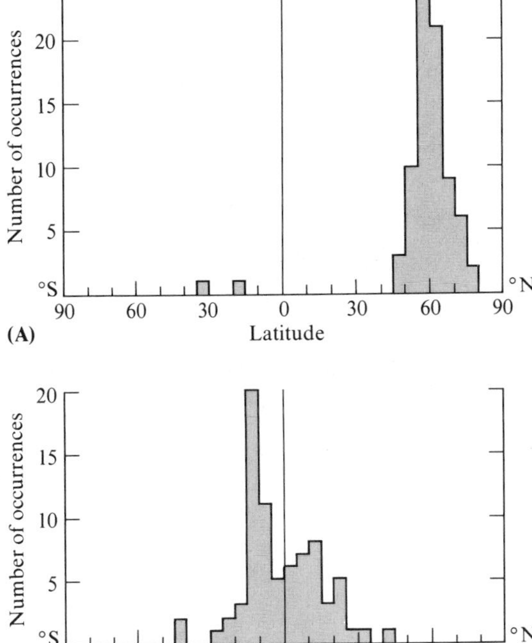

Figure 8–7
Latitudinal histograms for dolomites. (**A**) Present latitude. (**B**) Paleolatitude. [J. C. Briden and E. Irving, in A. E. M. Nairn (ed.), 1964, *Problems in Paleoclimatology* (New York: Wiley).]

habit only rarely. Rhombs of calcite in either limestones or dolomites are interpreted to indicate calcitization of dolomite crystals. Calcite has a very weak tendency to crystallize as euhedra.

Zoning

Dolomite crystals, particularly those of sand size or larger, commonly contain concentric, alternating zones of iron-rich (red) and iron-poor (clear) dolomite that mark stages of growth of the rhomb. Dolomite crystals can include significant percentages of ferrous iron in their crystal structure as a substituent for magnesium. Iron cannot substitute for calcium, so that calcite is never visibly zoned. Many calcitized dolomite rhombs retain the iron-enriched zones of the former dolomite. The ferrous iron must be oxidized to the ferric form (i.e., hematite) to be visible with standard petrographic techniques.

Zoning of dolomite euhedra can also occur without the presence of iron. Normally, this type of zoning is seen in thin section as a cloudy central zone surrounded by a rim of clear dolomite (see Figure 8–8A). The cloudy center contains inclusions of the original calcite crystals preserved during the earliest stage of dolomite crystal growth. The clear outer rim formed in empty pore space (see Figure 8–8B).

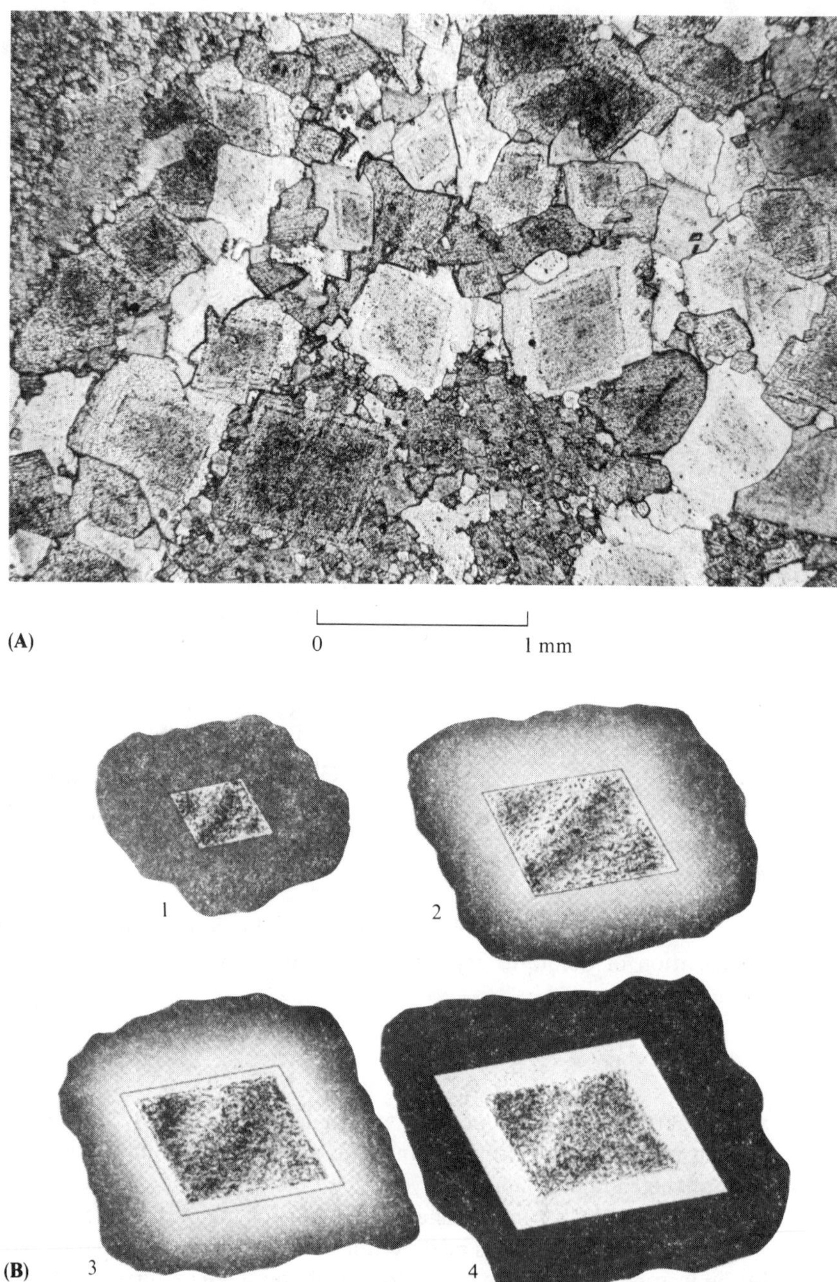

Figure 8–8
(**A**) Thin section of dolomite of Carboniferous age near Breedon, Leicestershire, England, showing zoned crystals formed by cloudy centers and clear rims. [Chilingar et al., 1979. Photo courtesy D. H. Zenger.] (**B**) Probable origin of cloudy-center/clear-rim texture of part A. (1) Dolomite begins growth by replacement. Dolomite requires more CO_3^{2-} than calcite, and excess is supplied by dissolution of $CaCO_3$ beyond limits of rhomb. (2) This creates higher porosity immediately around growing crystal. (3) Inclusion-free rim forms because of lack of $CaCO_3$ to be included. (4) Final result is dolomite rock with higher porosity than original limestone. [Murray, 1964.]

Twinning

Dolomite crystals are twinned only rarely in sedimentary rocks, although calcite is commonly twinned. This difference in frequency occurs because calcite has a plane of easy slip parallel to {0001}; dolomite has no plane of comparably easy slip. A corollary of this difference is that dolomite responds to stress more commonly by deforming plastically throughout the structure of the crystal (as does quartz). As a result, dolomite crystals of sand size may have visible undulatory extinction; calcite crystals do not.

Textures

Any texture that occurs in a limestone can occur in a dolomite, from essentially pure and homogeneous dolomicrites, to those that contain some allochems, to allochem-rich dolosparites, and finally to homogeneous, coarse-grained dolomites containing only spar crystals. In most dolomites the replacement origin of the mineral is clearly evident in the rock texture in thin section. Dolomite rhombs cut across the boundaries of allochems (see Figure 8–9) and may even preserve delicate structures within the allochem during the replacement process. In some partially dolomitized limestones,

Figure 8–9
Dolomite rhombs that have replaced ooliths, preserving fabric of ooliths as "ghosts." Diameter of large rhomb is 0.3 mm. [Photo courtesy R. C. Murray.]

an exquisite degree of selectivity occurs. Fossils (even specific kinds of fossils), coarse-grained lenses, fine-grained matrix, mud fragments, fossil burrows, or other inhomogeneities may be selectively composed of dolomite *or* preserved as calcite in contrast to the surrounding rock. This selectivity is typically interpreted to reflect control of dolomitization by slight differences in permeability, crystal size, or mineralogy of the original calcium carbonate host. In most dolomites the reason for the observed selectivity is unknown.

Dolomite rocks that contain replacement textures are, without doubt, formed after deposition of a sediment composed of calcite or aragonite. There are, however, many dolomites that lack clear evidence of replacement origin. Perhaps dolomitization completely destroyed the outlines and internal structures of preexisting allochems; perhaps the original sediment was a homogeneous calcitic or aragonitic micrite that did not contain diagnostic features; or, possibly, the dolomite formed as an original chemical precipitate from modified seawater. For many dolomites, none of these three possibilities can be eliminated on the basis of either field or thin-section observations.

Perhaps the most enigmatic are rhythmically alternating layers of calcite and dolomite, on a scale of millimeters or even micrometers. These layers commonly are sharply separated by a bedding plane or shaly parting and occur on a regional scale. It seems unlikely that the chemical composition of water at the seafloor could flip-flop so rapidly and completely. Might these alternating laminae represent times of subaerial exposure as the tides wax and wane adjacent to the mean high-tide line at the edge of the carbonate basin? On the basis of either field observations or thin-section studies, homogeneous dolomicrite is an enigmatic material.

ANCIENT DOLOMITES

Dolomites are associated almost exclusively with two other rock types: limestones and evaporites. Limestones, because of their biogenic origin, form largely in intertidal and shallow subtidal areas on the shallow seafloor. Evaporite formation requires large-scale evaporation of saline waters, either in an isolated arm of the sea or by subaerial exposure. What can we learn from regional field studies about the interrelationships among these three types of rocks—all of which can be formed within the proverbial "stone's throw" of a shoreline?

On a regional scale it is clear that when both dolomite and limestone occur within a carbonate sequence, the intensity of dolomitization increases in the direction of the craton (see Figure 8–6). Many examples are known, such as the Cambro-Ordovician rocks in Tennessee and Virginia (see Figure 8–10), Middle Ordovician rocks in southwestern Wisconsin (Badiozamani, 1973), and the entire Early and Middle Paleozoic carbonate rock section of the Cordilleran miogeosyncline in Nevada (Dunham and Olson, 1978).

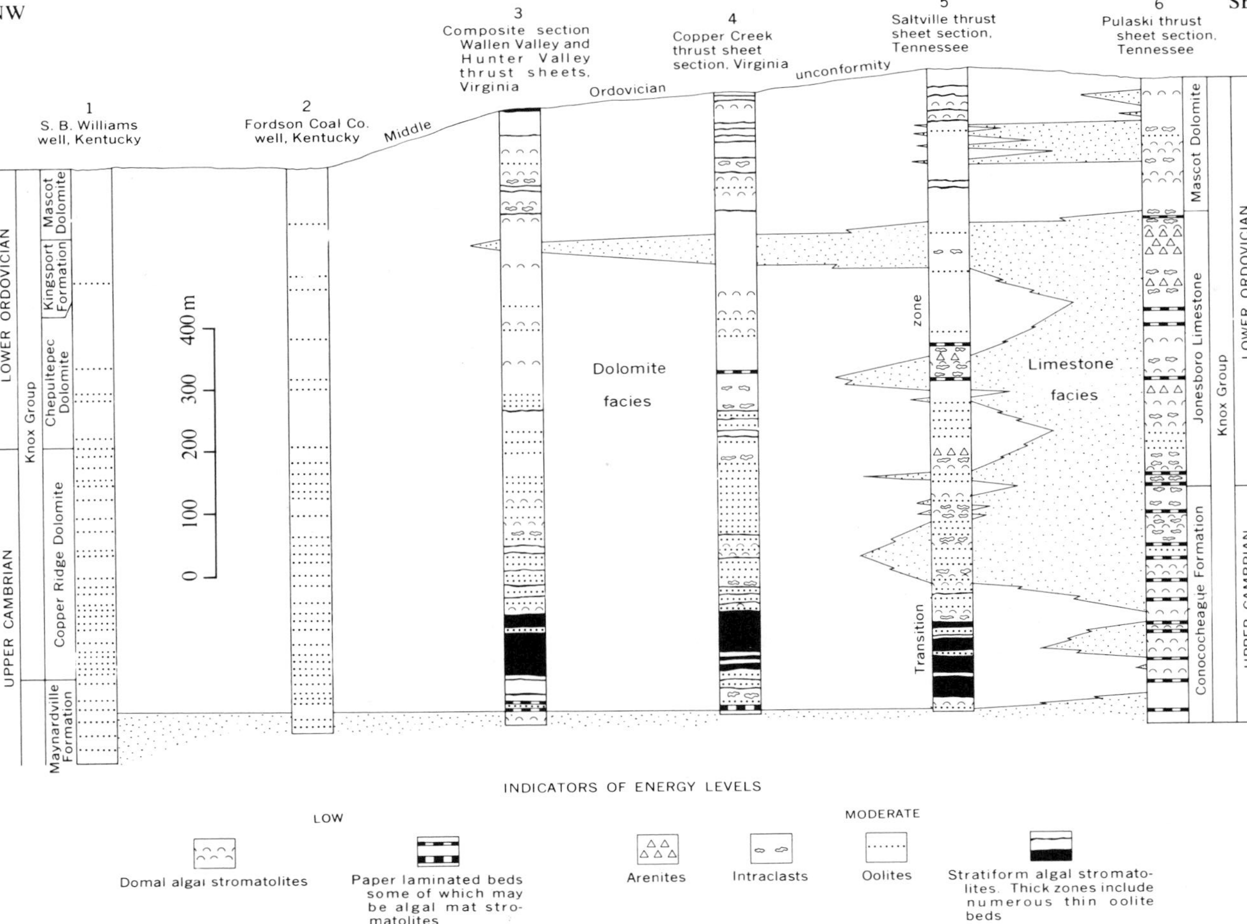

Figure 8–10
NW–SE cross section in central Appalachians, showing dominance of dolomite over limestone toward exposed craton (NW) and relationship between this and apparent "energy level" of depositional environment. [Harris, 1973.]

Hanson Creek Formation (Ordovician-Silurian), Nevada

The Hanson Creek Formation in northeastern Nevada is part of a regionally extensive Lower to Middle Paleozoic carbonate sequence that is interpreted to have formed on a broad, stable continental shelf. The craton lay to the east; deeper water to the west. The paleogeographic setting is in many respects a mirror image of many areas in the eastern United States at that time.

Stratigraphic sections of the Hanson Creek Formation in Eureka County in northeastern Nevada can be grouped into two categories, based on differing paleogeographic setting and diagenetic history (Dunham and Olson, 1980). Sections in northern and central Eureka County record deposition in shallow subtidal to peritidal environments. Replacement dolomite was the main diagenetic process affecting the original $CaCO_3$ skeletal grains, peloids, ooliths, and mud of these sections. In contrast, stratigraphic sections in southern Eureka County are undolomitized and consist of carbonate and clay mud succeeded by a shoaling-upward sequence capped by oolitic grainstone.

Lone Mountain

The dolomitic part of the Hanson Creek Formation is typified by the section at Lone Mountain (see Figure 8–11) in central Eureka County, which consists entirely of dolomite. Relic textures preserved within the dolomite prove that it formed as a replacement of $CaCO_3$. The lower part of the succession consists of dolomitized pelmatozoan columns that are mixed in a grain-supported texture with medium sand-size quartz grains derived from the underlying Eureka Quartzite. This part of the section is interpreted to represent a shallow subtidal shoal agitated by current and wave action. It is succeeded within 10 m upward by dolomitized coral and stromatoporoid colonies in growth position in a matrix of bioclastic debris and mud (see Figure 8–12), called *packstone* in the Dunham classification (see Figure 7–15). The presence of these organisms in growth position testifies to conditions of both normal salinity and dissolved oxygen. The packstone is overlain by another carbonate *grainstone* shoal, followed by a dolomicrite about 25 m thick that contains few fossils. The explanation for the scarcity of fossils is unclear. No evidence for hypersaline water (evaporites) or stagnant water (organic matter) is present. Perhaps the salinity was significantly above the normal value of 35,000 ppm, so that organic growth was discouraged, but not high enough (105,000 ppm) for evaporites (gypsum) to precipitate.

At 65 m above the base of the section, red-orange intraclasts of irregularly laminated dolomite are scattered in a matrix of gray-brown dolomite mudstone. The red color of the clasts indicates their thorough oxidation prior to deposition and burial in the contrasting gray-brown dolomite–mudstone matrix. The upper contact of the 1 m thick intraclast layer is riddled with fissures up to several centimeters deep, which are filled by dark-red to red-orange dolomite mud with scattered quartz-sand grains. The red-orange quartz-sandy dolomite deposit is approximately 1 m thick and is capped

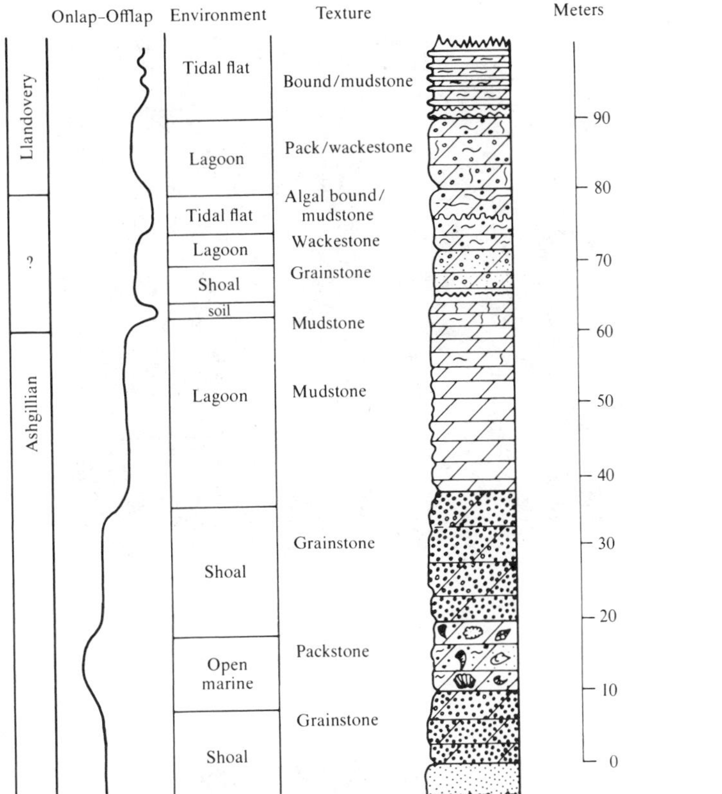

Figure 8–11
Interpretive stratigraphic section of Hanson Creek Formation (Ordovician-Silurian), at Lone Mountain. [Dunham and Olson, 1980.]

by a 1–2 cm thick laminated crust. This red zone is interpreted as an exposure surface, developed during a period of offlap on the platform (see Figure 8–13).

The sandy red zone is overlain at 67 m by medium-brown, well-sorted, quartz-sand–carbonate-pellet grainstone, recording renewed marine onlap onto the platform. The sandy zone is overlain by bioclastic *wackestone* to *mudstone,* containing a restricted fauna of brachiopods and ostracods and recording restricted lagoonal sedimentation. These lagoonal deposits are overlain at 76 m by a sequence of irregularly laminated, alternating light- and dark-gray dolomicrite bands (see Figure 8–14) containing stromatolites. From 77 m to 82 m, the rock is uniformly dark and contains stromatolites and channels up to several meters wide. Mud cracks are present, indicating periodic desiccation at this horizon. The coincidence of stromatolitic structures, channeling, fenestral fabrics, and mud cracks points to deposition in a peritidal mud-flat setting.

Overlying the tidal flat with a sharp contact is a lighter-colored bioclastic packstone that is extensively mottled with burrows, reflecting renewed subtidal deposition. Some early diagenetic chert occurs in the packstone; cross-cutting relationships indicate that most of the chertification preceded dolomitization.

(A)

(B)

Figure 8–12
Polished slabs of dolomitized stenohaline fossils at 15 m in Lone Mountain section. (**A**) Halysited coral. Note preservation of delicate corallites. (**B**) Massive stromatoporoid colony. Details of lamellar and pillar skeletal microstructure are well preserved. [Dunham and Olson, 1980. Photos courtesy J. B. Dunham.]

Figure 8–13
Red-soil horizon at 65 m in Lone Mountain section. Slab shows bright-red to red-orange quartz-sandy dolomite mudstone (1), overlain by thin, laminated crust (2), overlain by dark-brown dolomite mudstone (3). [Dunham and Olson, 1980. Photo courtesy J. B. Dunham.]

Figure 8–14
Polished slab from 76 m in Lone Mountain section, showing alternating light- and dark-gray dolomicrite bands with wavy, irregular lamination and distinct domal (stromatolitic) structures. [Dunham and Olson, 1980. Photo courtesy J. B. Dunham.]

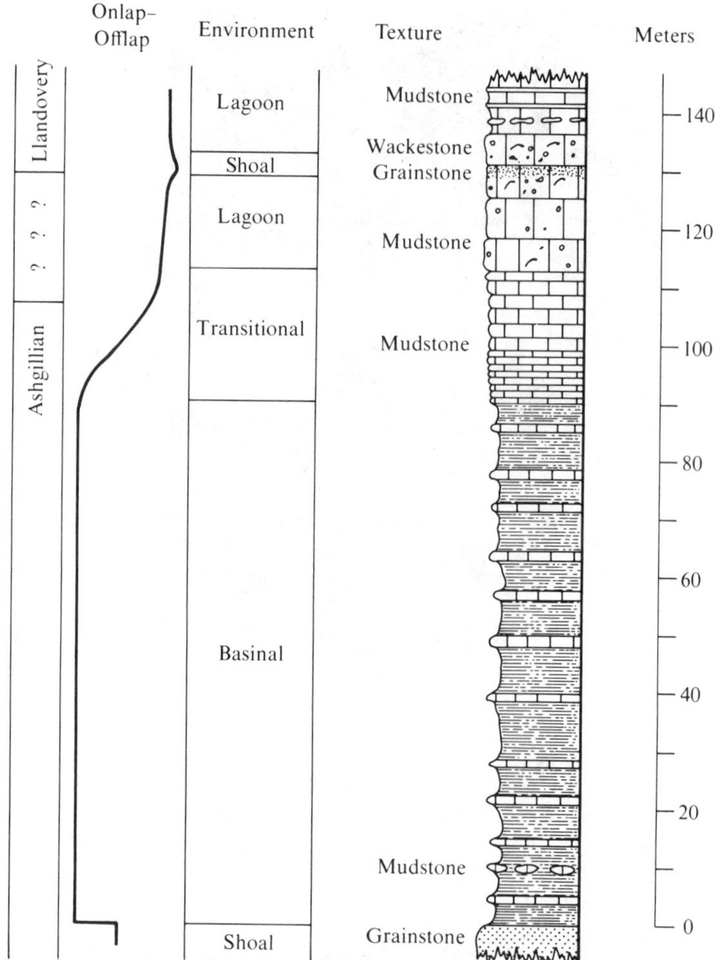

Figure 8–15
Hanson Creek Formation (Ordovician-Silurian) at Martin Ridge, showing inferred depositional environments. [Dunham and Olson, 1980.]

Martin Ridge

In marked contrast to the dolomitic section at Lone Mountain is the outcrop of the basal Hanson Creek Formation at Martin Ridge in southern Eureka County (see Figure 8–15). It consists of brownish-black, very finely laminated, shaly micrite (see Figure 8–16). Graptolites are common on bedding surfaces. Interbedded with this recessive-weathering, platy micrite are more resistant beds of gray-black, very finely laminated micrite 15–60 cm thick. Some well-preserved radiolarians occur, indicating early lithification of the sediment, that is, lithification before compaction and crushing of the delicate radiolarian shells. No benthic fauna and no evidence of burrowing are present. Hence, the bottom waters must have been quiet (planar, laterally extensive, unchanneled laminations) and inhospitable for living organisms.

Figure 8–16
Interbedded shaly limestone and thicker-bedded, purer micrite benches at Martin Ridge. [Dunham and Olson, 1980. Photo courtesy J. B. Dunham.]

Figure 8–17
Thin-bedded, dark-brown, laminated micrite at 90–113 m at Martin Ridge. [Dunham and Olson, 1980. Photo courtesy J. B. Dunham.]

At 90 m above the base of the section, a sequence of thin-bedded, dark-brown, laminated micrite occurs and persists to 113 m (see Figure 8–17). The only fossils present are conodonts and scattered laminae of transported shell debris. Above this unit is a lighter-colored, unlaminated, thicker-bedded micrite unit, and this is overlain by a pelmatozoan wackestone that persists to 130 m. The top of the wackestone unit is sharply overlain by oolitic grainstone containing well-sorted quartz grains and

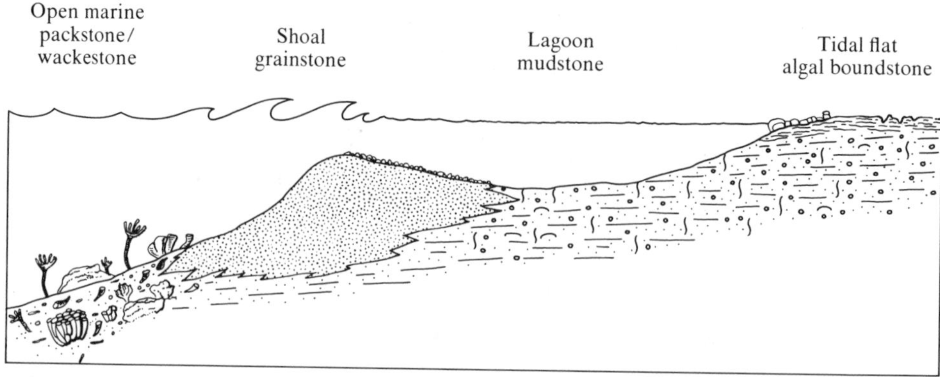

Figure 8–18
Lateral distribution of temporally equivalent depositional environments and rock textures in Hanson Creek Formation. Vertical scale in meters to tens of meters; horizontal scale in kilometers. [Dunham and Olson, 1980.]

rounded intraclasts. Overlying the grainstone is a wackestone unit containing scattered brachiopod fragments, pelmatozoan columnals, and rugose corals. At 137 m, darker-brown, parallel-laminated micrite with replacement chert stringers is present.

The integrated environmental interpretation of the Hanson Creek Formation is summarized in Figure 8–18. It was a quiet offshore setting at the onset of deposition, but the depth shallowed as deposition continued. The shallowing culminated with deposition of the oolitic grainstone, deposited in a very shallow-water setting of high kinetic energy, perhaps only 1 m or so deep. No dolomite is present in this section, demonstrating that this area was not influenced by the same set of conditions that led to the dolomitization of the formation at Lone Mountain.

The Origin of the Dolomite

The paleogeographic control of the distribution of limestone and secondary dolomite is definite and pronounced in the Hanson Creek Formation, as it is in most other epicontinental sea deposits. Subtidal deposition of the limestone is demonstrated by a variety of primary depositional fabrics such as ooliths and even, parallel laminations and by faunas that include pelmatozoans and corals. The dolomite in the Hanson Creek shows abundant evidence of having formed at or slightly above the normal high tide, but not in an area so desiccated that evaporites were able to form. Further, the dolomite clearly is secondary; it cuts across original calcium carbonate textures (fossils) and structures (laminations).

What was the salinity of the water from which the dolomite crystals precipitated? In thin section, it is evident that many of the dolomite crystals are large, some 150 μm long. Such large sizes indicate a slow rate of crystal growth and few crystal nuclei per unit volume of pore solution. This presumed slow growth rate is consistent with the determination from X-ray analysis that the crystals are structurally well-ordered; that is, they have compositions very close to the ideal 50 mol % each of calcium and magnesium. Chemical analysis reveals that the crystals of dolomite have low concentrations of foreign ions—an indication of precipitation from dilute solutions. Taken as a group, these data suggest that the dolomitizing solutions were dilute—probably more dilute than seawater—despite the fact that they formed in carbonate sediments located at the upper limit of onlap of normal high tides on the shore face—a location where intense evaporation might conceivably create waters of higher salinity than normal seawater. How might waters as dilute as suggested by the mineralogic evidence be generated?

The answer is provided by the fact that, during a change from calcite to dolomite, we are concerned with circulating pore waters, and not with surface seawater. Subaerially exposed carbonate terranes in humid climates are underlain by large lenses of fresh water. Below these lenses are even thicker lenses of brackish water, which grade downward into marine waters whose tops are roughly coincident with mean sea level. The lenses of fresh water can extend outward in the host rocks into the marine environment for distances of more than 100 km (Manheim, 1967). Is it possible that dolomite forms from seawater that has been diluted by mixing with fresh water? No dolomite precipitates from normal seawater, which contains much higher parts per million of calcium, magnesium, and carbonate ions. Why should it precipitate from a more dilute solution? For the moment, we will defer an answer to this question, but a satisfactory answer is known to this apparent geochemical dilemma. Land (1973) describes an apparent example of Holocene dolomitization caused by the mixing of fresh water and seawater in Pleistocene limestones in Jamaica (see Figure 8–19).

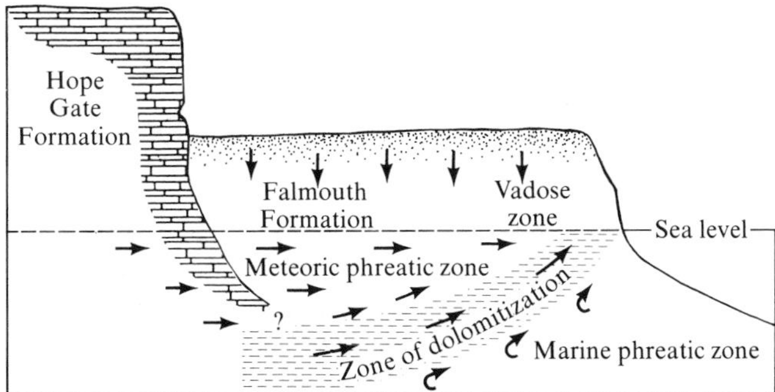

Figure 8–19
Schematic cross section showing contact relations between Hope Gate Formation and Falmouth Formation (Pleistocene), on northern side of Jamaica. Truncated marine terrace underlain by Falmouth Formation is capped by well-lithified, caliche-like caprock (dotted pattern), through which very little water percolates through vadose zone to water table. Most of the meteoric water in Falmouth is derived from water-table aquifer of island. [Land, 1973.]

If the model of dolomite formation by the mixing of seawater and meteoric waters is correct, several inferences can be made:

1. The mineralogic transition from limestone in a seaward direction to dolomite in a landward direction does not coincide with smaller-scale changes in depositional environment within the limestone section. It may, however, be closely related to differences in the predolomitization permeability of the limestone.

2. The position of the mineralogic transition is controlled by the position of exposed tracts of the carbonate sequence. The position of exposed tracts, and thus of freshwater lenses, is paleogeographically controlled.

3. Changes in regional paleogeography should produce changes in the position of the limestone–dolomite boundary. Regional regressions of the sea result in a seaward shift of the limestone–dolomite boundary because the position of the subaerially exposed carbonate terranes moves seaward. Conversely, regional transgressions would flood previously exposed tracts; this would push areas of freshwater recharge, and thus the limestone–dolomite boundary, landward.

4. If a close correlation exists in a stratigraphic section between transgressions and regressions and a shift in the limestone–dolomite boundary, it must be that dolomite formation occurred in the shallow subsurface soon after burial of the carbonate units. If dolomitization had occurred after deep burial of the limestone, the effect of surface paleogeography would not be present. Excellent correlation has been found between onlap–offlap and the movement of the limestone–dolomite boundary in many areas of Paleozoic rocks in the United States. For example, the Taghanic onlap, spanning the Middle–Late Devonian boundary, shifted the limestone–dolomite boundary several hundred kilometers eastward (shoreward) by Late Devonian time in east-central Nevada (Johnson, 1971).

The Dolomite–Evaporite Association

The stratigraphic association of dolomite with evaporites or pseudomorphs of evaporites is a very common one in the geological record; well-studied examples include the Duperow Formation (Devonian) in the Williston Basin (see Chapter 9), the Macumber Formation (Mississippian) in the Maritime Provinces of Canada (Schenk, 1967), and the Aghagrania Formation (Mississippian) in Ireland (West et al., 1968).

Aghagrania Formation (Mississippian), Ireland

The Aghagrania Formation in the Republic of Ireland has a total thickness of 71 m. West et al. (1968) divided the formation into two main groups on the basis of lithology, sedimentary structures, and fossils: (1) lithotypes with marine fossils, and (2) lithotypes with evidence of evaporites.

1. The fossiliferous rocks include both limestones and shales. The limestones are thin-bedded, unlaminated, and less than 0.3 m thick; they lack dolomite and evaporite minerals and contain goniatites and brachiopods. Rolled micrite-coated fossils (algal coatings?) suggest a shallow-water origin for this lithotype. The fossiliferous shales also lack dolomite and evaporites and are calcareous. Fissility is not always well developed, and the shales pass into mudrocks.

2. Underlying the fossiliferous, shallow-water marine rocks is a sequence of finely laminated, unfossiliferous limestones and dolomites (see Figure 8–20), usually in beds less than 1 m thick. No fossils are seen in thin section, and the laminae are seen to be thin, wavy, carbonaceous layers separating bands of dolomicrite or calcitic micrite (see Figure 8–21). They are apparently of algal-mat origin, and other algal textures are sometimes well developed. The lamination is sometimes marked by a variation in dolomite crystal size, by an alternation of dolomite and calcite, or by layers of pseudomorphs of algal structures. Desiccation cracks are common (see Figure 8–22), testifying to periods of subaerial exposure during the formation of this lithotype.

Figure 8–20
Unfossiliferous, laminated dolomicrite with algal fabric. [West et al., 1968. Photo courtesy I. M. West.]

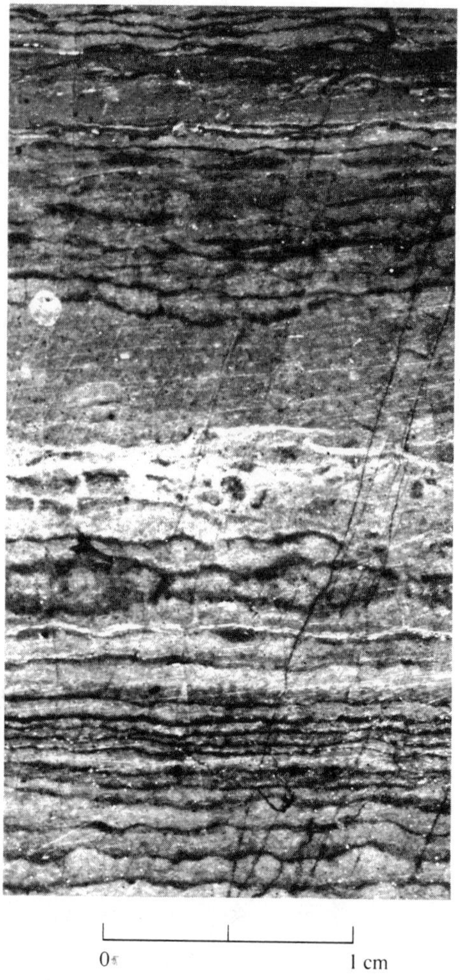

Figure 8–21
Unfossiliferous, finely laminated micrite, showing wavy carbonaceous (dark) layers. [West et al., 1968. Photo courtesy I. M. West.]

Figure 8–22
Small-scale desiccation cracks in laminated dolomicrite. [West et al., 1968. Photo courtesy I. M. West.]

Evidence of hypersaline conditions exists in this lithotype in the form of pseudomorphs of calcite after halite (see Figure 8–23), gypsum, and anhydrite. The most common pseudomorph is calcite after gypsum. These are sublenticular in form, and each is composed of several large sparry calcite crystals in a matrix of micrite. Some pseudomorphs consist of secondary anhedral gypsum after euhedral crystals of earlier, perhaps primary, gypsum. Pseudomorphs after a prismatic mineral, probably anhydrite, are also found frequently in thin sections of the unfossiliferous rocks. Celestite ($SrCO_3$) and length-slow chalcedony occur in the finely laminated, unfossiliferous limestones and dolomites (see Figure 8–24). Both minerals are common in evaporite deposits elsewhere. Macroscopic pseudomorphs of celestite after euhedral gypsum and anhydrite crystals reach a maximum size of 3 cm, but microscopic pseudomorphs are more frequent.

Small-scale, nontectonic folding and crumpling are conspicuous features of the laminated carbonates but are rarely present in other rock types of the formation. The folding seems to have occurred while the sediments were relatively plastic because fracturing is found only in very tight folds. West et al. suggest that the restriction of

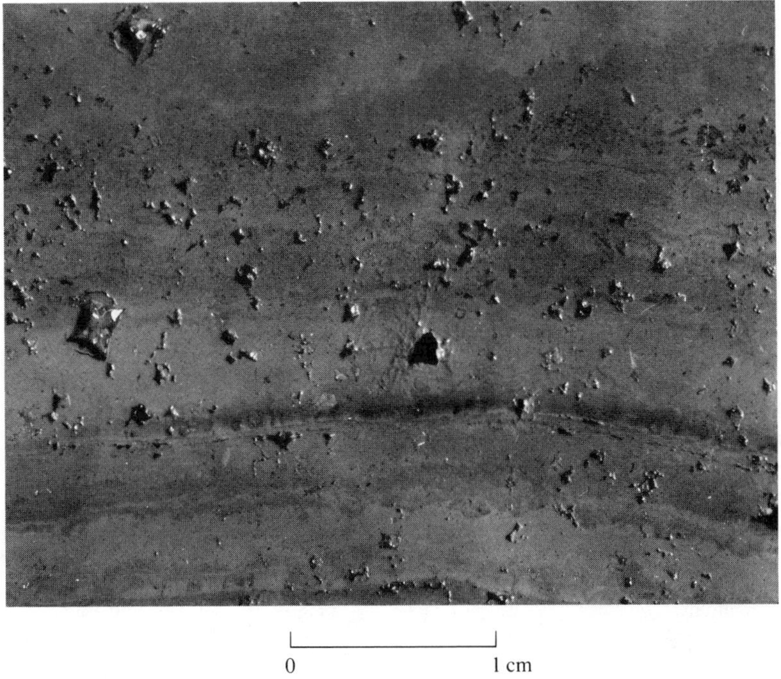

0 1 cm

Figure 8–23
Laminated dolomicrite with hollow molds and calcite pseudomorphs after halite. Original halite crystals have in many cases displaced sediment in which they developed. [West et al., 1968. Photo courtesy I. M. West.]

Figure 8–24
Laminated dolomicrite with crystals and globules of celestite. Possible dolomitized algal laminae above and below celestite-rich zone. [West et al., 1968. Photo courtesy I. M. West.]

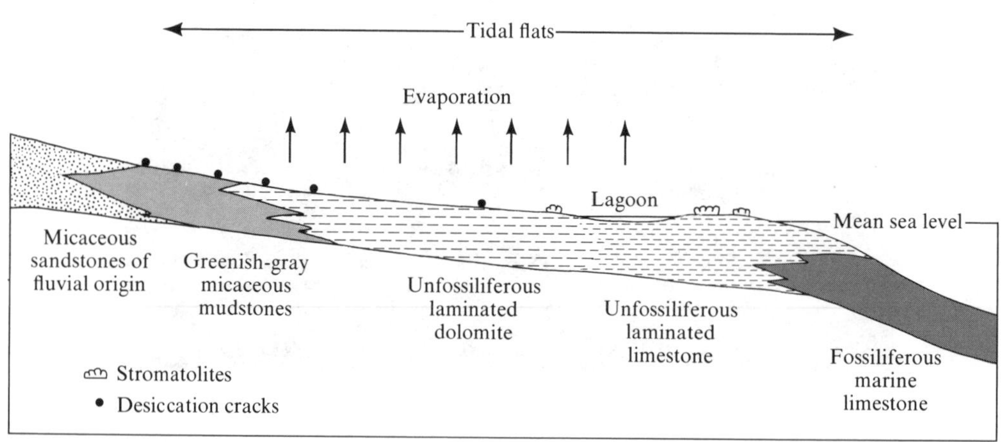

Figure 8–25
Diagrammatic sketch of arrangement normal to shoreline of rocks of Aghagrania Formation (Mississippian), Ireland. Width of tidal flat may have exceeded 30 km. [West et al., 1968.]

these distorted laminae to the laminated carbonate lithotype might have resulted from the original content of hydratable and soluble evaporites—a phenomenon known from Holocene sediments.

The stratigraphic associations, structures, textures, and mineralogy of Aghagrania sediments led West et al. to infer the facies relationships shown in Figure 8–25. The physiography and sequence of environments shown from shallow subtidal (fossils) to intertidal (algal stromatolites) to supratidal (dolomite and evaporites) are classic for rocks such as those in the Aghagrania Formation (Shinn et al., 1965; Lucia, 1972) and are duplicated in many ancient and modern settings.

Supratidal Dolomite Without Evaporite

Evaporites are very soluble minerals. Although they may form in association with dolomicrites, like those described from the Aghagrania Formation, they may subsequently have been dissolved. Sometimes casts or molds of evaporite minerals are preserved and can be recognized by their crystal habits. Some quartz geodes are pseudomorphous after evaporite nodules (Chowns and Elkins, 1974). Other times, the dolomite stands alone, and its environment of formation must be deciphered by the sedimentary textures and structures it contains. These are features that, as a group, are diagnostic of a supratidal environment: desiccation cracks; dolomicrite without fossils; laminations suggestive of replaced algal mats, such as wavy laminations and carbonaceous films; thin bedding; nontectonic, soft-sediment folding and brecciation (flat-pebble conglomerates); and very finely interlaminated dolomicrite and micrite. Examples of supratidal dolomites lacking preserved evaporites are described by Thompson (1970) and Laporte (1967).

Other Dolomite Occurrences

Nearly all dolomite occurrences are thought at present to have formed either by the mixing of fresh water and seawater in the shallow subsurface under a carbonate shoreline or by a chemical mechanism involving the formation of evaporites in a narrow zone immediately above mean high tide at the fringe of a carbonate basin. Other types of occurrences are known, however. Dolomite is sometimes found adjacent to and paralleling fault surfaces, clearly implying that the source of the water from which the dolomite crystallized was stratigraphically below the site at which the dolomite is found. The nature and origin of these waters are speculative. Dolomite is formed in some modern saline lakes (Clayton et al., 1968; von der Borch and Jones, 1976), and some is associated with unconformity surfaces. Some dolomite may be formed during relatively deep diagenesis of limestones. Probably these types of occurrences total less than 10% of the dolomite seen in the stratigraphic record.

MODERN DOLOMITES

Prior to 1964, dolomite was unknown as a significant deposit in Holocene sediments and a major concern of sedimentologists was "The Dolomite Problem." In the absence of a modern analog, generally accepted hypotheses concerning the origin of the extensive dolomites in the ancient record were of two types:

1. Seawater had a composition during the Precambrian and Paleozoic that was different from the composition today. In particular, magnesium was more abundant than at present. This hypothesis was purely ad hoc, as no reason for such a change in composition 250 million years ago was advanced and no other magnesium-rich minerals are found associated with most ancient dolomites, such as magnesite ($MgCO_3$) or brucite $Mg(OH)_2$.
2. Dolomite forms almost entirely during diagenesis below the water–sediment interface in the sea, perhaps at depths of hundreds of meters in most cases, so that it would be rare in Holocene carbonate deposits.

Then, in 1965, there appeared several publications describing a fairly common type of modern environment in which dolomite is forming at present (Deffeyes et al., 1965; Illing et al., 1965; Shinn et al., 1965). This environment is the low-lying, evaporative, supratidal flat that borders many areas where limestones are forming, such as the Bahamas and the Persian Gulf. From the vantage point of 1982, it seems difficult to imagine that dolomite had not previously been detected in these areas. But discoveries often seem "obvious" after someone has done the work. No doubt, large numbers of "obvious" facts about sedimentary rocks are still waiting to be recognized.

Six years after the discoveries of supratidal dolomite in association with evaporites, Hanshaw et al. (1971) proposed a method for achieving dolomitization of limestone by mixing meteoric or fresh waters with seawater, based on geochemical studies in the Florida aquifer where both calcite and dolomite occur in Tertiary carbonate rocks. The Hanshaw et al. concept was elaborated with a Holocene example by Land (1973; Land et al., 1975).

Hope Gate and Falmouth Formations (Pleistocene), Jamaica

The modern example described by Land (1973) of the nonevaporitic dolomite–limestone association so common in the geologic record was from Jamaica. On the northern coast of the island are two Pleistocene limestones: the Hope Gate Formation and the Falmouth Formation (see Figure 8–19). Using a portable core drill, Land drilled and cored five sites in Falmouth rocks, three of which were found to contain dolomite. All three were from the zone below the water table, the phreatic zone; no dolomite was found above the water table, the vadose zone. In thin section, the dolomite occurs in three forms:

1. As randomly oriented, euhedral rhombs 8–25 μm in diameter and scattered through a micrite matrix.

2. As a replacement of allochems of red algae.

3. As linings of dolospar in both primary and secondary pores. The dolomite crystals that line these pores are euhedral, and clearly formed as a chemical precipitate on the surfaces of the pore walls.

Chemical analyses revealed that the dolomite is nonstoichiometric (56 mol % $CaCO_3$) and contains a large amount of strontium (derived from the aragonite it replaced). It also contains an amount of sodium intermediate between the negligible amount typical of surface waters (less than 10 ppm) and the amount in the seawater (more than 10,000 ppm). Apparently, the dolomite forms in the zone of brackish, phreatic water in the shallow subsurface in a setting that is universal along coastlines: the zone of contact between fresh, essentially meteoric waters a few meters below the ground surface and seawater present below the freshwater lens. It is no surprise, then, that the dolomite–limestone association is so common in ancient rocks. As sea level rises or falls with respect to the land surface, the position of the brackish zone shifts laterally to produce dolomites that are very extensive over a low-lying land surface such as was characteristic of much of North America during Paleozoic time.

The Dolomite–Evaporite Association

One of the earliest descriptions of a modern example of the dolomite–evaporite association was made by Illing et al. (1965), based on observations in Qatar on the southern shore of the Persian Gulf (see Figure 8–26). The investigation was centered around Dohat (lagoon) Faishakh, but similar results have been obtained from other areas in the region. The peninsula of Qatar is a low-lying area that is nearly everywhere within 15 m of sea level. The land is extremely arid; years may pass without rain. The average daily temperature ranges between 23°C in January and 41°C in July. The salinity of the shallow seawater adjacent to Qatar can be as high as 60,000 ppm.

Dohat Faishakh (see Figure 8–27) has a maximum depth of 1.5 m and is floored by peloidal micrite containing small lamellibranchs, gastropods, and foraminifera. The northern shore of the lagoon is flanked by a flat, intertidal area about 300 m wide that is covered by dark-colored, rubbery sheets of blue-green algae that trap sediment carried in the tidal waters (see Figure 8–28). The resulting sediment is well laminated, and examples can be found resembling various types of stromatolites that have been described from ancient limestones (Kendall and Skipwith, 1968). These laminations are distinctive of the algal-flat environment when they are subsequently buried by supratidal sediment and dolomitized.

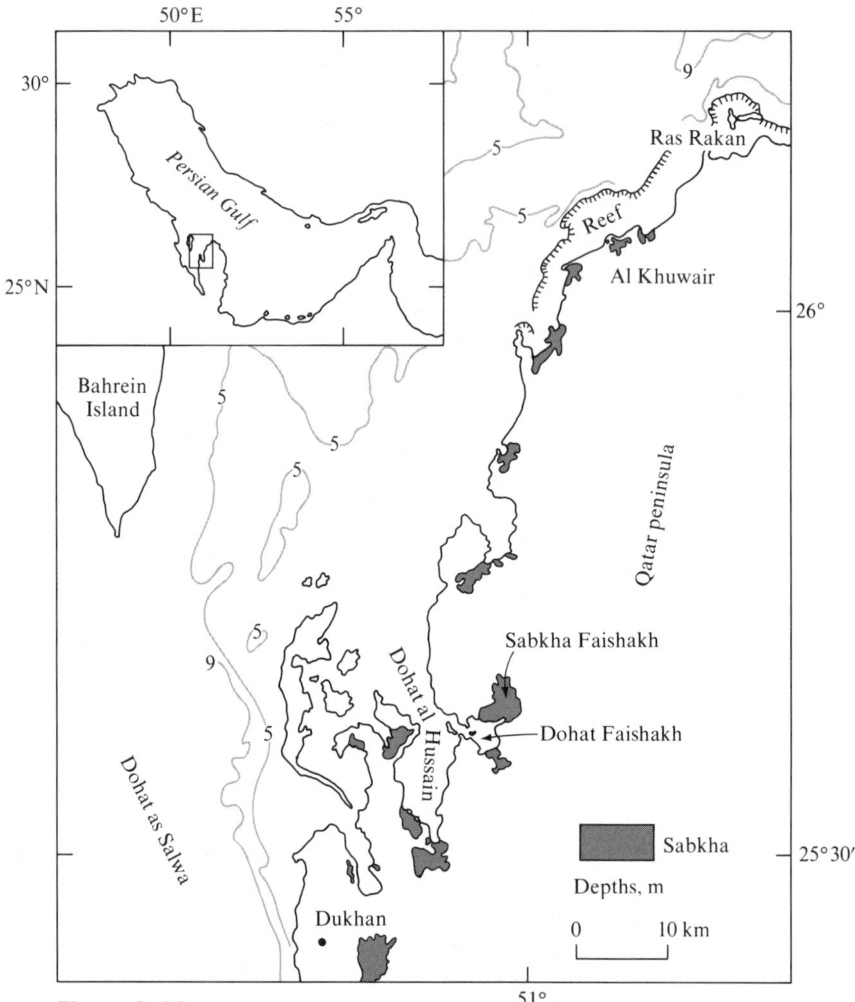

Figure 8–26
Location of supratidal areas (sabkhas) on northwestern side of Qatar peninsula, Persian Gulf. [Illing et al., 1965].

The Sabkha

Inland from the algal flats and above the level of normal high tides lies a halite-encrusted surface of salt flats called *sabkhas* (supratidal flats). The sabkha zone runs parallel to the coast and in places is 16 km in width. When strong offshore winds combine with high spring tides, very broad areas of sabkha are flooded. Subsequent intense evaporation causes upward movement of concentrated brine, with the resultant precipitation of halite (up to 8 cm thick), gypsum, aragonite, and celestite. Simultaneously, the underlying, dominantly aragonitic lagoonal sediment is dolomit-

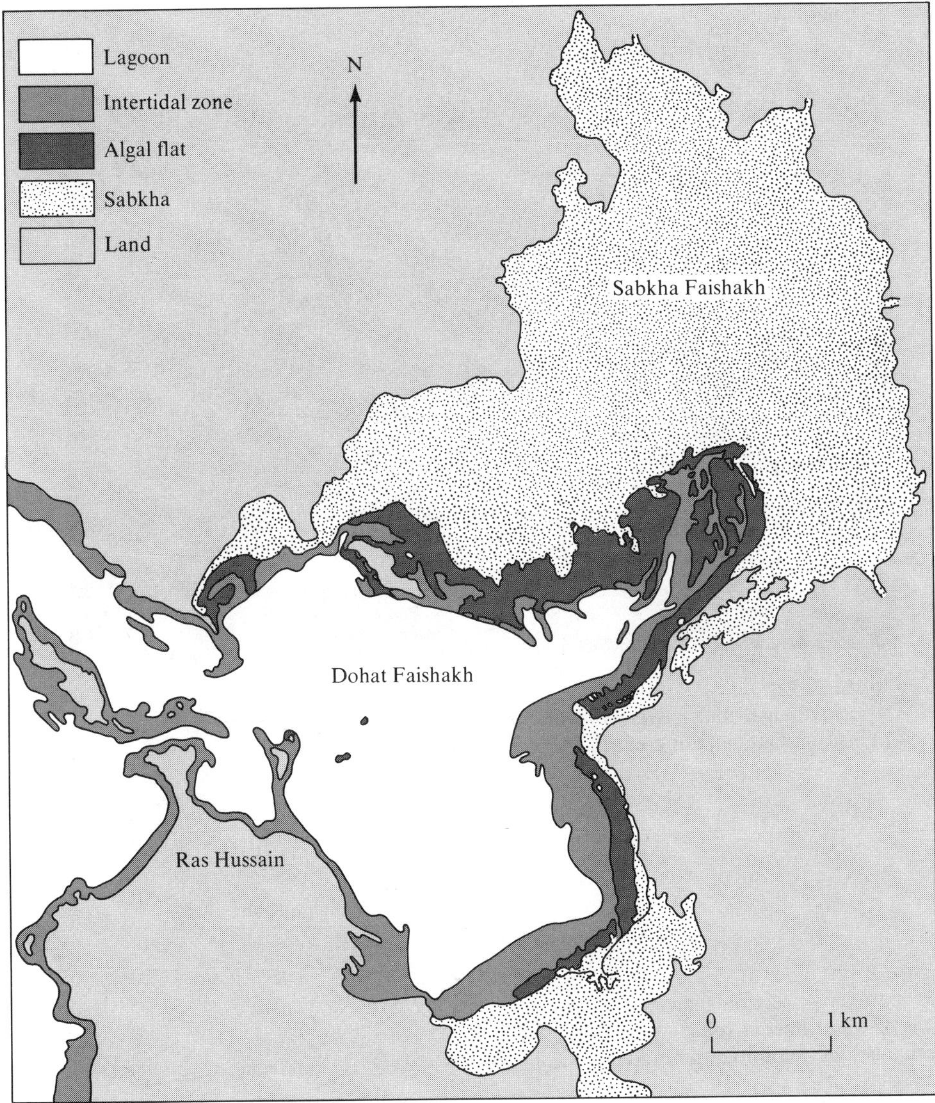

Figure 8–27
Sedimentary environments in Dohat Faishakh area, western coast of Qatar peninsula. [Illing et al., 1965.]

ized with the precipitation of unstoichiometric dolomite (protodolomite) containing up to 55 mol % $CaCO_3$. It is apparent from visual observation that, as water is lost from the sediment by evaporation, seawater moves in below the water table to replace it. Replacement of evaporated water is also effected by flooding. The depth of dolomitization in the sabkha may depend on the extent to which the heavier brines near the surface move downward as density currents in the pore system, a process termed *reflux* (see Figure 8–29).

Figure 8–28
Desiccated algal-mat polygons in shallow, intertidal drainage channel, Dohat al Hussain, Qatar. [Illing et al., 1965. Photo courtesy L. V. Illing.]

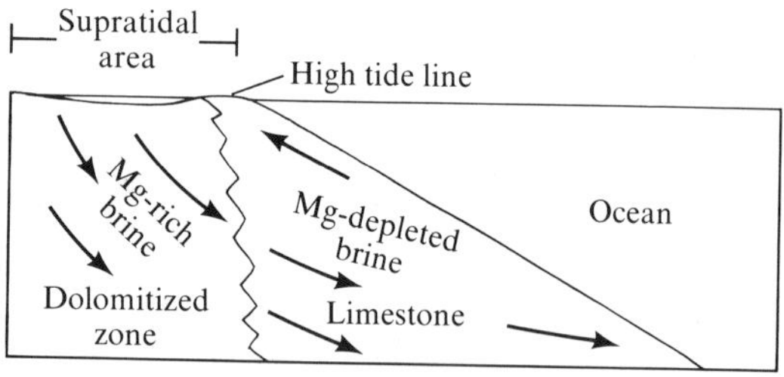

Figure 8–29
Schematic cross section showing paths of water flow in reflux mechanism of dolomitization under a sabkha.

Gypsum is common in the upper parts of the sediment; some parts contain nearly pure "gypsum crystal mushes" extending to 1 m below the surface. In the sabkha, gypsum can form more than 50% of the top meter of sediment, with layers of nearly pure gypsum alternating with gypsum-poor layers. Many gypsum crystals contain carbonate inclusions or traces of sedimentary lamination typical of the stromatolitic deposits of the algal flats. Crystal diameters range up to 25 cm. Secondary anhydrite after gypsum occurs sporadically but at some sites is as abundant in the sabkha sediment as is gypsum.

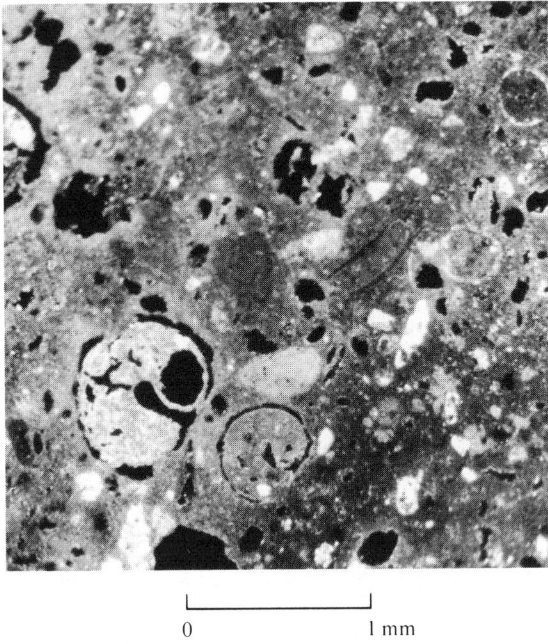

Figure 8–30
Dolomicrite from sabkha near Al Khuwair, showing floating quartz grains and molds of leached gastropods and other skeletal fragments. X-ray analysis reveals 75% dolomite, 15% calcite, and 10% quartz. [Illing et al., 1965. Photo courtesy L. V. Illing.]

Dolomite is found throughout the sabkha sediments (see Figure 8–30). The earliest crystals appear in the top 2–3 cm of sediment, at the contact between the algal flat and the sabkha, in the capillary zone above the water table. Inland, the depth at which the mineral occurs increases to perhaps 1 m and is present below the water table, in contrast to the more soluble evaporite minerals. In vertical section the amounts of dolomite and aragonite are inversely correlated, suggesting that most dolomite forms as a replacement of the needle-rich, peloidal aragonite mud. The dolomite seems to replace the interpeloidal aragonite cement first, then the peloids themselves, and finally the few fossils present. Just as the dolomitization in Jamaica that was described by Land (1973) is an analog of the ancient dolomite–limestone association, the sabkhas of the Persian Gulf region duplicate the features seen in the dolomite–evaporite association in ancient rocks. Present are algal laminations, dolomicrite, gypsum, anhydrite, halite, celestite; fossils are rare (see Figure 8–31). Further, the land surface adjacent to the supratidal flats is low-lying, so that relatively small changes in sea level cause very extensive areas of secondary dolomite to form.

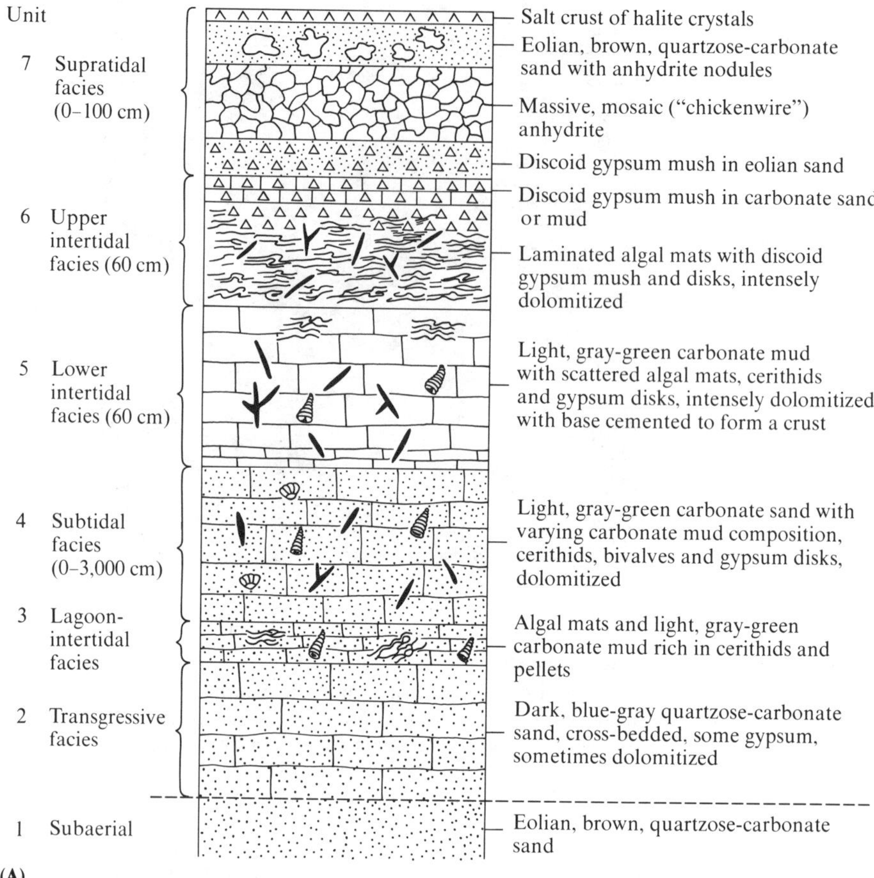

Figure 8–31
(A) Idealized vertical profile of sabkha sediments along southern shore of Persian Gulf.

Evaporative Pumping

McKenzie et al. (1980) have proposed a variant of the reflux model of dolomite formation based on studies in Abu Dhabi, along the southern shore of the Persian Gulf about 350 km ESE of the Qatar peninsula. Using several different geochemical approaches, they established a two-stage pattern of water movement under the sabkha surface.

The first stage is termed "flood recharge" and is initiated when the sabkha is flooded intermittently by seawater during unusually high tides or storms. Because of the flooding, seawater fills all available pore spaces, and the groundwater table is raised to the sabkha surface. Excess water flows back to the sea.

(B)

(**B**) Photograph of upper complex of sediments (units 5, 6, 7) from intermediate coastal sabkha of Abu Dhabi. Length of tape is 1.8 m. [McKenzie et al., 1980. Photo courtesy K. J. Hsü.]

The second stage entails evaporation from the capillary zone. Capillary evaporation begins immediately after the infiltration of floodwaters and lowers the water table below the sabkha surface. Drawdown continues until the water table falls below a level where the overlying sediments provide sufficient insulation against further capillary evaporation. The upward flow of groundwater in the zone of saturation to replace evaporative loss in the overlying capillary zone is termed "evaporative pumping," and this pumping process provides a continued supply of magnesium ions for the dolomitization of calcium carbonate. Calcium sulfate minerals are produced during the initial evaporative stages of capillary drawdown. They are also produced as a byproduct of subsequent dolomitization. As the evaporative-pumping process supplies magnesium ions to cause dolomite replacement of original calcium carbonate, the calcium ions released combine with sulfate ions in the pore waters to form additional gypsum and anhydrite.

MECHANISMS OF FORMATION

There appear to be two general models that can account for the vast bulk of dolomite seen in ancient rocks: one involving a mixing of fresh water with seawater, the other involving an evaporative area where seawater is concentrated to several times its normal salinity either by reflux or by evaporative pumping.

It is clear that a great deal of water movement is required to effect significant dolomite precipitation, irrespective of the mechanism of dolomitization. Is dolomitization in the deep subsurface uncommon because the permeability of limestones at depth is too low to permit much movement of subsurface water? How much supersaturated pore water must be moved through a rock to produce a microscopically visible amount of crystalline material? Such questions can be answered only by a careful consideration of the geochemistry of dolomite formation.

CHEMICAL CONSIDERATIONS

The chemical reaction from which dolomite is precipitated can be written as

$$CaMg(CO_3)_2 \rightleftharpoons Ca^{2+} + 2CO_3^{2-} + Mg^{2+}$$

(Reactions used to characterize equilibrium relationships and to determine equilibrium constants are, by convention, written as dissolution reactions, even though the reactions are described verbally as precipitation reactions.) Quantitative evaluation of this reaction and the related reaction in which calcite is replaced by dolomite

$$2CaCO_3 + Mg^{2+} \rightleftharpoons CaMg(CO_3)_2 + Ca^{2+}$$

reveals that for dolomite formation to be possible, three chemical requirements must be satisfied:

1. The ion activity product $(Ca^{2+})(Mg^{2+})(CO_3^{2-})^2$ in solution must exceed the solubility product of the mineral, approximately 10^{-17} mol/L. In normal seawater the product is equal to $10^{-15.1}$ mol/L, so that seawater is supersaturated by nearly two orders of magnitude with respect to dolomite. In average river water the ion activity product is close to 10^{-18} mol/L, less than the saturation level for dolomite.

2. The ion activity ratio Mg^{2+}/Ca^{2+} must be at least unity at 25°C; otherwise, $CaCO_3$ will form in preference to $CaMg(CO_3)_2$. In normal seawater this ratio is 5.6/1; in average river water, 0.5/1.

3. If the dolomite is to form by replacement of preexisting limestone, the limestone must be permeable and very large amounts of magnesium-rich water must percolate through the limestone. A single flushing of the original pore water by seawater can produce only an inconsequential amount of dolomite. The number of flushings (pore volumes) required to change a limestone into dolomite is very large but probably can be achieved when permeability is high and the replacement occurs very near the source of the flushing fluid, i.e., within 100 m or so of the surface.

Dolomite crystals are required to be rather precisely ordered, with planes of Ca^{2+} alternating with planes of CO_3^{2-} and planes of Mg^{2+}. No more than 5% of calcium ions are permitted as impurities in each layer of magnesium ions. The attainment of such near-perfection in cation ordering is a slow process at Earth surface temperatures and, as a result, abundant dolomite is difficult to form unless either condition 4a or 4b is met in addition to the three noted above:

4a. The Mg^{2+}/Ca^{2+} ratio is much in excess of the minimally required value of unity, perhaps 10/1 or more. If this condition prevails, the magnesium ions can displace the calcium ions more rapidly and effectively. In an ancient rock the evidence that this condition has been met is the presence (or former presence) of gypsum or anhydrite in association with the dolomite beds. This association indicates that a shallow pan of seawater having a restricted access to the sea was evaporated to the point of gypsum precipitation, which raised the Mg^{2+}/Ca^{2+} ratio by removing calcium from the solution, and made possible the rapid (10^3 yr) formation of dolomite.

4b. Or, alternatively, the solution has only a 1/1 ion activity ratio of Mg^{2+}/Ca^{2+}, but ions present in seawater that interfere with dolomite crystallization are not present. That is, the solution is dilute like fresh water but contains an Mg^{2+}/Ca^{2+} ratio higher than the value of 0.5/1 that characterizes fresh water. This condition can be met by a mixture of 5–30% seawater with 70–95% fresh water. Such a mixture is supersaturated with respect to dolomite, is undersaturated with respect to calcite, and has the required ratio of magnesium ions to calcium ions.

The postulated interrelationships among salinity, the Mg^{2+}/Ca^{2+} molar ratio, and the formation of dolomite are shown in Figure 8–32. Below an Mg^{2+}/Ca^{2+} ratio of 1/1, dolomite crystals cannot form at 25°C; calcite will precipitate instead. At ratios exceeding 1/1, dolomite can form but whether it does is determined by the relationship between salinity and the Mg^{2+}/Ca^{2+} ratio. If salinity is low, as in a mixture of mostly fresh water with some seawater, limpid dolospar and micrite crystals can form (Dorag dolomite); if salinity is high, as in a mixture of mostly seawater with little or no fresh water, dolomicrite forms only at high Mg^{2+}/Ca^{2+} ratios. Because such high ratios are produced as a consequence of gypsum precipitation, evaporites are normally associated with the dolomite in outcrop. Halite may occur in addition to a calcium sulfate mineral, but the halite is not essential to the formation of the dolomite.

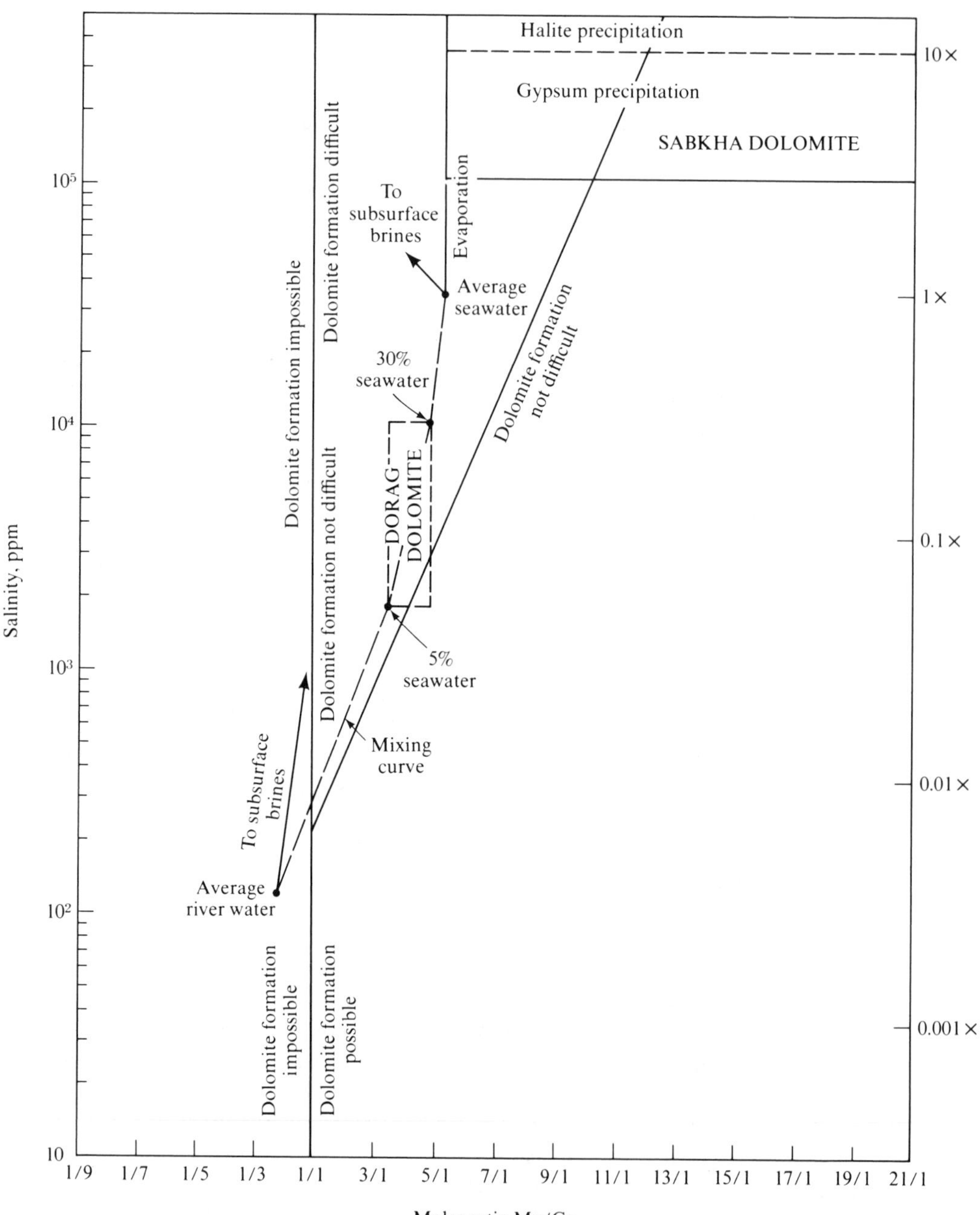

Figure 8–32
Relationships among total salinity, Mg/Ca ratio, and origin of dolomite, based on both field observations and laboratory experiments.

SUMMARY

Dolomite forms by replacement of limestone, most commonly very soon after the limestone has formed. This penecontemporaneous replacement can either retain or obliterate the textures and structures present in the limestone. Selectivity is common, with some features destroyed and others preserved, but no general explanation for selectivity is known. Sometimes it seems to be related to the amount of magnesium in the original calcareous shells; sometimes, to differences in permeability in the limestone facies; sometimes, to the crystal size of the calcareous particles; and sometimes, to none of these types of variations.

Occurrences of penecontemporaneous dolomitization can be grouped into two categories: those that are associated with evaporites and those that are not. Examples of these types are known from both ancient and modern settings. The two categories reflect complex chemical interactions within the fluid from which the dolomite precipitates. Fresh waters generally are undersaturated with respect to both dolomite and calcite; seawater is supersaturated with both minerals. Nevertheless, neither mineral normally precipitates inorganically from ordinary surface seawater unless uncommon conditions exist. In the case of dolomite, a mixture of 70–95% fresh water and 5–30% seawater results in a water that is supersaturated with dolomite, is undersaturated with calcite and aragonite, and is able to precipitate dolomite.

An alternative mechanism for the replacement of limestone by dolomite is intense evaporation of shallow pans of seawater on supratidal flats. The dolomite formed in these hypersaline environments is produced after the salinity has been tripled to more than 100,000 ppm and gypsum begins to form. The removal of calcium resulting from gypsum formation increases the Mg^{2+}/Ca^{2+} ion activity ratio to values greater than 10/1 so that dolomite forms rapidly.

FURTHER READING

Adams, J. E., and M. L. Rhodes. 1960. Dolomitization by seepage refluction. *Amer. Assoc. Petroleum Geol. Bull., 44,* 1912–1920.

Badiozamani, K. 1973. The Dorag dolomitization model—application to the Middle Ordovician of Wisconsin. *Jour. Sed. Petrology, 43,* 965–984. See also Discussion, by A. B. Carpenter, *46,* 254–258

Borch, C. C. von der. 1976. Stratigraphy and formation of Holocene dolomitic carbonate deposits of the Coorong area, South Australia. *Jour. Sed. Petrology, 46,* 952–966.

Borch, C. C. von der, and J. B. Jones. 1976. Spherular modern dolomite from the Coorong area, South Australia. *Sedimentology, 23,* 587–591.

Chilingar, G. V., D. H. Zenger, H. J. Bissell, and K. H. Wolf. 1979. Dolomites and dolomitization. In G. Larsen and G. V. Chilingar (eds.), *Diagenesis in Sediments and Sedimentary Rocks.* New York: Elsevier, pp. 425–536.

Chowns, T. M., and J. E. Elkins. 1974. The origin of quartz geodes and cauliflower cherts through the silicification of anhydrite nodules. *Jour. Sed. Petrology, 44,* 885–903.

Clayton, R. N., B. F. Jones, and R. A. Berner. 1968. Isotope studies of dolomite formation under sedimentary conditions. *Geochimica et Cosmochimica Acta, 32,* 415–432.

Deffeyes, K. S., F. J. Lucia, and P. K. Weyl. 1965. Dolomitization of Recent and Plio-Pleistocene sediments by marine evaporite waters on Bonaire, Netherlands Antilles. In L. C. Pray and R. C. Murray (eds.), *Dolomitization and Limestone Diagenesis.* Soc. Econ. Paleontol. Mineral. Spec. Pub. No. 13, pp. 71–88.

Dunham, J. B., and E. R. Olson. 1978. Diagenetic dolomite formation related to Paleozoic paleogeography of the Cordilleran miogeocline in Nevada. *Geology, 6,* 556–559.

Dunham, J. B., and E. R. Olson. 1980. Shallow subsurface dolomitization of subtidally deposited carbonate sediments in the Hanson Creek Formation (Ordovician-Silurian) of central Nevada. In D. H. Zenger et al. (eds.), *Concepts and Models of Dolomitization,* Soc. Econ. Paleontol. Mineral. Spec. Pub. No. 28, pp. 139–161.

Folk, R. L., and L. S. Land. 1975. Mg/Ca ratio and salinity: two controls over crystallization of dolomite. *Amer. Assoc. Petroleum Geol. Bull., 59,* 60–68.

Garrels, R. M., and F. T. Mackenzie. 1971. *Evolution of Sedimentary Rocks.* New York: Norton, 397 pp.

Hanshaw, B. B., W. Back, and R. G. Deike. 1971. A geochemical hypothesis for dolomitization of ground water. *Econ. Geol., 66,* 710–724.

Harris, L. D. 1973. Dolomitization model for upper Cambrian and lower Ordovician carbonate rocks in the eastern United States. *Jour. Res.* (U.S. Geol. Surv.), *1,* 63–78.

Illing, L. V., A. J. Wells, and J. C. M. Taylor. 1965. Penecontemporary dolomite in the Persian Gulf. In L. C. Pray and R. C. Murray (eds.), *Dolomitization and Limestone Diagenesis.* Soc. Econ. Paleontol. Mineral. Spec. Pub. No. 13, pp. 89–111.

Johnson, J. G. 1971. Timing and coordination of orogenic, epeirogenic, and eustatic events. *Geol. Soc. Amer. Bull., 82,* 3263–3298.

Johnson, J. G. 1974. Shorelines of epeiric seas. *Amer. Jour. Sci., 274,* 465–470.

Katz, A. 1971. Zoned dolomite crystals. *Jour. Geol., 79,* 38–51.

Kendall, A. C. 1977. Origin of dolomite mottling in Ordovician limestones from Saskatchewan and Manitoba. *Bull. Canad. Petroleum Geol., 25,* 480–504.

Kendall, C. G. St. C., and P. A. deE. Skipwith. 1968. Recent algal mats of a Persian Gulf lagoon. *Jour. Sed. Petrology, 38,* 1040–1058.

Land, L. S. 1973. Holocene meteoric dolomitization of Pleistocene limestones, North Jamaica. *Sedimentology, 20,* 411–424.

Land, L. S., M. R. I. Salem, and A. W. Morrow. 1975. Paleohydrology of ancient dolomites: geochemical evidence. *Amer. Assoc. Petroleum Geol. Bull., 59,* 1602–1625.

Laporte, L. F. 1967. Carbonate deposition near mean sea level and resultant facies mosaic: Manlius Formation (Lower Devonian) of New York State. *Amer. Assoc. Petroleum Geol. Bull., 51,* 73–101.

Lucia, F. J. 1972. Recognition of evaporite–carbonate shoreline sedimentation. In J. K. Rigby and W. K. Hamblin (eds.), *Recognition of Ancient Sedimentary Environments.* Soc. Econ. Paleontol. Mineral. Spec. Pub. No. 16, pp. 160–191.

McKenzie, J. A., K. J. Hsü, and J. F. Schneider. 1980. Movement of subsurface waters under the sabkha, Abu Dhabi, UAE, and its relation to evaporative dolomite genesis. In D. H. Zenger et al. (eds.), *Concepts and Models of Dolomitization.* Soc. Econ. Paleontol. Mineral. Spec. Pub. No. 28, pp. 11–30.

Manheim, F. T. 1967. Evidence for submarine discharge of water on the Atlantic continental slope of the southern United States, and suggestions for further research. *Trans. N.Y. Acad. Sci.,* Series 2, pp. 839–853.

Murray, R. C. 1960. Origin of porosity in carbonate rocks. *Jour. Sed. Petrology, 30,* 59–84.

Murray, R. C. 1964. Preservation of primary structures and fabrics in dolomite. In J. Imbrie (ed.), *Approaches to Paleoecology.* New York: Wiley, pp. 388–403.

Osmond, J. C. 1956. Mottled carbonate rocks in the Middle Devonian of eastern Nevada. *Jour. Sed. Petrology, 26,* 32–41.

Peterson, M. N. A., C. C. von der Borch, and G. S. Bien. 1966. Growth of dolomite crystals. *Amer. Jour. Sci., 264,* 257–272.

Sabins, F. F., Jr. 1962. Grains of detrital, secondary and primary dolomite from Cretaceous strata of the Western Interior. *Geol. Soc. Amer. Bull., 73,* 1183–1196.

Schenk, P. E. 1967. The Macumber Formation of the Maritime Provinces, Canada—a Mississippian analogue to recent strand-line carbonates of the Persian Gulf. *Jour. Sed. Petrology, 37,* 365–376.

Shinn, E. A., R. N. Ginsburg, and R. M. Lloyd. 1965. Recent supratidal dolomite from Andros Island, Bahamas. In L. C. Pray and R. C. Murray (eds.), *Dolomitization and Limestone Diagenesis.* Soc. Econ. Paleontol. Mineral. Spec. Pub. No. 13, 112–123.

Thompson, A. M. 1970. Tidal-flat deposition and early dolomitization in upper Ordovician rocks of southern Appalachian Valley and Ridge. *Jour. Sed. Petrology, 40,* 1271–1286.

West, I. M., A. Brandon, and M. Smith. 1968. A tidal-flat evaporitic facies in the Visean of Ireland. *Jour. Sed. Petrology, 38,* 1079–1093.

Zenger, D. H., et al. (eds.). 1980. *Concepts and Models of Dolomitization.* Soc. Econ. Paleontol. Mineral. Spec. Pub. No. 28, 320 pp.

9

Evaporites

Sometimes the sun, unkindly hot
My garden makes a desert spot.

PAUL LAURENCE DUNBAR

Evaporite minerals are defined as those minerals produced from a saline solution as a result of extensive or total evaporation of the water. Anhydrite (or its hydrated equivalent, gypsum) is the first evaporite mineral to precipitate from seawater, and its formation requires that 70% of the water be evaporated. Although evaporite minerals are very soluble, they have been found in rocks as old as 3.5 billion years. Their abundance is, however, very irregular (see Figure 9–1). Presumably, the times of great abundance of evaporites reflect times of concentration of drifting continents into the belts of descending, warm, dry air, centered at present at about 30°N and 30°S latitudes (see Figures 9–24 and 9–25). Differential preservation through geologic time may, however, modify original patterns of deposition.

The variety and abundance of minerals produced from an evaporating body of water depend on its initial composition. Nearly 70 evaporite minerals are known

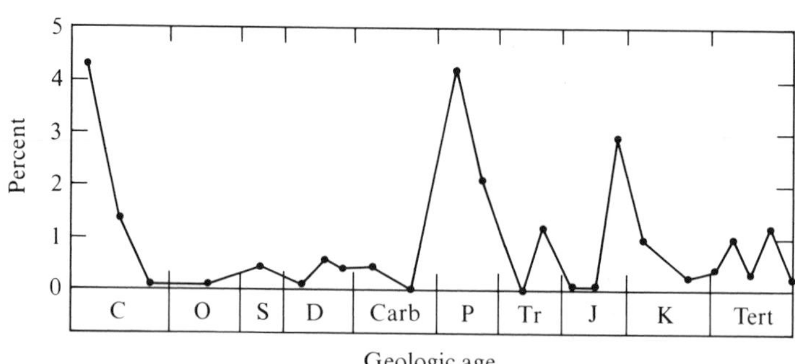

Figure 9–1
Variation in abundance of evaporite rocks through Phanerozoic time. [A. B. Ronov et al., 1980, *Sed. Geol., 25.*]

(Stewart, 1963), of which 27 are sulfates, 27 are borates, and 13 are halides. A single bed of evaporite is usually composed of more than one mineral, particularly when the deposit consists of the less common and more soluble evaporite minerals.

MINERALOGY

The bulk of evaporite deposits has seawater as its parent solution, and the most abundant anions in seawater are Cl^- (94.5% of anion molarity) and SO_4^{2-} (4.9% of anion molarity). As a result, the abundant evaporite minerals are chlorides (halite) and sulfates (gypsum and anhydrite). Table 9–1 lists the major minerals in marine evaporite deposits. Nonmarine, saline lake and playa evaporite waters commonly have initial compositions that differ significantly from the composition of seawater and, as a result, nonmarine evaporite deposits can contain large amounts of minerals rarely formed from seawater, such as the minerals trona [$Na_3H(CO_3)_2 \cdot 2H_2O$], mirabilite [$Na_2SO_4 \cdot 10H_2O$], glauberite [$Na_2Ca(SO_4)_2$], and borax [$Na_2B_4O_5(OH)_4 \cdot 8H_2O$]. Trona is used commercially in the manufacture of baking soda; borax is the raw material for household detergents.

FIELD OBSERVATIONS

Evaporites are, by definition, very soluble and rarely occur in outcrop except in arid areas. In these areas, gypsum is the most common and abundant mineral because (1) gypsum is the first evaporite mineral to precipitate from evaporating seawater (despite the fact that it is 15% more soluble than anhydrite), and (2) anhydrite uplifted to the surface is rapidly converted to gypsum because of the ease of hydration. Halite in outcrop is a poor second in abundance to gypsum because it is 150 times more soluble than gypsum.

Table 9–1
Major Minerals in Marine Evaporite Deposits

Mineral	Chemical structure
Chlorides	
Halite	$NaCl$
Sylvite	KCl
Carnallite	$KMgCl_3 \cdot 6H_2O$
Sulfates	
Anhydrite	$CaSO_4$
Langbeinite	$K_2Mg_2(SO_4)_3$
Polyhalite	$K_2Ca_2Mg(SO_4)_4 \cdot 2H_2O$
Kieserite	$MgSO_4 \cdot H_2O$
Gypsum	$CaSO_4 \cdot 2H_2O$
Kainite	$KMg(SO_4)Cl \cdot 3H_2O$

Despite the infrequent occurrence of evaporite beds in outcrop, they are common in the subsurface. Drilling for petroleum and natural gas has revealed that beds of evaporites underlie about one-third of the land surface of the United States (see Figure 9–2). Further, in the subsurface halite is more abundant than gypsum, as might be expected because of the dominance of sodium and chloride ions in seawater.

Evaporite rocks are most commonly interbedded with dolomite, limestone, and fine-grained detrital rocks, particularly thin, red shales. The association with dolomite reflects the sabkha origin of many evaporite deposits and the genetic association of dolomite with gypsum in the sabkha environment (see Chapter 8). The limestones, which are much less common than dolomite in sequences of evaporitic rocks, are the residual parent rocks of the dolomite. The fine-grained, red detrital beds reflect small amounts of mud carried into the evaporite basin from surrounding low-lying arid land areas. The red color reflects the general production of red pigment in arid climates through the leaching of ferromagnesian minerals in the vadose zone (see Chapter 6). The fissility of the mudrocks reflects the absence of burrowing, bottom-dwelling organisms in the highly saline waters from which evaporites are precipitated (see Chapter 3). Shelled fossils are rare in evaporitic rocks because few of them are able to tolerate hypersalinity for extended periods of time. The rare fossils that do occur in evaporite beds are planktonic forms swept into the evaporite basin from less saline lagoonal or open-sea areas adjacent to the basin.

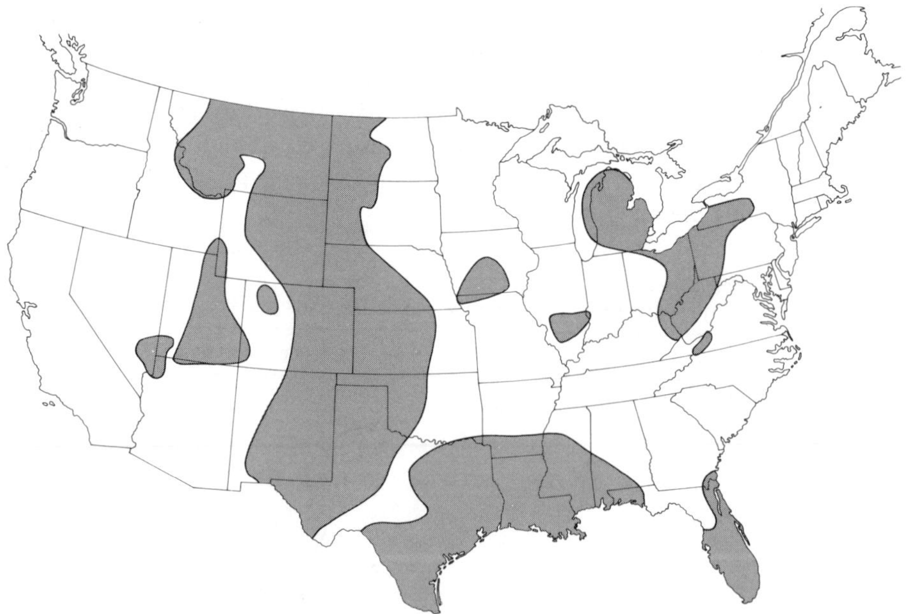

Figure 9–2
Areas of conterminous United States underlain by evaporites (shaded area).

Structures

Primary sedimentary structures described from evaporite units include lamination, cross bedding, graded bedding, ripple marks, mud cracks, and a variety of complex clastic and/or chemical structures. Several of these are illustrated in Figure 9–3. The most common structure is lamination. A typical laminated anhydrite or gypsum sequence consists of alternate laminae of white, nearly pure calcium sulfate mineral and gray-black lamellae rich in dolomite and organic carbon (see Figure 9–4). The lamellar pair is typically 0.2–2 mm thick but may be as thick as 10 mm, and the dark laminae may be one-fifth to one-tenth the thickness of the couplet. The thickness of the lamellae is usually within a range feasible for precipitation in one year, but a varve origin for the couplets is difficult to prove at most outcrops. An excellent description of the laminae in Figure 9–4 was given by Kaufmann and Slawson (1950, pp. 24–25):

> If the face of the salt is sprayed with water and slowly dissolved away, the core of the dark bands begins to protrude. The core consists of either a single layer or several layers of paper-thin lamellae of white or gray anhydrite, separated by narrow zones of salt of variable thickness. The lamellae are in some cases crowded so close together as to be almost indistinguishable. A group of four or five may occur in a centimeter of vertical distance; one or several such groups may make up a single dark band; or but one or two lamellae may occur within a dark band. In a few instances anhydrite lamellae were locally as thick as $\frac{1}{16}$ inch [1.6 mm]. The clear thick bands of white salt between the darker layers commonly contain these same thin lamellae of anhydrite, but these are not observable until solution causes the edges of the anhydrite to project from the surface.

The authors point out that, although each pair of light and dark bands may be an annual varve, the presence of numerous thin, dark lamellae within each dark band indicates that environmental changes were much more frequent than would be suggested by the dark bands alone. Furthermore, the presence of several lamellae and groups of lamellae in a single dark band may indicate cyclic disturbances lasting several years.

Many laminated gypsum and/or anhydrite sequences are remarkably uniform. Individual laminae have sharp contacts, are perfectly flat, and are continuous across a core, a mine face, or an outcrop. The vertical sequence of thicknesses seems to be cyclic, and detailed statistical analyses of the thicknesses of laminae in several areas have revealed a cyclic pattern that can be correlated for tens and even hundreds of kilometers across the sedimentary basin. In these basins the laminar calcium sulfate sequences are underlain by finely laminated carbonate and clastic rocks and overlain by nodular anhydrite and banded (coarsely laminated) halite beds.

These characteristics have led to the conclusion that the laminated rock was sedimented in a very quiet-water, marine environment, either a shallow-water setting protected by a barrier of some sort or a "deep-water" environment below wave base. The remarkable uniformity of the laminae certainly attests to quiet waters, but in an

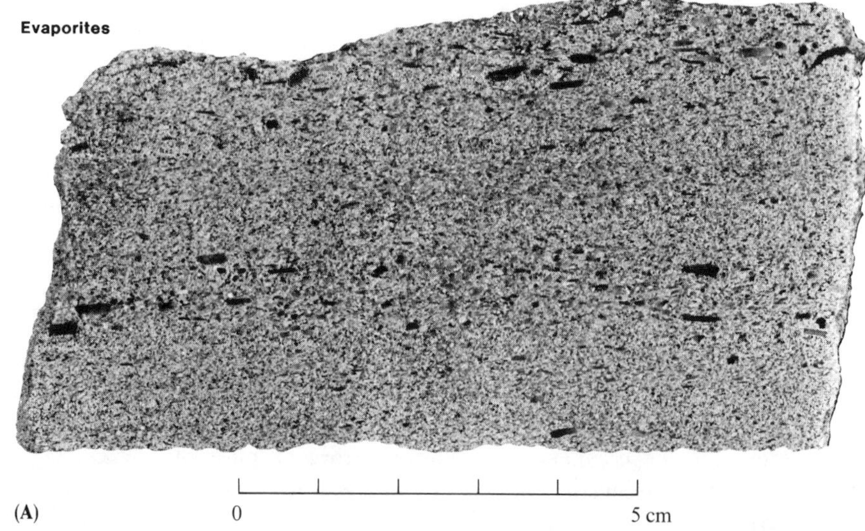

(A)

(B)

Figure 9–3
(**A**) Laminated evaporite rock composed of large, tabular selenite gypsum fragments (dark), fragments of planktonic foraminifera *Globigerina,* and pieces of microcrystalline carbonate and shale. Scale in millimeters. (**B**) Nodular gypsum strung out along bedding and coalescing to gypsum bands. Pencil shows scale. (**C**) Bedded selenite with interbeds of fine gypsum sand. (**D**) Selenite layers interbedded with laminated gypsum. Gypsum–carbonate laminae are draped over large selenite crystals with pockets between selenite crystals filled by fine gypsum sand. All photos from Solfifera Series (Miocene), Sicily. [Hardie and Eugster, 1971. Photos courtesy L. A. Hardie.]

evaporite setting such undisturbed waters can be attained at shallow depths. As Holser (1979) has pointed out, sharp brine stratification in an evaporite basin might dampen wave motion at depths less than would be normal in a more uniform water column such as normal seawater.

The vast majority of sedimentary structures seen in outcrops of evaporite beds is of diagenetic origin. This is to be expected because of the extreme solubilities of evaporite minerals. Recall that even aragonite is rare in pre-Pleistocene rocks, and arago-

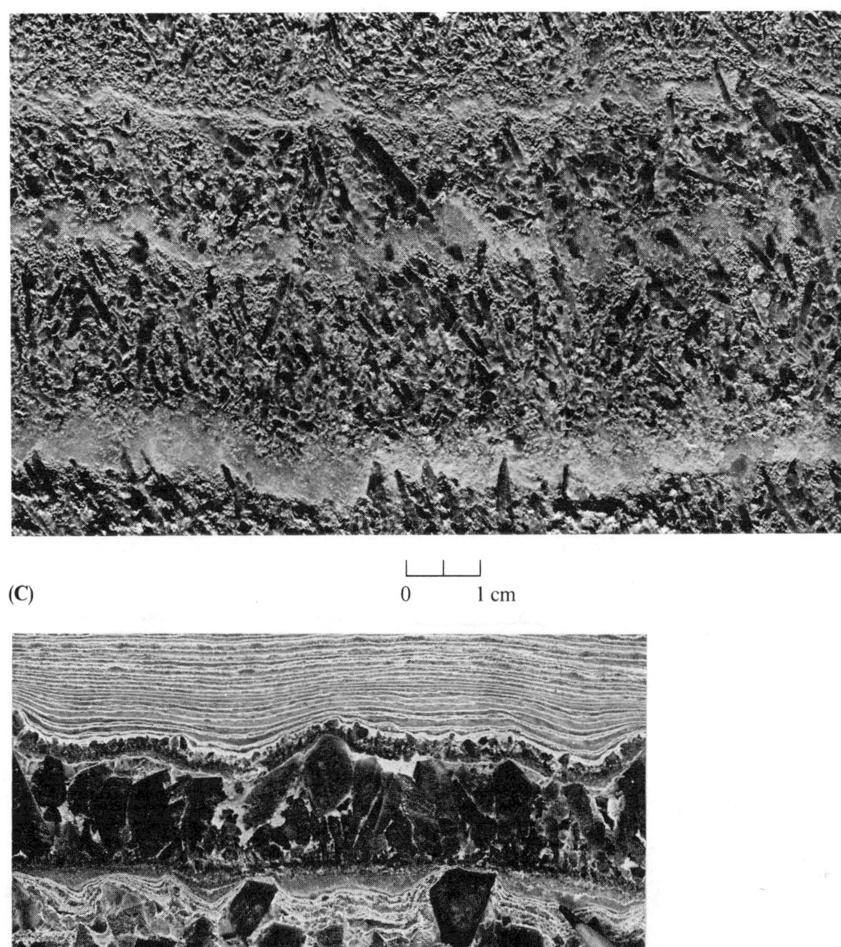

(C)

(D)

nite is 140 times less soluble than anhydrite, 160 times less than gypsum, and 24,000 times less than halite. Maiklem et al. (1969) devised a useful classification of anhydrite structures and textures that are seen commonly in hand specimens, based on occurrences in Devonian evaporites of the Elk Point Basin in northern Alberta, Canada. They illustrate their classification with 44 high-quality photographs of slabbed cores, two of which are shown as Figures 9–5 and 9–6. Figure 9–5 illustrates nodular or chicken-wire structure—a striking structure formed of elongate gobs

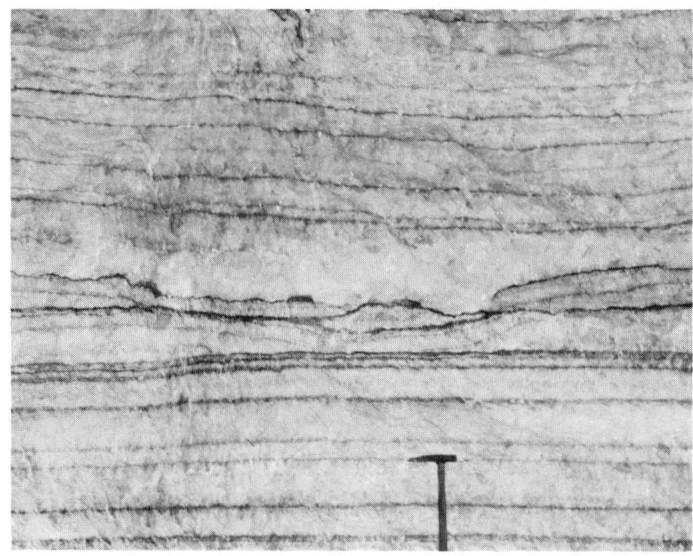

Figure 9–4
Bands of white halite and laminae of dark, clayey anhydrite, Salina Formation (Silurian), New York. One or two unconformities, probably resulting more from solution than physical erosion, are present in center. Width of photo is 3.5 m. [L. F. Dellwig and R. Evans, 1969, *Amer. Assoc. Petroleum Geol. Bull., 53.* Photo courtesy L. F. Dellwig.]

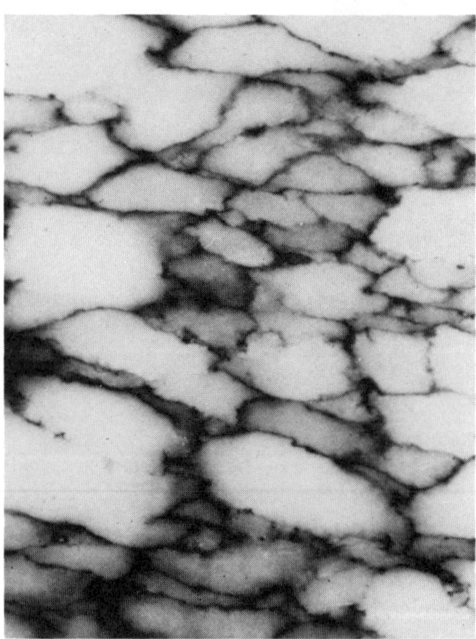

Figure 9–5
Mosaic anhydrite (chicken-wire structure). Actual size. [Maiklem et al., 1969. Photo courtesy R. P. Glaister.]

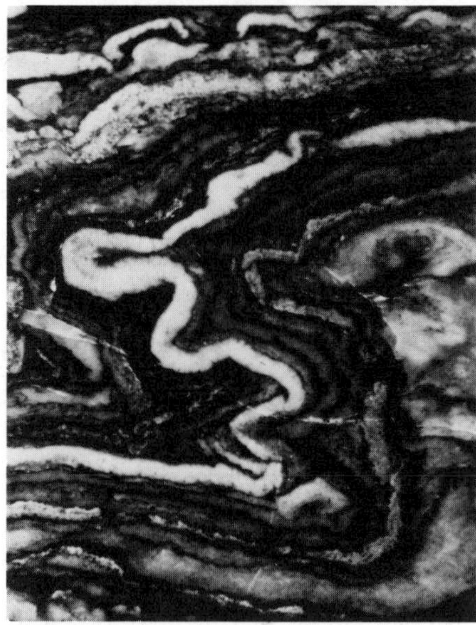

Figure 9–6
Ropy, bedded anhydrite (enterolithic folding). Actual size. [Maiklem et al., 1969. Photo courtesy R. P. Glaister.]

of anhydrite set in a darker matrix of microcrystalline anhydrite. This anhydrite mosaic forms from separate crystals and aggregates of gypsum that originated within a few meters of the ground surface by displacement of the surrounding sediment (Shearman, 1981). The gypsum subsequently dehydrated (normally with burial to a few hundred meters) with a loss of 38% of the solid volume of the gypsum. The resulting watery mush of anhydrite nodules compacted as the water escaped. Anhydrite nodules tend to be initiated in layers parallel to bedding. If the rate of the dehydration is sufficiently slow, the sediment may remain rigid enough to permit the formation of pseudomorphs of anhydrite after gypsum, on either a phaneritic or an aphanitic scale (see Figure 9–7).

Nodular structure in gypsum–anhydrite sequences is commonly observed as displaced growth in the carbonate muds of modern supratidal environments (sabkhas), and for this reason the structure is widely believed to be evidence of sabkha deposition. However, nodular structure has also been found in laminated evaporite sequences overlain by varve-banded halite. Indeed, the main requirement for the formation of gypsum or anhydrite nodules is growth under a mud in contact with high-salinity brines. Consequently, nodular structure should be used with caution in drawing environmental conclusions.

Figure 9–7
Thin section of aphanitic anhydrite grading upward into coarser crystals of anhydrite pseudomorphous after gypsum, Blaine Formation (Permian), Kansas. Black area in upper half is halite. [Holdaway, 1978. Photo courtesy Kansas Geol. Surv.]

Figure 9–6 illustrates intralayer crumpling (enterolithic folding) of laminated anhydrite. The same structure occurs in laminated gypsum. Laminated gypsum forms in quiet, standing bodies of water in which there are no bottom-dwelling, burrowing organisms—as do all finely laminated sediments. As noted earlier, the laminations in some gypsum units may be varves. There are several possible ways by which ropy anhydrite can form:

1. The original gypsum dehydrates to anhydrite on burial, releasing the water of hydration in the process. Crumpling of laminae may occur during dehydration.
2. Crumpling may occur as a near-surface, soft-sediment deformational feature.
3. Crumpling may occur during the hydration of diagenetic anhydrite back to gypsum, a process involving a 64% volume increase.

In most cases it is not possible to distinguish among these three alternatives.

The production of many sedimentary structures and textures in evaporite units can be attributed to the enormous volumes of water released by the dehydration of primary gypsum during relatively shallow burial. A gypsum bed 10 m thick will convert to 6.2 m of anhydrite, releasing 4.9 m^3 of water per square centimeter of cross section. At 30°C, this volume of $CaSO_4$-saturated water can dissolve 0.8 m^3 of halite or 5.4 m^3 of carnallite. Alternatively, the water can cause 8.1 m^3 of carnallite ($KMgCl_3 \cdot 6H_2O$) to be replaced by sylvite (KCl). There may or may not be primary porosity and permeability in a particular evaporite bed, but the great solubility and hydrated nature of many evaporite minerals assure that fluid flow will always occur within the bed and among adjacent beds. Most evaporite units are as completely metamorphosed as a kyanite schist or an amphibolite; the only difference is in the temperature required to accomplish the metamorphism.

Evaporite-Solution Breccias

The great solubility of evaporites and the ease with which evaporite beds can be plastically deformed commonly cause the development of solution breccias after burial (Stanton, 1966). These breccias can develop on a very large scale, sometimes extending over several hundred square kilometers (DeMille et al., 1964; Smith, 1972). When seen in their entirety in outcrop, evaporite-solution breccias have a well-defined base that coincides with the top of the strata underlying the evaporites and typically extend upward through two or more formations of differing lithology, perhaps for several hundred meters. Their upward extent depends on the amount of evaporite removed and on the competence of the overlying rocks.

The shape and size of the breccia fragments are determined by the lithology of the nonevaporite rocks within and above the evaporite beds. Competent rocks such as thickly bedded dolomites tend to fracture during foundering into large, rectangular joint blocks with little fine debris. Less competent beds such as shales deform initially by buckling and only later break into small slabs and angular fragments separated by much fine-grained debris (see Figure 9–8).

Evaporite-solution breccias can form in two ways. The evaporite may simply dissolve to produce large voids and cause the collapse of the less soluble beds within and above it. Alternatively, during evaporite-solution brecciation, gradual subsidence, evaporite flowage, and small-scale precipitation may occur simultaneously with gross evaporite removal; thus cavern formation and roof collapse may be relatively unimportant. As evaporite dissolves, overlying and interbedded nonevaporite strata subside

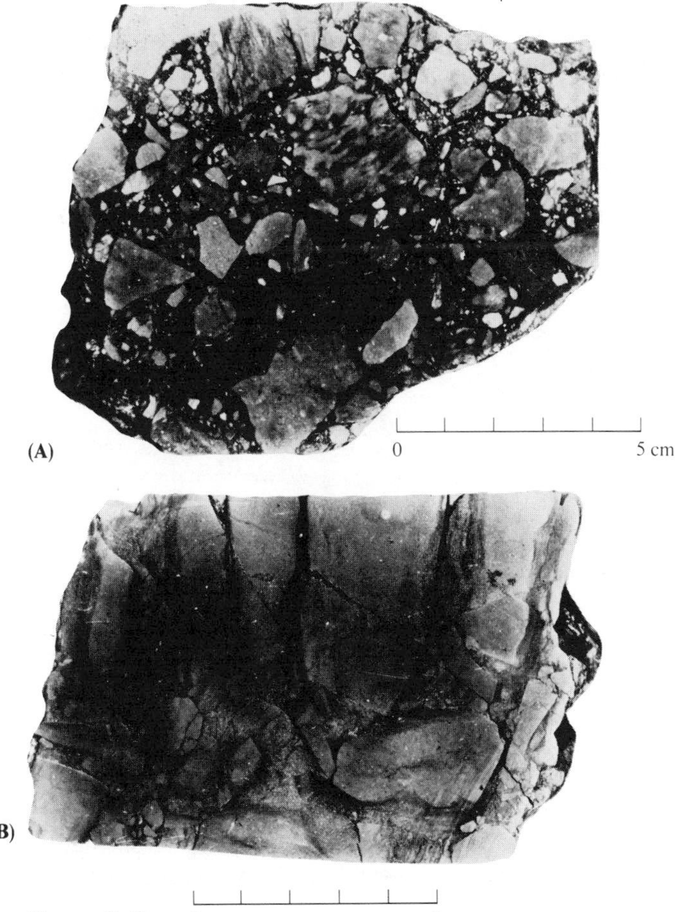

Figure 9–8
Two stages in formation of evaporite-solution collapse breccia, Prairie Formation (Devonian), Saskatchewan, Canada. (**A**) Lower part of breccia, in which much of the material is aphanitic. Some particles are insoluble residue present within original salt; other particles are fragments of overlying beds. (**B**) Higher parts of breccia, where fragments are larger and show less relative separation. [DeMille et al., 1964. Photos courtesy G. DeMille.]

Figure 9–9
Limestone fragments floating in secondary gypsum during early stage of evaporite-solution brecciation, Middle Magnesian Limestone (Permian), Durham, England. Bedding planes of limestone are still concordant, although separated by gypsum flowage. [Smith, 1972. Photo courtesy D. B. Smith.]

and fragment; evaporite deforms plastically and flows, so that the insoluble rock fragments are floating in it (see Figure 9–9). A grain-supported mass gradually forms and, finally, an evaporite-free breccia. Depending on the stage to which the process has progressed, an evaporite-solution breccia may or may not contain residual evaporite minerals.

LABORATORY STUDIES

In thin sections of ancient evaporites, both apparently primary and clearly secondary textures can be observed, although the distinction between the two is not always clear-cut. Figure 9–10 is a photomicrograph of normally graded beds of gypsum.

Long (0.3–0.6 mm), tabular, twinned gypsum grains, with their long axes parallel to bedding, and trains of equant sand-size gypsum are concentrated at the base of each lamination. Small equant grains (15–30 μm) are packed between the large tablets. Fining upward is shown mainly by a rapid decrease in the proportion of large to small grains, so that the mean grain size at the top of a layer is only about 20 μm. The grain boundaries show minor interlocking and suturing—clearly a diagenetic feature superimposed on the primary graded bedding.

Figure 9–10
Normally graded beds of gypsum in thin section, Solfifera Series (Miocene), Sicily. Lower lamination of coarse crystals is 2 mm thick. [Hardie and Eugster, 1971. Photo courtesy L. A. Hardie.]

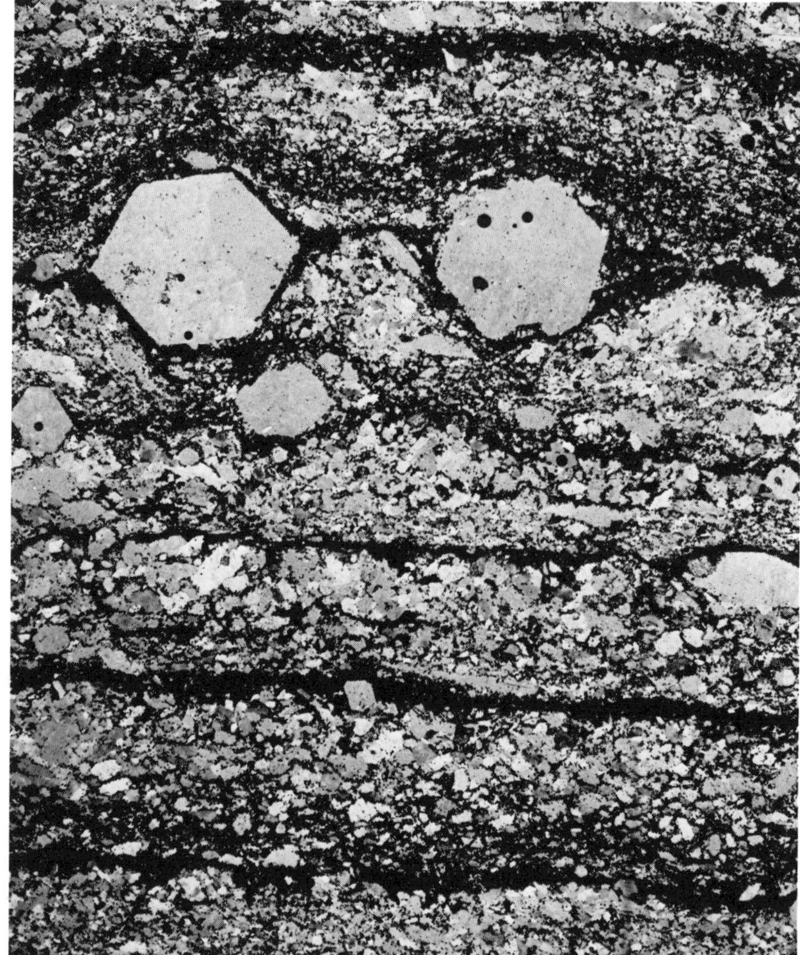

Figure 9–11
Thin section showing gypsum–dolomite triplets and large selenite crystals draped by dolomite mud, Solfifera Series (Miocene), Sicily. Gypsum is different shades of gray; dolomite is black. Large crystal in upper left is 0.8 mm across. [Hardie and Eugster, 1971. Photo courtesy L. A. Hardie.]

Figure 9–11 shows complex laminations produced by alternations of dolomite-rich layers and gypsum-rich layers. The base of each lamination is marked by a thin concentration of very even-sized dolomite silt averaging 7 μm in diameter, which passes upward into a coarse gypsum mosaic peppered with dolomite silt, which in turn is capped by dolomite mud, making each lamination a distinct triplet. It is difficult to distinguish primary from secondary fabric in this thin section.

Thin sections of halite commonly reveal the crystals to contain an internal chevron structure formed by zones of fluid inclusions (see Figure 9–12A). This pattern reflects a peculiar growth habit of halite crystals at the surface of an evaporating brine pan as the small crystals are held at the water surface by surface tension. As the crystals grow or are disturbed by winds blowing over the water surface, the crystals fragment and sink to the floor of the basin. Subsequent secondary growths occur on the crystal surfaces, analogous to the secondary growths on calcite and quartz crystals, but these new growths contain no fluid inclusions and are clear as a result (see Figure 9–12B).

Many unusual textures and structures seen in evaporitic rocks are believed to have resulted from dehydration–hydration reactions among evaporite minerals. Holliday (1970), however, has pointed out that there is abundant textural evidence in some evaporite beds that these mineral replacements can occur without a change in volume. Observations used to support this view include:

1. Bedding structures in unreplaced parts of the rock pass without deflection through replaced parts.
2. Anhydrite fabrics passing undisturbed through secondary gypsum (see Figure 9–13).
3. Optically continuous anhydrite and secondary gypsum.
4. Corroded remnants of single anhydrite crystals, now separated by secondary gypsum, but still perfectly aligned and in the same orientation.

It certainly is possible, therefore, that gypsum can replace anhydrite without volume change, with the excess sulfate carried away by moving waters. There is even evidence that enough sulfate might be removed in solution to result in a net volume *decrease* during gypsification. Holliday (1970) has observed that secondary gypsum rocks are usually more porous than their parent anhydrite.

Evidence of replacement textures is common in evaporite rocks. Pseudomorphs are widespread and reaction rims are common, as are relics of an earlier evaporite mineral enclosed within a mineral that replaced it. Well-known examples are pseudomorphs of anhydrite after swallowtail twins of gypsum, reaction rims of polyhalite around earlier anhydrite, relics of carnallite in secondary sylvite, pseudomorphs of sylvite plus kieserite after earlier langbeinite, and many others. The equations below describe these types of transformations recognized in thin section:

$$\underset{\text{gypsum}}{CaSO_4 \cdot 2H_2O} \rightleftharpoons \underset{\text{anhydrite}}{CaSO_4} + 2H_2O$$

$$\underset{\text{anhydrite}}{2CaSO_4} + 2K^+ + Mg^{2+} + 2SO_4^{2-} + 2H_2O \rightleftharpoons \underset{\text{polyhalite}}{K_2Ca_2Mg(SO_4)_4 \cdot 2H_2O}$$

$$\underset{\text{carnallite}}{KMgCl_3 \cdot 6H_2O} + 4H_2O \rightleftharpoons \underset{\text{sylvite}}{KCl} + Mg^{2+} + 2Cl^- + 10H_2O$$

$$\underset{\text{langbeinite}}{K_2Mg_2(SO_4)_3} + MgCl_2 + 3H_2O \rightleftharpoons \underset{\text{kieserite}}{3MgSO_4 \cdot H_2O} + \underset{\text{sylvite}}{2KCl}$$

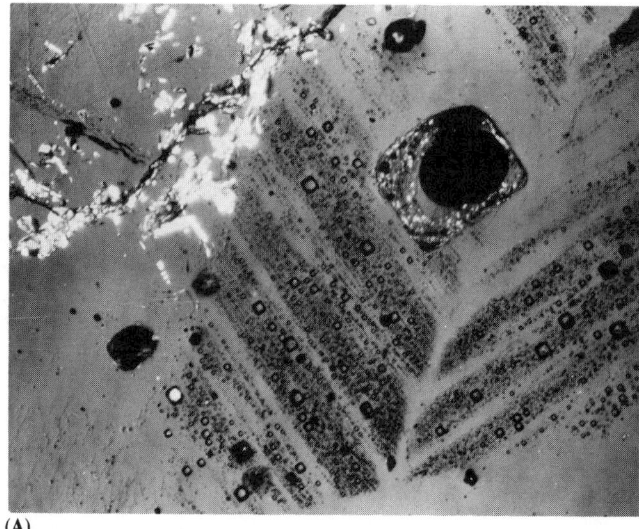

(A)

(B)

Figure 9–12
Thin section showing chevron structure within halite crystals. (**A**) Close-up, with white anhydrite along crystal boundaries. Width of photo is 3.4 mm. (**B**) Broader field of view, showing chevrons within crystals surrounded by clear halite overgrowths that meet along subplanar boundaries. Width of photo is 4.2 mm. [Holdaway, 1978. Photos courtesy Kansas Geol. Surv.]

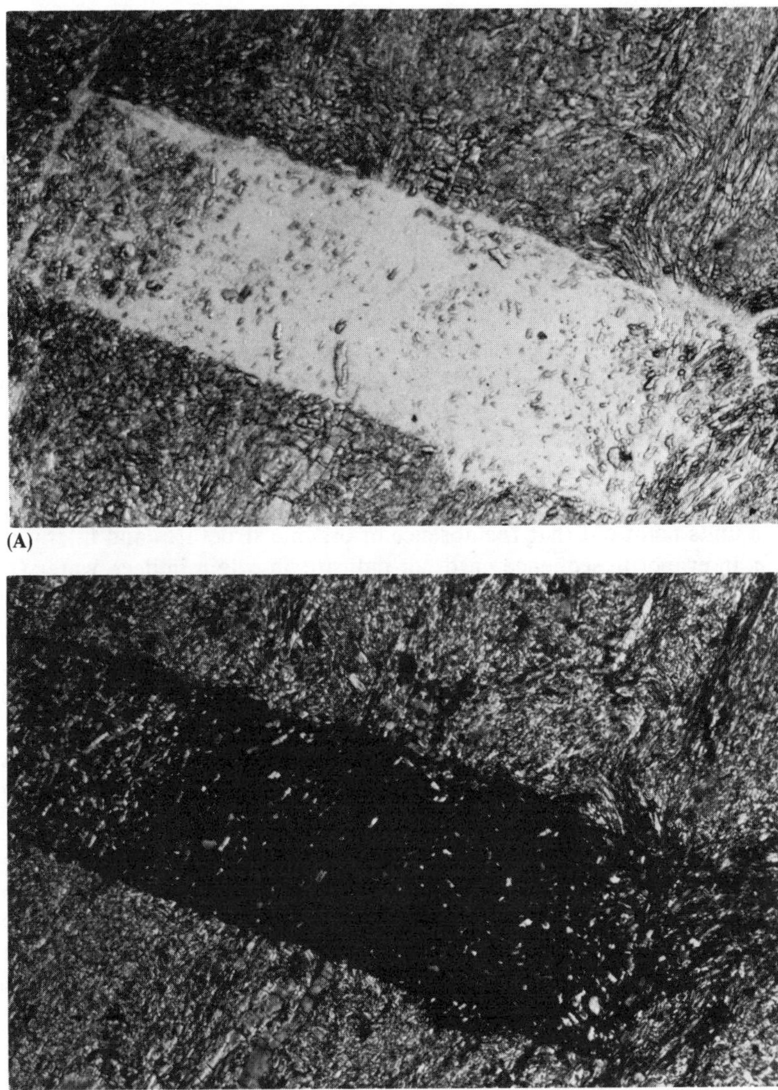

Figure 9–13
Subhedral gypsum porphyroblast replacing anhydrite, Ebbadelen Formation (Carboniferous), Spitsbergen. Porphyroblast contains numerous corroded relics of anhydrite and sharply truncates but does not disrupt anhydrite fabric, demonstrating that gypsification took place without an increase in volume. (**A**) Uncrossed nicols. (**B**) Crossed nicols with porphyroblast in extinction position. Length of porphyroblast is 2 mm. [Holliday, 1970. Photos courtesy D. W. Holliday.]

WATER DEPTH AND DEPOSITION

In recent years there has arisen a dispute concerning the depth of water in which some laminated gypsum–anhydrite sequences were formed. The question was posed by the textural and sedimentary structural characteristics of the beds as seen in the field. Laminated sulfate units thought to be of deep-water origin are characterized by greater vertical and lateral uniformity than those formed in shallow water. For example, in the Delaware Basin of western Texas, the laminite section of the Castile Formation and lower Salado Formation is about 400 m thick and contains 260,000 anhydrite–calcite couplets. Although some structures are present that might be indicative of shallow water (breccias and nodular anhydrite), the overall impression is of physical and chemical constancy over a very extensive flat area. There is a conspicuous absence of normal shallow-water features such as desiccation cracks, rip-up conglomerates, ripple marks, algal-growth features, and discontinuous laminae resulting from laterally varying physical and chemical conditions.

Opponents of the deep-water hypothesis of origin of the laminated gypsum–anhydrite units point out that the absence of organic structures and lateral discontinuities in an evaporite sequence need not indicate an origin in deep water (meaning water at least 200 m deep). For example, as noted above, Holser (1979) has pointed out that a density-stratified water mass can effectively dampen the effects of surface waves at shallow depths. Also, there are many shallow-water evaporite sequences that lack an obvious organic influence, as might be expected given the very high salinities of brines from which sulfates and halite precipitate.

Shaw (1977) has made a quantitative analysis of the meteorologic (evaporation rate, air-mass movements) and chemical requirements for a deep and laterally extensive water mass to become saturated with respect to the calcium sulfates. He concluded that deep-water evaporite formation is physically impossible, noting, for example, that ionic diffusion within a water mass is far more rapid than evaporative concentration, so that a deep-water basin could neither develop evaporite-producing salinities nor maintain them even if it were filled originally with brine of sufficient salinity derived from some other source. The deep-water enthusiasts dispute some of Shaw's assumptions or calculations. The question is still unresolved.

ANCIENT EVAPORITES

Castile Formation (Permian), Texas and New Mexico

Perhaps the most spectacular evaporite deposit is the anhydrite–halite sequence in the Castile Formation (Upper Permian) in the Delaware Basin of western Texas and southeastern New Mexico (see Figure 9–14). The Castile crops out only in the western part of the Delaware Basin (as 250 m of gypsum) but in the subsurface underlies an area more than 100,000 km^2 in extent. It has a maximum thickness of 1,200 m and a volume of approximately 25,000 km^3. The original depositional volume of the

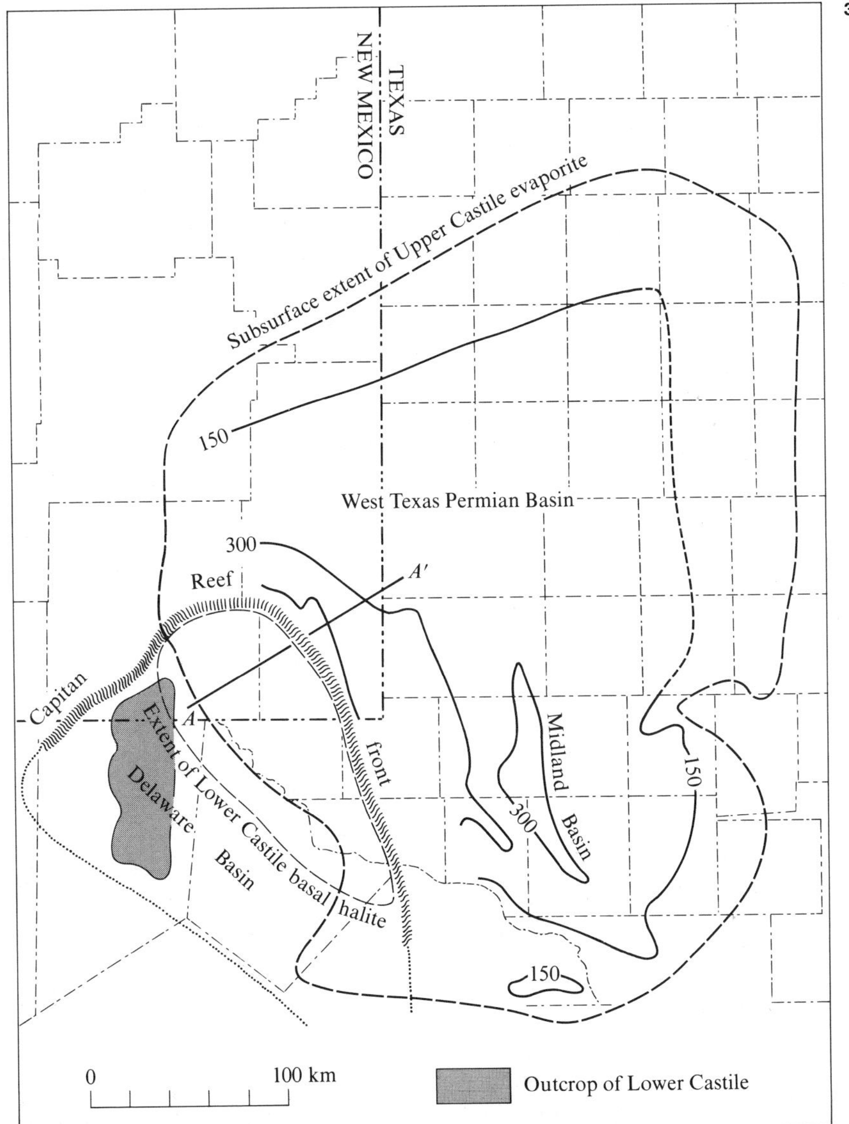

Figure 9–14
Map of western Texas and southeastern New Mexico, showing location and thickness of upper Castile evaporite (meters). *A–A′* is line of cross section in Figure 9–15. [G. A. Kroenlein, 1939, *Amer. Assoc. Petroleum Geol. Bull., 23.*]

evaporites was even greater than at present, as evidenced by erosional contacts laterally and many anhydrite-solution breccias caused by the removal of halite. Near the surface the anhydrite is converted to gypsum. The present ratio of halite to the calcium sulfate minerals is estimated to be about 1/1.

The depositional environment of this giant evaporite deposit was a steep-sided, deep basin (but not necessarily deep water) bordered by a carbonate platform that

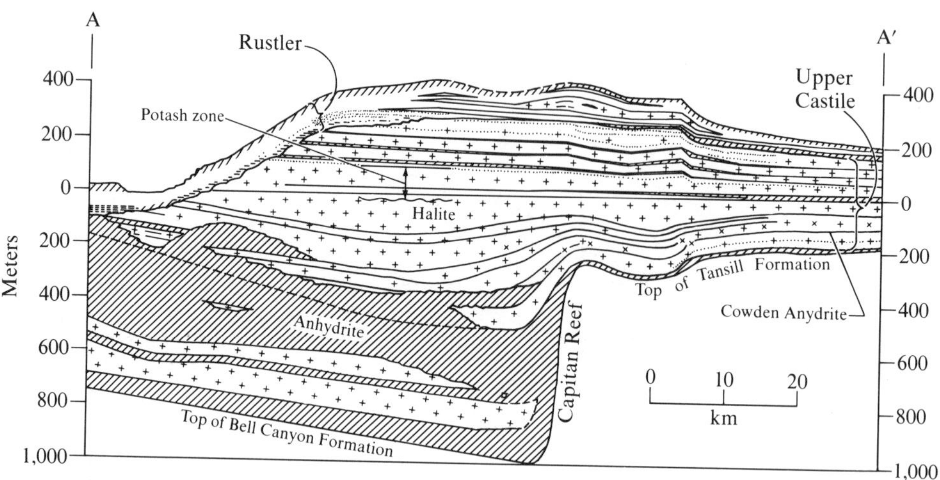

Figure 9–15
Generalized cross section of Castile Formation normal to reef front, showing changes in evaporite composition and thickness between Delaware Basin and backreef area. [G. A. Kroenlein, 1939, *Amer. Assoc. Petroleum Geol. Bull., 23.*]

stood more than 500 m above the basin floor (see Figure 9–15). In the basin the evaporites are underlain by organically laminated quartz siltstone, fine sandstone, and carbonates of the Bell Canyon Formation; on the platform the beds below the evaporites are shallow-water carbonates of the Tansill Formation. Above the Castile in both areas is the Rustler Formation, mostly anhydrite with lesser amounts of fine-grained red clastics, dolomite, and halite. Evaporite deposition closed in the region with an influx of fluvial muds and fine sands derived from the foreland of the Ouachita–Marathon orogen to the south and east.

The lower portion of the Castile is composed of alternating thin laminae of calcite, anhydrite, and organic matter. The anhydrite layers are 2–3 times thicker than the calcite layers. In addition to many variations in the character of these primary sedimentary structures, many diagenetic features are present. Calcite laminae disappear into nodular masses of anhydrite. Concretionary anhydrite lenses grow between the partings and disrupt them. Bands of crinkled laminae occur between undisturbed beds. Sometimes the anhydrite is interlaminated with dolomite rather than calcite.

The calcite crystals are subhedral to ovoid in outline and uniformly 25 μm in size—possibly a result of recrystallization. The anhydrite crystals range up to 100 μm in diameter and have interlocking rectangular outlines. The laminae in the calcite-laminated anhydrite average 1.6 mm in thickness and are remarkably persistent laterally, having been traced for distances up to 113 km. Each lamina is synchronous, and each couplet has been interpreted by Anderson et al. (1972), as representing an annual layer of sedimentation—a varve. Anderson et al. measured and correlated a total of 260,000 of these couplets within the basin.

The upper Castile is dominated by halite, and the onset of halite precipitation seems to have been synchronous throughout the basin. The halite, however, is not as easily traceable in detail as is the anhydrite below it. The halite typically is thoroughly recrystallized and occurs in laminae up to 10 cm thick, alternating with anhydrite. The halite is composed of nearly pure NaCl, the only impurities being blebs and crystals of anhydrite, laminated calcite, and a few euhedral quartz crystals.

The distribution of sedimentary rock types in the Delaware Basin shows a crude concentric zonation (see Figure 9–16), characteristic of a desiccating basin. Along the outer fringe are either fine-grained clastics or limestones, depending on the location of nearby land areas. Within this outer fringe are gypsum and/or anhydrite, followed by halite, and finally by the more soluble salts in the center of the desiccated area—salts such as polyhalite, langbeinite, carnallite, and sylvite. Polyhalite is the

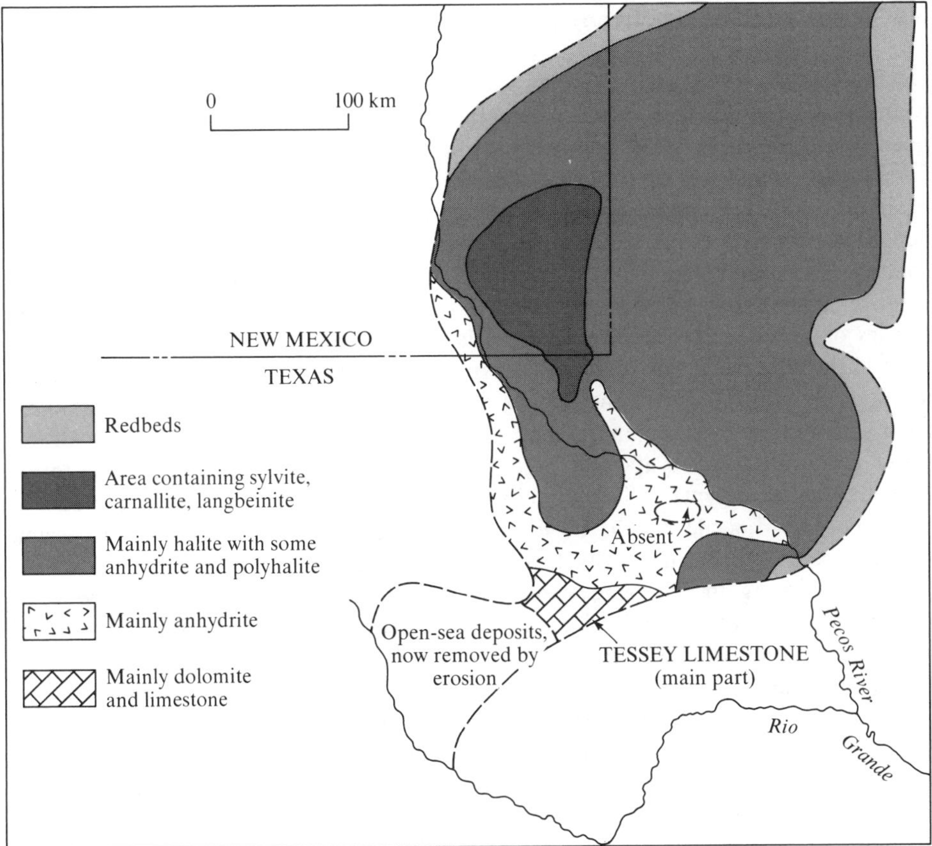

Figure 9–16
Map showing distribution of sedimentary rock types during deposition of upper Castile Formation in Texas and New Mexico. [Stewart, 1963.]

least soluble of the posthalite salts, occurs throughout a greater stratigraphic interval, and is more widespread in distribution than the more soluble potassium salts. Deposition of the potassium minerals started first on the shelf area during Tansill time. With basinward changes in chemical and physical conditions in later Tansill and upper Castile time, the beginning of potassium-mineral deposition moved basinward and progressively higher in the evaporite sequence. The very soluble potassium salts persist into the lower part of the Rustler Formation, where they are finally terminated by the influx of red muds. The potassium minerals occur in the evaporite deposits as accessory minerals, as stratified deposits in the sulfate units, as bedded deposits in mixed halite–clastic units in the upper part of the Castile, and as vein or lenticular deposits that have replaced or displaced the host evaporite beds.

Mediterranean Sea (Miocene)

During the late 1960s, seismograms were taken of several areas of the Mediterranean Sea basin, and they revealed structures resembling salt diapirs under the seafloor. In 1970, these structures were drilled, and the presence of Late Miocene (Messinian) evaporites up to 2 km thick was discovered in several areas within the basin (Hsü, 1972). The evaporites are sharply underlain by pelagic muds that were deposited in water about 1,000 m deep, based on the benthonic foraminifera they contain. The sparse fauna in the overlying Lower Pliocene muds suggests a water depth of 200–600 m. Apparently, prior to Late Miocene time the input–output dynamics of water in the Mediterranean basin were similar to that existing today, a deep basin filled with fairly normal ocean water. Then, in the Late Miocene, plate interactions associated with the closing of the Tethys Ocean isolated the Mediterranean (2.5 million km^2) to the extent that within about 1,000 years nearly all the water evaporated and evaporites precipitated. The eastern closing (Turkey, Syria, Lebanon, Israel, Egypt) still exists today, but approximately 7 million years ago, at the end of the Miocene Epoch, the western end at Gibraltar opened slightly to permit the refilling of the basin with waters from the Atlantic Ocean. Pelagic muds were deposited once again. The rate of deposition of this mud was sufficiently rapid, so that most of the salt was effectively isolated from the water in the filling basin and was not dissolved. An empty bathtub 4,000 m deep and with a volume of 4×10^6 km^3 was refilled almost instantaneously.

The topography of the floor of the Mediterranean Sea divides it into four basins (see Figure 9–17). The relief between the basin lip and its bottom ranges from 1,000–2,500 m among the basins. In each basin is a sequence of Miocene evaporites 0.3–2.0 km in thickness that includes gypsum, halite, and the more soluble salts and, in each, the areal distribution resembles a bull's-eye pattern (see Figure 9–18). It looks much like the distribution in the Castile Formation or in a desert playa, with the least soluble salt (gypsum) at the fringes and the more soluble salts toward the center as the amount of water decreases and the salinity increases. Evaporation of 1,000 m of seawater produces only 15.9 m of evaporite deposit. Therefore, more than

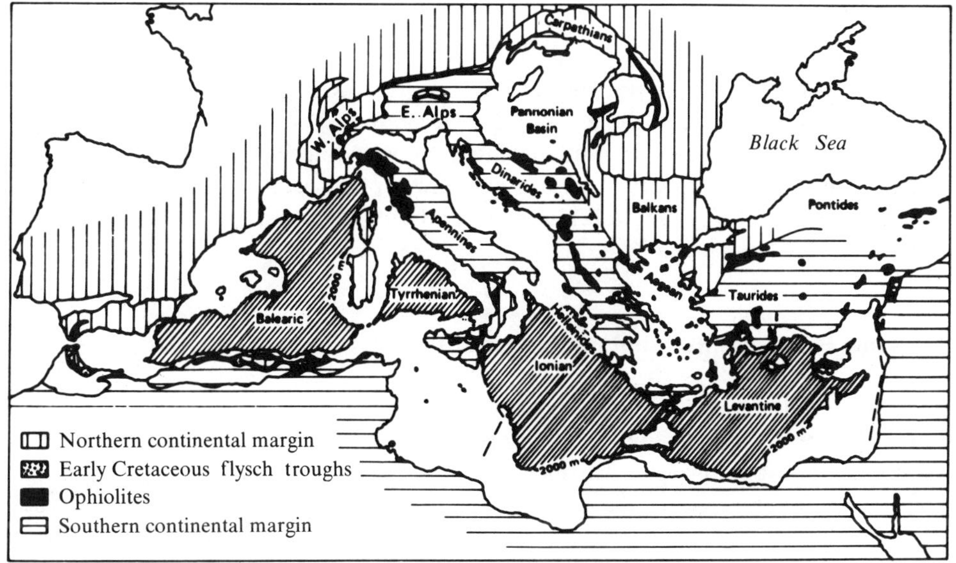

Figure 9–17
Tectonic setting of Mediterranean basins in which thick evaporites accumulated during Miocene time. [K. J. Hsü, 1979, Deep Sea Drilling Project Report, *42A*.]

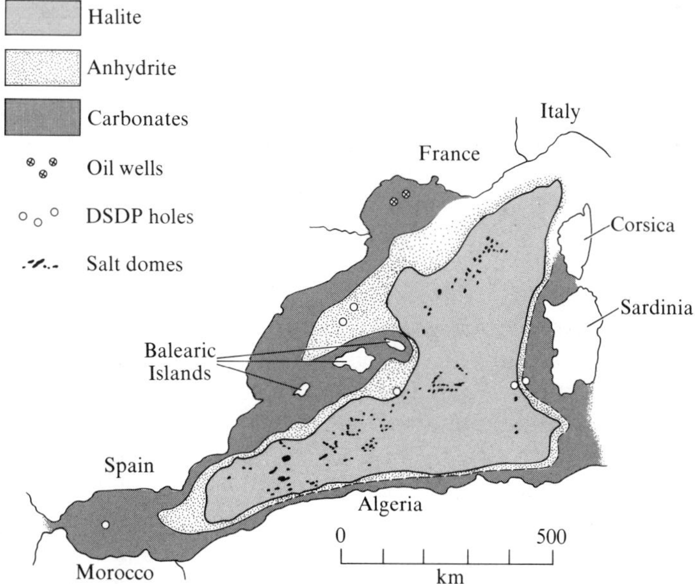

Figure 9–18
Evaporite distribution in Miocene rocks of Balearic Basin of Mediterranean Sea immediately east of Strait of Gibraltar. Balearic was one of several topographically low salt pans on floor of desiccated Mediterranean. Potash salts (not shown) occur in middle of halite area. [Hsü, 1972.]

125 km of water must have been evaporated during Messinian time (2 million years) to produce the maximum thickness of evaporite present in the Mediterranean basins. Apparently, the small amount of inflow into each basin during the Late Miocene was adequate to maintain the salt concentration between that needed for gypsum to form (3 times normal salinity) and total dryness. Certainly, there was never water 125 km deep in the Mediterranean area.

Duperow Formation (Devonian), Williston Basin

The Castile Formation (Permian) in western Texas and southeastern New Mexico and the Miocene evaporites in the Mediterranean Sea basin exemplify the first of two main types of evaporite deposits: thick accumulations of essentially pure evaporites initiated in the central parts of a desiccating, standing body of water. The second type of evaporite accumulation forms at the margins of a marine basin on a sabkha, typically interbedded with dolomitic carbonate rocks (see Chapter 8). The Duperow Formation is one of several evaporite–carbonate sequences that have been studied in detail (Wilson, 1967).

The Duperow is part of a great sheet of Upper Devonian strata that extend beneath the Prairie Provinces of Canada southeastward from the Canadian Rockies and arctic Canada into the United States as far south as the transcontinental arch of South Dakota and Nebraska. Its maximum thickness is about 250 m in southern Saskatchewan. It thins southeastward to 150 m in northwestern North Dakota and disappears because of erosion in south-central South Dakota. The formation is composed of about twelve regular carbonate–evaporite cycles. Each cycle consists of a lower member of burrowed, fossiliferous micrite containing a normal marine, shallow-water fauna of brachiopods, crinoids, and stromatoporoids (a group of extinct calcareous coelenterates); a middle member of brown micrite with a restricted microfauna of ostracods interbedded with unfossiliferous, peloidal beds or laminated micrite; and an upper member of bedded anhydrite with gray-green, silty dolomicrite displaying intertidal and supratidal sedimentary structures. The anhydrite member is typically the thinnest of the three members (see Figure 9–19). Duperow cycles are exceedingly widespread, and constituent beds only 3–5 m thick can be traced for several hundred kilometers across the Williston Basin.

Deposition occurred in a vast backreef lagoon south of a reef belt in Alberta and stretched to a sandy shore in South Dakota and Wyoming. This very shallow basin was periodically and apparently rapidly flooded with marine water, permitting certain benthonic organisms to flourish and even small reefs to grow at times. Gradual shallowing as sedimentation filled the basin resulted in extensive tidal flats and evaporitic sabkhas; extensive dolomitization occurred on the peripheral shelves. The time required for the deposition of each cycle can be estimated as from 500,000 to a million or so years, assuming a constant rate of sedimentation through the Late Devonian.

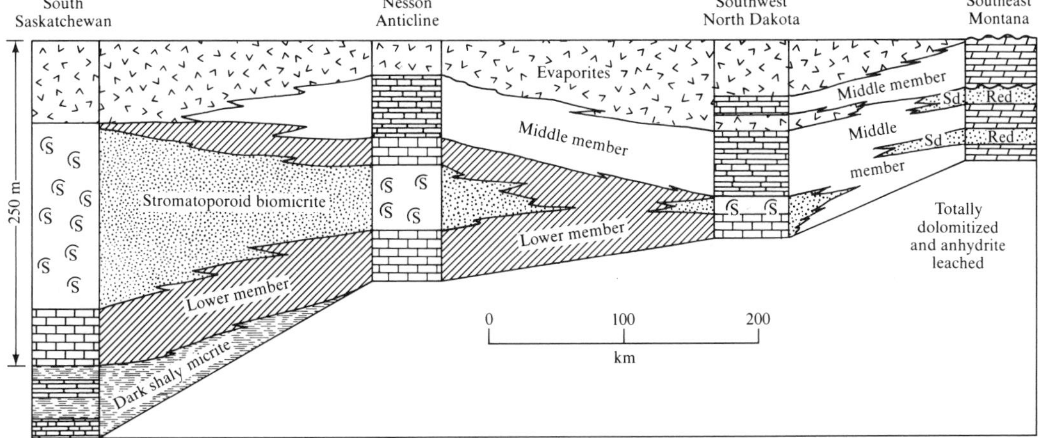

Figure 9–19
Idealized stratigraphic diagram showing interpreted Duperow Formation (Devonian) facies belts from north to south across Williston Basin. Normal marine, burrowed biomicrite and stromatoporoid reefs grade into restricted marine, laminated micrite with microfauna (white area). Microfaunal micrite grades upward into sabkha dolomicrites and evaporites. [Wilson, 1967.]

The bedded anhydrite of the upper member of the Duperow Formation occurs in beds 0.3–3.0 m thick that have a fibrous texture and ubiquitous lamination. The laminae are of two types: horizontal laminae about 1 mm thick and secondarily distorted laminae. Some beds contain "slump-and-crumple structure" or "ball-and-flow structure," discernible when the anhydrite is interlaminated with microcrystalline dolomite. In these disturbed beds, local areas of breccia and conglomerate occur, probably a result of early solution of sulfate and consequent breakage of the thin layers of carbonate. Postdepositional flowage of anhydrite layers is responsible for other localized conglomerates. Mud cracks and arched laminae, indicative of algal stromatolites, are also present in the bedded anhydrites, particularly where dolomites are associated.

Rarely, the anhydrite is interlaminated or otherwise closely associated with lime mud instead of cryptocrystalline dolomite. Except where associated with cryptocrystalline dolomite, the anhydrite is without detrital impurities and, of course, is without fossils or any trace of life, unless some of the rounded grains of the microconglomerate can be interpreted as fecal pellets.

The abundance of sedimentary structures, such as mud cracks and microbreccias, is taken as evidence that the evaporites were formed on periodically exposed tidal flats.

Green River Formation (Eocene), Western United States

The Green River Formation, located in southwestern Wyoming and adjacent parts of Utah and Colorado, is the best-studied ancient nonmarine evaporite deposit. It is of

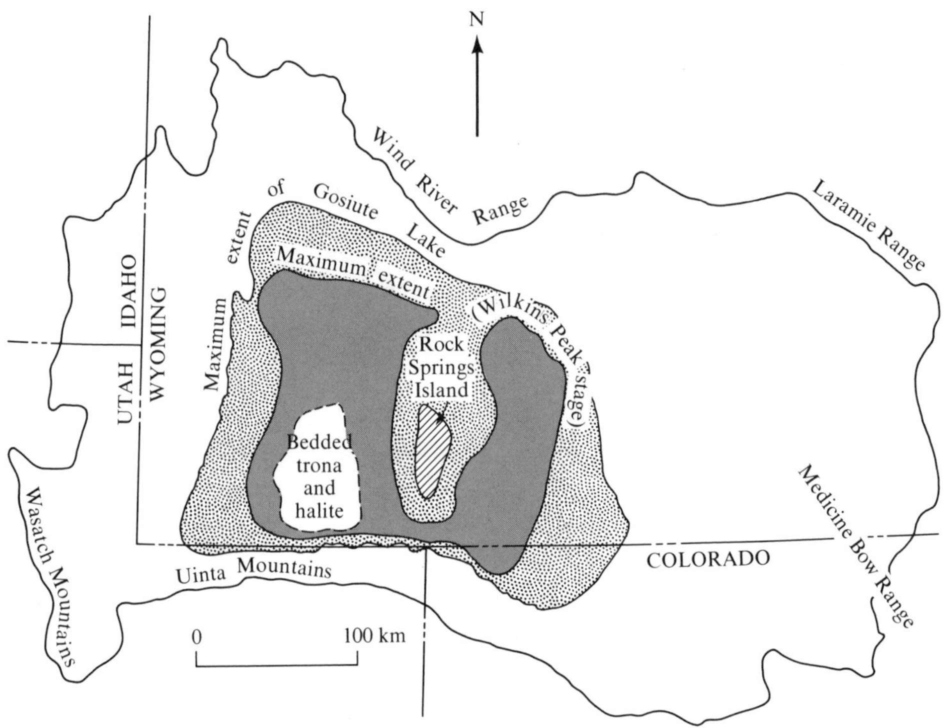

Figure 9–20
Outline of drainage basin of Gosiute Lake, showing maximum extent during third stage of its development, probable extent of Wilkins Peak stage, and area underlain by beds of trona and halite. [Bradley and Eugster, 1969.]

lacustrine origin and has the form of a great lens, or pile of lenses, within an enormous volume of fluvial sediments. Gosiute Lake, in which the Green River Formation was deposited, changed considerably and repeatedly in size and character during the Green River epoch (see Figure 9–20). But it had three major stages, each of which lasted a million years and corresponds to a member of the Green River Formation:

1. During the first stage, the lake was a large freshwater lake that had an outlet.
2. In the second stage, the climate became more arid and the lake shrank, becoming a playa lake with no outlet. It was during this period that the evaporites were deposited in the Wilkins Peak member.
3. In the third stage, the climate become humid once again, the lake expanded to its maximum size (about 40,000 km^2), and a second nonevaporitic member was deposited.

The Wilkins Peak Member has a maximum thickness of about 400 m and is dominantly carbonaceous dolomicrite, containing some illite and quartzo-feldspathic

Figure 9–21
Large crystals of trona in dolomitic mudstone of Wilkins Peak Member. [Bradley and Eugster, 1969. Photo courtesy W. H. Bradley.]

silt. It is sometimes tuffaceous and contains many beds of economically valuable oil shale (Bradley and Eugster, 1969). Within it are at least 25 beds of trona [$Na_3H(CO_3)_2 \cdot 2H_2O$] 1–13 m in thickness, locally with an aggregate thickness of 50 m, and extending over several thousand square kilometers. Halite generally is subordinate but, in the southern part of the Green River Basin, predominates over trona in most of the evaporite beds. Between the major trona beds are locally few to many thin trona layers, some no thicker than films. Bedded trona consists of interlocking crystals of the mineral, which differ greatly in size from place to place (see Figure 9–21). Most of the trona beds are brownish because of their content of carbonaceous matter, but fibrous, recrystallized trona layers are free of impurities of any kind.

In addition to trona and halite, the Wilkins Peak contains locally abundant shortite [$Na_2Ca_2(CO_3)_3$] (see Figure 9–22), northupite [$Na_3Mg(CO_3)_2Cl$], and much smaller amounts of other complex carbonates. Many of these minor evaporitic carbonate minerals have formed by the diagenetic alteration of shortite (see Figure 9–23).

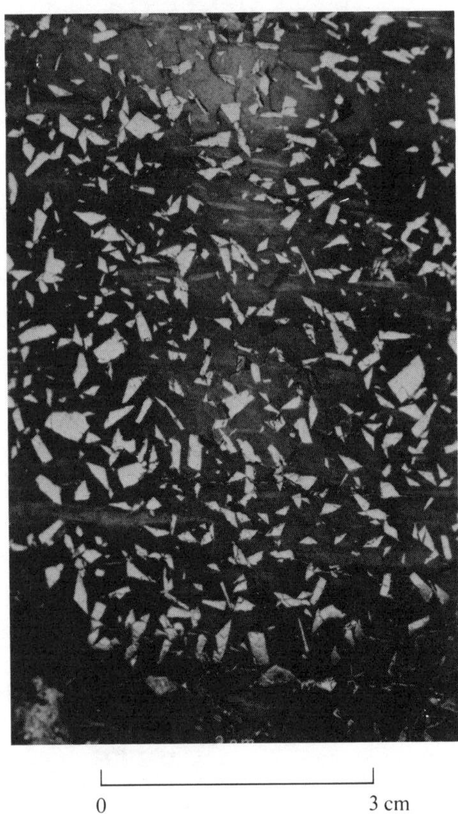

Figure 9–22
Thin section showing shortite crystals uniformly dispersed in fairly homogeneous low-grade oil shale. [Bradley and Eugster, 1969. Photo courtesy W. H. Bradley.]

THE ORIGIN OF THE GIANT MARINE DEPOSITS

Giant evaporite deposits such as the Castile appear to have been formed in standing bodies of water that were isolated or nearly isolated from the sea by reef growth, tectonic activity, volcanic eruptions, or some other cause. Therefore, the simplest model to consider for the origin of thick, pure evaporite deposits is a straightforward evaporation-to-dryness of a standing body of seawater. If the rate of potential evaporation exceeds the rate of inflow of water over a prolonged period of time, the concentration of dissolved salts in the water increases and an evaporite mineral is precipitated.

Laboratory experiments performed by numerous scientists during the past 130 years have revealed the order in which we should expect the evaporite minerals to precipitate. If we start with normal seawater, gypsum will precipitate when the water has been reduced to one-third its original volume; halite, one-tenth the original volume. The other 70 evaporite minerals (sulfates, borates, halides, and carbonates) do not appear until the volume is less than one-twentieth the original.

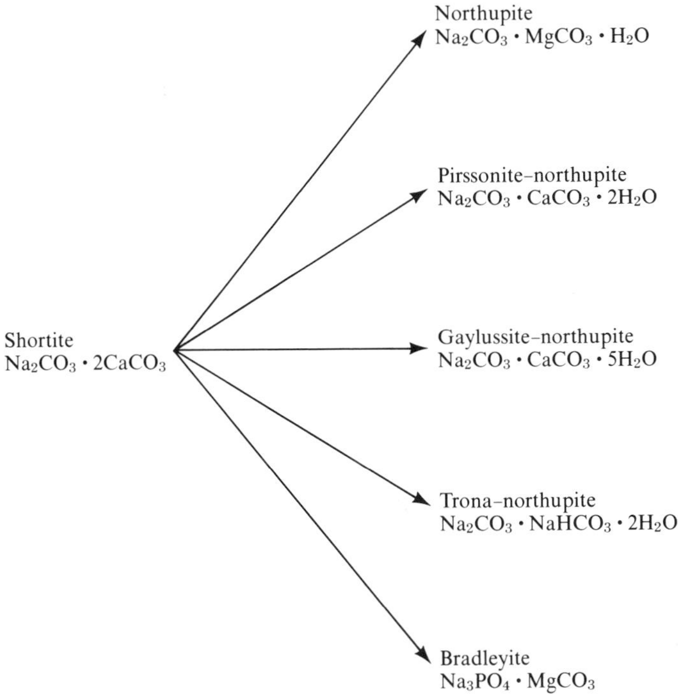

Figure 9–23
Diagrammatic representation of paragenetic sequence of six evaporite minerals in Wilkins Peak Member. [J. J. Fayey, 1962, U.S. Geol. Surv. Prof. Paper No. 405.]

What thickness of evaporite rock might we expect from the complete evaporation of seawater? Suppose we assume a column of water 1,000 m deep with 3.5% dissolved salts. Complete evaporation would yield a nonporous precipitate of sediment with a thickness of 0.035 (1,000)/specific gravity of the mineral precipitated. The average evaporite mineral has a specific gravity of 2.2, so that the total thickness of the deposit will be 15.9 m. If we assume that all the sodium in seawater is precipitated as NaCl (sp. gr. = 2.16), then Table 9–2 indicates that 41.80×2 or 83.6% of the evaporite will be halite. This is 13.5 m $[(35/2.16) \times 0.836]$. If all the calcium comes out as gysum (sp. gr. = 2.32), Table 9–2 indicates that 0.91×2 or 1.82% of the evaporite will be gypsum. This is 0.27 m $[(35/2.32) \times 0.018]$. The remaining 2.1 m of evaporite will be composed of some of the 70 or so other evaporite minerals in appropriate proportions.

Our calculation indicates that the evaporation of a standing body of water as much as 1,000 m deep can yield only 15.9 m of evaporite deposit. As we have seen in the Castile Formation, however, the thicknesses of gypsum–halite sequences can be several hundreds of meters. Most evaporite deposits are several meters to several tens of

Table 9–2
Relative Amounts of Most Abundant Dissolved Chemical Constituents in Seawater

Dissolved species	Molarity	Percentage
Cl^{1-}	0.535	48.72
Na^{1+}	0.459	41.80
Mg^{2+}	0.052	4.74
SO_4^{2-}	0.028	2.50
Ca^{2+}	0.010	0.91
All others	0.014	1.33
	1.098	100.00

meters thick. The amount of evaporation this requires seems staggering (Think of the humidity!) but that is only because of the brief span of human life as compared to the thousands of years during which uninterrupted evaporation can proceed in an isolated body of water.

The simple evaporation-to-dryness model we tested with our calculation has defects in addition to the fact that it could not supply the great thicknesses of evaporites seen in the field. For example, our model produced 50 times as much halite as gypsum because the dissolved material in seawater is mostly sodium and chloride ions. How are we to explain halite/gypsum ratios that are greatly different from 50/1, such as the Castile Formation, in which the ratio is about 1/1? Study of ancient evaporites reveals there is no norm for this ratio. Some deposits are all gypsum; others, all halite.

The mechanism that permits this to occur is called *reflux*. The requirements for reflux are:

1. Constant influx of seawater.
2. Constant evaporation within a restricted basin.
3. Loss of the heavy brine produced, either by seepage down through the underlying sediments (subsurface reflux) or by flow of the brine out of the basin beneath the inflow of new seawater.

Under these conditions a constant salinity can be maintained and a single evaporite mineral deposited. The main factor controlling what the constant salinity will be is the rate of brine loss by reflux. Greater reflux produces lower salinity; lesser reflux, high salinity.

Applying this model to the Castile Formation, we would infer that the salinity achieved during evaporation and reflux was sufficient initially to cause the precipitation of calcium sulfate (lower Castile); but as reflux flow decreased, halite was precipitated. These evaporite accumulations will be somewhat younger in age than the rocks of the basin walls and will not grade laterally into contemporaneous sedimentary facies.

If the rate of evaporation and reflux is greater than the rate of inflow, the water level drops (evaporative drawdown). The Great Salt Lake is a familiar example, a remnant of the much larger glacial Lake Bonneville; Death Valley in southern California was the site of glacial Lake Manly, as evidenced by the present salt flat and bordering wave-cut benches marking the levels of former shorelines.

CLIMATE

We have seen that the formation of extensive evaporite deposits requires a relatively high rate of evaporation coupled with a low rate of precipitation. In what areas of the present Earth's surface do such conditions exist? To find the answer to this question, we must examine the gross circulation pattern of the atmosphere (see Figure 9–24). The intensity of solar radiation reaching the Earth's surface is determined primarily by the angle of incidence of the sun's rays: the more nearly perpendicular to the Earth's surface, the greater the intensity. As a result, the area around the equator is heated more than at higher latitudes, so that this zone is characterized by rising air. As the air rises, it cools, causing its moisture to be released and forming the rainy, tropical jungles. The air continues to cool and compress as it moves away from the equator until it sinks back toward the surface at about 30° latitude. At this point it is a dry air mass because it has dropped nearly all its moisture as it cooled moving toward the poles; and as a dry air mass, it has the capability to absorb surface moisture as it descends. Thus, the belts of desert originate at 30° latitude on both sides of the equator. As shown in Figure 9–24, modern evaporite deposits center at these 30° latitude locations, as do belts of high average salinity in the oceans.

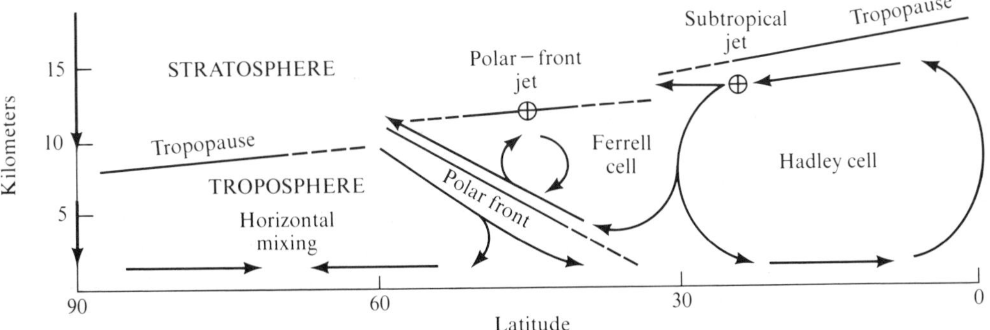

Figure 9–24
Longitudinal cross section through Earth's atmosphere, showing circulation between Earth's surface and tropopause. Surrounding 30° latitude is belt of descending, dry air that causes arid belt at Earth's surface.

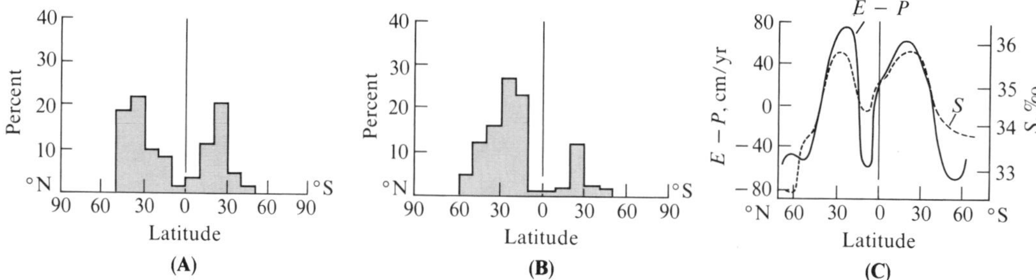

Figure 9–25
Distribution by latitude. (**A**) Modern evaporites. (**B**) Areas with less than 25 cm annual precipitation, excluding high-latitude deserts. (**C**) Average evaporation (E) minus precipitation (P) and surface salinity (S) of modern oceans. Evaporite precipitation centers around 30° latitude on both sides of the equator. [Gordon, 1975. Reprinted by permission of The University of Chicago Press. Copyright © 1975 by The University of Chicago.]

The control of evaporite occurrence by atmospheric circulation suggests that the locations of ancient evaporite deposits might reflect paleolatitudes of ancient landmasses (possibly modified by topographic controls). At present, the latitudinal distribution of ancient evaporites is haphazard (see Figure 9–25A) because of differences in the amount and direction of drift of the continents. But if the deposits are placed in their paleolatitudes, as determined from paleomagnetic measurements (see Figure 9–25B), some semblance of order appears. In the Northern Hemisphere the pattern is still very diffuse. When the occurrences are grouped irrespective of hemisphere (see Figure 9–25C), not much improvement is obtained. Probably the lack of the perfect pattern found for modern evaporite locations results from changes in solar radiation through time. For example, in nonglacial times the intensity of solar radiation was generally greater, and latitudinal belts were less clearly defined. This would cause the belts of evaporite formation to be spread out over a greater latitudinal range (see Figure 9–26).

SUMMARY

Evaporite rocks are not widely exposed on the Earth's surface because of their high solubility but underlie about one-third of the land surface and perhaps one-half of the land area covered by sedimentary rocks. In surface outcrops, gypsum is most abundant; in the subsurface, halite probably dominates. Because they are so soluble, evaporites are generally recrystallized at least once during their geologic history and are best viewed as metamorphic rocks, with textures and structures produced by such "metamorphic" processes as plastic deformation, recrystallization, and relative rates of growth of different minerals. However, as is true of metamorphic rocks produced from silicates at high temperatures and pressures, primary sedimentary features are sometimes preserved, such as lamination and cross bedding.

Many ancient, thick evaporite deposits appear to have formed by reflux of shallow, hypersaline brine in basins largely surrounded by considerable topographic relief.

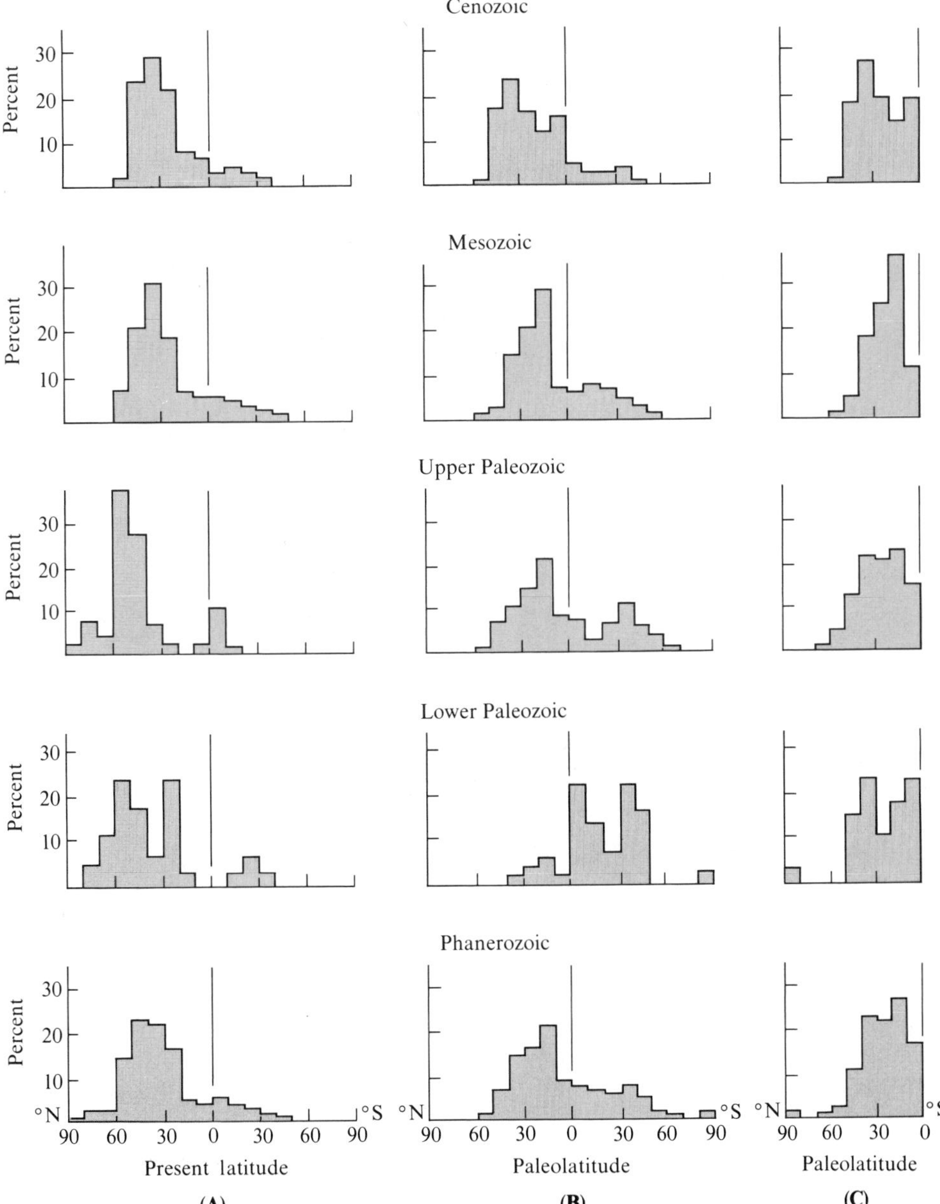

Figure 9–26
Distribution of ancient evaporites by latitude. (**A**) Present latitude. (**B**) Paleolatitude based on paleomagnetic data. (**C**) Paleolatitude without respect to hemisphere. [Gordon, 1975. Reprinted by permission of The University of Chicago Press. Copyright © 1975 by The University of Chicago.]

The relative rates of evaporation and reflux determine the salinity of the brine and the types of evaporites that will be precipitated. Most other marine evaporites are formed in sabkha settings and are interbedded with dolomite as a result. Nonmarine, lacustrine evaporites are less abundant than marine types and contain many minerals not found in marine types.

FURTHER READING

Anderson, R. Y., et al. 1972. Permian Castile varved evaporite sequence, west Texas and New Mexico. *Geol. Soc. Amer. Bull., 83,* 59–85.

Borchert, H., and R. O. Muir. 1964. *Salt Deposits: The Origin, Metamorphism and Deformation of Evaporites.* New York: Van Nostrand, 338 pp.

Bosellini, A., and L. A. Hardie. 1973. Depositional theme of a marginal marine evaporite. *Sedimentology, 20,* 5–27.

Bradley, W. H., and H. P. Eugster. 1969. *Geochemistry and Paleolimnology of the Trona Deposits and Associated Authigenic Minerals of the Green River Formation of Wyoming.* U.S. Geol. Surv. Prof. Paper No. 496B, 71 pp.

Braitsch, O. 1971. *Salt Deposits: Their Origin and Composition.* New York: Springer-Verlag, 297 pp.

Briggs, L. I., D. Gill, D. Z. Briggs, and R. D. Elmore. 1980. Transition from open marine to evaporite deposition in the Silurian Michigan Basin. In A. Nissenbaum (ed.), *Hypersaline Brines and Evaporitic Environments.* New York: Elsevier, pp. 253–270.

Dean, W. E., and B. C. Schreiber (eds.), 1978. *Marine Evaporites.* Soc. Econ. Paleontol. Mineral. Short Course No. 4, 188 pp.

DeMille, G., J. R. Shouldice, and H. W. Nelson. 1964. Collapse structures related to evaporites of the Prairie Formation, Saskatchewan. *Geol. Soc. Amer. Bull., 75,* 307–316.

Eugster, H. P. 1980. Geochemistry of evaporitic lacustrine deposits. *Ann. Rev. Earth Planet. Sci., 8,* 35–63.

Fisher, J. H. (ed.). 1977. *Reefs and Evaporites—Concepts and Depositional Models.* Amer. Assoc. Petroleum Geol. Studies in Geol. No. 5, 196 pp.

Gill, D. 1977. Salina A-1 sabkha cycles and the Late Silurian paleogeography of the Michigan Basin. *Jour. Sed. Petrology, 47,* 979–1017.

Gordon, W. A. 1975. Distribution by latitude of Phanerozoic evaporite deposits. *Jour. Geol., 83,* 671–684.

Gornitz, V. M., and B. C. Schreiber. 1981. Displacive halite hoppers from the Dead Sea: some implications for ancient evaporite deposits. *Jour. Sed. Petrology, 51,* 787–794.

Hanford, C. R. 1981. Coastal sabkha and salt pan deposition of the lower Clear Fork Formation (Permian), Texas. *Jour. Sed. Petrology, 51,* 761–768.

Hardie, L. A., and H. P. Eugster. 1971. The depositional environment of marine evaporites: a case for shallow, clastic accumulation. *Sedimentology, 16,* 187–220.

Holdaway, K. A. 1978. *Deposition of Evaporites and Red Beds of the Nippewalla Group, Permian, Western Kansas.* Kansas Geol. Surv. Bull., *215,* 43 pp.

Holliday, D. W. 1970. The petrology of secondary gypsum rocks: a review. *Jour. Sed. Petrology, 40,* 734–744.

Holser, W. T. 1979. Mineralogy of evaporites. In P. H. Ribbe (ed.), *Marine Minerals.* Mineral. Soc. Amer. Short Course Notes, *6,* 211–294.

Hsü, K. J. 1972. Origin of the saline giants: a critical review after the discovery of the Mediterranean evaporite. *Earth-Sci. Rev., 8,* 371–396.

Kaldi, J. 1980. The origin of nodular structures in the Lower Magnesian Limestone (Permian) of Yorkshire, England. In H. Füchtbauer and T. Peryt (eds.), *Zechstein Basin.* Stuttgart: E. Schweizerbart'sche Verlagsbuchhandlung, pp. 45–60.

Kaufmann, D. W., and C. B. Slawson. 1950. Ripple marks in rock salt of the Salina Formation. *Jour. Geol., 58,* 24–29.

Kendall, A. C. 1979a. Continental and supratidal (sabkha) evaporites. In R. G. Walker (ed.), *Facies Models.* Geol. Assoc. Canada Reprint Series No. 1, pp. 145–157.

Kendall, A. C. 1979b. Subaqueous evaporites. In R. G. Walker (ed.), *Facies Models.* Geol. Assoc. Canada Reprint Series No. 1, pp. 159–174.

Kinsman, D. J. J. 1969. Modes of formation, sedimentary associations and diagnostic features of shallow-water and supratidal evaporites. *Amer. Assoc. Petroleum Geol. Bull., 53,* 830–840.

Maiklem, W. R., D. G. Bebout, and R. P. Glaister. 1969. Classification of anhydrite—a practical approach. *Bull. Canad. Petroleum Geol., 17,* 194–233.

Mattox, R. B. (ed.). 1968. *Saline Deposits.* Geol. Soc. Amer. Spec. Paper No. 88, 701 pp.

Milliken, K. L. 1979. The silicified evaporite syndrome—two aspects of silicification history of former evaporite nodules from southern Kentucky and northern Tennessee. *Jour. Sed. Petrology, 49,* 245–256.

Murray, R. C. 1964. Origin and diagenesis of gypsum and anhydrite. *Jour. Sed. Petrology, 34,* 512–523.

Neev, D., and K. O. Emery. 1967. *The Dead Sea.* Israel Geol. Surv. Bull. No. 41, 147 pp.

Schreiber, B. C., et al. 1976. Depositional environments of Upper Miocene (Messinian) evaporite deposits of the Sicilian Basin. *Sedimentology, 23,* 729–760.

Shaw, A. B. 1977. A review of some aspects of evaporite deposition. *Mountain Geol., 14,* 1–16.

Shearman, D. J. 1981. Displacement of sand grains in sandy gypsum crystals. *Geol. Magazine, 118,* 303–306.

Smith, D. B. 1971. Possible displacive halite in the Permian Upper Evaporite Group of northeast Yorkshire. *Sedimentology, 17,* 221–232.

Smith, D. B. 1972. Foundered strata, collapse-breccias and subsidence features of the English Zechstein. In G. Richter-Bernburg (ed.), *Geology of Saline Deposits.* Paris: UNESCO, pp. 255–269.

Stanton, R. J. 1966. The solution brecciation process. *Geol. Soc. Amer. Bull., 77,* 843–848.

Stewart, F. H. 1963. *Data of Geochemistry,* Chap. Y: Marine evaporites. U.S. Geol. Surv. Prof. Paper No. 440–Y, 52 pp.

Wilson, J. L. 1967. Carbonate–evaporite cycles in lower Duperow Formation of Williston Basin. *Bull. Canad. Petroleum Geol., 15,* 230–312.

Worsley, N., and A. Fuzesy. 1979. The potash-bearing members of the Devonian Prairie Evaporite of southeastern Saskatchewan, south of the mining area. *Econ. Geol., 74,* 377–388.

10

Cherts

Rocks whereon greatest men have oftest wreck'd.

JOHN MILTON

Chert is a microcrystalline sedimentary rock composed largely or entirely of quartz that was precipitated from an aqueous solution. The sizes of the quartz crystals are typically 5–20 μm, but cherts that consist of coarser or finer crystals are not uncommon. Most chert beds are quite pure, in concert with other chemically formed sedimentary rocks, such as evaporites, limestones, and dolomites, and for many of the same reasons. Rock units composed of chemical or biochemical precipitates cannot form in areas of significant influx of detrital sediment. Impurities in chert normally total less than 5% and consist of calcite, dolomite, clay minerals, hematite, and organic matter.

FIELD OBSERVATIONS

Megascopically visible chert occurs in three different types of stratigraphic and tectonic settings:

1. As nodules in cratonic carbonate rocks.
2. As pure, bedded chert in geosynclinal tectonic regions.
3. In association with hypersaline lacustrine deposits.

Almost all occurrences of chert are in groups 1 and 2. The relative volumes of chert in these two types of occurrences are uncertain, and estimates are strongly biased by the geologic upbringing of the geologist. Geologists whose training and experience have been largely in mid-continental North America tend to emphasize the fact that carbonate rocks form 10–15% of all sedimentary rocks, and many of these contain visible bodies of chert. Geologists whose experience is dominated by studies in areas of modern or ancient convergent plate margins (for example, Tertiary rocks around the rim of the Pacific Ocean basin, alpine areas of southern Europe, mid-Paleozoic rocks of the Ouachita Mountains–Marathon Basin trend of the southwestern United States) generally believe the volume of bedded chert exceeds the volume of chert nodules. In either event, however, chert *in toto* forms less than 1% of the volume of sedimentary rocks.

Nodules in Carbonate Rocks

Chert nodules are irregularly shaped, usually structureless, dense masses of microcrystalline quartz, and occur most commonly in carbonate rocks, although they have also been reported in mudrocks and sandstones. The nodules range in size and shape from egg-shaped spheroids a few centimeters in length to large, highly irregular, tuberous bodies 30 cm in length. The larger ones typically have warty or knobby exteriors. The outer few centimeters of a nodule may show desiccation cracks filled with secondary chert probably formed penecontemporaneously with the formation of the nodule. Nodule exteriors are commonly bleached, soft, and porous because of alteration at some stage in the history of the nodule. The interior of the nodule may show concentric color zoning of translucent shades of gray, caused either by carbonaceous matter or by zones of different water content. Occasionally, the bedding, texture, or color shading in the host carbonate rock is seen to pass without interruption through the chert nodule, but this is uncommon. Some nodules contain visible calcareous fossils that are partially or completely silicified, as well as the remains of siliceous organisms such as sponges, radiolarians, and diatoms. In some occurrences, it is clear that the siliceous organisms have supplied dissolved silica for the formation of the nodules. As described by Lowenstam (1942, p. 828):

> The nodules contain layer-like masses of sponge spicules, indicating that originally continuous layers of spicules were present. The dolomites and dolomitic limestones in which the nodules are enclosed are free from visible siliceous material. It is therefore evident that a considerable portion of the original spicule layers has been dissolved, the nodules being local remnants of the former spicule layers which were cemented and enveloped by silica. The latter was derived from the desilicified parts of the former skeleton layers by local solution and lateral migration.

Most chert nodules in carbonate rocks formed very early during diagenesis. This is indicated by laminae in the carbonate rock that curve around the nodules, showing that compaction of the host rock postdated nodule formation.

In most occurrences the nodules are elongated parallel to the bedding in the host rock and are commonly concentrated along certain horizons to the exclusion of others (see Figure 7–13). In some carbonate rocks, nodules are numerous enough to coalesce into nearly continuous, irregular or wavy layers several hundred centimeters in thickness. It is possible that these layers or zones of chert concentration may be rhythmically spaced, reflecting either depositional or diagenetic control. In some limestones the zones are sufficiently close to form bridges of chert between them, cutting through intervening bedding.

Geosynclinal Bedded Cherts

Stratigraphic sections that contain bedded cherts are composed of several types of rocks, among which the chert layers are most prominent. Individual chert layers range in thickness from a few centimeters to a meter or more (see Figure 10–1), are evenly bedded, thinly laminated to massive, and typically green or black in color, although white bedded cherts (novaculite) are common in some areas. The upper and lower surfaces of chert layers can be either smooth or wavy, and many layers pinch

Figure 10–1
Bedded chert in Caballos Formation (Devonian-Mississippian), Texas, showing pinch-and-swell bedding, typical bed thicknesses, and minor partings of shale. [E. F. McBride and A. Thomson, 1969, in *Dallas Geol. Soc. Guidebook to Marathon Basin, Texas*. Photo courtesy E. F. McBride.]

and swell while others bifurcate. Both ripple marks and cross bedding have been observed in bedded chert layers, but at most outcrops no sedimentary structures other than bedding are evident.

Interbedded sedimentary layers are typically thin, green or dark-colored shales, frequently siliceous, and contain small amounts of pyrite. Less commonly the interbeds are micrite, graded bioclastic limestone and fine sandstone, and fine-grained conglomerate lenses.

On a regional scale, bedded chert sequences most commonly occur with graded turbidites, ophiolites, and mélanges in tectonic settings associated with convergent plate margins (Garrison, 1974). Examples include the Franciscan Formation (Jurassic) in California, many Mesozoic occurrences in the Mediterranean–Himalayan (Tethys) region, Ordovician cherts in Scotland, and Jurassic cherts in New Zealand. Bedded chert deposits *not* associated with ophiolites and mélanges include Carboniferous–Jurassic rocks in Japan, the Caballos Formation of Devonian–Mississippian age in Texas, and the Arkansas Novaculite of the same age in Arkansas and Oklahoma.

Cherts in Hypersaline Lakes

Thin beds and nodules of chert occur in many Cenozoic and Mesozoic nonmarine sequences in units interpreted to be lacustrine, based on criteria such as facies relationships, the absence of marine fossils, and the presence of evaporite minerals (Surdam et al., 1972). The chert beds are characterized by soft-sediment deformation features such as slump structures, intraformational breccias, various types of surface ornamentation resulting from dehydration, and casts of mud cracks (see Figure 10–2). In addition, casts of crystals such as pirssonite [$Na_2Ca(CO_3)_2 \cdot 2H_2O$], gaylussite [$Na_2Ca(CO_3)_2 \cdot 5H_2O$], and trona occur on chert surfaces. These minerals form only in hypersaline, alkaline lakes. The assemblage of sedimentary and mineralogic features associated with these nonmarine cherts has not been described either from chert nodules and beds in cratonic carbonate rocks or from the bedded cherts in geosynclinal sequences.

LABORATORY STUDIES

Chert in thin section appears as a colorless, microcrystalline aggregate (see Figure 10–3), similar in appearance to micrite and microsparite in carbonate rocks but with much lower birefringence. In some bedded cherts, circular or elliptical clear areas are seen with uncrossed polarizers and are interpreted to be "ghosts" of diatoms, radiolarians, or siliceous sponge spicules (see Figure 10–4). Cherts of Tertiary age may contain isotropic (amorphous) material among the microquartz crystals, remnants of an opaline precursor of the chert. Opal is chemically unstable relative to quartz.

(A) (B)

(C)

Figure 10–2
Textural features on surfaces of nonmarine chert nodules. Magnification 3×. (**A**) Surface cracks, Green River Formation (Eocene), Wyoming. (**B**) Surface ornamentation, Morrison Formation (Jurassic), Wyoming. (**C**) Crystal casts of trona or gaylussite, Moonstone Formation (Pliocene), Wyoming. [Surdam et al., 1972. Photos courtesy H. P. Eugster.]

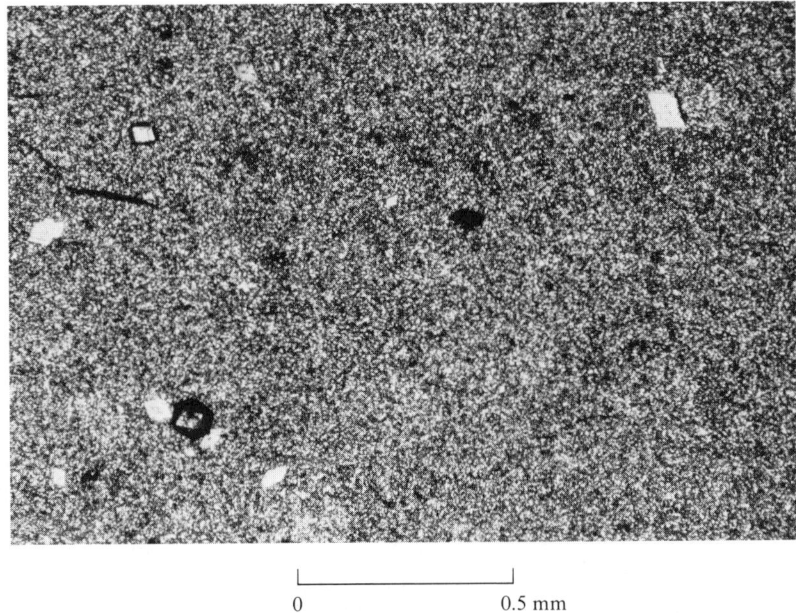

Figure 10–3
Typical appearance of chert in thin section, showing microcrystallinity and black-gray speckled character. Rhombs are manganocalcite. [McBride and Thomson, 1970. Photo courtesy E. F. McBride.]

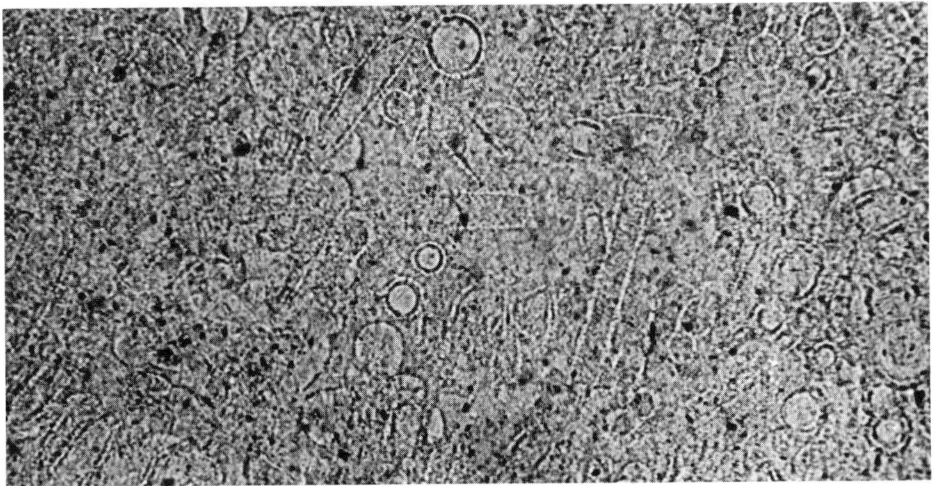

Figure 10–4
Photomicrograph of novaculite in Caballos Formation in uncrossed nicols, showing "ghosts" of sponge spicules about 0.03 mm in diameter. [McBride and Thomson, 1969. Photo courtesy E. F. McBride.]

Many cherty rocks contain not only the microcrystalline equant quartz crystals but also an elongate, fibrous variety of microquartz called chalcedony (see Figure 10–5). When both chert and chalcedony are present, as in chert-cemented sandstones, the chert is always closest to the detrital grain surface, with the chalcedonic quartz filling cavities among neighboring detrital grains. The reason a silica-saturated solution sometimes precipitates fibrous quartz rather than the equant variety is not known but is thought to result either from a difference in degree of supersaturation or from the presence of other ions in the solution.

Most chalcedonic fibers have negative elongation (length-fast), the *c* axis being perpendicular to the length of the fibers. In some chalcedony, however, the *c* axis is parallel to the fibers (length-slow), and Folk and Pittman (1971) have shown that there is a strong correlation between the occurrence of this variety of chalcedony and the presence of evaporites. The chalcedony occurs commonly as pseudomorphs of gypsum and anhydrite or in nodules mimicking the appearance of chicken-wire structure. Sometimes, unreplaced remnants of evaporite minerals are still present in the rock. The correlation between length-slow chalcedony and evaporites is not perfect but is common enough to be very useful in petrologic investigations of ancient rocks.

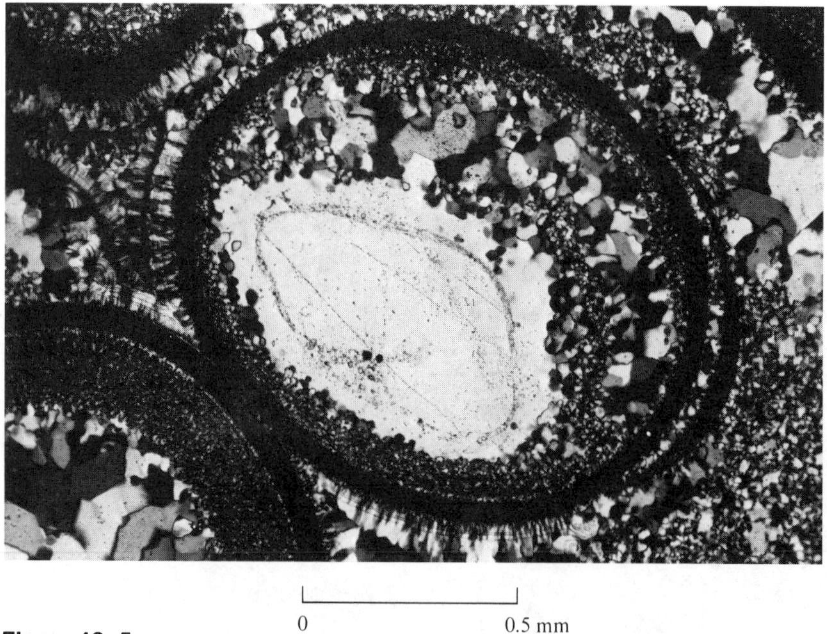

0 0.5 mm

Figure 10–5
Siliceous oolith showing geopetal structure, State College oolite (Cambrian), Pennsylvania. Secondarily enlarged detrital quartz grain serving as nucleus of oolith is displaced to lower part of oolith. Several generations of fibrous quartz (chalcedony) can be seen in upper left. [F. J. Pettijohn, 1975, *Sedimentary Rocks,* 3rd ed. (New York: Harper & Row). Photo courtesy F. J. Pettijohn.]

Many cherts, particularly those in cratonic carbonate rocks, contain shreds of calcite, indicating incomplete replacement of limestone by the chert. Rhombic molds may also occur and are interpreted as evidence of partial dolomitization of the preexisting limestone, the dolomite having weathered out after chertification. Cherts with carbonate inclusions exhibit all gradations, from chert that contains a few scattered mottles of calcite or dolomite to rocks in which the carbonate is so abundant that it forms a spongelike network whose interstices are filled with opal and chalcedony. The carbonate may be uniformly distributed throughout the chert or may be concentrated in laminae.

A rather uncommon type of chert is the oolitic variety (see Figure 10–5), described by Choquette (1955). The ooliths are about 1 mm in diameter, contain a nucleus composed of either detrital quartz or carbonate, and are composed of microcrystalline quartz of different crystal size and shape arranged concentrically around the nucleus. The nucleus is not necessarily in the center of the oolith but may be displaced to one side. This feature is interpreted as a geopetal fabric resulting from the removal in solution of the central core of the oolith, leaving the insoluble material displaced to the lower side, followed by infilling with precipitated, coarsely crystalline quartz. Some oolitic cherts contain chert pseudomorphs of dolomite rhombs.

The textures of cherts as seen in thin sections are complex because of variations in size, shape, and arrangement of crystals and because of replacement and recrystallization phenomena. As is true of the other monomineralic rocks such as limestones and gypsum beds, simple mineralogy is not accompanied by simple textures and structures. As seen in thin sections, the common textures can be accurately described in terms borrowed from igneous and metamorphic petrology: seriate, merocrystalline, granoblastic, and porphyroblastic. Unfortunately, it is generally not possible to distinguish an "igneous" primary crystallization texture from a "granoblastic" recrystallization texture, or an "igneous" porphyritic texture from a "porphyroblastic" recrystallization texture. Euhedral quartz crystals do not occur in cherts, probably because of mutual interference among adjacent crystals during primary crystallization and recrystallization.

With an electron microscope, two types of textures are seen in ancient cherts: equigranular and spongy (see Figure 10–6). They correspond to the normal, polyhedral, equant microquartz and to the chalcedonic (fibrous) microquartz, respectively, as seen in thin section. The interpretation of the various textures seen in cherts, either in thin section or with the electron microscope, is uncertain.

CHEMICAL CONSIDERATIONS

Pure chert contains nothing but SiO_2 and H_2O and, as is true for calcite and aragonite, there are two possible ways to remove the dissolved silica from solution: inorganic precipitation and biochemical removal as part of the life processes of shelled organisms.

Figure 10-6
Transmission electron micrographs of chert, showing the two main textural types. (**A**) Equigranular texture, Phosphoria Formation (Permian), Montana. (**B**) Spongy texture of chalcedony, locality unknown. [R. L. Folk and C. E. Weaver, 1952, *Amer. Jour. Sci., 250.* Photos courtesy R. L. Folk.]

The solubility of quartz in either fresh or saline water is 6 ppm ($10^{-4.2}$ mol/L). Therefore, an aqueous solution with a greater concentration has the chemical potential to release silica and to form quartz. Despite this fact, quartz does not precipitate at surface temperatures unless the silica content greatly exceeds 6 ppm because of difficulties in the process of nucleation. For most geochemical considerations the solubility of amorphous silica controls the precipitation of solid silica, and this solubility is about 120 ppm ($10^{-2.9}$ mol/L). Normal surface seawater contains less than 1 ppm dissolved silica (H_4SiO_4), deep-sea water up to 6 ppm, and average river water 13 ppm, so that chert does not precipitate from them. A more concentrated source of silica is required.

For chert nodules in carbonate rocks and for bedded cherts in geosynclinal sequences, the silica is supplied by the dissolution of siliceous sponges (marine, Cambrian-Holocene; nonmarine, Jurassic-Holocene), radiolarians (Ordovician-Holocene, marine only, planktonic), and diatoms (Jurassic-Holocene, mostly marine, planktonic). During their life processes, these organisms are able to remove molecules of silica from undersaturated solutions—an impossible occurrence by purely inorganic means. The silica is used to construct opaline shell material (see Figure 10–7). After the death of the organism and during diagenesis, the opal dissolves to produce aqueous solutions with contents of dissolved silica of 6–120 ppm, so that chert can be precipitated.

In the shallow, nearshore marine areas in which carbonate sediments are produced, these siliceous organisms are known to be present (e.g., Land, 1976) and are buried with the hard parts of the carbonate-secreting organisms with which they live. The opal subsequently dissolves and, by an unknown mechanism, the uncharged H_4SiO_4 molecules migrate to centers of crystallization to form nodules. As indicated by textural evidence described earlier, the silica is not deposited in cavities within the limestone but makes its own space by replacing existing carbonate material, which is then carried away in the moving pore solution.

The sites of deposition of siliceous shells in geosynclinal settings are quiet, deep waters—locations inaccessible to detrital sediment (except for wind-blown fine silt and clay). Silica-secreting organisms are not so numerous as planktonic carbonate secretors, so that a further restriction applies to sites of accumulation of post-Paleozoic bedded cherts. The sites of accumulation must be below the carbonate-compen-

Figure 10–7
Siliceous skeletons in the >44 μm size fraction in modern siliceous ooze from northern equatorial Pacific Ocean. [Photo courtesy J. R. Hein, in *Jour. Sed. Petrology, 50* (1980).]

sation depth. Planktonic, carbonate-secreting organisms (e.g., *Globigerina*) did not evolve until the Mesozoic Era, so this depth restriction does not apply to the mid-Paleozoic bedded cherts in the Ouachita–Marathon orogenic belt in the southwestern United States.

The chemistry of chert formation in hypersaline lakes is considerably more complex than that of the nodules in carbonate rocks or bedded cherts. The chemical setting is a lake in which hypersalinity is caused either by an excess of evaporation over inflow (e.g., Gosiute Lake in Wyoming during Eocene time) or by the presence of exceptionally soluble rocks in the drainage basin surrounding the lake (e.g., many lakes in the rift basins of Kenya and Tanzania). Seasonally high pH values are created in the lake either by high rates of algal photosynthesis or by an unusual chemical composition of the lake waters. High concentrations of silica are dissolved in the waters at high pH values and, when the seasonally high values are lowered, the silica precipitates.

ANCIENT AND MODERN CHERTS

Caballos Formation (Devonian), Western Texas

One of the best-studied bedded chert deposits is the Caballos Formation in the Marathon Basin of western Texas (see Figure 10–8) and its correlative in the Ouachita Mountains (Arkansas–Oklahoma), the Arkansas Novaculite. The two units are believed to be correlative because the total Paleozoic sequence in the two areas is very similar and both areas were at the continental margin during Paleozoic time. The two areas are connected structurally in the subsurface through central Texas.

The Caballos Novaculite is composed of 90% bedded chert and 10% interbeds of green and red clay shale and siliceous shale. The formation is lens-shaped and ranges from 30 to 225 m in thickness. Underlying the Caballos is the Maravillas Chert, the uppermost unit of an Ordovician succession of limestone, shale, chert, and conglomerate with a total thickness of 600 m. The nature of the contact between the two formations is uncertain and may be unconformable. Overlying the Caballos is the Tesnus Formation, a Mississippian turbidite unit about 2,000 m thick.

There are few sedimentary structures in the Caballos, other than bedding surfaces, most of which are featureless. The structures that are seen occasionally include slump structures, stylolites, trace fossils, geopetal fabric, and iron oxide nodules. In thin section, there occur laminae of coarse silt and fine sand less than 1 mm thick, ghosts of sponge spicules and radiolarians (see Figure 10–4), brownish specks of organic matter, rhombs of carbonate minerals, traces of illite, and specks of hematite. The microquartz occurs mostly as subequant crystals 5–25 μm in diameter, but a variety of intercrystal textural patterns occurs within most thin-section slides. Much of the variation seems to result from the deformation and smearing of siliceous skeletons during their crystallization from opal to microquartz.

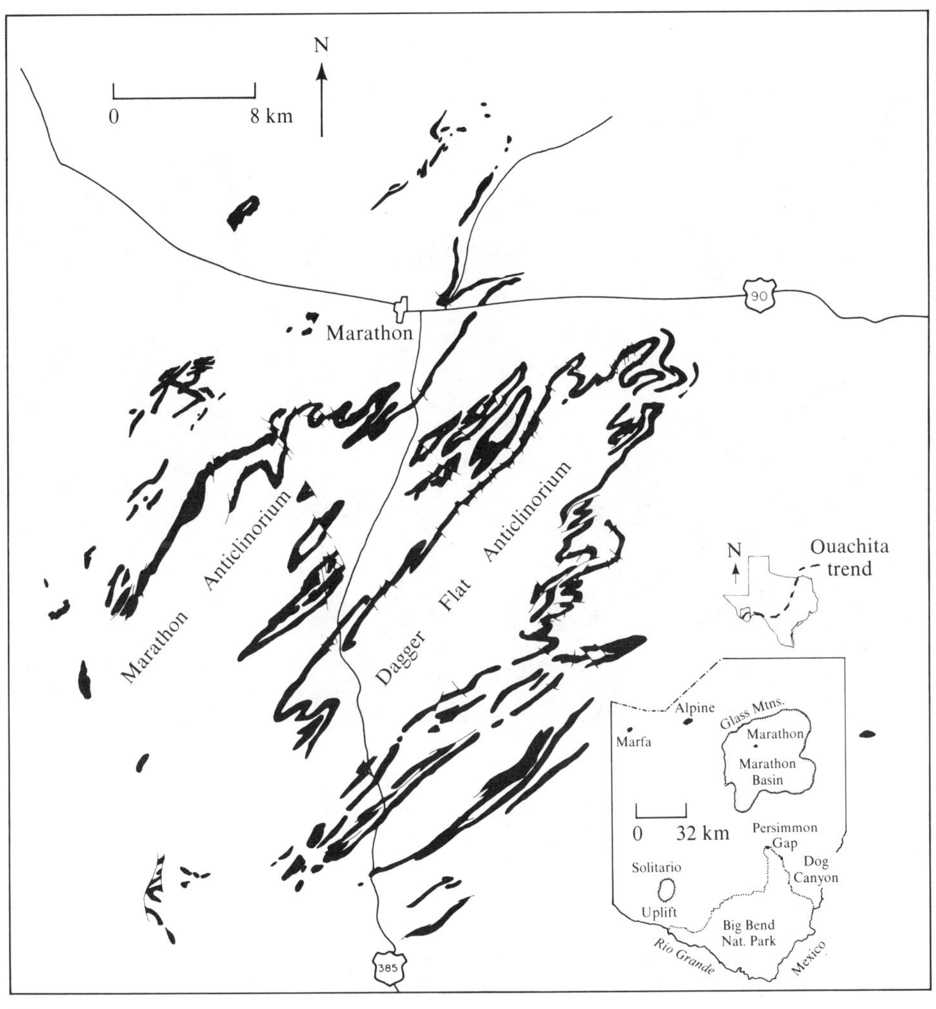

(A)

Figure 10–8
(A) Locality and outcrop map of Caballos Formation. [E. F. McBride and R. L. Folk, 1977, *Jour. Sed. Petrology, 47.*]

The depositional environment of the Caballos has traditionally been thought to be deep water because of the following evidence:

1. The cherts contain no shallow-water fossils such as corals, bryozoa, or crinoids—all of which are characteristic of nearshore Paleozoic deposits. The only fossils are radiolarians, pelagic organisms characteristic of the open ocean, and siliceous sponge spicules, whose abundance testifies to the lack of detrital sediment from a nearby land area.

(B)

Figure 10–8 *(continued)*
(B) Aerial view of northeastern part of Marathon Basin, showing Caballos Formation cherts forming resistant ridges. [Photo courtesy E. F. McBride, on cover of *Dallas Geol. Soc. Guidebook to Marathon Basin, Texas,* 1969.]

2. The tectonic setting of the formation is that of a continental margin, with the thick units both above and below containing features characteristic of deposition in deep water, such as repeated turbidite sequences, graded bedding, organic burrows only along bedding planes, and the absence of exclusively shallow-water sedimentary structures such as reefs, intertidal features, or mud cracks.

As is true of nearly all environments interpreted to be in deep water, the evidence is largely negative rather than positive. The conclusion regarding depth of water relies heavily on the absence of indigenous shallow-water structures and fossils rather than on the presence of deep-water features. This is typical of such interpretations because there are few bathymetric indicators known below the photic zone, where ecologic data can indicate water depth. In most geologic reports the term "deep water" is as likely to mean 100 m as 1,000 m.

The deep-water interpretation of the Caballos Formation has been challenged by Folk (1973), who has found features in parts of the formation that he interprets as evidence of deposition in a supratidal–intertidal environment. These features include stromatactis (?) and bird's-eye structures, chert breccias he believes resulted from evaporite solution and collapse, jasperized (red chert) areas resembling silicified fossil soils, and traces of fossils that are thought to be freshwater fungi. The controversy is still unresolved. Folk's observations, however, point out once again the importance of careful and detailed field observations.

Modern Analogs of Ancient Bedded Cherts

Field evidence such as the typical association of bedded cherts with graded sandstones, mélanges, and ophiolites indicates that cherts are marine, were associated with active tectonism, were at one stage located at a continental margin, and were probably formed in deep water. Where on the present Earth do such conditions exist today? The clear answer is: in the oceanic basin. The presence of siliceous fossils in thin sections of ancient bedded cherts suggests that the source of at least some of the silica was skeletons of organisms such as radiolarians, siliceous sponges, and diatoms. All these groups occur in marine waters, and diatoms occur in fresh waters as well.

The type and distribution of sediment currently accumulating on the deep oceanic floor were not well known until relatively recently. Numerous "grab samples" from the ocean floor have been available since the globe-circling voyage of the ship *Challenger* in the 1870s, but the number of samples was small and all were necessarily from the surface layer; nothing was known of the sediment more than a meter or so below the surface. Then in 1964, four of this country's major oceanographic institutions formed a working group they named JOIDES (Joint Oceanographic Institutions for Deep Earth Sampling) whose objective was to construct a ship capable of drilling holes in the deep oceanic floor, core the sediment and rocks encountered, and bring the cores to the surface. The idea was sold to the U.S. government for financing and a ship, the *Glomar Challenger,* was launched in 1968. To date, the vessel has drilled in water up to 7,044 m in depth. The maximum depth penetrated below the seafloor so far is 1,741 m, and the location of the coring site can be determined by satellite fixing to within ± 1.6 km. So far, more than 600,000 km of ship track has been logged in the world's oceans and seas (Mediterranean, Black, Red, etc.), and about 60 thick volumes of results have been published as *Initial Reports of the Deep Sea Drilling Project* (DSDP reports). The *Glomar Challenger* continues to operate and accumulate geological, geophysical, and oceanographic data about the 70% of the Earth's surface that is covered by marine waters.

The cores recovered by JOIDES commonly contain carbonate oozes, clays, and thick sequences of nearly pure siliceous sediment (which covers 14% of the ocean floor) and rock. At and near the seafloor, the sediment consists of accumulations of skeletons of diatoms and radiolarians. At depth, all stages of transition are observed, from unlithified aggregates of amorphous opaline shells through various stages of recrystallization of these materials to the final product: microcrystalline quartz or chert. Beds of chert underlie many areas of the ocean floor. Apparently, the shells can be dissolved and the silica reprecipitated within a few million years after burial. Below the chert layers are often found mafic and ultramafic igneous rocks, the essential ingredients of the ophiolite suite. The overall picture seems to be that the igneous rocks are extruded from rift zones on the ocean floor and are overlain by carbonate and siliceous skeletons raining down from the ocean surface. The carbonate sediment may be subsequently dissolved as it is carried into deeper water on the moving plate. The ocean plate converges toward a continental margin (as the southeastern Pacific does with South America); the igneous rocks are altered to serpentine, spilite, and related rocks by seawater, and the accumulations of skeletons are converted to chert. At the site of convergence, some of these rocks are subducted into the mantle, and some are accreted onto the edge of the continent; subsequently, they are raised above sea level, where we find them. The graded sandstone turbidites typically associated with the bedded cherts are formed adjacent to the continental edge and are in position for accretion as the ophiolite–chert complex arrives from the areas farther offshore.

Unfortunately, the deep-sea cherts now forming in the Pacific Ocean basin are markedly different in character from most cherts in ancient, orogenic belts. The modern cherts in the Pacific are all replaced chalks, limestones, or pelagic clays; they are not rhythmically bedded as the ancient cherts are; their rates of accumulation are much less; and their burrowing characteristics are different. Perhaps most chert beds in orogenic belts were deposited near shore in deep water, such as in the Gulf of California or adjacent to island arcs. That is, they were not carried to the edge of the continent from far offshore. It seems clear that the origin of bedded cherts is still enigmatic.

Lake Magadi, Kenya

The Magadi Basin lies in the eastern Rift Valley of southern Kenya, at 2°S latitude. The lowest part of the basin is occupied by Lake Magadi, an intermittently dry, alkaline lake 30 km in length and 5 km in width that contains an extensive deposit of trona. Exposures of Quaternary chert beds are abundant throughout the Magadi Basin. Chert beds are usually 2–20 cm thick, though accumulations up to 3 m are known. They may form small, isolated knobs, or they may cover the flat valley floors for several square kilometers. The total area covered by chert is about 30 km^2.

Natural outcrops and trenching into the lake beds reveal that the chert beds are laterally contiguous with horizons of a sodium silicate mineral, magadiite [$NaSi_7O_{13}(OH)_3 \cdot 3H_2O$], with the transition taking place over a distance of no more than 1 m (see Figure 10–9). Chert plates in the vicinity of the magadiite–chert transition are coated on both sides with magadiite and, commonly, kenyaite [$NaSi_{11}O_{20.5}(OH)_4 \cdot 3H_2O$].

The Magadi cherts show a great variety of textures and structures that indicate soft-sediment deformation. As the magadiite has a soft, puttylike consistency, these features are explained readily if it is assumed that they formed prior to or during the

Figure 10–9
Magadiite–chert conversion at Lake Magadi, Kenya. Chert plate on left (under knife) lies on surface and is continuous with buried magadiite bed on right. White magadiite layers below chert plate show slump folds. Ruler (in inches) lies on surface strewn with chert plate fragments. [Eugster, 1969. Photo courtesy H. P. Eugster.]

magadiite–chert transformation. The largest of these features are four-, five-, and six-sided polygonal ridges of chert 10 cm to 2 m high and extending for 2–50 m. Apparently, they are very large-scale desiccation polygons—gigantic mud cracks. The conversion of magadiite to chert involves a shrinkage of at least 25%. In places, magadiite was extruded like toothpaste between already chertified edges of neighboring polygons (see Figure 10–10). Chert plates and nodules display the same surface features described from the nonmarine cherts in the Green River (Eocene) lake beds: surface ornamentation, desiccation cracks, and casts of evaporite minerals.

The conversion of magadiite or kenyaite to chert is described by

$$\underset{\text{magadiite}}{NaSi_7O_{13}(OH)_3 \cdot 3H_2O} + H^+ \rightleftarrows \underset{\text{chert}}{7SiO_2} + Na^+ + 5H_2O$$

and

$$\underset{\text{kenyaite}}{NaSi_{11}O_{20.5}(OH)_4 \cdot 3H_2O} + H^+ \rightleftarrows \underset{\text{chert}}{11SiO_2} + Na^+ + 5.5H_2O$$

Figure 10–10
"Toothpaste" chert, apparently extruded as magadiite between already chertified polygon edges. These cherts are laminated internally, with laminations strongly contorted. [Eugster, 1969. Photo courtesy H. P. Eugster.]

Precambrian Cherts

The oldest known nonmarine siliceous organisms are the diatoms, which evolved during the Jurassic Period. The oldest recognized marine siliceous organisms are the sponges, which first appeared in the Cambrian Period. It is the great number of marine siliceous organisms that keeps the sea perpetually undersaturated with respect to quartz. But what of Precambrian time, in which bedded cherts are very common and total about 15% of the entire stratigraphic column in rocks about 2 billion years old? How might these rocks have formed? Presumably, the answer is that they formed as precipitates of silica gel from seawater, which would have had a concentration of silica of some tens of parts per million. Some workers, however, believe that they have found scattered remnants of siliceous organisms in some Precambrian

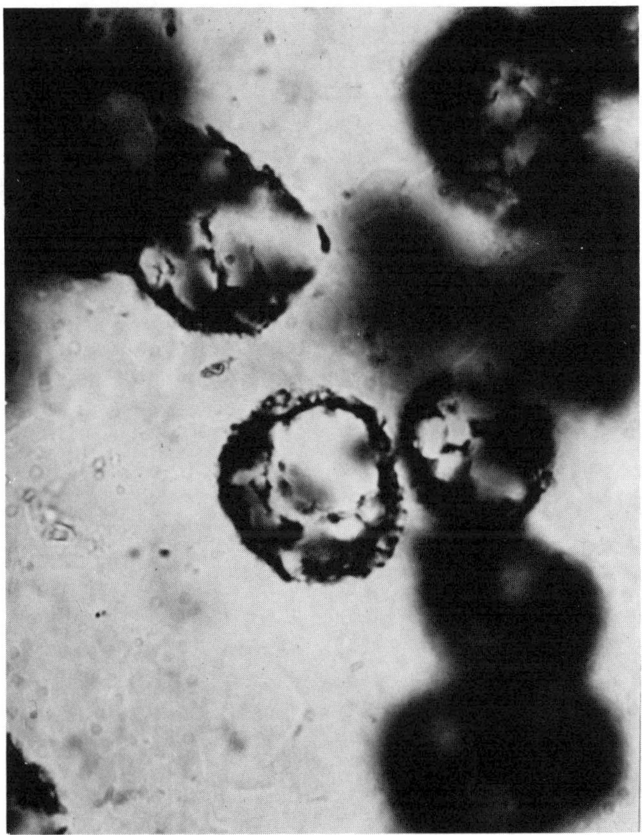

Figure 10–11
Possible recrystallized radiolarian (?) shells, Brockman Iron Formation, Hammersley Range, Western Australia. Structures are about 25 μm in diameter. [H. Blatt et al., 1980, *Origin of Sedimentary Rocks,* 2nd ed. (Englewood Cliffs, NJ: Prentice-Hall, Inc.). Photo courtesy G. L. LaBerge.]

bedded cherts (see Figure 10–11; LaBerge, 1973), and these cherts may have the same bioclastic origin as do Phanerozoic cherts. Other workers doubt the authenticity of the "fossils" in these cherts, believing instead that they are structures formed inorganically in silica gel. The question is still unresolved.

SUMMARY

Chert is a mineralogically simple rock whose origin can be quite complex. Most Phanerozoic cherts were formed in two stages: (1) removal of dissolved silica from undersaturated seawater by siliceous sponges, radiolarians, and diatoms and its solidification as opal; and (2) dissolution of the opaline skeletons during shallow burial after death of the organisms, the silica in solution subsequently crystallizing as microcrystalline quartz (chert). In carbonate environments and at sites where the number of shells per unit volume is low, the crystallization results in scattered nodules of chert. At sites where the number per unit volume is large, as in some areas in the deep oceanic basins at present, crystallization results in beds of chert that can be hundreds of meters thick. Cherts in both types of settings can show replacement features, such as undigested remnants of carbonate material.

Many nonmarine cherts of Mesozoic and Cenozoic age may have been formed in hypersaline lakes by the leaching and dehydration of a sodium silicate precursor such as magadiite or kenyaite. In pre-Mesozoic times, no nonmarine siliceous organisms are known, so that all nonmarine cherts must have been formed by this or a similar mechanism.

The origin of Precambrian cherts is uncertain. No siliceous organisms are known to have existed at that time, and the majority opinion among petrologists favors an origin for Precambrian marine cherts by precipitation from normal seawater, which would have been supersaturated with respect to quartz.

FURTHER READING

Biggs, D. L. 1957. *Petrography and Origin of Illinois Nodular Cherts.* Ill. Geol. Surv. Circ. No. 245, 25 pp.

Calvert, S. E. 1974. Deposition and diagenesis of silica in marine sediments. In K. J. Hsü and H. C. Jenkyns (eds.), *Pelagic Sediments: On Land and Under the Sea.* Internat. Assoc. Sedimentol. Spec. Pub. No. 1, pp. 273–299.

Choquette, P. W. 1955. A petrographic study of the "State College" siliceous oolite. *Jour. Geol., 63,* 337–347.

Cressman, E. R. 1962. *Nondetrital Siliceous Sediments.* U.S. Geol. Surv. Prof. Paper No. 440-T, 22 pp.

Eugster, H. P. 1969, Inorganic bedded cherts from the Magadi area, Kenya. *Contrib. Mineral. Petrology, 22,* 1–31.

Folk, R. L. 1973. Evidence for peritidal deposition of Devonian Caballos Novaculite, Marathon Basin, Texas. *Amer. Assoc. Petroleum Geol. Bull., 57,* 702–725.

Folk, R. L., and E. F. McBride. 1976. The Caballos Novaculite revisited. Part I: Origin of novaculite members. *Jour. Sed. Petrology, 46,* 659–669. See also *47,* 1261–1286, for Part II.

Folk, R. L., and J. S. Pittman. 1971. Length-slow chalcedony: a new testament for vanished evaporites. *Jour. Sed. Petrology, 41,* 1045–1058.

Garrison, R. E. 1974. Radiolarian cherts, pelagic limestones, and igneous rocks in eugeosynclinal settings. In K. J. Hsü and H. C. Jenkyns (eds.), *Pelagic Sediments: On Land and Under the Sea.* Internat. Assoc. Sedimentol. Spec. Pub. No. 1, pp. 367–399.

Grunau, H. R. 1965. Radiolarian cherts and associated rocks in space and time. *Eclogae Geologicae Helvetiae, 58,* 157–208.

LaBerge, G. L. 1973. Possible biological origin of Precambrian iron formations. *Econ. Geol., 68,* 1098–1109.

Land, L. S. 1976. Early dissolution of sponge spicules from reef sediments, North Jamaica. *Jour. Sed. Petrology, 46,* 967–969.

Lowenstam, H. A. 1942. Geology of the Eastern Nazareth Mountains, Palestine. *Jour. Geol., 50,* 813–845.

McBride, E. F. (ed.). 1979. *Silica in Sediments: Nodular and Bedded Chert.* Soc. Econ. Paleontol. Mineral. Reprint Series No. 8, 184 pp.

McBride, E. F., and A. Thomson. 1970. *The Caballos Novaculite, Marathon Region, Texas.* Geol. Soc. Amer. Spec. Paper No. 122, 129 pp.

Murata, K. J., and R. R. Larson. 1975. Diagenesis of Miocene siliceous shales, Temblor Range, California. *Jour. Res.* (U.S. Geol. Surv.), *3,* 553–566.

Surdam, R. C., H. P. Eugster, and R. H. Mariner. 1972. Magadi-type chert in Jurassic and Eocene to Pleistocene rocks, Wyoming. *Geol. Soc. Amer. Bull., 83,* 2261–2266.

11

Iron-Rich Rocks

It is a riddle wrapped in a mystery
inside an enigma.

WINSTON CHURCHILL

Bedded iron-rich rocks, like bedded phosphate rocks, are defined by their chemical composition rather than by mineral composition, texture, or structure. Iron-rich deposits are defined as those that contain at least 15% Fe. This is either 21.4% Fe_2O_3, 19.3% FeO, or some combination of the two totaling 15% Fe. These amounts are much greater than the average for mudrocks (4.8% Fe), sandstones (2.4% Fe), or limestones (0.4% Fe). All three groups of rocks can, however, grade laterally into an iron-rich deposit. The iron deposits are termed either *ironstone* or *iron formation*; in the Lake Superior region the unweathered, cherty iron formation that is mined for the metal is known as *taconite.*

The iron-rich rocks have generally been neglected by sedimentary petrologists. The impetus for studying iron-rich rocks has come from economic geologists rather than from petrologists or sedimentologists. In part, this results from the fact that a larger proportion of the iron ores is Precambrian and is metamorphosed, typically to the greenschist facies and frequently with the destruction of original sedimentary depositional features. It is only within the last 20 years that sedimentary petrologists have begun intensive study of the bedded iron-rich rocks.

The iron-rich rocks are of two distinctly different types: Precambrian banded mixtures of chert and iron-bearing minerals, and Phanerozoic oolitic ironstones. The banded iron formations are much more abundant.

BANDED IRON FORMATIONS

Outcrop Characteristics

Cherty iron formations are composed of quartz, iron oxides (hematite and magnetite); carbonates (siderite, ankerite, dolomite, and calcite); iron silicates such as greenalite [(Fe, Mg)$_3$Si$_2$O$_5$(OH)$_4$], minnesotaite [Fe$_3$Si$_4$O$_{10}$(OH)$_2$], stilpnomelane (hydrated Mg, Fe, Mn, Al silicate of uncertain composition), and riebeckite [Na$_2$Fe$_5$Si$_8$O$_{22}$(OH)$_2$]; and iron chlorite. Terrigenous components such as sand and silt are present locally, and transitions from ferruginous cherts into slightly cherty ferruginous sandstones, siltstones, and shales occur. The characteristic layering is formed by alternating layers rich in cryptocrystalline iron oxide with layers rich in microcrystalline quartz (chert). The layers may be lenticular, anastomosing, or nodular, and the appearance and thickness of laminations can change over short distances because of the solution and precipitation of silica during diagenesis and low-grade metamorphism. The original banding, however, is of depositional origin. It is present in all of the unmetamorphosed units and occurs as penecontemporaneously reworked, already banded boulders in some banded iron formations. Some of the depositional bands have been traced laterally over hundreds of kilometers and may represent annual layers (varves).

Most of the sedimentary structures in banded iron formations are not easily seen in outcrop, except in rare cases where they have been accentuated by differential weathering (Dimroth, 1976). The structures are clearly visible, however, on polished slabs, and include many of those prominent in sandstones and carbonate rocks (Gross, 1972). Among those described from banded iron formations (see Figure 11–1) are cross bedding, graded bedding, load casts, ripple marks, stylolites, erosion channels, shrinkage cracks, intraformational breccias, and slump structures. Among organically produced features, structures resembling stromatolites have been described, as have some ovoid, gravel-sized grains that may be oncolites.

The textural characteristics of these rocks commonly resemble those of carbonate rocks (see Figure 11–2). Intraclasts (normally termed *granules* by workers in iron formations), peloids, ooliths, and pisoliths seem to be common in banded iron formations, and problematic fossils have been described (see Figure 10–11). Matrix chert and femicrite (composed of microcrystalline iron silicate and/or carbonate) are equivalents of micrite in limestones. The fact that some iron formations are loadcasted indicates their depositional origin and original particulate character. It is certainly possible that some banded iron formations are replaced limestones.

Regional Characteristics

The initial investigations of iron-rich rocks were made in the latter part of the nineteenth century by American and Canadian geologists in the Lake Superior region. They described linear belts of Precambrian rock composed of banded mixtures of chert and a variety of iron-bearing minerals (see Figure 11–3). Subsequent dating of such banded rocks has established that they are all Precambrian in age. They typically occur in belts hundreds to thousands of kilometers long, up to several kilometers wide, and with thicknesses as great as 600 m. The rock successions in which these iron formations occur usually lie above high-grade metamorphic rocks, but the iron formations are generally not the basal units in the succession.

(A)
0
5 cm
(B)
0
1 cm
(C)
0
5 cm

(D)

(E)

Figure 11–1
Examples of sedimentary structures seen in banded iron formations. (**A**) Cross bedding. (**B**) Scour-and-fill. (**C**) Intraformational breccia. (**D**) Syndepositional folds. (**E**) Boudinage. [A–D from Gross, 1972. Photos courtesy G. A. Gross. E from R. W. Bayley and H. L. James, 1973, *Econ. Geol., 68*. Photo courtesy H. L. James.]

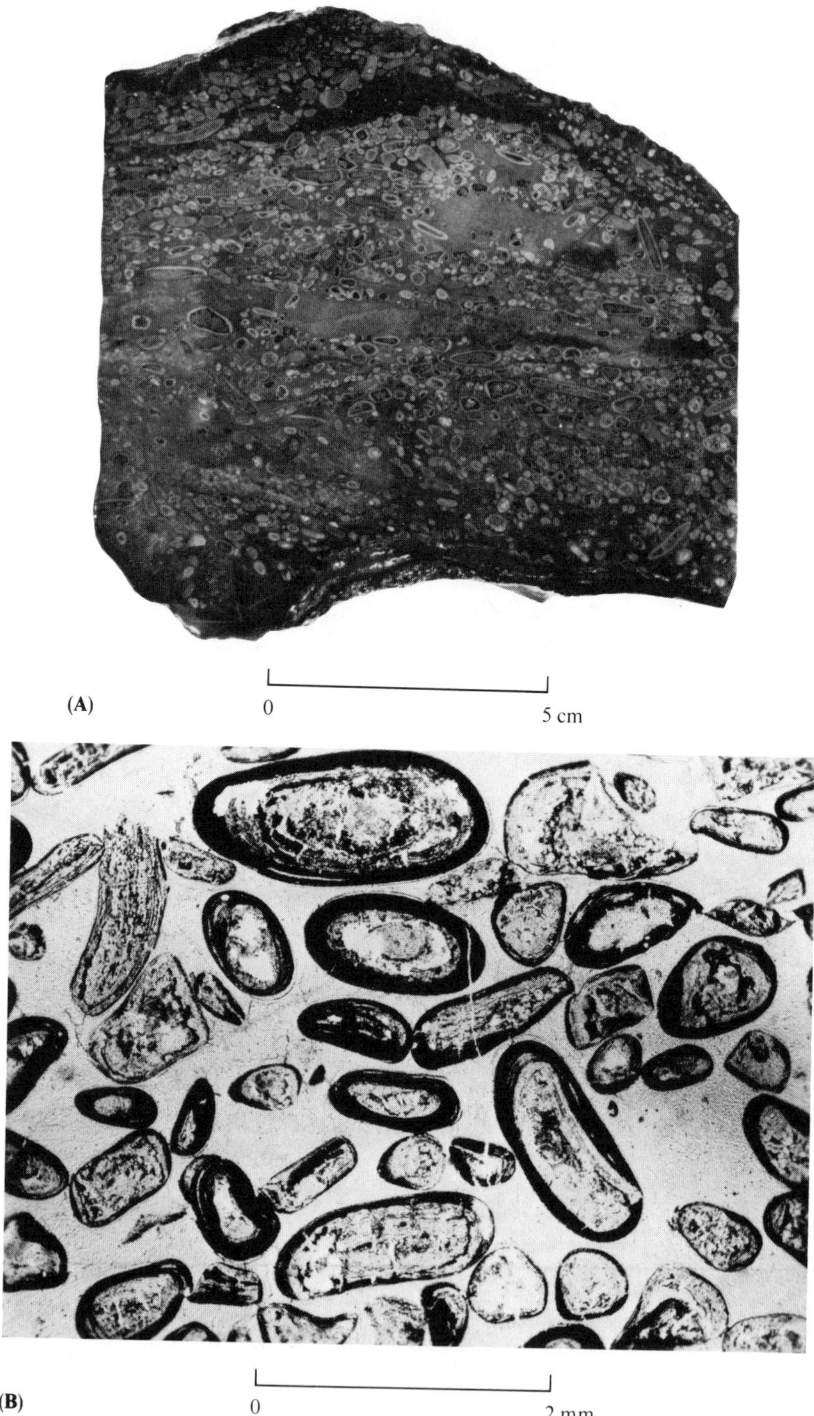

Figure 11–2
Examples of textural features seen in banded iron formations. (**A**) Chert–hematite pebble conglomerate, showing scour of dark microcrystalline layer near base of specimen. (**B**) Thin section showing concentric rings of magnetite and chert in ooliths. (**C**) Thin section showing intraclasts 2–4 mm long with thin oolithic coatings. (**D**) Thin section showing chert spherulites in chert–hematite (black) rock.

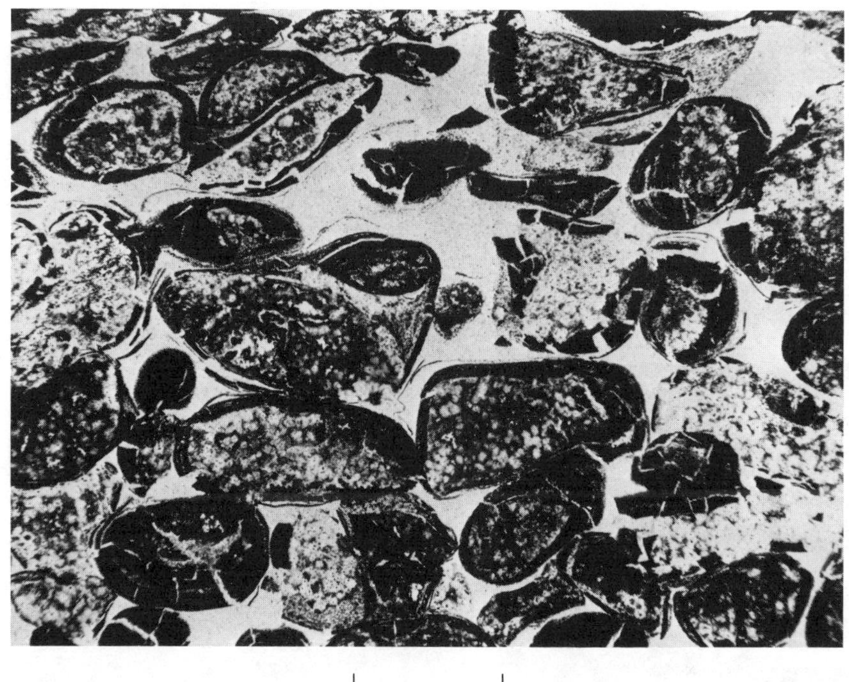

(C)

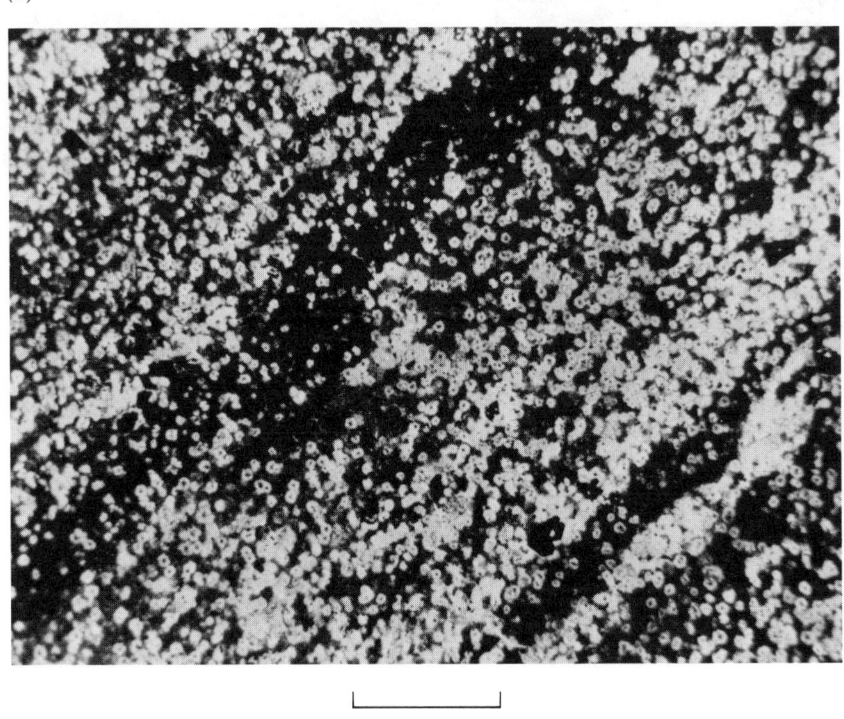

(D)

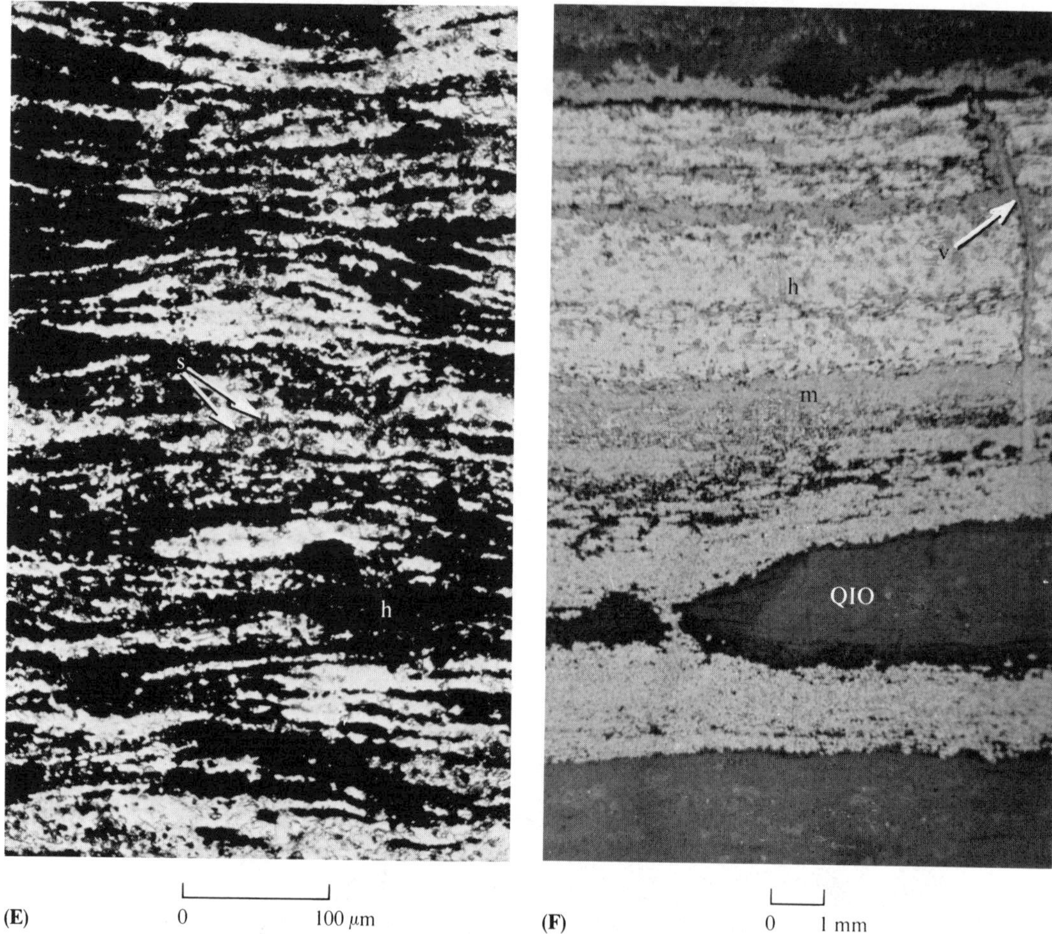

Figure 11–2 (*continued*)
(**E**) Polished section, viewed in reflected light, showing anhedral, lenticular hematite (h) aggregates defining banding in chert–hematite rock and siderite (s) crystals. (**F**) Polished section, viewed in reflected light, showing magnetite (m) bands alternating with hematite (h) bands and elongate pods of chert–iron oxide mixtures. Boundaries between hematite and magnetite bands are irregular, inclusions of magnetite are abundant within hematite bands, and vein (v) of recrystallized magnetite transgresses both hematite and magnetite bands. QIO = quartz plus iron oxide. [A and D from Zajac, 1974. Photos courtesy I. S. Zajac, Geol. Surv. Canada. B from Gross, 1972. Photo courtesy G. A. Gross. C from Dimroth, 1976. Photo courtesy E. Dimroth. E and F from D. E. Ayers, 1972, *Econ. Geol., 67*. Photo courtesy D. E. Ayers.]

Figure 11–3
Outcrop in upper part of Negaunee Formation, Marquette District, Michigan. Red (dark) bands are hematitic chert (jasper); gray (light) bands are specular hematite. [H. Blatt et al., 1980, *Origin of Sedimentary Rocks,* 2nd ed. (Englewood Cliffs, NJ: Prentice-Hall, Inc.). Photo courtesy H. L. James.]

Banded iron formations occur in the oldest rocks still preserved on Earth, about 3.8 billion years old. Those older than 2,600 million years are different in appearance (see Table 11–1) and tectonic setting from those of younger Precambrian age and have been termed the *Algoma type* (Gross, 1965). They characteristically are thinly banded or laminated with interbands of ferruginous gray or red chert and hematite and magnetite. Siderite and iron silicate or sulfide minerals are present locally or in small amounts. Occasionally, massive siderite and pyrite–pyrrhotite beds form part of the formation. Single iron formations range from millimeters to perhaps 100 m thick and are rarely more than a few kilometers long. Usually, a number of lenses of iron formation are distributed *en êchelon* within a volcanic belt. They are intimately

Table 11–1
Characteristic Features of Algoma- and Superior-Type Banded Iron Formations

Feature	Algoma-type (Archean-type)	Superior-type (Animikie-type)
Age	pre-2,600 m.y. (also Proterozoic and Phanerozoic)	pre-1,800 m.y.
Sedimentary environment	eugeosynclinal tectonic basins of several 100 km diameter; iron formation in marginal parts in connection with greenstone belts	miogeosynclinal; iron formation along margins of stable continental shelves; shallow water; restricted intracratonic basins
Extent	commonly lenticular bodies of a few kilometers	extensive formations persistent over some 100 km to more than 1,000 km
Thickness	0.1–10 m	several meters to more than 100 m (1,000 m)
Location in sedimentary sequence	irregular, lenticular bodies within Archean "basement" rocks	in bottom and middle parts of sedimentary sequences as sheet deposits, transgressive on older "basement" rocks
Associated rocks	graywackes and shales; carbonaceous slates; mafic volcanic; felsic pyroclastics; ryolithic flows; pillowed andesites	coarse clastics; quartzites, conglomerate dolomites, black shales (graphitic)
Volcanics	close association to volcanism in time and space	no direct association with contemporaneous volcanism; volcanics normally absent
Sedimentary facies	oxide facies predominant; carbonate and sulfide facies thin and discontinuous; silicate facies; all facies frequently closely associated	oxide facies most abundant; silicate and carbonate facies frequently intergradational
	sulfide and carbonate facies near the center of volcanism; oxide facies on the margins	sulfide facies insignificant or absent
	heterogeneous lithological assemblages with fine grained clastic beds	more homogeneous (especially oxide facies); little or no detritus
	granular and oolitic textures	granular and oolitic textures
Examples	Canada: Archean basins (e.g., Michipicoten)	Labrador Trough
	U.S.A.: Vermilion District, Minnesota	Lake Superior region
	S. Africa: greenstone belts, Kaapvaal and Rhodesia cratons	Transvaal and Witwatersrand supergroups
	Brazil: Rio das Velhas series	Minas Series (itabirites); Carajás/Pará
	India: southern Mysore	Bihar, Orissa, Goa, Mysore
	Australia: Yilgarn and Pilbara blocks	Hamersley Group
	U.S.S.R.: Taratash/Urals	Krivoy Rog, Kursk; Ukrainian Shield

Source: Eichler, 1976.

associated with various volcanic rocks, including pillowed andesites, tuffs, pyroclastic rocks, or rhyolitic flows, and with graywacke, gray-green slate, or black carbonaceous slate (see Figure 11–4). Tuff and fine-grained clastic beds or ferruginous cherts are interbedded in the iron formation, and detailed stratigraphic successions show heterogeneous lithologic assemblages. These iron formations have streaky lamination or layering, and oolithic or granular textures are generally absent or inconspicuous. The associated rocks indicate a eugeosynclinal depositional environment of formation and a close relationship in time and space to volcanic activity (see Figure 11–5).

Banded iron formations 2,600–1,800 million years in age form the bulk of the Precambrian banded iron formations. They are clearly different in appearance and lithologic association from their older relatives and are termed the *Superior type* (Gross, 1965). In addition to the ubiquitous banding of cherty and iron-rich layers, granules and ooliths composed of both chert and iron minerals are a typical textural feature. The uniformity of the layering may be disrupted by nodular or stubby lenses of chert, by cross bedding, or by cherty forms that appear to be silicified stromatolites (see Figure 11–6). Shrinkage and desiccation cracks are present in some chert granules and nodules, and stylolites are common. The types of textures and sedimentary structures in the Superior type of iron formation are remarkably alike in detail wherever examined, although the relative prominences of the features may differ from one unit to another.

The stratigraphic succession from bottom to top is generally dolomite, quartzite, red and black ferruginous shale, iron formation, black shale, and finally argillite, although variations in the succession do occur. Volcanic rocks, either tuffs or flows, are not always directly associated with the Superior type of banded iron formation, but they are nearly always present somewhere in the succession.

Banded iron formations become noticeably less common in rocks younger than 1,800 million years; the youngest is 800 million years old. Iron formations younger than 800 million years, the Phanerozoic ones, are very different in character from their older relatives, as discussed later in this chapter.

Worldwide Distribution

The Precambrian iron formations occur in the shield areas on all the present continental masses (see Figure 11–7), and it is obvious that many of them lie close to the borders of the continents or on the margins of the cratonic masses now surrounded by younger fold belts and platform sediments. It seems likely that some of the iron formations now on the borders of the different continents were originally part of a single depositional basin before being separated by continental drift. Where the general geologic setting has been studied in more detail, a comparison of iron formations on opposite continents fits fairly well with the assumption of originally coherent iron-formation belts. The type and depositional environment of these iron formations with their associated rocks are similar, if not identical, and so are their ages and metamorphic histories.

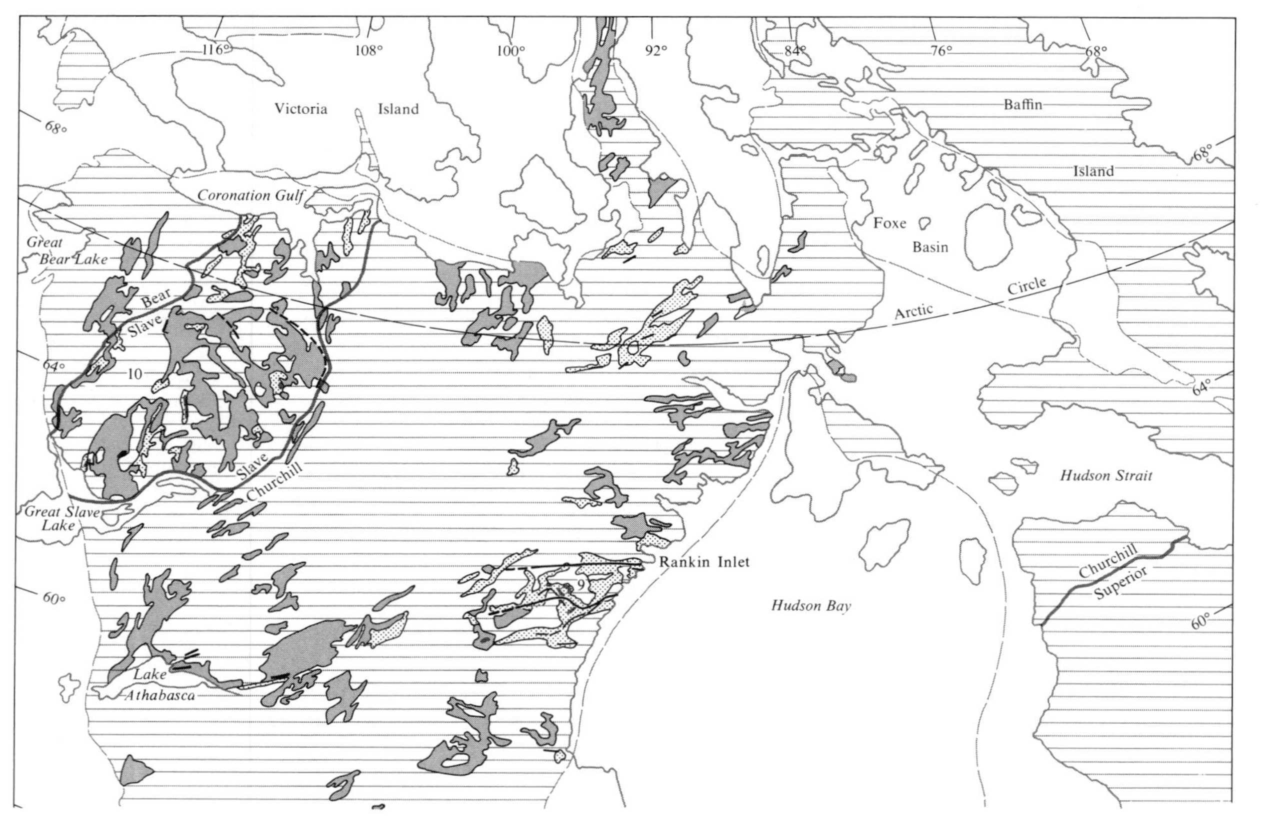

116°
108°
100°
92°
84°
76°
68°
68°
64°
60°
68°
64°
60°
Victoria Island
Baffin Island
Coronation Gulf
Great Bear Lake
Bear
Slave
10
Slave
Churchill
Great Slave Lake
Lake Athabasca
Foxe Basin
Arctic Circle
Hudson Strait
Rankin Inlet
9
Hudson Bay
Churchill
Superior

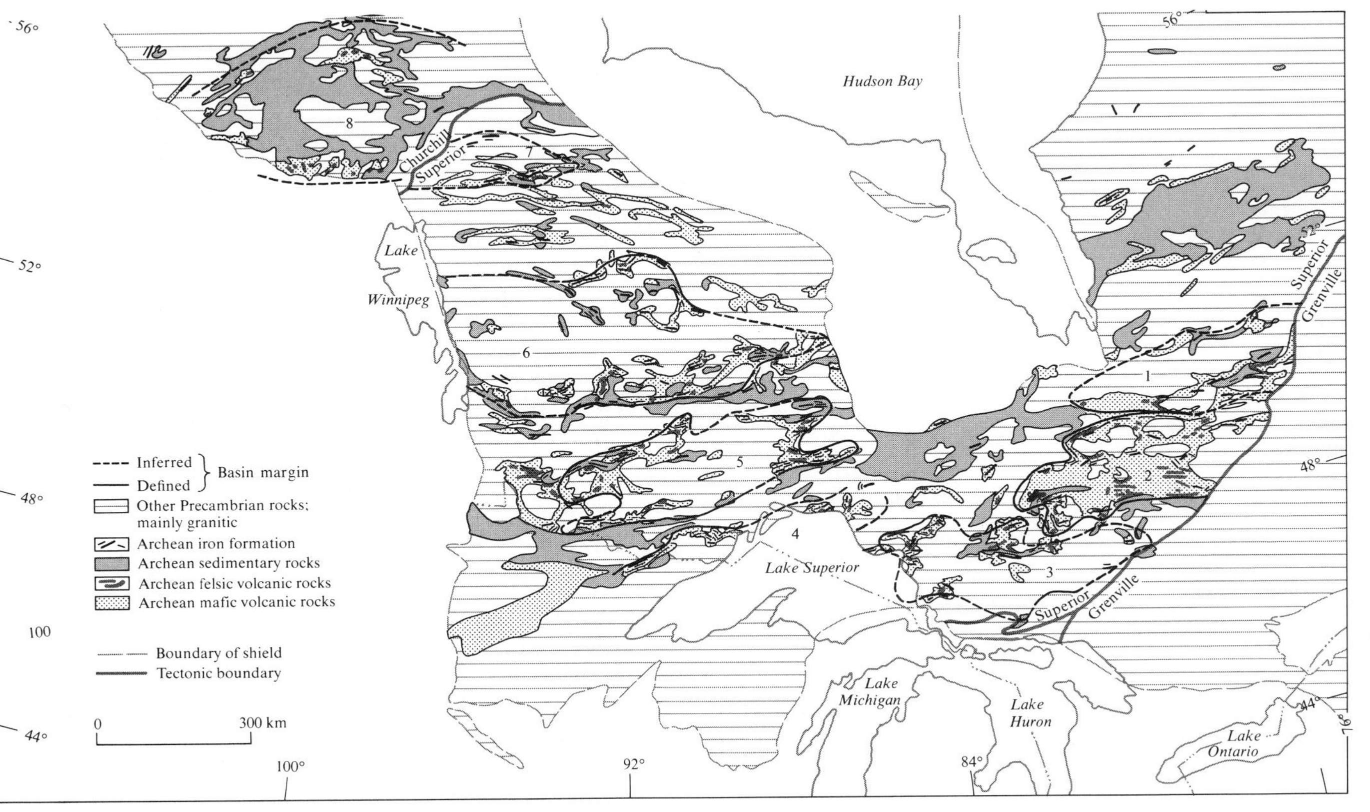

Figure 11–4
Distribution of Archean depositional basins surrounding shield area of Hudson Bay, Canada. Much of the apparent separation among basins may result from later tectonic activity; some of the ovoid areas may have formed a continuous belt around older shield area centered in Hudson Bay. [A. M. Goodwin, 1973, *Econ. Geol., 68.*]

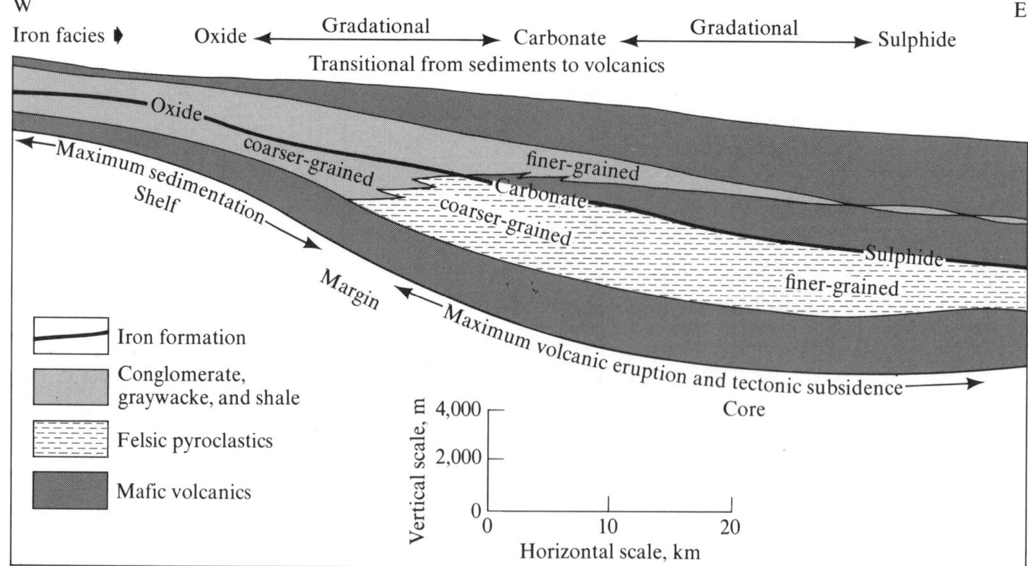

Figure 11–5
Reconstructed stratigraphic and facies relationships in Archean Michipicoten Basin (located along western edge of tectonic basin 3 in Figure 11–4. In oxide facies, hematite and magnetite predominate; in carbonate facies, siderite; in sulfide facies, pyrite and pyrrhotite. [A. M. Goodwin, 1973, *Econ. Geol., 68.*]

Sources of the Silica and Iron

The derivation of iron and silica from deep weathering of a landmass is considered unlikely for the Algoma type of iron formation. The associated sedimentary rocks are graywackes, fine-grained slates, and other clastic rocks derived from the mechanical disintegration of volcanic rocks and are typical eugeosynclinal assemblages. Sediments such as orthoquartzites, limestones, dolomites, alumina-rich mudstones, and beds rich in the well-rounded, heavy detrital minerals typically derived from maturely weathered areas are generally absent. The thin layers of sedimentary rock interbedded with chert and iron formation have a normal amount of iron, except for those that show evidence of postdepositional alteration as noted above, and are similar to other sediments in the rock sequence that lie some distance from the iron formation. If an abundance of iron were being supplied to the sea from a land source, a gradual increase would be expected in the amount of iron oxide and carbonate in the sedimentary beds as the iron formation is approached. As only clastic beds containing a normal amount of iron distributed in the common rock-forming silicate minerals are interlayered with iron formation and chert, a more direct and possibly intermittent source of iron is suggested. Volcanic emanations could provide such a direct source with rapid precipitation of the iron as the thermal waters mix with the seawater. Iron derived from weathering of a landmass is likely to be precipitated in the shallow-water parts of a basin near shore. The Algoma type of iron formation is characteristically present in eugeosynclinal volcanic-sedimentary rock assemblages, and not with continental shelf or deltaic rocks.

(A)

(B)

Figure 11–6
Stromatolitic structures in Superior type of iron formation. Note striking similarity between these structures and those in shallow-water carbonate rocks, as shown in Figure 7–7B. (**A**) Outcrop appearance. [Zajac, 1974. Photo courtesy I. S. Zajac, Geol. Surv. Canada.] (**B**) Thin-section appearance. [Gross, 1972. Photo courtesy G. A. Gross.]

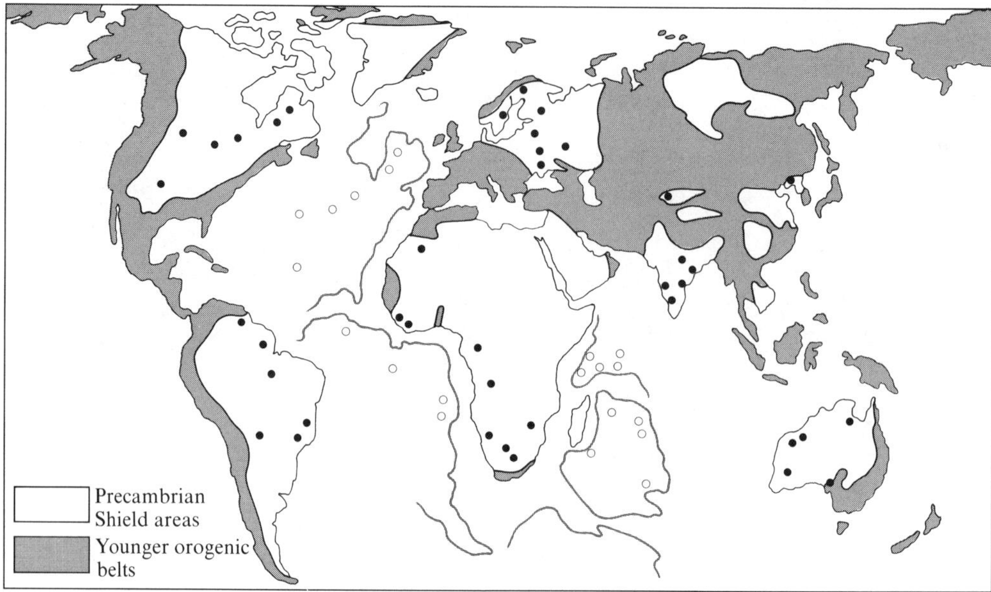

Figure 11–7
Distribution of prominent Precambrian banded iron formations and reconstruction of continental locations prior to continental drift. [Eichler, 1976.]

The source of the iron in the Superior type of banded iron formation is less clear. These units are associated with continental-shelf and miogeosynclinal sediments composed of well-sorted detrital and chemical constituents derived from a maturely weathered landmass. The presence of ooliths and stromatolites in the iron formations further supports the idea that the iron was deposited in nearshore environments. In the Labrador Geosyncline the iron formation occurs as a continuous stratigraphic unit more than 1,000 km long before being disrupted by erosion or structural features. Apparently, a single iron-formation unit was once present around the entire shoreline of the Ungava craton for a distance of more than 3,000 km. It is evident that the factors that led to the deposition of the iron existed on a very broad scale and were a major part of the tectonic framework of the time. What is the mechanism for the episodic generation of intensely iron-rich laminae over such extensive areas? Why has this not occurred during the past 800 million years? As yet there are no satisfactory answers to these questions.

The source of the silica in all banded iron formations is probably the same. There is abundant evidence found in the thick-banded chert masses to indicate that the silica was precipitated as a gel that later hardened and crystallized to form chert. Syneresis or shrinkage cracks, septarian structures, lenticular cracks, or desiccation cracks that have been distorted during compaction all indicate that the silica was deposited as a gel and later dehydrated and compacted. Many slump structures, folds, and slips confined to single beds or to a restricted group of beds indicate penecontemporaneous deformation of very soft and incompetent beds. Structures identical to those visible in Phanerozoic chert formations are found in Precambrian cherty iron formations.

Sokoman Iron Formation, Labrador Trough

Regional Extent

Perhaps the best-studied banded iron formation is the Sokoman Formation (Dimroth and Chauvel, 1973; Zajac, 1974), located in northwestern Labrador and northeastern Quebec, Canada (see Figure 11–8). The Labrador Trough, in which the Sokoman is located, is the preserved part of an Early Proterozoic geosyncline and contains a belt of Proterozoic rocks. The trough extends for about 1,100 km across the center of the Labrador peninsula; its maximum width is 100 km near Knob Lake.

The eastern half of the Labrador Trough consists mainly of mafic volcanic and intrusive rocks; the western half, almost entirely of sediments. In the west, the practically unmetamorphosed sediments rest unconformably on the Archean basement. In the east, the rocks become progressively more metamorphosed and merge almost imperceptibly with older schists and gneisses.

Stratigraphy

Zajac (1974) has described the Knob Lake area, in the western part of the trough, in some detail. His SW–NE cross section through this area is shown in Figure 11–9. The base of the Attikamagen Formation is not exposed in the Knob Lake area but may be 700 m or more in thickness. It consists of thin-bedded to laminated, cherty, greenish-gray shales and slates. The overlying Denault is a light-gray, cherty dolomicrite, averaging 17% chert. In the Knob Lake area, it is 60–200 m thick but thins to the east and west (see Figure 11–9).

The Fleming Chert Breccia is very limited in distribution, approximately 13 km long, 2.5 km wide, and 100–130 m thick. Its lower contact is gradational into cherty and sandy shales of the Denault Dolomite; its upper contact with the Wishart Quartzite is usually sharp. The breccia consists of angular chert fragments up to 10 cm in size, cemented by chert and coarser quartz crystals. Some of the chert fragments are banded. The breccia is clearly intraformational, but the origin of the chert and the manner in which brecciation occurred are still uncertain.

The Wishart Quartzite is a blanket of quartz-cemented, supermature quartz arenite with minor amounts of other types of sandstones. It is well-bedded and medium- to coarse-grained. Some boulder- to pebble-sized conglomerates composed of quartz fragments occur in the upper and lower parts of the Wishart. The blanket shape of the Wishart Quartzite, together with its end-product mineralogy and texture, strongly suggests an origin in a shallow marine and beach environment. This interpretation is further supported by the occurrence of stromatolites in a dolomite lens within the Wishart 15 km from Knob Lake. The contact between the Wishart and the overlying Sokoman Formation appears to be conformable, although the presence of a minor, local unconformity cannot be ruled out.

The contact between the Sokoman Formation and the overlying Menihek Formation is similar to that between the Sokoman and the underlying Wishart—generally

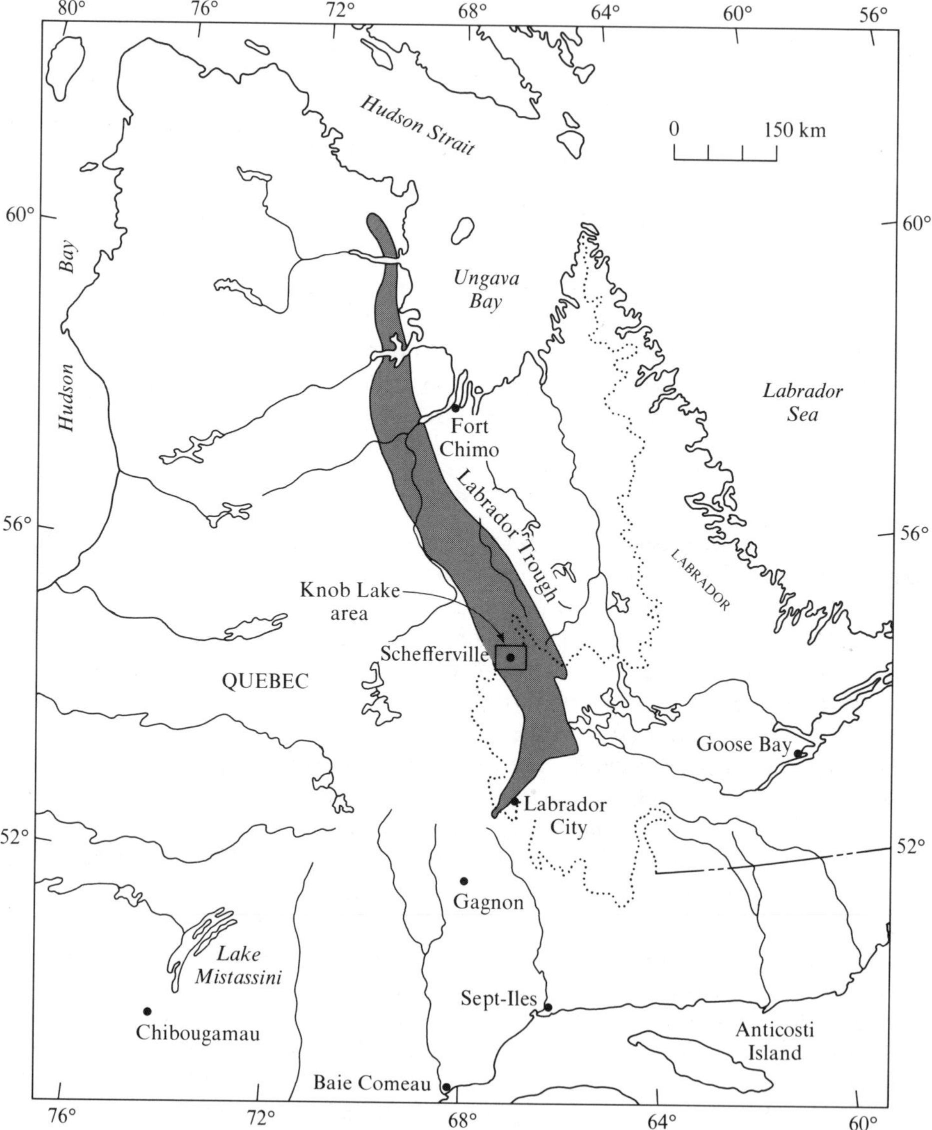

Figure 11–8
Location of Labrador Trough. [Zajac, 1974.]

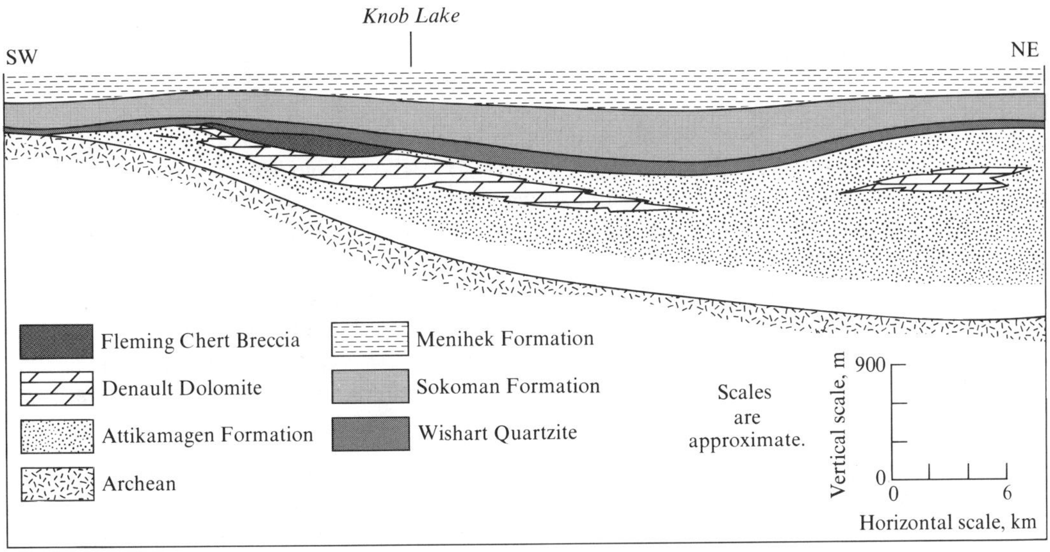

Figure 11–9
Cross section through central part of Knob Lake area before deformation. [Zajac, 1974.]

conformable but with the presence of very local unconformities. The Menihek Formation is composed almost entirely of gray–black carbonaceous shales, slates, and minor siltstones. In the eastern part of the Knob Lake area, feldspathic graywackes are common.

In summary, the stratigraphic section in which the Sokoman Formation lies consists of shales (or slates), dolomites, and pure quartz sandstones, interrupted by a local chert breccia unit of unknown origin. This rock sequence is identical to many cratonic sequences of Paleozoic and later age. It shows no peculiarities that would supply clues to the reason for the existence of a banded iron formation in its midst.

Petrology

The Sokoman Formation in the Knob Lake area is 300–500 m thick and is composed mainly of alternating silica-rich and iron-rich beds. It is a typical cherty iron formation of the Superior type. Chert, magnetite, minnesotaite, and siderite are the most common minerals. The granules, ooliths, and predominantly thin but commonly irregular bedding—features that are characteristic of the Superior type of iron formation—are typical of the Sokoman. The intraformational conglomerates, cross bedding, and abnormally thick beds, which are also present in many iron formations in the Lake Superior region, are particularly common in the Sokoman Formation. All these features, with the exception of the abnormally thick beds, suggest a shallow, agitated environment of deposition. On the other hand, the even and continuous laminations in the Sokoman and the other banded iron formations suggest quiet waters. The apparent paradox is unresolved.

Banded iron formations are commonly classified on the basis of their mineralogic facies: oxide, silicate, carbonate, and sulfide. This system of classification is chemical in character and is an attempt to distinguish environments of deposition on the basis of pH and Eh. The oxide facies (hematite) is commonly interpreted to represent the facies nearest to shore; the sulfide facies with its carbonaceous shales would be the environment farthest from shore and in deepest water. It seems doubtful, however, that the oxide–silicate–carbonate–sulfide classification can be so simply interpreted. In modern environments it is quite common to find oxygenated environments in deeper water and black, fine-grained muds accumulating close to shore. No doubt, environments in which hematite is forming are oxygenated and those in which carbon is accumulating are not oxygenated. But to correlate these facts with either distance from shore or depth of water is probably erroneous. As James (1966, p. 15) has observed:

> No example is known of a complete array of major facies grading laterally one into the other . . . . The main reason for this is that the shore-to-depth profile is merely a device for indicating a range in environmental conditions, particularly oxygen availability.

Dimroth and Chauvel (1973; Chauvel and Dimroth, 1974; Dimroth, 1975) have made a detailed study of the sedimentary facies and petrography of the Sokoman Formation in the Labrador Trough and have interpreted the depositional history of the unit. The basis of their interpretation is the similarity in texture among the Superior type of iron formation, Phanerozoic limestones, and modern carbonate sediments. Because of the textural similarities, they assume that the iron formations formed under the same hydrodynamic conditions as did Holocene limestones with equivalent textures. The only differences between the two types of rocks are the absence of metazoan skeletons and reefs and the preservation in the iron formations of delicate bedding structures because of the absence of bottom-stirring organisms. Chauvel and Dimroth (1974) recognized shallow-water zones in which there were shifting banks of oolithic or intraclastic sand (granules), shifting mud banks (femicrite), and lagoonal sediments. Based on a synthesis of many measured stratigraphic sections, they inferred the environmental setting for the Sokoman Formation shown in Figure 11–10. The major unanswered question with regard to the Sokoman—and, indeed, the Superior type of iron formation in general—is whether the striking similarities between the cherty iron formations and the modern carbonates means that the iron-rich rocks are simply replaced carbonates. And if at least some rocks of the Superior type were originally limestones, why did this massive replacement process stop about 800 million years ago?

IRONSTONES

The ironstones are mostly Phanerozoic unbanded, oolitic rocks (see Figure 11–11) composed of goethite, hematite, and chamosite (iron-rich chlorite); siderite is not uncommon. Locally, magnetite or pyrite may predominate. Common nonferriferous minerals include detrital quartz, calcite, dolomite, and nondetrital phosphorite. Authigenic chert is commonly present but is never equal in abundance to the ferriferous minerals.

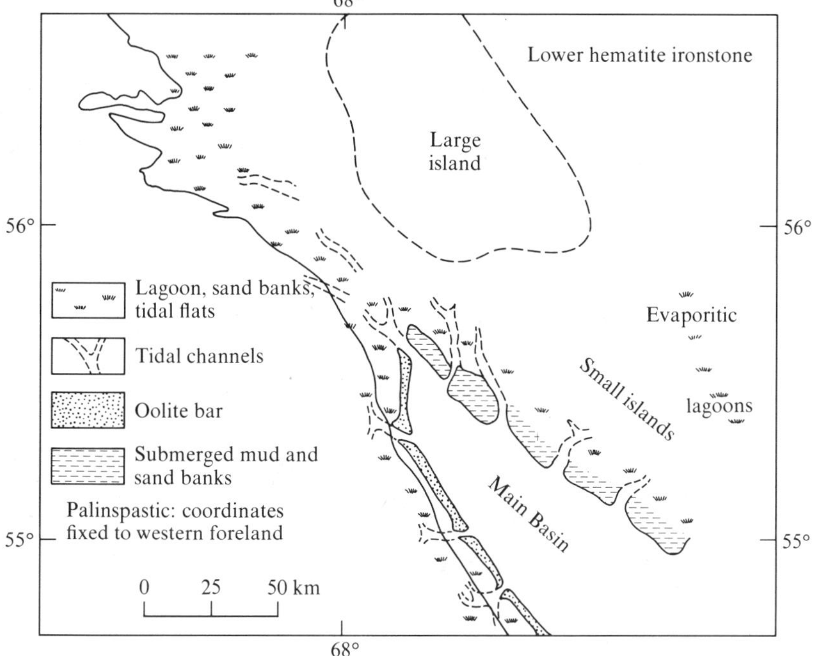

Figure 11–10
General paleogeography of Sokoman Iron Formation in central Labrador Trough. [Dimroth, 1975.]

Figure 11–11
Fractured surface of fresh oolithic ironstone (Cretaceous), Nigeria. Ooliths and matrix consist of siderite, magnetite, and chlorite, with subordinate limonite and clay minerals. Width of photo is 30 mm. [H. A. Jones, 1965, *Jour. Sed. Petrology, 35*. Photo courtesy H. A. Jones.]

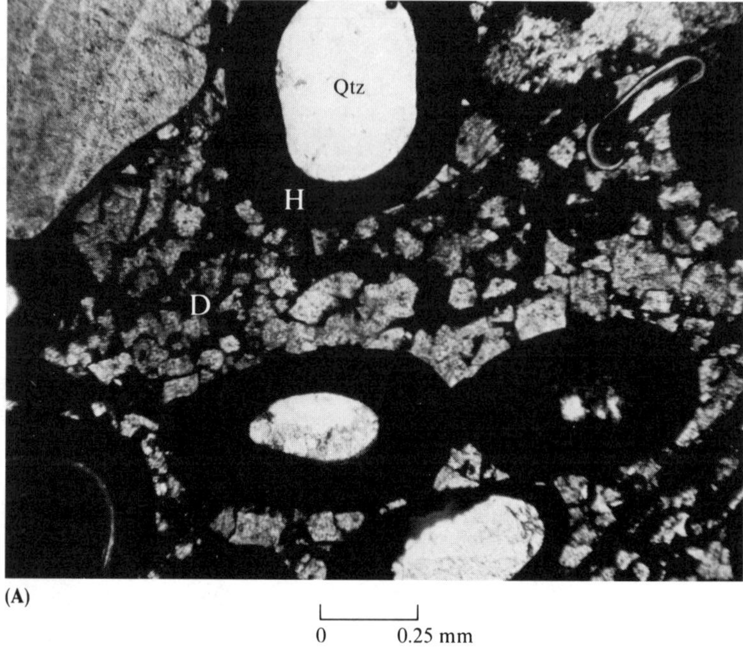

(A)

Figure 11–12
(A) Photomicrograph (crossed nicols) of Clinton Formation (Silurian), Clinton, New York, showing detrital quartz (Qtz) nucleus surrounded by coating of hematite (H) to form oolith. Dolomite (D) rhombs were formed by replacement of original calcium carbonate. [Schoen, 1964. Photo courtesy R. Schoen.]

The areal distribution of the ironstones is much more restricted than that of the Precambrian banded iron formations, with individual basins not exceeding 150 km in length, and the thickness of the units ranges from less than 1 m to a few tens of meters. Interbedded and interfingering sedimentary rocks are carbonates, mudrocks, and fine-grained sandstones of shelf to shallow-marine origin.

Sedimentary structures in oolitic ironstones include large cross beds, ripples, scour-and-fill structures, ripped-up clasts of penecontemporaneously lithified oolitic sediment, and worm burrows. In thin section, fossils of shallow-water origin are as common as they are in oolitic limestones. Many of the fossils are abraded. Commonly, the fossils in ironstones are partly or completely replaced by ferriferous minerals. The irregularly shaped burrows that occur in many ironstones reveal the presence of abundant shallow-water life forms even where shells are lacking. Tertiary ironstones locally contain abundant wood and tree seeds. Clearly, the oolitic ironstones are near-shore, shallow-water deposits.

The nuclei of the ooliths may be quartz grains (angular or rounded), fossil fragments, peloids, intraclasts, or other sediment locally available. The oolithic coatings may be monomineralic or not. For example, some consist of alternating rings of hematite and chamosite. Commonly, individual ooliths are broken and regrown. Sometimes, delicate, well-preserved algal borings cut across some of the oolith rings.

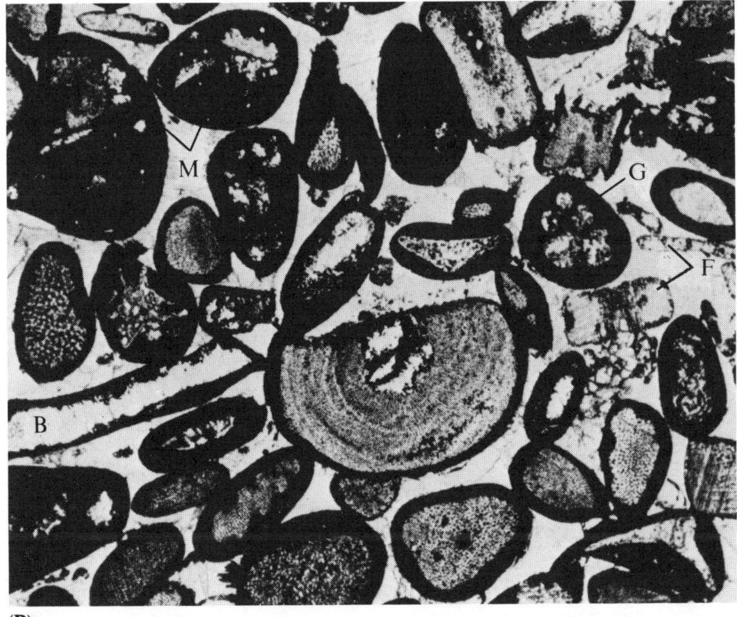

(B)

(B) Photomicrograph of oolithic hematite, Keefer Sandstone (Silurian), Pennsylvania. Most oolith nuclei are crinoid columnals. Others are brachiopod (B) shell fragments, small gastropods (G), and intraclasts composed of small fossil fragments in hematite matrix (M). Hematite permeates fossil nuclei and their pore systems, but a few fossils (F) scattered among ooliths are only slightly replaced by hematite and lack oolithic coatings. Cement is calcite. Large, broken, coated crinoid in center is 1 mm in long dimension. [Photo courtesy R. E. Hunter.]

Clinton Group (Silurian), Appalachia

The ironstones of the Clinton Group in the Appalachian Mountain belt are the best-studied oolitic iron ores in North America. They have been studied most recently by Hunter (1970) and Schoen (1964). The ores crop out sporadically over a distance of 1,500 km from New York southward to Alabama and form part of a sequence of fluvial to shallow-water, marine quartz sandstones, illitic shales, and carbonate rocks in the Appalachian miogeosyncline.

Most of the facies of the Clinton Group are not unusually rich in iron, and they resemble rocks in other parts of the Valley and Ridge Province of the central and southern Appalachians. Ironstones occur in the hematite-cemented sandstone facies, green shale facies, carbonate rock facies, and carbonate-cemented sandstone facies of the Clinton Group. The ironstones are up to 8 m thick, and most are oolitic hematites. Most oolith nuclei in ironstones associated with the hematite-cemented sandstone facies are quartz grains (see Figure 11–12A); most oolith nuclei in the other ironstones are carbonate fossil grains (see Figure 11–12B). The fossils are typically partially replaced by iron minerals, and their internal cavities are filled by iron minerals. All of the oolitic ironstones are cemented mainly by calcite and dolomite.

Thin beds of oolitic chamosite are commonly associated with the oolitic hematite in the Clinton Group. The chamositic beds occur mainly as transitional units between oolitic hematite and overlying or underlying shale. In a few localities, oolitic chamosite is interbedded with shale and is not associated with oolitic hematite.

In general aspect, the Clinton seems identical to sediments formed in modern and ancient beach–nearshore neritic settings, except that the ooliths are composed of hematite rather than calcite or aragonite. For this reason, it is generally believed that the ooliths were formed by the replacement of calcium carbonate, probably soon after deposition, which permits iron-rich solutions to completely permeate the unconsolidated sediment.

SUMMARY

Bedded iron ores are enigmatic rocks of two different types. The more abundant type is the Precambrian bedded iron ore or cherty ironstone. These rocks are typically composed mostly of chert but contain at least 15% iron in the form of hematite, magnetite, siderite, or other mineral. These minerals are concentrated in varvelike layers within the chert. The iron-bearing minerals were precipitated from marine water, lake water, or pore solutions during very early diagenesis. No modern counterparts of these rocks are known, and so their origin remains uncertain.

The second type of ironstone is the Phanerozoic oolitic iron ore. These units are much thinner than their Precambrian relatives and much less extensive. The ooliths are composed mostly of goethite and contain a nucleus of detrital quartz or a carbonate shell fragment; the units grade laterally into shallow-water limestone deposits. It is believed that the iron oxide ooliths were originally composed of calcium carbonate but were replaced by goethite during early diagenesis.

FURTHER READING

Chauvel, J., and E. Dimroth. 1974. Facies types and depositional environment of the Sokoman Iron Formation in the central Labrador Trough, Quebec. *Jour. Sed. Petrology, 44,* 299–327.

Dimroth, E. 1975. Paleo-environment of iron-rich sedimentary rocks. *Geologische Rundschau, 64,* 751–767.

Dimroth, E. 1976. Aspects of the sedimentary petrology of cherty iron-formation. In K. H. Wolf (ed.), *Handbook of Strata-Bound and Stratiform Ore Deposits.* New York: Elsevier, 7, 203–254.

Dimroth, E., and J. Chauvel. 1973. Petrography of the Sokoman Iron Formation in part of the Central Labrador Trough, Quebec, Canada. *Geol. Soc. Amer. Bull., 84,* 111–134.

Eichler, J. 1976. Origin of the Precambrian iron-formations. In K. H. Wolf (ed.), *Handbook of Strata-Bound and Stratiform Ore Deposits.* New York: Elsevier, *7,* 157–202.

Gross, G. A. 1965. *Geology of Iron Deposits in Canada. I: General Geology and Evaluation of Iron Deposits.* Geol. Surv. Canada Econ. Geol. Report No. 22, 181 pp.

Gross, G. A. 1972. Primary features in cherty iron-formations. *Sed. Geol., 7,* 241–261.

Gross, G. A. 1980. A classification of iron formations based on depositional environments. *Canad. Mineral., 18,* 215–222.

Hunter, R. E. 1970. Facies of iron sedimentation in the Clinton Group. In G. W. Fisher et al. (eds.), *Studies of Appalachian Geology.* New York: Wiley, pp. 101–124.

James, H. L. 1966. *Chemistry of the Iron-Rich Sedimentary Rocks.* U.S. Geol. Surv. Prof. Paper No. 440–W, 60 pp.

James, H. L., and P. K. Sims (eds.). 1973. Precambrian iron formations of the world. *Econ. Geol., 68,* 913–1179.

Kimberley, M. M. 197. Origin of oolitic iron formations. *Jour. Sed. Petrology, 49,* 111–132. See also Discussion, *50,* 295–304.

Mellon, G. B. 1962. Petrology of Upper Cretaceous iron-rich rocks from northern Alberta. *Econ. Geol., 57,* 921–940.

Schoen, R. 1964. Clay minerals of the Silurian Clinton ironstones. *Jour. Sed. Petrology, 34,* 855–863.

Van Houten, F. B., and D. P. Bhattacharyya. 1982. Phanerozoic oolitic ironstones—facies model and distribution in space and time. *Rev. Earth Planet. Sci., 10,* in press.

Van Houten, F. B., and R. M. Karasek. 1981. Sedimentary framework of Late Devonian oolitic iron formation, Shatti Valley, west-central Libya. *Jour. Sed. Petrology, 51,* 415–427.

Zajac, I. S. 1974. *The Stratigraphy and Mineralogy of the Sokoman Formation in the Knob Lake Area, Quebec and Newfoundland.* Geol. Surv. Canada Bull., *220,* 159 pp.

12

Phosphorites

Attempt the end, and never stand to doubt;
Nothing's so hard but search will find it out.

ROBERT HERRICK

Phosphorites are sedimentary rocks that contain at least 20% P_2O_5 by laboratory chemical analysis—an enormous enrichment over the amounts in other types of rocks. For example, mudrocks average only 0.17% P_2O_5; limestones, 0.04%. The phosphorous is present in ancient phosphorites in the mineral fluorapatite, $Ca_5(PO_4)_3F$, typically with several percent of CO_3^{2-} substituting for PO_4^{3-} in the crystal structure. Most commonly, the apatite is cryptocrystalline; phosphorites are not concentrations of detrital apatite grains. The term *collophane* is commonly used to describe cryptocrystalline to X-ray amorphous phosphatic material.

FIELD OBSERVATIONS

The total thickness of dominantly phosphatic sequences can be several hundred meters, although generally they are much thinner. Most studies of phosphorites have been made for economic purposes (fertilizer and byproduct uranium, vanadium, and rare-earth elements), so that the abundance of thin phosphorite sequences or individual beds in otherwise nonphosphatic sections is unknown. The phosphorites occur in outcrop as black layers ranging in thickness from laminae of 1 mm to beds of a few meters. Commonly, they are interbedded with carbonates, dark-colored chert (phosphatic or organic), and carbonaceous mudstone. Coarse detrital sediments are rare.

Sedimentary structures visible in outcrop include bedding, lamination, scour structures, and cross bedding (see Figure 12–1). Phaneritic textural features include grain rounding and ooliths. Calcareous fossils include mollusks, brachiopods, sponges, and foraminifera; phosphatic fragments of fish bones and teeth can also be seen with a hand lens.

Figure 12–1
Sedimentary structural and textural features visible in outcrops of phosphorite, as seen in sawed and polished hand specimens. (**A**) Finely laminated bed from Areyonga Formation (Late Precambrian), central Australia. Darker patches are rounded grains of chert. (**B**) Dark-colored, ovoid phosphate grains cemented by light-colored dolomite, Meade Peak Member of Phosphoria Formation (Permian), Idaho. Note cross bedding and parallelism of long axis of grains to bedding.

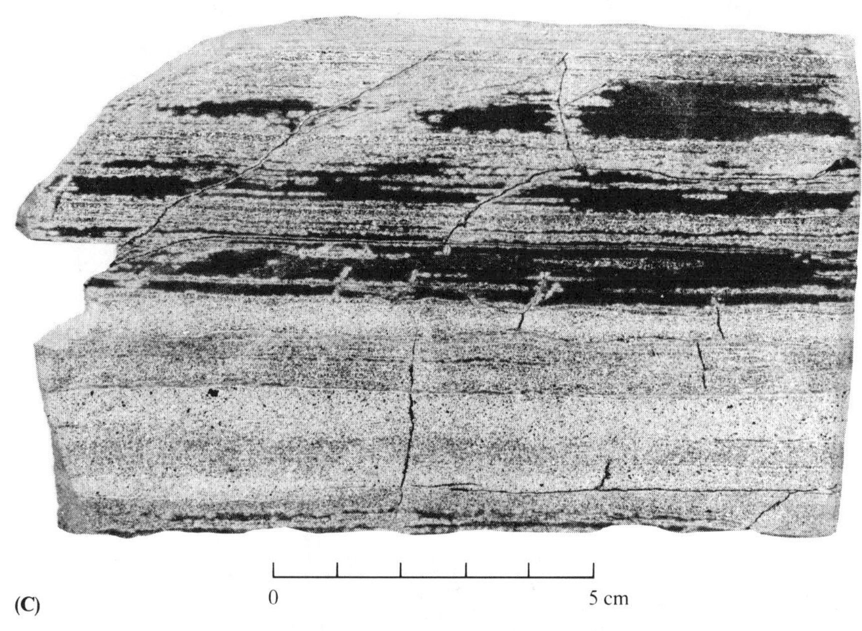

(C)

(D)

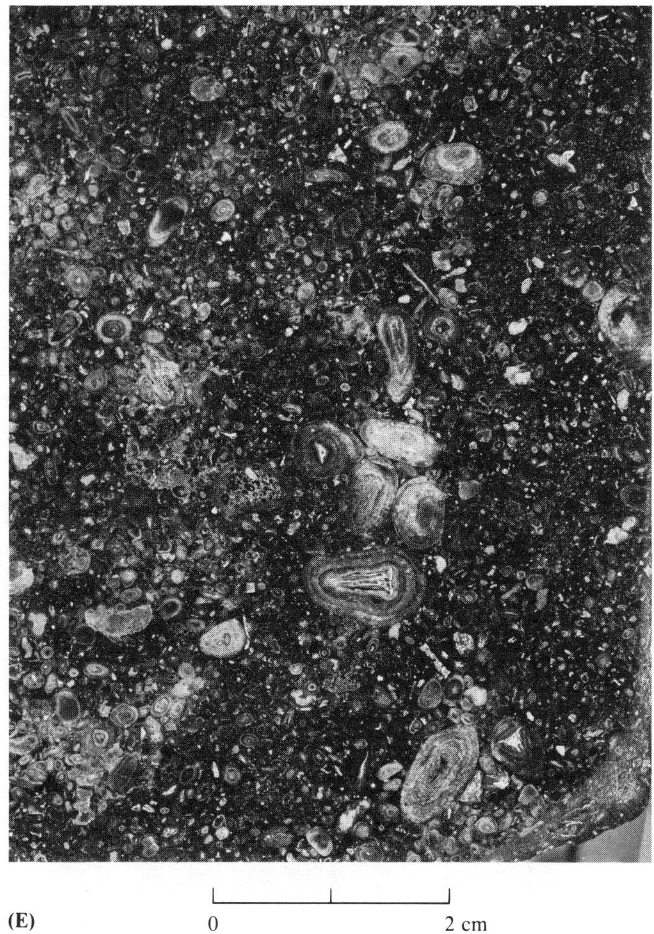

(E)

Figure 12–1 (*continued*)
(**C**) Phosphatic shale, Meade Peak Member. Where unweathered, it is a hard, black, bituminous phosphatic dolomitic shale. With weathering, carbonates and hydrocarbons leave. Note how black organic matter has left most of this specimen and how remaining organic matter is leaving along bedding and fracture surfaces. Bed contains 25% P_2O_5. (**D**) Aphanitic phosphate in Meade Peak, containing calcitic shell fragments and phosphatic fish scales visible only in thin section. Darker layer at bottom was apparently thinned by erosion. Bed contains 32% P_2O_5. (**E**) Polished surface of oolitic phosphomicrite in Phosphoria Formation (Permian), Wyoming. Some ooliths show nuclei of shark's teeth, as in left center. [A from Cook, 1976. Photo courtesy P. J. Cook. B–D from G. D. Emigh, 1967, in L. A. Hale, ed., *Intermountain Assoc. Geol. 15th Annual Field Conference Guidebook*. Photos courtesy G. D. Emigh. E from G. R. Mansfield, 1927, U.S. Geol. Surv. Prof. Paper No. 152. Photo courtesy G. R. Mansfield.]

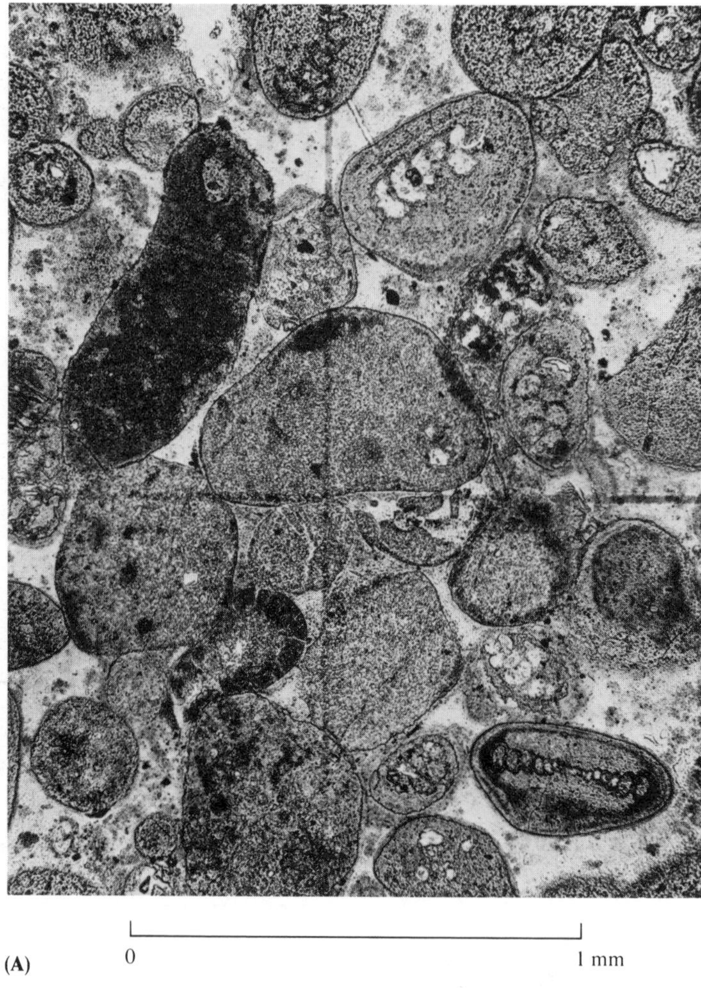

(A)

Figure 12–2
(A) Thin section of cherty phosphorite containing many foraminiferal tests that form nuclei of ooliths, Phosphoria Formation (Permian), Montana. Clear chert fills interstices between grains, cracks in grains, and chambers of fossils.

LABORATORY STUDIES

Except for the difference in mineral composition (apatite rather than calcite), thin sections of phosphorites look very much like thin sections of limestones. The terminology used to describe carbonates in the laboratory can also be used to describe phosphorites, although this has not been done by phosphophiles. There occur phosphomicrites, phosphosparites, and varieties of both that contain variable percentages of fossils, peloids, ooliths, and phosphoclasts. (Specialists in phosphorites refer to all clastic grains of phosphorite coarser than 2 mm as "nodules"; grains finer than 2 mm,

(B)

(**B**) Thin section of plastically deformed peloids composed of cryptocrystalline apatite (collophane), Phosphoria Formation. Deformation of peloids resulted from compaction while peloids were soft. [Cressman and Swanson, 1964. Photos courtesy E. R. Cressman, U.S. Geol. Surv.]

"pellets.") Examples of the textures of phosphorites are shown in Figures 12–2, 12–3, and 12–4.

The bulk of phosphorite grains and matrix is cryptocrystalline (crystal diameters smaller than 1 μm) rather than microcrystalline and, as a result, phosphorite appears as a brownish, isotropic material in thin section. Within many particles, there are areas that are more coarsely crystalline and have visible birefringence. Typically, these areas are not random within the particle but parallel internal structure, such as concentric zones within ooliths (see Figure 12–4) or thin zones paralleling the shell outline in linguloid brachiopods.

Mabie and Hess (1964) present a comprehensive classification of textures commonly seen in phosphorites. Their system is based on the internal structure of the particle, the apatite crystal size, and the degree of compaction of the rock as determined by the types of grain-to-grain contacts. Only some of the suggestions made by Mabie and Hess have been widely adopted by phosphophiles. Because of the textural similarities between limestones and phosphorites, it seems preferable to describe

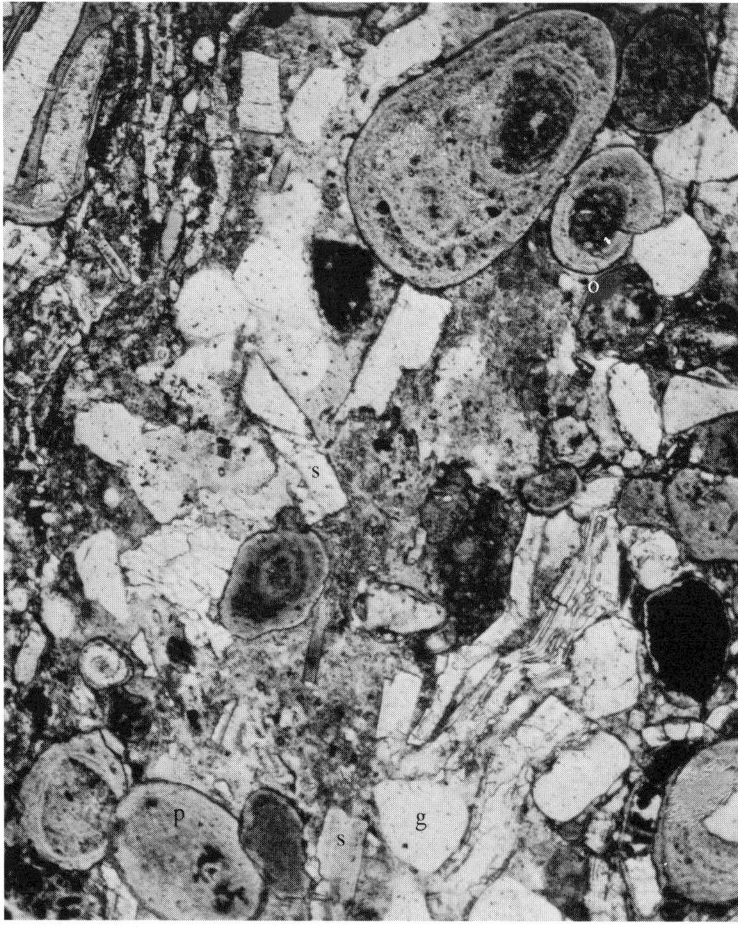

Figure 12–3
Photomicrograph of phosphorite in Phosphoria Formation, containing quartz sand grains (g) and apatite ooliths (o), "pellets" (p), and skeletal (s) fragments. Note oolith nuclei consisting of quartz sand grains and apatite coating. Uncrossed polarizers. [Cressman and Swanson, 1964. Photo courtesy E. R. Cressman, U.S. Geol. Surv.]

phosphorites in terms applied to limestones, such as phosphomicrite (cryptocrystalline or microcrystalline), phosphosparite, biophosphomicrite, intraphosphosparite, etc. This type of terminology is appropriate not only because of the textural similarities between limestones and phosphorites, but also because many phosphorites are clearly phosphatized limestones. Some specialists in phosphorites believe that *all* phosphorites are replaced limestones. Even the common impurities in phosphorites are identical to those in carbonate rocks: clay minerals, chert, glauconite, organic matter, iron oxide, pyrite, and fine-grained detrital quartz.

CHEMICAL CONSIDERATIONS

Phosphorite beds occur in rocks of all ages, from Precambrian to Holocene, and on all continents (see Figure 12–5). Therefore, the conditions in which they form have not been uncommon through geologic time. Nevertheless, phosphorites are peculiar rocks, consisting as they do of concentrations of a minor element (phosphorous) more than 100 times the amount in other types of sedimentary rocks.

Where are phosphorites forming today? When we asked that question about limestones, which are similar in many respects to phosphorites, we were able to look around and find limestone reefs and calcium carbonate shells to examine. Using the principle that the present is the key to the past, we studied these materials and applied the information gained to the interpretation of the ancient limestones. In the case of phosphorites, however, this approach is inadequate. No phosphatic organisms build reefs. Only a single brachiopod group among the invertebrates in the sea secretes a phosphatic skeleton (*Lingula*), and it is not abundant. So we must use a different approach.

It is clear that phosphorites are chemical or biochemical rocks rather than terrigenous detrital rocks. We also know, from the fossils they contain, that nearly all phosphorites are marine. Therefore, it may be informative to determine the character of the waters in which phosphorous can be concentrated. The first person to consider this question was a Russian, A. V. Kazakov, in 1937. From oceanographic data he determined that the P_2O_5 content of marine waters is at a maximum at depths about 30–500 m. At depths less than 30 m, the phosphorous is consumed by phytoplankton during photosynthesis. At depths greater than 500 m, the content of carbon dioxide in the water is so great that the waters cannot become saturated with respect to apatite because apatite, like calcite, is soluble in acid. As we have seen, ancient phosphorites formed in fairly shallow waters (fossils, cross bedding), so that depths greater than 100 m or so are not of much interest to us in any event—at least not so far as the major phosphorite deposits are concerned.

Where in the modern ocean is phosphorite forming today? Kazakov pointed out that modern deposits are located in low latitudes on the eastern sides of marine basins where deep, phosphate-rich waters are upwelling adjacent to a shallow shelf. In most locations the upwelling is caused by the Trade Winds blowing surface waters off-

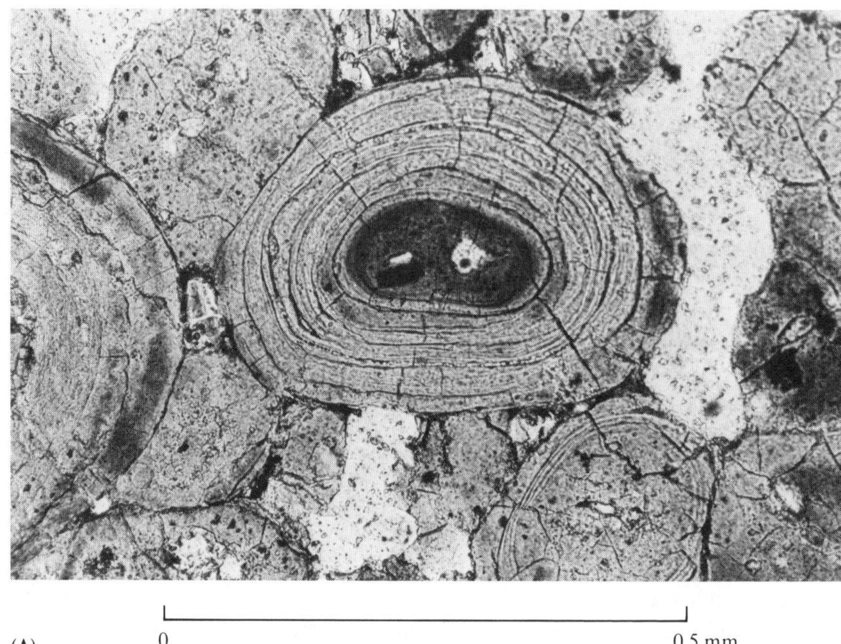

(A) 0 0.5 mm

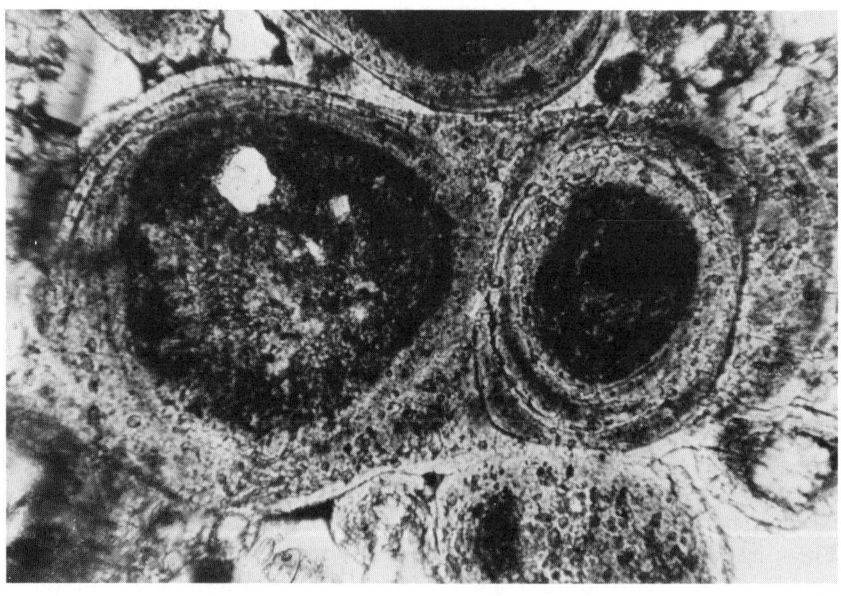

(B) 0 0.5 mm

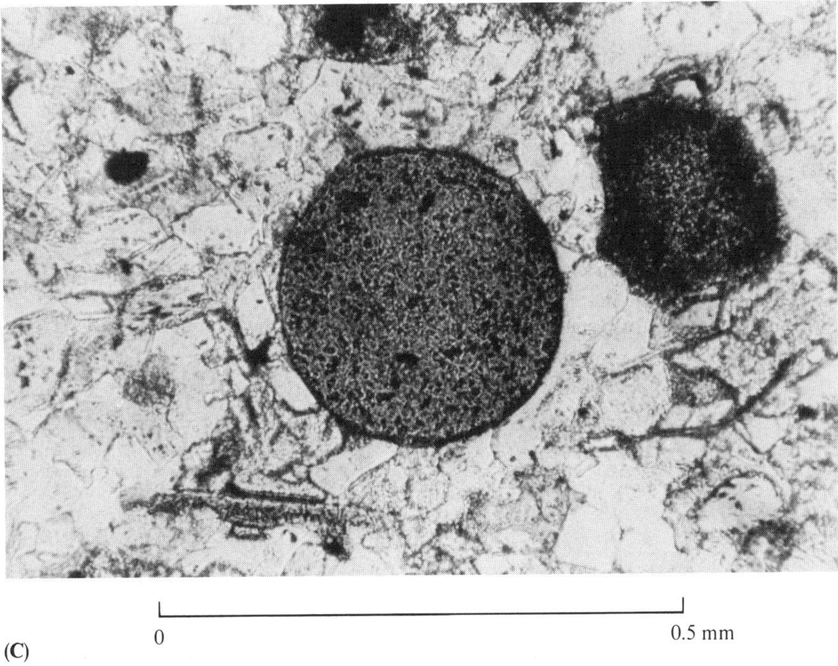

(C)

Figure 12–4
(**A**) Thin section of oolith ("oolitic pellet") showing well-developed concentric banding. (**B**) Thin section of compound ooliths or grapestone (type of "pellet"). (**C**) Thin section of perfectly oval peloid ("ovule pellet") composed of phosphomicrite. [Cook, 1976. Photos courtesy P. J. Cook.]

shore, so that they are replaced at the surface by the deeper waters. Has this preferred location been characteristic of ancient phosphorites as well (see Figure 12–6)? Once again, paleomagnetic data are required, so that the present locations of ancient phosphorites can be changed to the locations they occupied when they formed (see Figure 12–6C). It is clear that ancient phosphorites formed in the Trade Wind belt in low latitudes, apparently in latitudes even lower than the Holocene average of 25°. The reason for this difference is not clear. It is clear, however, that the upwelling of phosphate-rich, cold, bottom waters is critical to phosphorite formation. The largest and best-known phosphorite deposit, the Phosphoria Formation of Permian age in the western United States, was located along the western margin of the North American continent at the time it was formed and within 5° of the Permian equator. (We consider the Phosphoria Formation in more detail a bit later in this chapter.)

Although it is generally agreed among sedimentary geochemists that upwelling waters stimulate the production of phosphorite, the mechanism by which the solid phosphate is produced remains unclear. No one has seen a precipitate of collophane or microcrystalline apatite form in clear, marine water in the way that gypsum, for

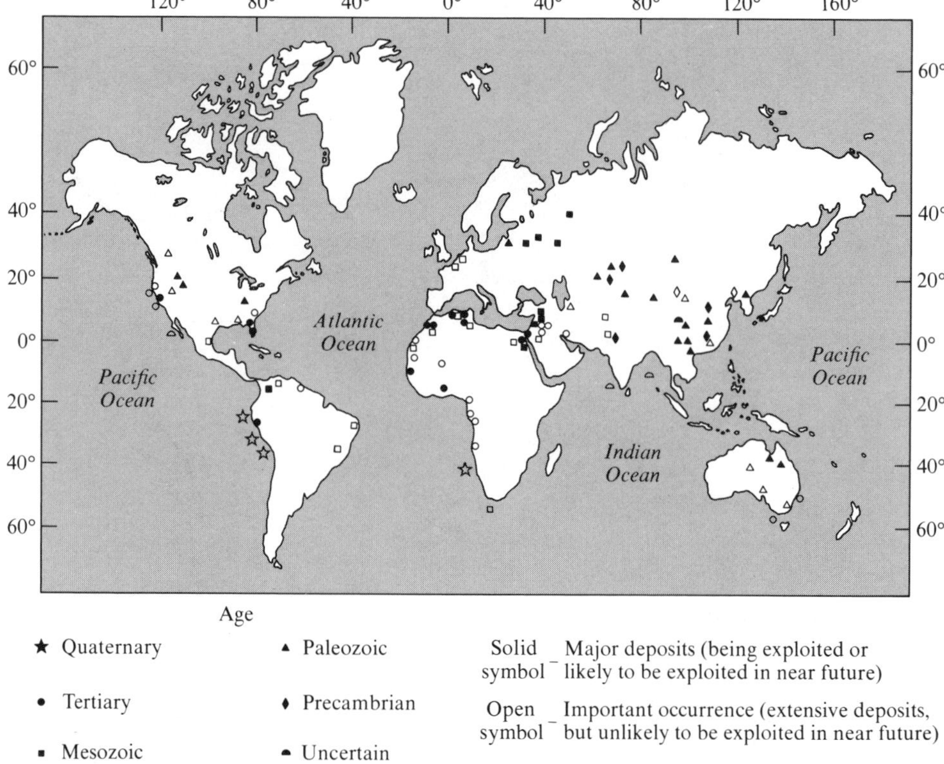

Figure 12–5
Geographic and stratigraphic distribution of major sedimentary phosphorite deposits. [Cook, 1976.]

example, forms in more saline waters. Apparently, the phosphate material crystallizes interstitially among particles at and immediately below the water–sediment interface. In the process of formation, the phosphorite commonly replaces existing materials. Replacements of calcareous foraminifera and siliceous diatoms by phosphorite on the modern ocean floor are well documented.

Metasomatism of Limestone

As noted earlier, the stratigraphic relationships, structures, and textures of many phosphorite deposits resemble those of limestones. In addition, some phosphorites contain originally calcareous fossils that have been phosphatized or ooliths that have been partially converted to apatite after burial. We are reminded of dolomite replacing limestone. In these cases, we are certain that the apatite was not a primary precipitate but is diagenetic in origin, as is the dolomite. Unfortunately, many phosphorites show no evidence of a replacement origin.

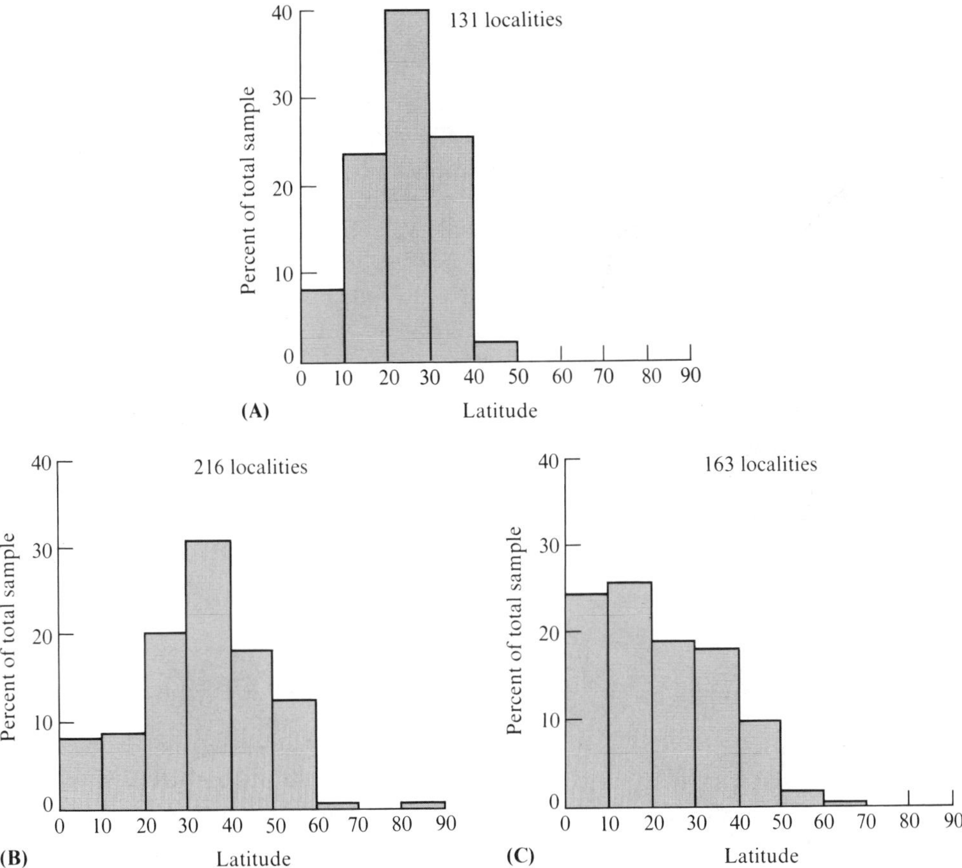

Figure 12–6
Latitudinal distribution of phosphorite deposits. (**A**) Quaternary and Late Tertiary deposits, unaffected by plate movements. (**B**) Present latitudes of older deposits. (**C**) Original latitudes of ancient deposits. [Cook and McElhinny, 1979.]

Nevertheless, as geologists, we are interested in defining the conditions in which the phosphatization of calcium carbonate can occur, as we are in the case of dolomitization. To study the phosphatization process, Ames (1959) conducted laboratory experiments in which he passed a basic solution of sodium phosphate and sodium hydroxide through a tube packed with fragments of calcite. From time to time he sampled the composition of the solution from the bottom of the tube and studied samples of the "sediment" in the tube to determine its composition. He found that the original calcite was replaced and perfectly pseudomorphed by apatite and that the apatite contained a few percent of CO_3^{2-} substituting for PO_4^{3-} in its crystal structure, just as occurs in most ancient phosphorites. Ames' apatite was a hydroxyl apatite rather than a fluorapatite because he had no fluorine in his phosphatizing solution.

$$3Na_3PO_4 + NaOH + \underset{\text{calcite}}{5CaCO_3} \rightarrow \underset{\text{apatite}}{Ca_5(PO_4)_3OH} + 5Na_2CO_3$$

Ames found that the phosphatization of calcite would occur in PO_4^{3-} concentrations as low as 0.1 ppm. In areas of oceanic upwelling, the PO_4^{3-} content is usually 0.3–0.8 ppm—more than adequate to cause replacement of calcite by apatite. In reducing environments immediately below the seawater–sediment interface, it is common to find 1.0 ppm PO_4^{3-} in solution because of the decay of phosphate-containing organic tissue in such an environment. It certainly seems possible that a significant amount of phosphorite can be produced by very early diagenetic replacement just below the seafloor and, just as is true in many instances of early replacement of calcite by dolomite, the replacement process may not be detectable in the resulting rock. It may be significant, as pointed out by Russell and Trueman (1971), that, whereas granular phosphorites occur in Phanerozoic sediments, almost all known Precambrian phosphorites are cryptocrystalline, collophane mudstones without clastic phosphorite grains. Might this fact be related to the fact that calcareous organisms were not present in Precambrian time?

ANCIENT PHOSPHORITES

Phosphoria Formation (Permian), Northwestern United States

The Phosphoria Formation is the best-studied marine phosphorite unit in the world. For this reason it is often used as a "type section" of phosphorite occurrence and, based on what is known of marine phosphorite units in other parts of the world, it is well suited for this use.

The Phosphoria Formation has a maximum thickness of 420 m and was deposited during a period of 15 million years over an area of about 350,000 km^2. Phosphorites occur in both the platform and the geosynclinal parts of the Paleozoic Cordilleran structural belt (see Figure 12–7). The regional stratigraphy of the Phosphoria reflects its structural setting not only by the great increase in thickness toward the geosyncline but also by lithologic changes. Spiculiferous chert and cherty, carbonaceous mudstone dominate the stratigraphic section in the deeper parts of the depositional basin; they are succeeded successively eastward by dolomitic limestones and calcareous sandstones and ultimately by continental red shales and sandstones. Highly phosphatic beds occur in both the geosynclinal and the platform facies and appear to reach a maximum thickness near the "hinge line" separating the two facies.

In the vicinity of its type locality in southeastern Idaho, the formation has a thickness of 80–150 m and consists mainly of dark chert, phosphatic and carbonaceous mudstone, phosphorite, cherty mudstone, and minor amounts of dark carbonate rock. Among the four members into which the Phosphoria has been subdivided, two are phosphorites: the Meade Peak and the Retort. The maximum combined thickness of the two members is 30–35 m.

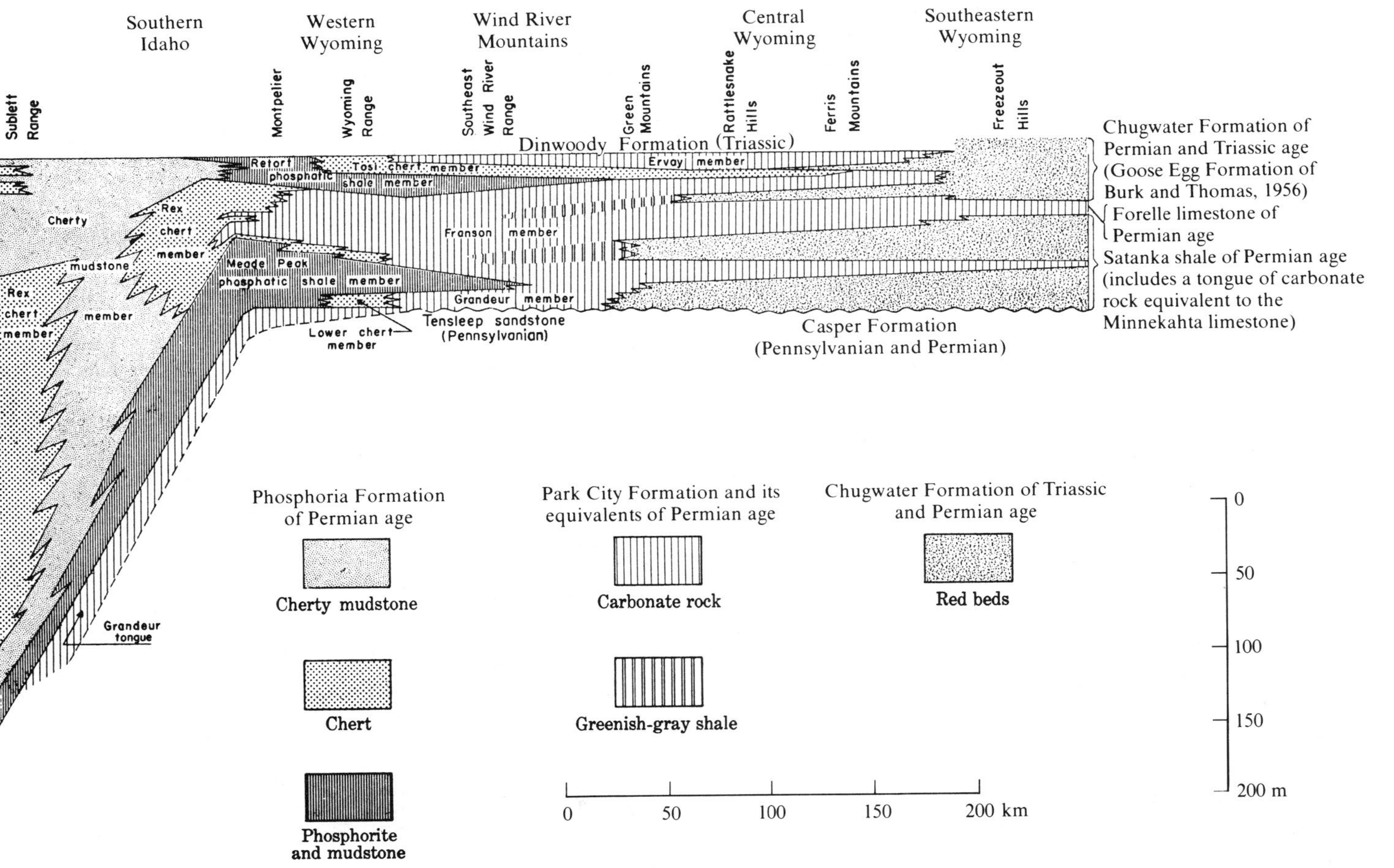

Figure 12–7
Stratigraphic relations of Phosphoria Formation (Permian), Park City Formation (Permian), and Chugwater Formation (Permian) in Idaho and Wyoming. [V. E. McKelvey et al., 1959, U.S. Geol. Surv. Prof. Paper No. 313A.]

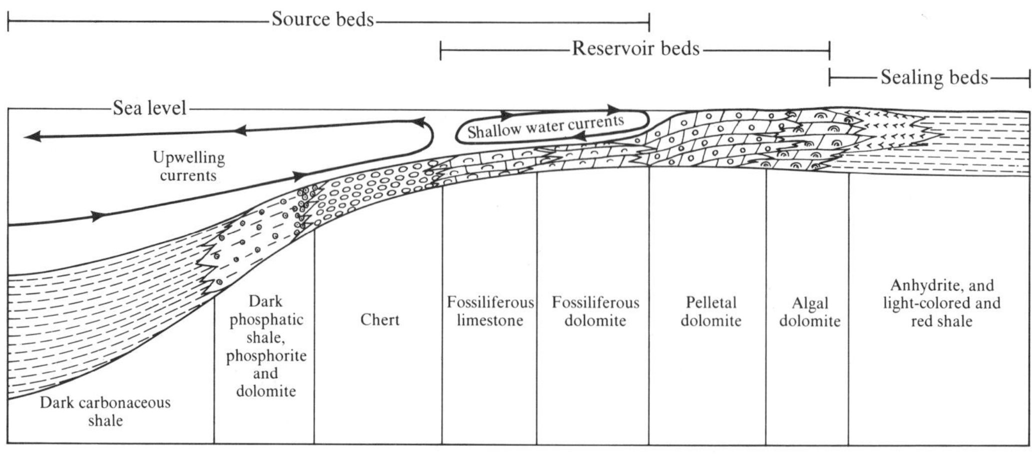

Figure 12–8
Idealized model of sedimentation in Phosphoria sea. The Phosphoria is believed to be the source rock for major petroleum accumulations; zones of source rock, reservoir rock, and impermeable seal are also shown. [Sheldon, 1963.]

The thickest and most extensive phosphorite unit in the Phosphoria Formation is the Meade Peak Member, composed of phosphatic shale, mudstone, and carbonate rocks. The phosphorites are mostly thin-bedded, laminated, dark-colored, well-sorted, granular rocks consisting of about 50% apatite. The grains include peloids (possibly fecal pellets), ooliths, and intraclasts. Linguloid brachiopods, phosphatic fish scales, and phosphatized gastropods, cephalopods, bryozoans, echinoid spines, and sponge spicules also occur in the allochemical fraction.

In the nearshore area, well-aerated waters are suggested by the noncarbonaceous, cross-bedded, well-sorted, granular phosphorites that contain abraded fossils and lineated, elliptical grains. Contrasting with this lithology are the offshore carbonaceous, pyritic, phosphatic mudstones, whose few fossils show no evidence of current transport. The change in lithologic character is strikingly similar to that encountered in limestones as the rocks are followed offshore to a micritic area farther from the shoreline. Sheldon (1963) has proposed the cross section in Figure 12–8 as an idealized model of sedimentation in the Phosphoria sea.

Bone Valley Formation (Pliocene), Florida

The Bone Valley Formation is an economically important phosphorite unit of Pliocene age exposed in west-central peninsular Florida. It is representative of the "land pebble" deposits that supply the bulk of the production of phosphate rock in the United States (Altschuler et al., 1956). The deposits occur in an irregularly shaped area about 110 km in length and 65 km in width, mostly in the Bone Valley Formation but subordinately in the underlying Miocene Hawthorn Formation (see Figure 12–9). The surface of the Hawthorn is irregular, its carbonate rock is solution-pitted,

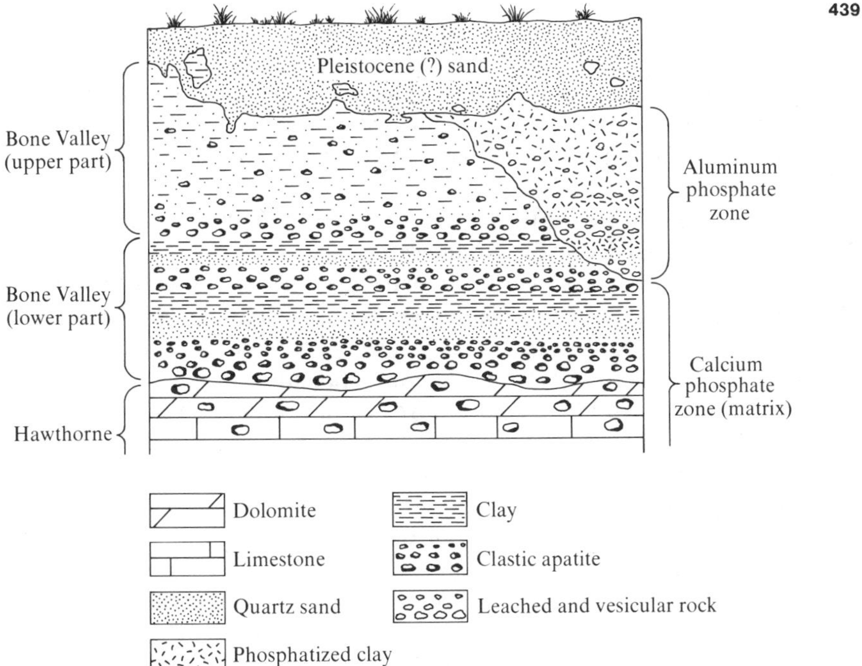

Figure 12–9
Stratigraphic relations in aluminum phosphate zone of Bone Valley Formation (Pliocene), Florida. [Altschuler et al., 1956.]

and it contains many small and large slumps, within which the overlying Bone Valley thickens. The Hawthorn was exposed and weathered during Late Miocene time, developing an irregular topography and accumulations of phosphatic residue containing primary apatite clasts inherited from the Hawthorn and secondarily phosphatized limestone pebbles. During a period of Pliocene marine transgression, this residuum of Hawthorn was reworked into the unconformably overlying Bone Valley Formation, with additions of quartz, clay, and probably phosphate.

The Bone Valley is 8–10 m thick and, where unweathered, is composed of iron-rich smectite, quartz, and carbonate fluorapatite. In its lower two-thirds the formation is graded from phosphatic pebbly sand to fine-grained phosphatic sand to argillaceous phosphatic sand to clay. The upper one-third consists of thick-bedded, argillaceous quartz sand containing only minor amounts of the clastic phosphate rock so abundant in the lower part of the formation. Also, the clay in the upper part is kaolinite—a result of postdepositional alteration of the smectite.

Generally, the Bone Valley is compact, although not indurated, and either is gray or shows the reddish-brown or orange color caused by release during weathering of the ferric iron from the clay. However, in some areas a white zone of leaching and alteration irregularly transgresses the upper part of the section (see Figure 12–9). In this zone, quartz sand is cemented and indurated by secondary phosphate occurring as the aluminum phosphate minerals wavellite, crandallite $[CaAl_4(OH)_8(PO_4)_2 \cdot H_2O]$, and, locally, millisite $[Na\ Ca\ Al_6\ (OH)_9(PO_4)_4 \cdot 3H_2O]$. Where the altered zone transgresses the apatite-rich, pebbly layers at the base of the graded beds, the calcium aluminum phosphates crandallite and millisite occur. Where the alteration is restricted to the clay-rich, apatite-poor upper part of the Bone Valley, however, the aluminum phosphate wavellite $[Al_3(PO_4)_2(OH)_3 \cdot 5H_2O]$ is the dominant mineral. The aluminum in all three of the secondary phosphate minerals was derived from the decomposition of clay by acidic groundwaters.

Secondary precipitates of apatite, limonite, and chert underlie the aluminum phosphate zone in irregular and discontinuous seams of hardpan that cement layers of sand and encrust upper surfaces of clay. The chert represents silica released during the phosphatization of clay to wavellite and crandallite. Seams of silica are closely associated with clay in intermediate stages of such alteration higher in the section. These hardpans are formed by material leached from above and redeposited at the groundwater table during the early stages of leaching; hence, apatite, rather than aluminophosphates, is present as the precipitate.

Braithwaite (1968) has described Quaternary phosphatization of Pleistocene calcareous sediment caused by the leaching of surficial guano deposits. Excellent pseudomorphing of shell fragments by apatite was reported. Petrographic study of the phosphatic rocks reveals the presence of whitlockite $[Ca_3(PO_4)_2]$ and suggests the former presence of brushite $[CaHPO_4 \cdot 2H_2O]$. Both of these minerals are unstable relative to apatite; the conditions under which they may form and persist are unknown.

MODERN PHOSPHORITES

Isolated, large nodules of phosphorite many centimeters in diameter are commonly reported to occur in sediments dredged from modern marine environments. However, the dense concentrations of sand-size phosphorite grains that characterize ancient deposits such as the Phosphoria Formation are reported only rarely. One of these rare reports is by D'Anglejan (1967), who described modern sediments on an extensive, shallow marine shelf that contain phosphatic grains as a major constituent. The shelf, located offshore of Baja California (see Figure 12–10), is approximately 13,000 km^2 in surface area, is as much as 80 km in width, and has only slight relief on its surface. The shelf is bounded on the west by a down-faulted basin whose bottom is more than three times the depth of the shelf. Most of the phosphorite grains occur at depths less than 100 m.

The phosphorite occurs as well-sorted, sand-size particles (see Figure 12–11) within a fine sand to silt matrix. As in ancient marine deposits, the phosphorite is carbonate fluorapatite, and the apatite forms 90% by weight of the particles; the remainder is composed of detrital inclusions, syngenetic sulfides, organic matter, opaline silica, and disseminated calcite. Chemical analyses of the phosphorite indicate a P_2O_5 content of 30%. The grains of fine sand size are black, are subspherical to elongate, and have knobby to botryoidal surfaces. In thin section, they are normally

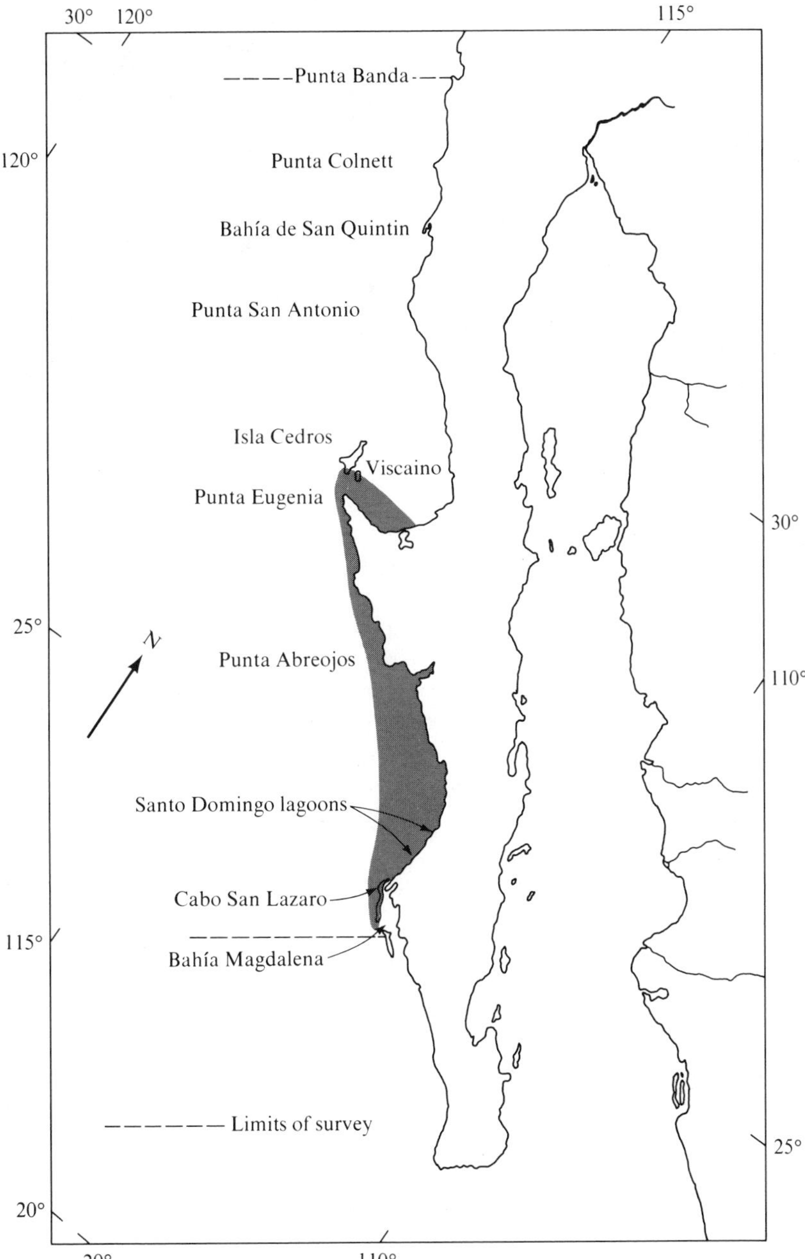

Figure 12–10
Extent of fine-grained phosphorite in surficial sediments along western coast of Baja California, Mexico. [D'Anglejan, 1967.]

Figure 12–11
Grains of phosphorite of fine sand size from continental shelf, Baja California, Mexico. [D'Anglejan, 1967. Photo courtesy B. F. D'Anglejan.]

structureless, with only occasional apparent concentricity due to zonal staining by diffuse organic matter. A superficial reduction layer imparts a black color to the grains collected in the reducing zone of the middle shelf, but it is absent in the reddish-brown grains from the oxidizing environment of the beach.

The apatite within the peloids occurs both as anhedral, equant crystals a few microns in length and as a cryptocrystalline matrix surrounding these crystals. The crystals are sometimes grouped as circular mosaics around black, organic inclusions.

Opaline silica is an important constituent of the pellets. It occurs as a whitish filling in surface cavities and as a very thin surface film that appears to be formed of an intergrowth of about 10% silica and 90% apatite. Diatom fragments are present as inclusions in the matrix of some phosphorite grains, and probably supply the silica for the surface films.

Thus, apart from the associated silica, these fine-grained phosphorites are similar in mineralogy, composition, and general lack of internal structure of the grains to those of ancient marine deposits. The opaline coatings on the modern phosphorite grains are probably removed during diagenesis. The association of ancient, bedded phosphorites with cherts, siliceous mudstones, and calcareous black shales seems compatible with the modern deposit near Baja California.

The environment of peloid formation is adjacent to a zone of seasonal upwelling of phosphate-rich waters, and it is possible that the peloids are concretionary, developing through surface accretion from phosphate colloids in the supersaturated seawater. The accretion may be a slow, continuous process occurring randomly at the periphery of the particle, so that no visible concentricity is produced. Alternatively, the peloids may have formed by phosphatization of originally structureless particles such as fecal pellets or aggregates of mud.

SUMMARY

Phosphorites are enigmatic rocks defined by chemical composition. They must contain at least 20% P_2O_5. Rocks that contain this much phosphatic material are composed largely of cryptocrystalline to microcrystalline apatite, are black in color, and have structures and textures similar to those found in limestones. Phosphorites can be described in the same terms used for describing limestones, micrite, and sparite by adding the prefix *phospho-* to generate descriptions, such as *phosphopelmicrite.*

Phosphorites form in many of the same depositional environments as do limestones, and many phosphorites may be replaced limestones. The origin of most phosphorites, however, is uncertain.

FURTHER READING

Altschuler, Z. S., E. B. Jaffe, and F. Cuttitta. 1956. The aluminum phosphate zone of the Bone Valley Formation, Florida, and its uranium deposits. U.S. Geol. Surv. Prof. Paper No. 300, pp. 495–504.

Ames, L. L., Jr. 1959. The genesis of carbonate apatites. *Econ. Geol., 54,* 829–841.

Bentor, Y. K. (ed.). 1980. *Marine Phosphorites.* Soc. Econ. Paleontol. Mineral. Spec. Pub. No. 29, 249 pp.

Braithwaite, C. J. R. 1968. Diagenesis of phosphatic carbonate rocks on Remire, Amirantes, Indian Ocean. *Jour. Sed. Petrology, 38,* 1194–1212.

Brown, C. E. 1974. Phosphatic zone in the lower part of the Maquoketa Shale in northeastern Iowa. *Jour. Res.* (U.S. Geol. Surv.), *2,* 219–232.

Cook, P. J. 1970. Repeated diagenetic calcitization, phosphatization, and silicification in the Phosphoria Formation. *Geol. Soc. Amer. Bull., 81,* 2107–2116.

Cook, P. J. 1976. Sedimentary phosphate deposits. In K. H. Wolf (ed.), *Handbook of Strata-Bound and Stratiform Ore Deposits.* New York: Elsevier, *7,* 505–536.

Cook, P. J., and M. W. McElhinny. 1979. A reevaluation of the spatial and temporal distribution of sedimentary phosphate deposits in the light of plate tectonics. *Econ. Geol., 74,* 315–330.

Cressman, E. R., and R. W. Swanson. 1964. Stratigraphy and petrology of the Permian rocks of southwestern Montana. U.S. Geol. Surv. Prof. Paper No. 313-C, pp. 275–569.

D'Anglejan, B. F. 1967. Origin of marine phosphorites off Baja California, Mexico. *Marine Geol., 5,* 15–44.

Emigh, G. D. 1958. *Petrography, Mineralogy, and Origin of Phosphate Pellets in the Phosphoria Formation.* Idaho Bur. Mines and Geol. Pamphlet No. 114, 60 pp.

Jour. Geol. Soc. 1980. Vol. 136, Part 6. An issue devoted to phosphatic and glauconitic sediments.

Mabie, C. P., and H. D. Hess. 1964. *Petrographic Study and Classification of Western Phosphate Ores.* U.S. Bur. Mines Report of Investigations No. 6468, 95 pp.

Riggs, S. R. 1979. Petrology of the Tertiary phosphate system of Florida. *Econ. Geol., 74,* 195–220.

Russell, R. T., and N. A. Trueman. 1971. The geology of the Duchess phosphate deposits, northwestern Queensland, Australia. *Econ. Geol., 66,* 1186–1214.

Sheldon, R. P. 1963. Physical stratigraphy and mineral resources of Permian rocks in western Wyoming. U.S. Geol. Surv. Prof. Paper No. 313–B, pp. 49–273.

13

Coal

Coal is a portable climate.

RALPH WALDO EMERSON

Coal is a combustible, stratified organic rock composed largely of altered and/or metamorphosed plant remains mixed with a variable but subordinate amount of inorganic material. Coal is brittle, has a hackly to conchoidal fracture, and is noncrystalline. It is dull to brilliant in luster and, with increasing degree of metamorphism, the color changes from light brown to black, the specific gravity increases from 1.0 to 1.7, and the Mohs hardness increases from less than 0.5 to 3.0 (anthracite coal).

Coal is one of the more important energy resources for the present operation and future development of industrial civilization. It has been used as a fuel for at least 3,000 years and was the foundation for the industrial revolution that began in England in the latter part of the eighteenth century. In North America, coal was discovered in Illinois in 1673 by Joliet and Marquette during the early explorations of the continent by the French. Major production began in the 1800s and increased steadily with demands brought on by the emergence of the steel industry, the development of the railroad steam engine, increased use for general heating, and the growth of electric power. The increase in the rate of use lessened in the twentieth century, largely because oil and gas supplanted coal for many purposes. Since 1970, coal has supplied about 20% of America's energy needs, despite a continual growth in the total amount of energy consumed. Because of the decline in oil reserves of the United States, it seems probable that coal will become more important as an energy fuel in this country during the final part of the twentieth century. It has been estimated that nearly 90% of the world's recoverable reserve of energy from fossil fuels lies in coal and lignite.

Coal-bearing rocks underlie 13–14% of the land area of the United States and are present in 39 states (see Figure 13–1). The United States contains about 20% of the world's coal reserves, enough to supply our energy needs for several hundred years at the present rate of usage (Hubbert, 1971).

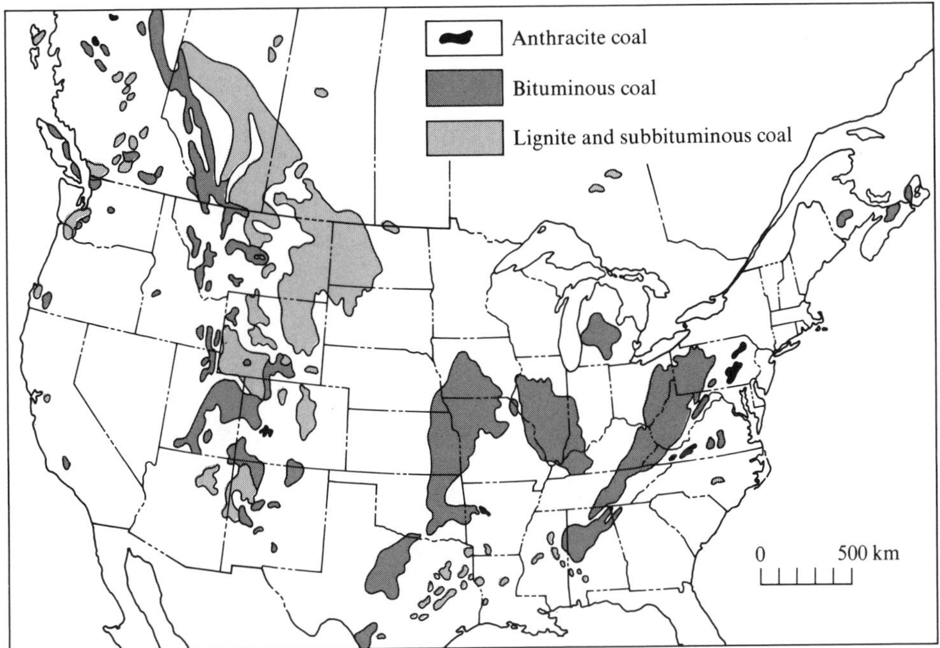

Figure 13–1
Coalfields of conterminous United States and southern Canada. [R. F. Flint and B. J. Skinner, 1977, *Physical Geology,* 2nd ed. (New York: Wiley).]

FIELD OBSERVATIONS

Stratigraphy

Most coal, but not all, is of continental origin so that, although coal occurs in rocks as old as early Proterozoic, it did not become widespread until the development of woody land plants in the Devonian Period. Pre-Devonian coals are formed of algal remains. Coal-bearing rocks are most abundant in sediments of Carboniferous age (see Figure 13–2), but the reasons for this are unknown. Presumably tectonic, eustatic, climatic, and biological factors all played a part.

Coal beds are commonly found associated with shales and fine-grained sandstones of continental to coastal depositional environments and represent poorly drained swampy to deltaic settings, based on associated sedimentary facies (see Figure 13–3). Nonmarine sandstones that may be associated with the coal consist of angular, fine- to medium-grained sand that is moderately to poorly sorted and cross-bedded, texturally submature to immature. Indigenous marine fossils are rarely present. The sandstones are lenticular in cross section and shoestring-shaped in plan view, as is normal for stream-channel deposits. Accompanying shales are interpreted as slack-water or floodplain deposits.

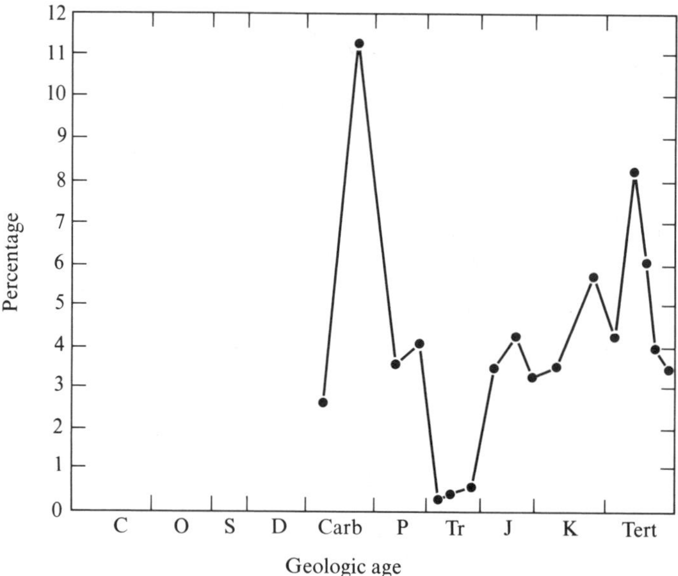

Figure 13–2
Variation in abundance of coal-bearing rocks through Phanerozoic time. [A. B. Ronov et al., 1980, *Sedimentary Geol., 25*.]

Some coal beds are associated with deltaic and shoreline sandstones. The sandstones in such cases have features produced in an environment of high kinetic energy, such as rounding of quartz grains, high percentages of quartz, fossils, or distinctive types of sedimentary structures (Fassett, 1977).

The thickness of a coal seam and the thickness and stratigraphic position of its interleaved shale and sandstone layers are closely related seam properties. Both are determined largely by the degree of subsidence that took place during the accumulation of peat in the basin of deposition. Seam splitting into two or more coal layers tends to occur away from the margin of the basin in response to differential subsidence, either of the basin floor or of the organic overburden. At the margin, seams are thin and commonly contain an appreciable amount of mineral-rich detritus; at the center of the basin, the number of seams and the total thickness of sediment are maximal.

Seam splitting is a common feature of coal-bearing stratigraphic sections (see Figure 13–4). Seams bifurcate into two or more seams, possibly rejoining several meters or kilometers away. This phenomenon can result from several causes, including irregularities in local topography within the swamp, differential compaction of the peat layer, differential subsidence of the depositional basin, transgressions, and regressions. For example, a layer of peat is laid down on a gently sloping surface; the lower part is then covered by mud from a sluggish stream winding its way slowly through the swamp. The stream meanders away, vegetation grows and decays on the

(A)

(B)

Figure 13–3
(**A**) Discontinuous layers of coal about 5–10 cm thick interbedded with sandstone, Petersburgh Formation (Pennsylvanian), Indiana. Continuity of coal seams is frequently interrupted by lenticular sandstones, e.g., upper right. Environment of deposition of coal was apparently a floodplain adjacent to a meandering, low-gradient stream. [F. J. Pettijohn and P. E. Potter, 1964, *Atlas and Glossary of Primary Sedimentary Structures* (New York: Springer-Verlag). Photo courtesy P. E. Potter.] (**B**) Upper Freeport Coal (Pennsylvanian), near Homer City, Pennsylvania. Constant thickness and evenly bedded character of coal indicate undisturbed swamp lacking through-cutting stream channels at this location. Field and petrologic characteristics of Upper Freeport Coal are discussed in E. F. Koppe, 1963, *Penn. Topographic and Geol. Surv. Mineral Resources,* Report M48, 43 pp. [Photo courtesy P. C. Lyons.]

Figure 13–4
Splitting of one coal bed into two by intervening sandstone bed, about 10 m thick, Emery coalfield (Cretaceous), Utah. Splitting of coal seam originates to left, and divergence increases in magnitude toward right. Roadside railing at base shows scale. [Photo courtesy T. A. Ryer, U.S. Geol. Surv.]

newly exposed muddy surface, and the peat that forms is joined on its high-relief end to the original peat deposit. The entire process may occur in a geologic instant, perhaps within a few thousand years.

Banded Coal

The coal seams themselves normally consist of banded or laminated organic materials of differing texture or luster. This layering is attributable to variations in local environments, plant communities, and processes of plant destruction. True bedding features, such as partings formed of inorganic detritus or placers of inorganic impurities, are usually not readily visible. In describing banded coal, the term *banding plane* may be more appropriate than bedding plane.

One of the remarkable features of a coal seam is its wide lateral extent compared to its thinness. A seam less than 1 m thick may extend over many hundreds of square kilometers. Hardly any seam consists of pure coal; nearly everwhere the coal is interlayered with stratified impurities, usually shale partings.

Banded coal seams rarely form more than 1–2% of the coal-bearing section. The seams vary in thickness from paper-thin to tens of meters. In the eastern United States most are less than 3 m thick; an average thickness is about 0.5 m. In the western United States coal beds are typically thicker. Lateral extents of seams also vary. Some have essentially constant thicknesses for thousands of meters before gradually pinching out; others pinch out within a meter or less, and others are truncated laterally by penecontemporaneous stream channels (see Figure 13–5). Sometimes small, rafted fragments of the coal seam can be found within the detritus of the channel fill. Sometimes the coal bed is underlain by a "seat earth" or underclay, a penecontemporaneous kaolinitic soil zone that may be produced either by leaching resulting from root penetration or by downward percolation of meteoric water. This

Figure 13–5
Edge of "washout" of Indiana Coal III (Pennsylvanian), Indiana. Coal bed (1) and underclay (2) terminating abruptly against channel sandstone (3). Rafted stringers of coal (4) at base of sandstone. [F. J. Pettijohn and P. E. Potter, 1964, *Atlas and Glossary of Primary Sedimentary Structures* (New York: Springer-Verlag). Photo courtesy P. E. Potter.]

water was rich in humic and other organic acids derived from its percolation through the decaying plant material that subsequently became the coal bed.

Megascopic description of coal is uncommon in North America although it is used extensively elsewhere (Marchioni, 1980). Although coal-bearing stratigraphic units are described in North American publications, the descriptions emphasize the thickness and position of partings, the thickness and number of coal interlayers, and the lithology of the beds directly above and below the coal seam. Few publications describe the coal itself.

There is a widespread feeling among coal geologists, both industrial and academic, that megascopic descriptions are simply not worth the effort in terms of useful information. Most petrologists place reliance only on petrographic data obtained microscopically. There are several reasons for this tendency:

1. One is the difficulty experienced by petrographers in describing the megascopic variations (lithotypes) even when there is general agreement on the characteristics of these lithotypes. There is a large degree of subjectivity in the identification of

lithotypes and, as a result, reproducibility among various researchers may be poor. However, automated methods of image analysis with reflected-light microscopes promise to be of value in this area.

2. In addition, there are difficulties that involve the condition of the coal being described. Thus, one may deal with fresh coal exposed in a mine or a weathered outcrop or with broken coal in a laboratory from a core or a polished surface. These represent quite different conditions for observation and control the detail in which a description can be made.

3. Yet another problem is the lack of agreement on the physical limitations of the lithotypes or layers that one describes. The International Committee for Coal Petrology (ICCP) suggests a minimum band thickness of 3 mm but also indicates that this minimum thickness may vary from 3 to 10 mm in different countries. For example, in West Germany a minimum band thickness of 10 mm is presently used.

The most commonly used system of megascopic description of coal beds was devised by Stopes (1919), who recognized four types of appearance of fresh coal surfaces (see Table 13–1; Figure 13–6). Examination of the adjectives used to describe (1) vitrain, (2) clarain, (3) durain, and (4) fusain clearly shows the subjectivity and, therefore, lack of reproducibility inherent in the definitions. Words such as "brilliant" and "glossy" (vitrain) must be contrasted with "pronounced gloss or shine" (clarain). To recognize durain, one needs a universally accepted definition of "hard, with a close, firm texture."

In North America, the terms clarain and durain are not always used. Extremely glossy, vitreous fresh surfaces are termed vitrain; dull coal with a silky luster that powders immediately on being touched and soils the fingers is termed fusain. Other types of surfaces (clarain, durain) with varying degrees of luster or brilliance (reflec-

Table 13–1
Lithotypes of Coal as Defined by Megascopic Textural Appearance

Type	Description
Vitrain	Is "a coherent and uniform whole, brilliant, glossy, indeed vitreous in its texture"; occurs in thin, horizontal bands; friable, breaks with a clean, conchoidal fracture.
Clarain	Has a "definite and smooth surface when broken at right angles to the bedding plane and these faces have pronounced gloss or shine: the surface lustre is seen to be inherently banded."
Durain	Is "hard, with a close, firm texture, which appears rather granular to the naked eye; a broken face is never truly smooth but always has a fine lumpy or mat surface."
Fusain	Occurs "chiefly as patches or wedges; it consists of powdery, readily detachable, somewhat fibrous strands"; this is the constituent in coal that soils the fingers and has been described as "mineral charcoal."

Source: Stopes, 1919.

Figure 13–6
Natural fracture surface (cleat) on bituminous coal from Harbour coal seam, Sydney coalfield, Nova Scotia. [Photo courtesy A. R. Cameron, Geol. Surv. Canada.]

tivity) are termed bright or dull *attrital coal.* Attrital coal constitutes the microfragmental matrix that occupies the interstices between visibly discrete coal ingredients. The differences in appearance of the coal lithotypes defined by Stopes (1919) result from differences both in composition of the parent plant material and in degree of diagenetic change (metamorphism). These relationships are poorly understood but are being actively studied by coal researchers.

Nonbanded Coal

Nonbanded coal is less common than banded coal. It may occur as layers within a thick banded coal, or it may compose the entire bed but rarely exceeds 1 m in thickness. Nonbanded coal is completely attrital in origin, having been derived almost entirely from water-transported and size-sorted plant debris (Schopf, 1960). In this respect it differs from banded coal, which is formed from unsorted plant debris of more heterogeneous sizes. For this reason, nonbanded coal is notably uniform in texture and is compact and massive; in outcrop it usually has widely spaced joints (*cleats* in mining terminology) and breaks with a broadly conchoidal fracture. Surfaces may be minutely rough, greasy or satiny, or visibly smooth in texture, and they may show dull to moderately bright luster. On rare occasions, nonbanded coal may contain a large erratic plant fragment transformed to vitrain.

Nonbanded coal is generally very heterogeneous in composition, but the particles of which it is composed are all microscopic in size. This is the cause of the massive, structureless appearance—quite analogous to the appearance of micrite in stratigraphic sections of limestone or dolomite. Because of their very small size, the botanical affinities of the organic particles in nonbanded coals usually can be identified only with ultraviolet fluorescence microscopy but are sapropelic in origin. These coals are termed either *boghead* or *cannel* coals, depending on the dominant microscopic constituent. Boghead coals are composed largely (80–90%) of nonspore algal remains (sapropelic coals), with a noticeable amount of the cutinous or waxy envelopes of a unique type of colonial alga, *Botryococcus brauni* Kuetzing. Ordinarily, the nonwaxy matrix predominates, however. Cannel coals contain a conspicuous percentage of spores. In thin or polished sections of nonbanded coal, the spores are conspicuous because of their distinctive shapes, even though they may form only a small percentage of the coal. Their appearance contrasts sharply with the morphologically unidentifiable groundmass of the coal.

Minerals

Relatively few minerals (contained in the "ash" in a coal analysis) can be considered common or abundant in coal beds, although Stach (1975) lists 45 that have been reported to occur occasionally. Common minerals are the clays illite, kaolinite, and montmorillonite; carbonates siderite, ankerite, calcite, and dolomite; evaporites halite, sylvite, and bischofite; and quartz, pyrite, and various iron sulfate hydrates such as melanterite ($FeSO_4 \cdot 7H_2O$) that form from the pyrite on exposed surfaces. These hydrates form a whitish, bluish, or pinkish stain on the coal surfaces. Some of the quartz may have resulted from crystallized opal of silica-secreting plants such as *Calamites,* but the amount of such quartz in coal has not been determined.

LABORATORY STUDIES

Petrographic studies of coal were first made in 1883 from thin sections, but today are made most commonly from polished sections of crushed coal particles that have been mounted in an epoxy cement. The technique of examination is very similar to that used in the study of opaque minerals in silicate or sulfide rocks; a qualitative or quantitative evaluation is made in reflected light of the polished surface or parts of the surface. From the reflectivity of the surface and the morphology of the constituents that compose the coal (the *macerals*), plus chemical analysis, three determinations are made: (1) rank, (2) grade, and (3) type.

1. The *rank* of the coal is its position in the series: peat, lignite, subbituminous coal, bituminous coal, anthracite coal, and meta-anthracite coal. Rank is established by determining the reflectance or the chemistry of the coal. Coals of higher rank have higher reflectivities.
2. The *grade* of the coal is determined by chemical analysis and refers to its content of impurities such as ash; higher grades of coal have fewer impurities.
3. The *type* of coal refers to the kinds of particles (macerals) of which the coal is composed. The macerals are the equivalent of minerals in inorganic rocks.

Thin sections are not so versatile as polished sections for petrographic studies and are difficult to prepare because of the high degree of opacity and brittleness of coal. Coal must be ground to a thickness of about 5 μm before it will transmit a usable quantity of light for petrographic purposes. Special equipment and great care are required to manufacture such thin slices of a brittle rock. A further disadvantage of thin sections is that they are inappropriate for the determination of reflectivity. However, both thin sections and polished sections do reveal the types of macerals in the coal and contribute toward the determination of the environment of deposition and facies relationships.

Macerals

The term *maceral* (from the Latin *macerare,* meaning "to soften") was suggested by Stopes (1935) as a parallel to the term *mineral,* used for inorganic rocks. Macerals are organic units, and because some macerals consist of more than one type of plant tissue, they are sometimes more analogous to lithic fragments in silicate rocks than to mineral grains. Stopes also proposed that the names of macerals end with the suffix *-inite;* for example, vitrinite is normally the most abundant maceral in the coal lithotype vitrain.

The International Committee for Coal Petrology (1971, unpaginated looseleaf book) states:

> Coal macerals evolve from the different organs or tissues of the initial coal-forming plant materials during the course of the first stage of carbonification. However, because of variable but severe alteration, it is not always possible to recognize the starting materials.

Macerals are defined by their colors in reflected, transmitted, and fluorescent light and by their morphology, size, shape, botanical affinity, and occurrence. The ICCP

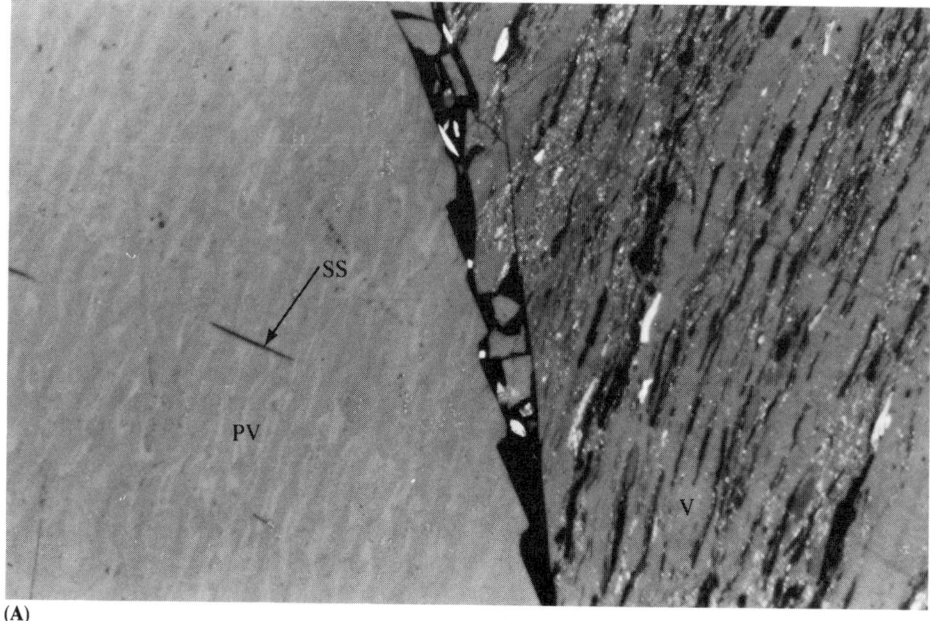

(A)

Figure 13–7
Vitrinite-group macerals. (**A**) Polished fragments of coal seen in reflected light, showing contrast in reflectance and internal structure between vitrinite (V) and pseudovitrinite (PV). Pseudovitrinite has higher reflectance and well-developed cellular structure and shows characteristic serrated edges that result from fracture of very brittle pseudovitrinite. It also

concept implies that the properties (reflectance, hardness, chemistry) of a maceral change with coal rank although the name of the maceral does not.

A significantly different definition for the term maceral was proposed by Spackman (1958) and is used in North America. According to Spackman:

> [Macerals] are organic substances, or optically homogeneous aggregates of organic substances, possessing distinctive physical and chemical properties, and occurring naturally in the sedimentary, metamorphic, and igneous materials of the earth.

In this definition, macerals are substances with distinctive physical and chemical properties. Thus, vitrinite in bituminous coal is a different maceral from vitrinite in anthracite coal because the two have different sets of properties, e.g., reflectances and contents of fixed carbon. In the ICCP system, these two macerals might well be given the same name.

Coal petrographic analyses consisting of a maceral (compositional) analysis and a reflectance (rank) analysis are routinely performed in academic, industrial, and commercial laboratories. Such analyses permit a coal to be characterized well enough

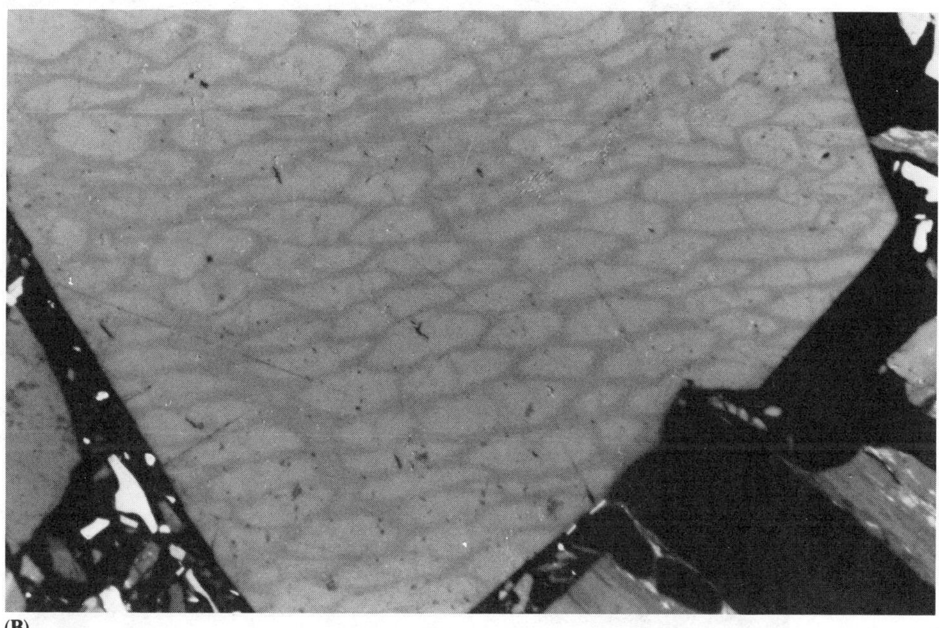

(B)

frequently contains slitted structures (SS), oriented normal to long cell dimension in this specimen. Vitrinite (right half) is full of sporinite, the dark, elongate material (see detail in Figure 13–8C.) (**B**) Pseudovitrinite, showing large-scale cell structure only faintly visible in part A. Width of photos is 0.34 mm. [Photos courtesy J. C. Crelling.]

that its behavior in some industrial processes can be predicted. For practical work (industrial or academic) only approximately 8–9 macerals are identified in a maceral analysis.

Macerals are divided into three groups (see Figures 13–7, 13–8, and 13–9) based largely on the part of the plant from which the maceral originated and the degree of degradation of the plant material.

Vitrinite Group

The individual macerals in this group originate in woody and cortical tissues (wood and bark). The vitrinite group forms 50–90% of the macerals in most North American coal seams and is the primary constituent of bright coal. There are two major types: If the cellular structure of the woody tissue can be distinguished in a polished section or by etching techniques, the maceral is called *telinite*. If structureless, it is called *collinite*. True collinite is the colloidal material that dominates in all bright coals; however, collinite may also be composed of compressed cell walls. Fragments of vitrinite smaller than 10 μm in diameter are called *vitrodetrinite*.

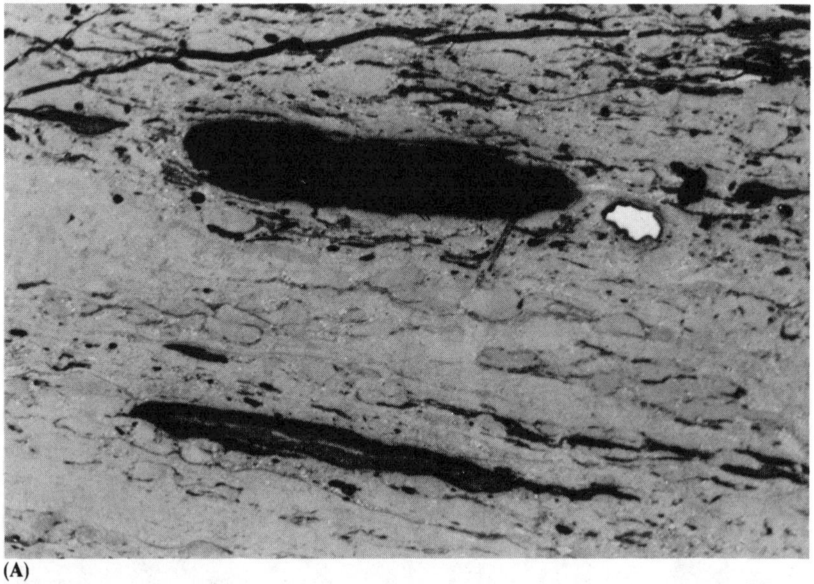

(A)

(B)

Figure 13–8
Exinite-group macerals. (**A**) Polished fragments of coal seen in reflected light, showing resinite (dark, oval body in upper half) and sporinite (crushed, elongate body with central light-gray silt) in vitrinite matrix. Small fragment of highly reflectant macrinite occurs immediately to right of resinite maceral. Width of photo is 0.34 mm. [Photo courtesy J. C. Crelling.] (**B**) Resinite in vitrinite. White material in lower left is fusinite (see Figure 13–9A, B). Width of photo is 0.34 mm. [Photo courtesy J. C. Crelling.]

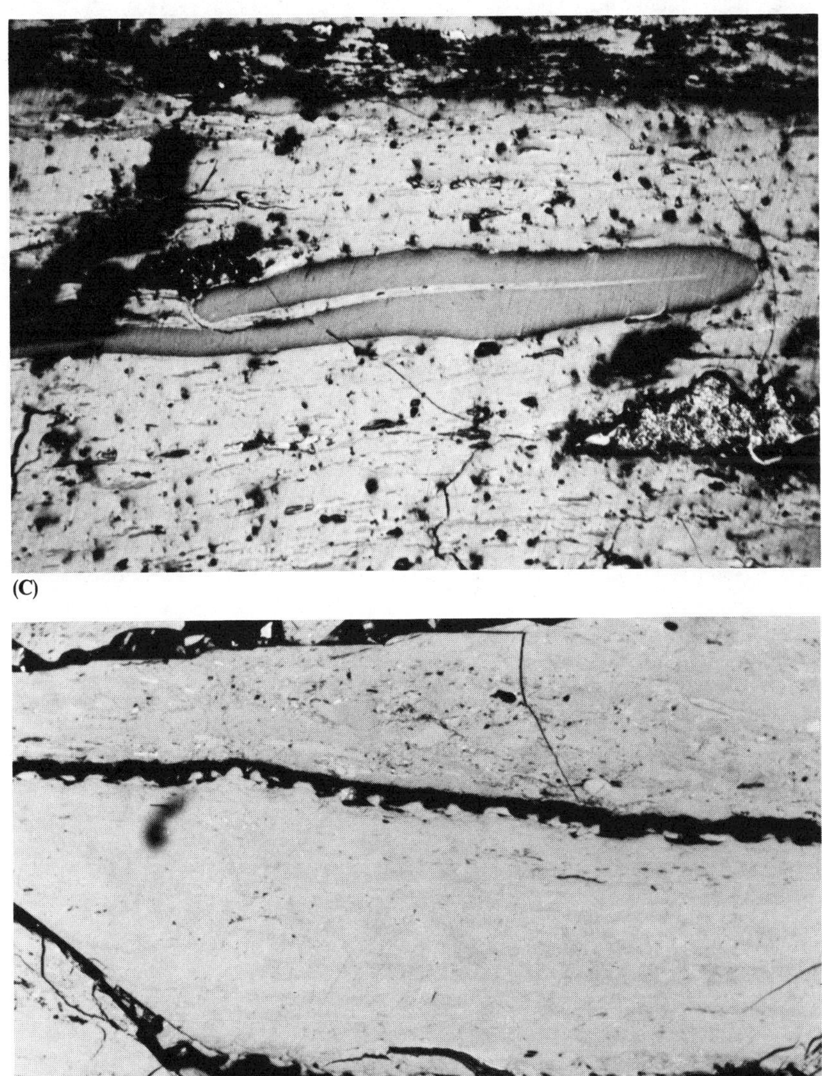

(C)

(D)

(**C**) Sporinite (center) and fusinite (left, upper half) in vitrinite matrix. Black material is epoxy; its formless character indicates that it is not resinite or cutinite (see part D), which are also black in reflected light. Width of photo is 1.0 mm. [Photo courtesy P. C. Lyons.] (**D**) Cutinite (dark, serrated stringers) in vitrinite. Cutinite is on outer surface of a leaf, root, or stem cross section. Width of photo is 0.34 mm. [Photo courtesy J. C. Crelling.]

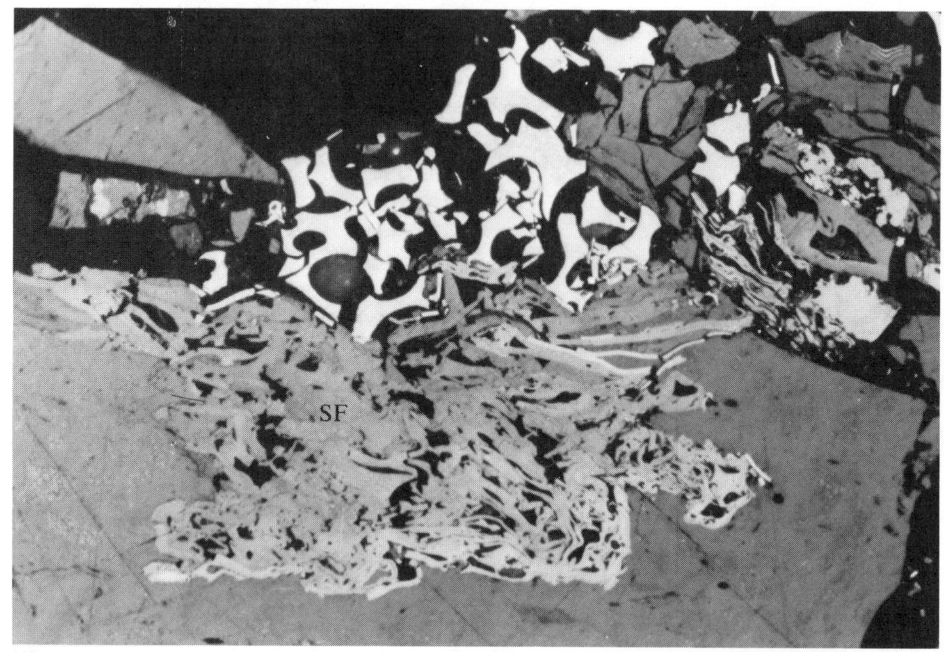

(A)

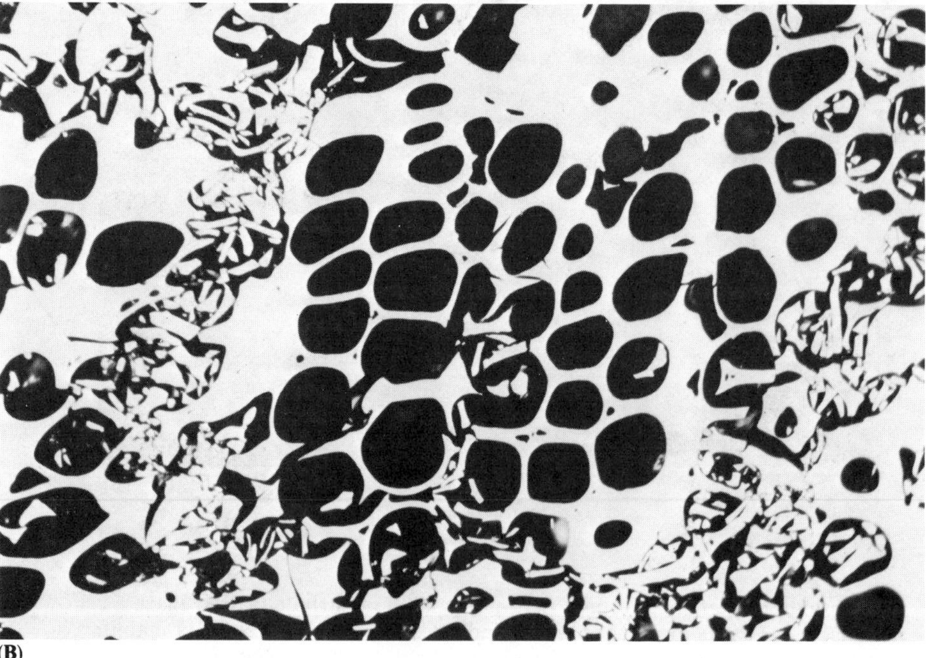

(B)

Figure 13–9
Inertinite-group macerals. (**A**) Broken fragments of fusinite (white in upper center) underlain by semifusinite (SF) within larger vitrinite fragment. (**B**) Large, unbroken, cellular fusinite in center, bordered on left and right by granulated fusinite and other, unbroken fusinite macerals.

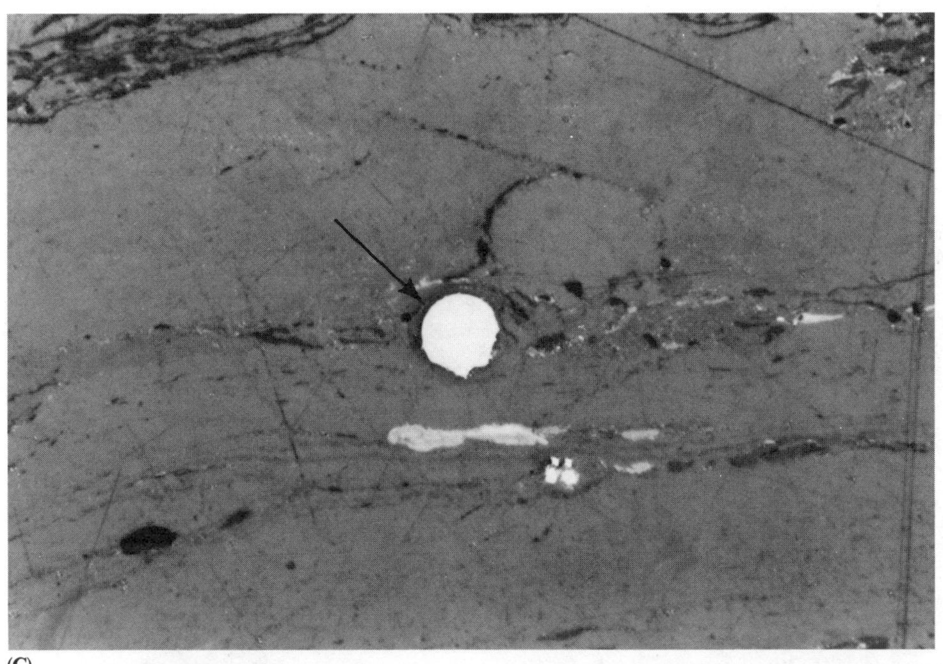

(C)

(D)

(**C**) Macrinite maceral (white) in vitrinite. Note absence of internal structure in macrinite. (**D**) Large fragment in upper center contains light-gray, finely granular micrinite and darker-gray vitrinite. Macrinite occurs in extreme upper right (white). Most of the smaller fragments are vitrinite or noncellular pseudovitrinite.

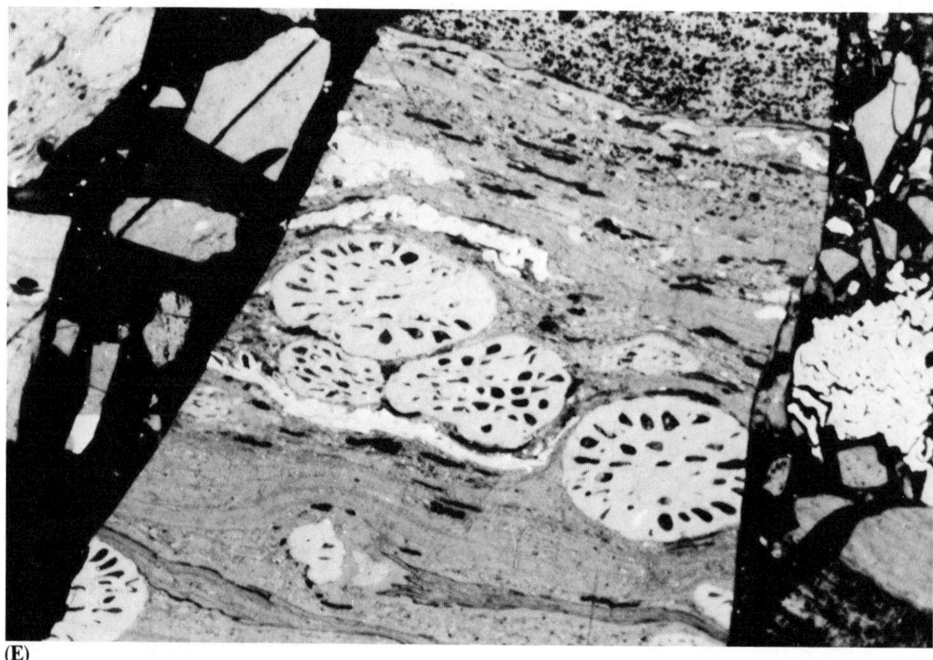

(E)

Figure 13-9 (*continued*)
(E) Round and cellular sclerotinite in vitrinite. White fusinite in center right edge; other fragments are vitrinite. Width of photos is 0.34 mm. [Photos courtesy J. C. Crelling.]

Exinite or Liptinite Group

The macerals in this group originate in the waxy and resinous parts of plants, such as spores, needle and leaf cuticles, and wound resins. Members of this group form 5–15% of most North American coal seams. The major categories of the exinite group are:

1. *Sporinite*. The fossil remains of spore exines, flattened parallel to the stratification; normally the interior of the spore can be recognized only as a very thin slit. Spores range in size from a few micrometers (microspores) to about 5,000 μm (megaspores).

2. *Cutinite*. Formed from cuticles. They occur as narrow bands frequently having one serrated edge.

3. *Resinite*. The remains of plant resins and waxes. This maceral occurs as round and oval bodies and as inclusions in collinite and as filling of cell lumens and fractures.

4. *Alginite*. The remains of algal bodies, the principal exinite constituent of boghead coal. They have a serrated oval shape.

5. *Liptodetrinite*. A structureless member of the exinite group too small in size to be assigned with certainty to one of the other four maceral types.

Inertinite Group

These macerals are fungal remains, oxidized wood or bark, plant fragments of uncertain origin, or fungal remains. Most are biochemically altered or fire-charred, woody, cell wall tissue.

1. *Fusinite.* The maceral that forms fossil charcoal (the coal lithotype fusain). It always shows a cellular structure composed of carbonized or oxidized cell walls and hollow lumens. The lumens are commonly mineral-filled.
2. *Semi-Fusinite.* The transition stage between fusinite and vitrinite; has a lower reflectance than vitrinite.
3. *Sclerotinite.* The remains of fungal sclerotia or altered resins. Its oval shape and internal complexity make this maceral easy to recognize.
4. *Micrinite and Macrinite.* These macerals are completely structureless. Micrinite is granular with particles smaller than 10 μm in size and occurs as inclusions in dull coal and in bright coal. Macrinite particles are larger than 10 μm and may be the principal component of durain.
5. *Inertodetrinite.* Finely comminuted, structureless inertinite.

More complete descriptions of the petrographic appearance and physical properties of these macerals, as well as further classification of these maceral types into submacerals, varietal macerals, and cryptomacerals, are given by Crelling and Dutcher, 1980; Stach, 1975; Van Krevelen, 1961; and the International Committee for Coal Petrology, 1963.

COAL RANK

As noted earlier, rank refers to the position of a coal within the metamorphic series: peat, lignite, subbituminous coal, bituminous coal, anthracite coal, and meta-anthracite coal. The petrographic and chemical characteristics of these six diagenetic or metamorphic stages of plant alteration are not defined very precisely, and the boundaries between successive stages in the higher ranks are commonly given by the percentages of fixed carbon or volatile matter in the rock. The Btu content is commonly used for low-rank coals. Volatile matter in a coal analysis is defined as those substances, other than moisture, that are given off as a gas during heating in the absence of air (baking or carbonization) in a standardized laboratory analysis. This gas contains some carbon (as carbon dioxide and other gases) so that a distinction is made between the terms *carbon* and *fixed carbon.* Fixed carbon is a value calculated from a chemical analysis and is the carbon *not* driven off during the volatile-matter analysis; it is the carbon remaining in the macerals. Fixed carbon plus volatile matter plus ash plus moisture totals 100% of the coal material. Considerable care is required in the interpretation of *x–y* graphic plots of the results of coal analyses. Sometimes the axes are, for example, "carbon" and "volatile matter," both of which contain the element carbon as a component. Sometimes the axes give ash-free percentages; other times the percentages refer to the amounts in the whole coal. And, of course, the analysis may be reported on either a moisture-free or a moisture-containing basis. Further, sometimes the analysis is of the megascopic lithotype vitrain and sometimes of the microlithotype vitrite.

Microlithotype is a term applied in microscopic studies. All microlithotype terms end in *-ite*. *Vitrite* is composed almost entirely of the maceral group vitr*inite*. There also exist *clarite,* which consists predominantly of vitrinite-group and exinite-group macerals; *durite,* consisting largely of exinite-group and inertinite-group macerals; and *fusite,* composed largely of the fusinite maceral. Clearly, similar-sounding terms do not have identical meanings, and some terms used by coal petrologists have partially overlapping meanings. Pitfalls in the usage of coal terminology in petrologic studies are abundant, and great care is required to avoid falling into a trackless morass.

Living terrestrial plants contain about 50% carbon, the remainder being oxygen, hydrogen, nitrogen, sulfur, and other elements in lesser amounts. Peat contains 50–60% carbon, lignite 60–70%, subbituminous coal 70–78%, bituminous coal 78–90%, semianthracite coal 90–92%, anthracite coal 92–98%, and meta-anthracite coal > 98%—all on a dry, ash-free basis.

Because coal is a complex material, it is best if the rank of a coal can be measured on one specific entity. This avoids the interactions that are bound to occur when "whole coal" is examined. Petrographically, the changes during coalification of bituminous and anthracitic coals are mainly characterized by an increasing similarity in appearance of the various macerals, which originally differ markedly in their microscopic appearance. This convergence in aspect is difficult to quantify, however. One feature that is easily quantifiable petrographically is the reflectance of macerals in a polished section. Vitrinite undergoes a gradual and progressive increase in reflectivity during coalification in bituminous coal and, for this reason, vitrinite reflectance is used routinely by coal petrographers to estimate the rank of coal. As noted earlier, vitrinite is by far the most abundant maceral in most coals and coaly materials, so that it is nearly always present in such rocks. The increase in reflectivity of liptinites and inertinites is less regular than that of vitrinites, and both groups of macerals are less abundant than vitrinites in rocks. The relationship between vitrinite reflectance and coal rank is shown in Figure 13–10.

Causes of Increase in Coal Rank

There are several generalizations that have been made from field observations concerning the causes of increase in coal rank. These are that older coals are higher in rank (see Figure 13–11), coals at greater depth are higher in rank—Hilt's rule, and coals in orogenic belts or near igneous intrusions are higher in rank (see Figure 13–12). From these observations it has been concluded, and subsequently confirmed in laboratory experiments, that temperature and time are the main factors that cause

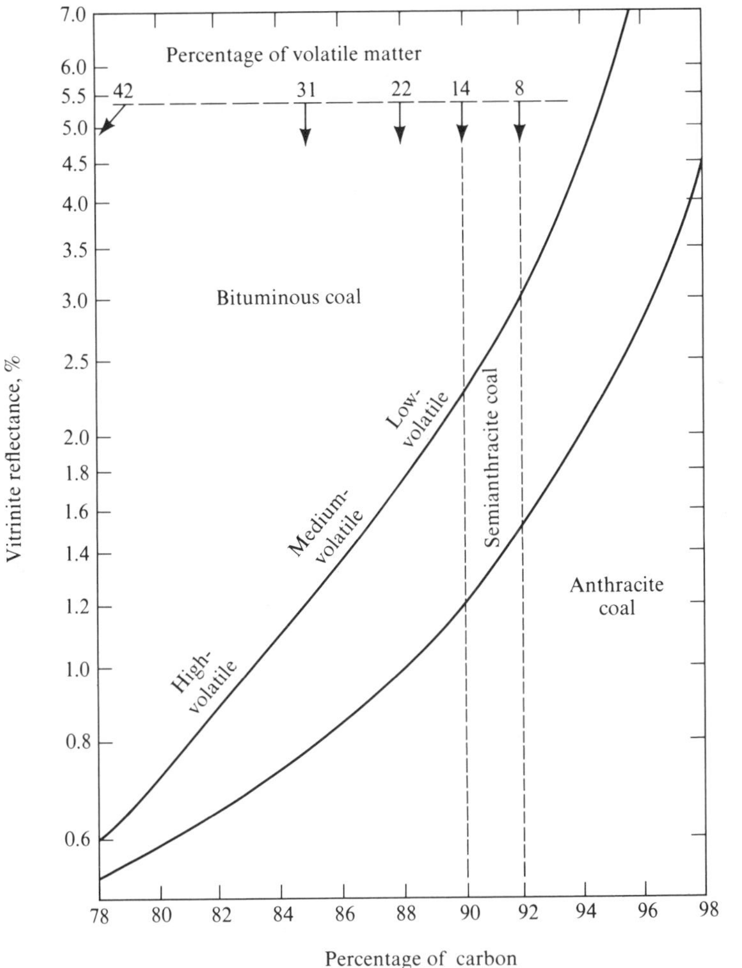

Figure 13–10
Correlations among percentage of carbon and volatile matter (dry, ash-free basis), vitrinite reflectance, and coal rank. Solid lines are boundaries of coal ranks. In terms of percentage of *fixed* carbon, lower limit of semianthracite coal is 86%; low-volatile bituminous coal, 78%; medium-volatile bituminous coal, 69%; and high-volatile bituminous coal, 54% (Amer. Soc. for Testing and Materials standards). Solid lines indicate range in percentage of carbon and vitrinite reflectance reported by various investigators.

increase in coal rank. Time is important because a suitable temperature must be maintained for extensive periods before coalification is achieved (see Figure 13–13). The temperatures necessary for the formation of bituminous coal are commonly no higher than 100–150°C. Thus the rank of coal may serve as a relatively sensitive geothermometer in a low temperature range. This has great practical importance in prospecting for hydrocarbons because the generation and maturation of petroleum and

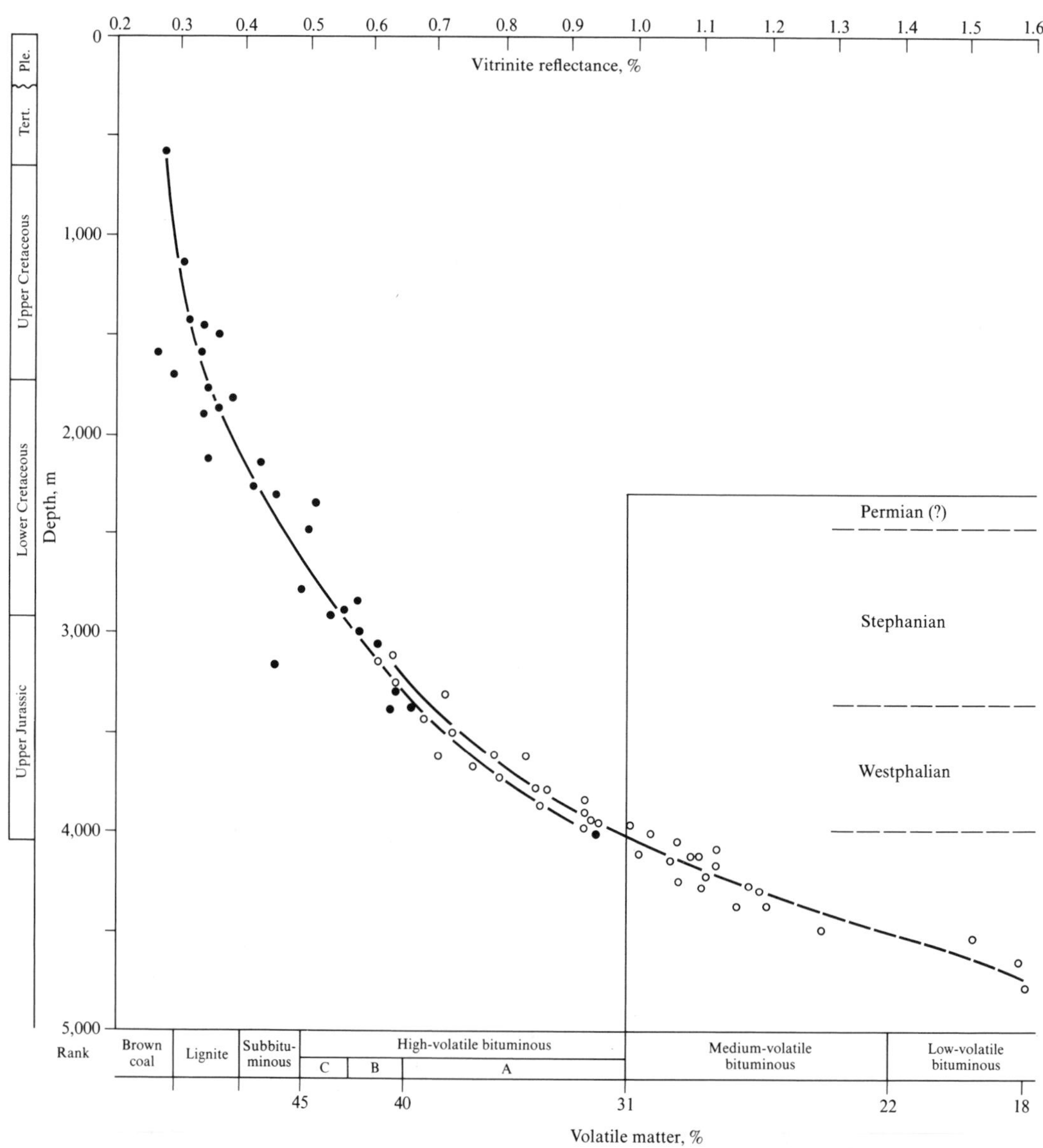

Figure 13–11
Progressive coalification with depth (dry, ash-free basis) in Mesozoic and Carboniferous coal sequences in Maritime Provinces, Canada. "Brown coal" is term used for transitional stage of coalification between peat and lignite. [P. A. Hacquebard, 1975, *Colloque International de Pétrographie de la materière organique et potentiel petrolier* (Paris: Centre National de la Recherche Scientifique).]

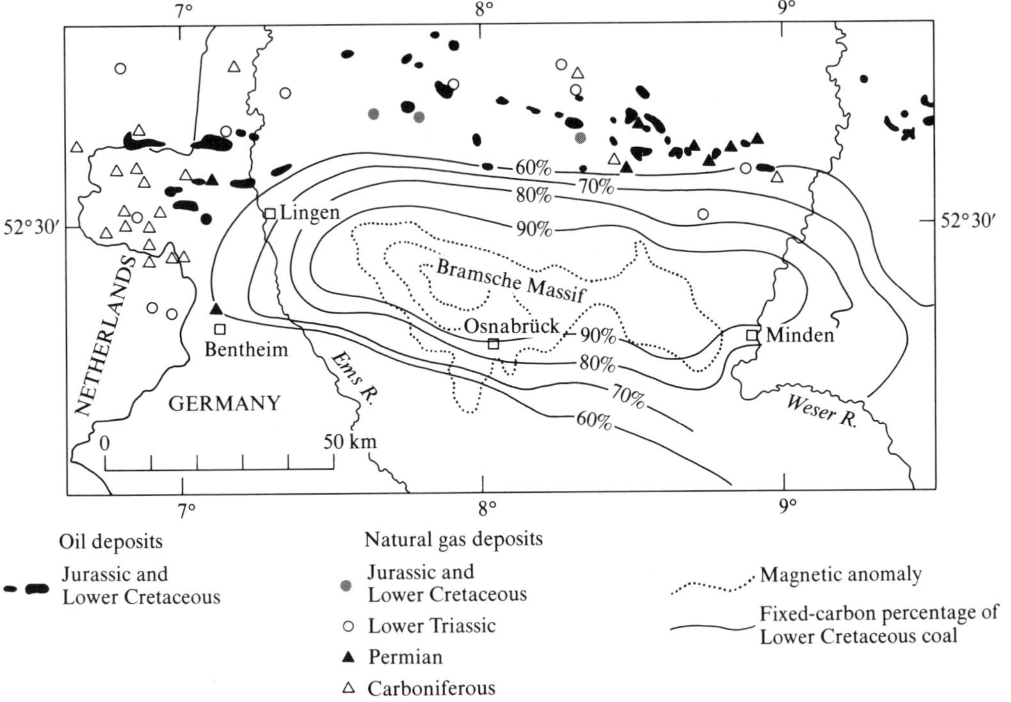

Figure 13–12
Brahmsche Massif of northwestern Germany, a deeply buried mafic igneous intrusion, is well delineated by a magnetic anomaly. Paralleling magnetic contours are contours of coal rank, which increase toward center of anomaly. Directly above center of intrusion, coals are anthracitic. Oil fields are restricted to coal-carbon percentages of 60 or less, a result of upward migration after generation. Isoranks of coal are interpreted as isotemperature surfaces during cooling of intrusion. [Teichmüller and Teichmüller, 1968.]

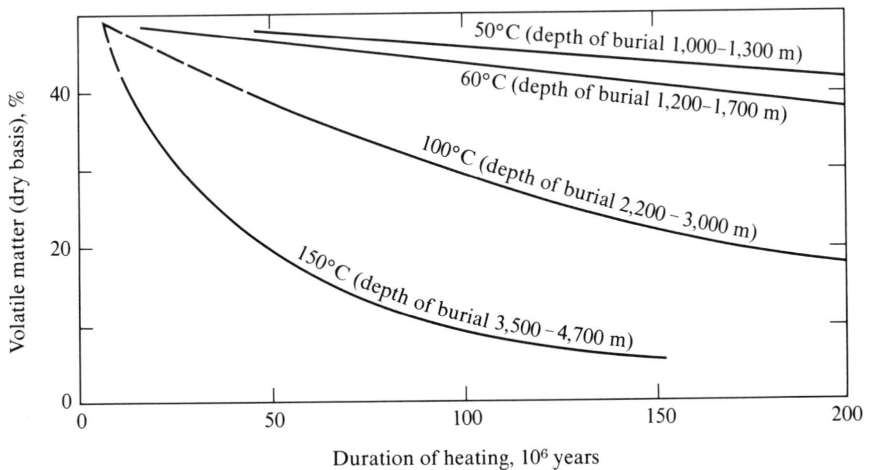

Figure 13–13
Relationship between coal rank and coalification time at four temperatures. [Teichmüller and Teichmüller, 1968.]

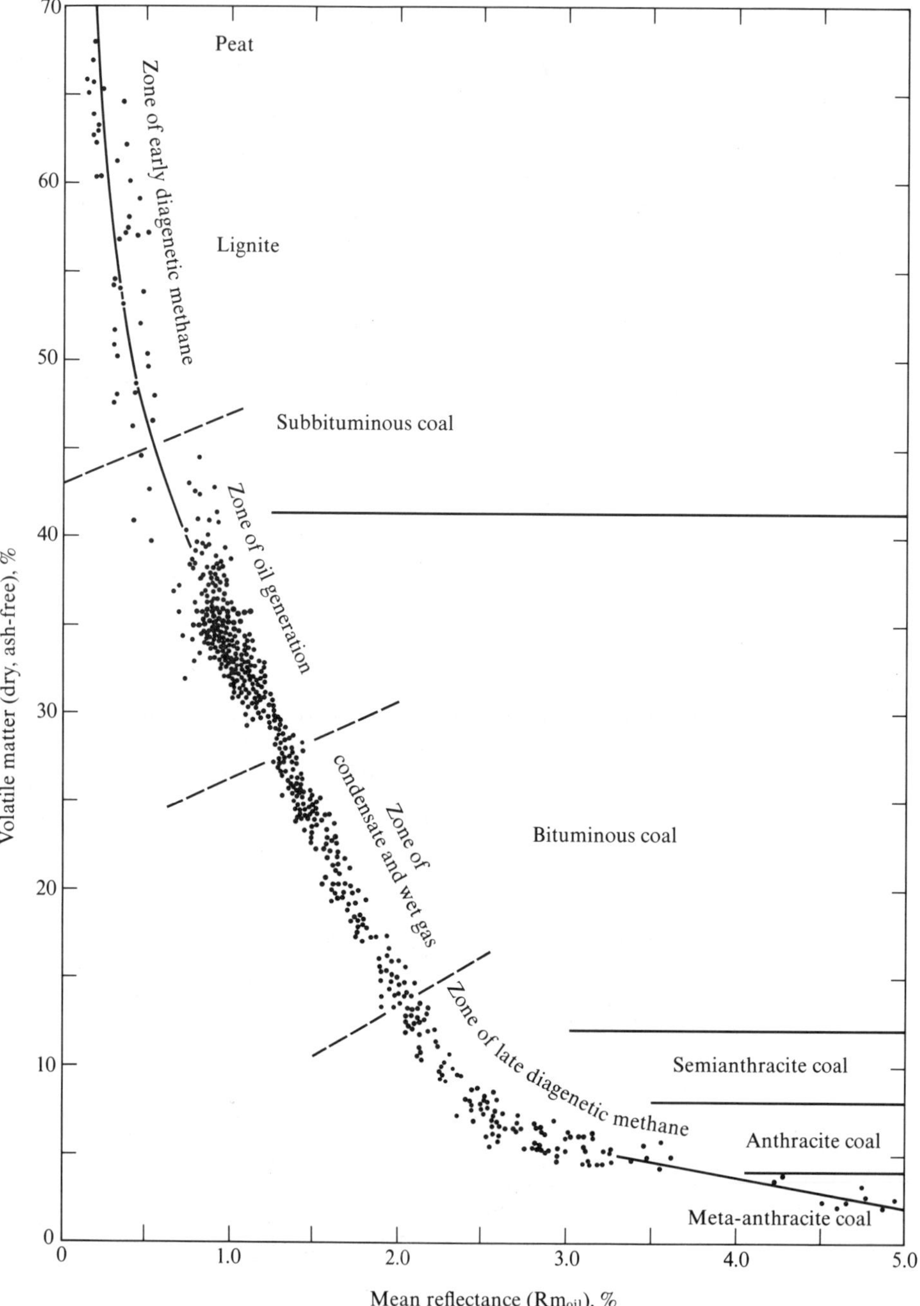

Figure 13–14
Relationship between volatile matter, coal rank, vitrinite reflectance, and generation of oil and gas. Data points are for coal deposits in West Germany. [Stach, 1975; Teichmüller and Teichmüller, 1979.]

natural gas also depend on the temperature to which the source rocks have been subjected (see Figure 13–14).

The rate at which coal rank increases with depth depends on the geothermal gradient and on the heat conductivity of the surrounding rocks. For example, in the flat-lying Tertiary sediments of the Upper Rhine graben, the geothermal gradient is very variable locally. In areas where the gradient is 7–8°C/100 m, bituminous rank is attained at a depth of 1,500 m; while in areas where the gradient is only 4°C/100 m, bituminous rank is not attained until a depth of 2,600 m (Stach, 1975). In stratigraphic sections characterized by alternations of predominantly sandy and shaly formations, the gradient of rank variation is less pronounced in the sandy formations. This results from the fact that the sands conduct heat better than the shales, so that the thermal gradient is less steep in sandy than in shaly formations (see Figure 13–15).

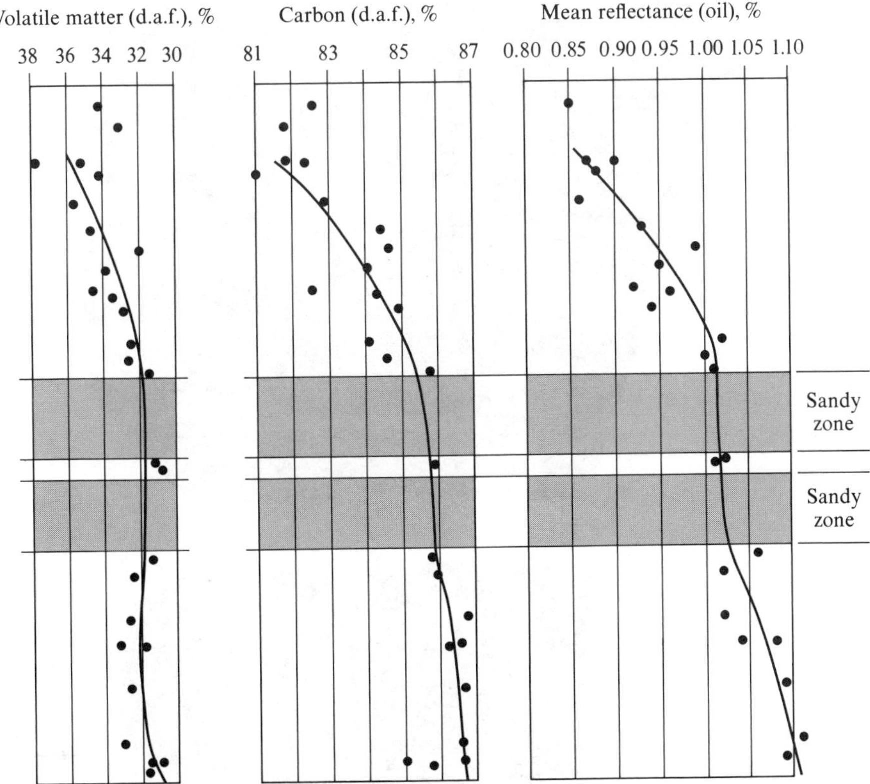

Figure 13–15
Increase in coal rank with depth in Tuefelspforte borehole, Saar district, West Germany, based on analyses of microlithotope vitrite. Note break in trend of coalification tracks in sandy zone. d.a.f. = dry, ash-free. [Teichmüller and Teichmüller, 1968.]

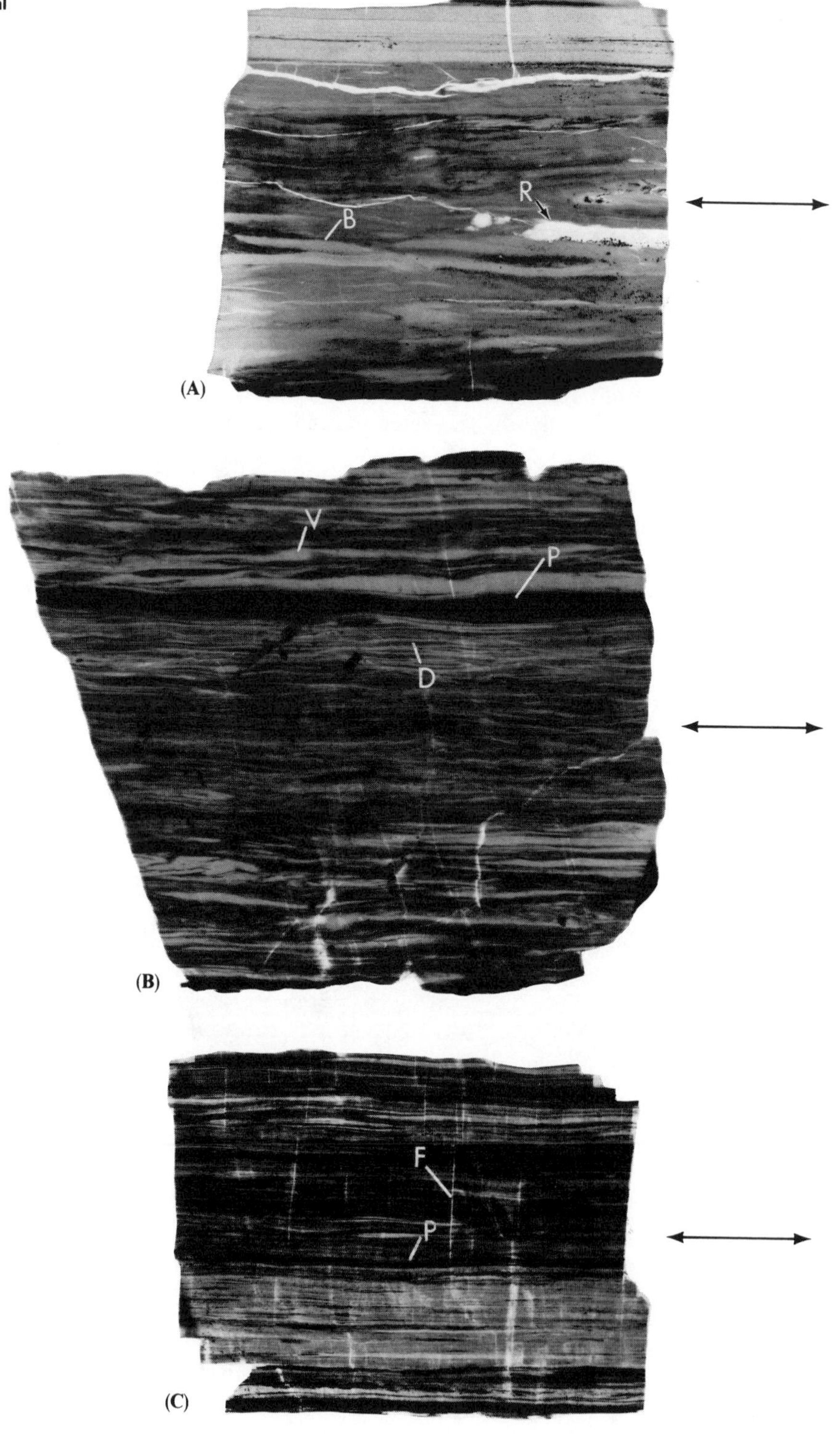
B
R
(A)
V
P
D
(B)
F
P
(C)

(D)

Figure 13–16
Effect of pressure on perfection of orientation in lignite, bituminous coal, and anthracite coal, as seen in X-radiographs of thin slabs. (**A**) Tertiary lignite with roots or branches (B) in huminite-rich matrix with varying admixtures of mineral matter producing darker tones. Large resin body (R) is transparent to X-rays. (**B**) Pennsylvanian high-volatile bituminous coal, showing greater degree of compaction than lignite, with pronounced banded fabric defined by vitrain (V) and durain (D) bands. Note difference in coal fabric above and below clay parting (P). Subvertical fractures are face cleats. (**C**) Pennsylvanian low-volatile bituminous coal with numerous minor clay partings (P) and well-developed face-cleat system (F). Lighter-tone lower portion of sample contains less mineral matter. (**D**) Pennsylvanian banded anthracite possessing numerous spherical mineral-matter bodies (S), which are randomly distributed. Large vitrain band (V) is relatively free of mineral matter and is likely a branch or root. Scale bar is 1 cm. [Vinopal, 1979. Photos courtesy R. Vinopal.]

Tectonic pressure seems to be of minor importance as a cause of increase in coal rank. For example, the more highly folded parts of coal basins do not invariably contain coals of higher rank than the less folded parts. In some basins, where a correlation between folding and rank has been observed, the rank variation can be explained more plausibly as a result of greater temperature than as a result of tectonic pressure. In other basins, it can be demonstrated that the rank of the coal was established before the folding took place. The major effect of pressure on coalification occurs in the transition from peat to lignite, where overburden pressure rapidly reduces porosity and moisture content by perhaps 50%—Schürmann's rule. For example, in a Macedonian peat profile, approximately 200 m thick, it was found that moisture decreases from 89% to 69%. In the peats and lignites in boreholes near Cologne, West Germany, the moisture content of the organic accumulation decreases smoothly from 65% with an overburden thickness of 20 m to 50% at a depth of 330 m. The increase in amount of overburden and pressure that accompanies the transition from lignite to anthracite causes the other half of the total porosity and moisture loss, together with an increase in the perfection of orientation of the layers of inequant organic particles (see Figure 13–16). Bituminous and anthracite coals contain less than 10% moisture (Teichmüller and Teichmüller, 1968).

Rapid tectonic movements occurring along faults may occasionally cause a local rise in coal rank because of a concentration of frictional heat along the fault surface. More generally, however, tectonic movements take place so slowly that frictional heat is lost before it can affect the coal. The effect of intense shearing stress is clearest during the metamorphism from anthracite coal to graphite. Although this occurs at 600–800°C (depending on the duration of heating) in the absence of shear, the transformation may occur at temperatures as low as perhaps 300°C under shear (Stach, 1975).

Diagenetic Stages of Siliceous Rocks

Many sandstones and mudrocks contain coaly particles (i.e., vitrinite), and these can be used to determine the maximum temperature to which the siliceous rock unit was subjected (Bostick, 1979). The correlation between coal rank and specific minerals that form during the diagenesis of silicate rocks is, however, very imprecise. The reason is that, in contrast to coalification, the recrystallization of minerals depends not only on temperature, pressure, and time, but also on the total salinity and specific ion content of pore waters. As a result, most chemical reactions involving silicate minerals in sedimentary rocks can occur at any diagenetic temperature, for example, cementation of sandstones or intrastratal solution of heavy minerals. The most reliable correlations between coal rank and silicate-mineral diagenesis are between the formation of subbituminous coal and the illitization of interlayered smectite–illite clays (50–100°C) and the initiation of the zeolite facies (above 150°C) after anthracite has formed but before graphitization.

PLATE TECTONICS

Both temperature and precipitation are strongly affected by latitudinal factors, so that it is possible that the presence of coal beds in ancient rocks might give information about former positions of the continents. Coal deposits originating from terrestrial plants are initiated in deltaic environments, coastal swamp–marsh complexes, and alluvial flood basins separated by major river courses. All of these settings either are covered with very shallow water or are peat bogs in which the groundwater table is at the surface. Clearly, a humid climate is required.

As a temperature indicator, coal is useful only if supplemented with botanical evidence. Modern peat is generated in high latitudes (e.g., Alaska) as well as in temperate and tropical climates. The character of the wood, the distinctness of growth rings, the types of leaves, and the types of spores may be used as temperature indicators. By the use of such evidence it has been established, for example, that the Carboniferous and Permian coals of the Northern Hemisphere, formed within 15° of the equator according to paleomagnetic evidence, indicate a tropical climate. Growth rings are indistinct or absent, reflecting a warm and uniform climate. The Permian Gondwana coals, on the other hand, contain abundant gymnosperm woods with distinct growth rings, indicating a cool climate with marked seasons. Paleomagnetic data indicate these coals formed at latitudes 50–75°S. Modern peat accumulations in Alaska occur in the same latitude belt in the Northern Hemisphere and in association with glacial deposits—an association very similar to the Gondwana deposits. It is apparent, therefore, that the stereotypical hot, steamy climate is not required for coal formation, although such a climate may well result in thicker and more extensive coals.

A further complication in the interpretation of ancient global temperature patterns is the fact that the latitudinal widths of the temperature belts can vary without the need for continental drift. This is well shown by the repeated advances and retreats of glaciers within the past few million years, during which time latitudinal continental drift has been insignificant. At present, the mean annual temperature range between the equator and the polar regions is about 50°C, from +30°C to −20°C. During past geologic periods, however, the gradient was apparently much smaller, and subtropical climates such as those in the southeastern United States extended into New England and western Canada, with no help from continental drift. For example, bauxite pebbles of Early Tertiary age have been found in Pleistocene deposits in Massachusetts, and crocodile fossils of the same age have been found in New Jersey. The flora of the London Clay (Eocene) in England is that of a tropical rainforest, and the Late Eocene to Early Oligocene forests of British Columbia and southeastern Alaska contain palms, which indicate a subtropical climate. These data make it abundantly clear that coal-forming swamps may have existed at almost any latitude during past geologic periods, regardless of the positions of the landmasses during the time (see Figure 13–17).

ANCIENT COAL DEPOSITS

Fruitland Formation (Cretaceous), New Mexico and Colorado

One of the more extensive areal investigations of a coal sequence was made by Fassett and Hinds (1971) in their study of the Fruitland Formation in New Mexico and Colorado. The Fruitland Formation is contained within the San Juan Basin (see Figure 13–18) and is part of an Upper Cretaceous section about 2,000 m thick composed of shallow marine and continental rocks. The Fruitland Formation is in the upper half of this section and ranges from zero to 160 m in thickness, averaging about 100 m. Its lower contact with the Pictured Cliffs Sandstone is conformable and partially intertonguing; the lowermost coal bed in the Fruitland is used as the base of the unit. The contact of the Fruitland with the overlying Kirtland Shale is mostly conformable and is taken as the top of the highest coal bed or carbonaceous shale (see Figure 13–19).

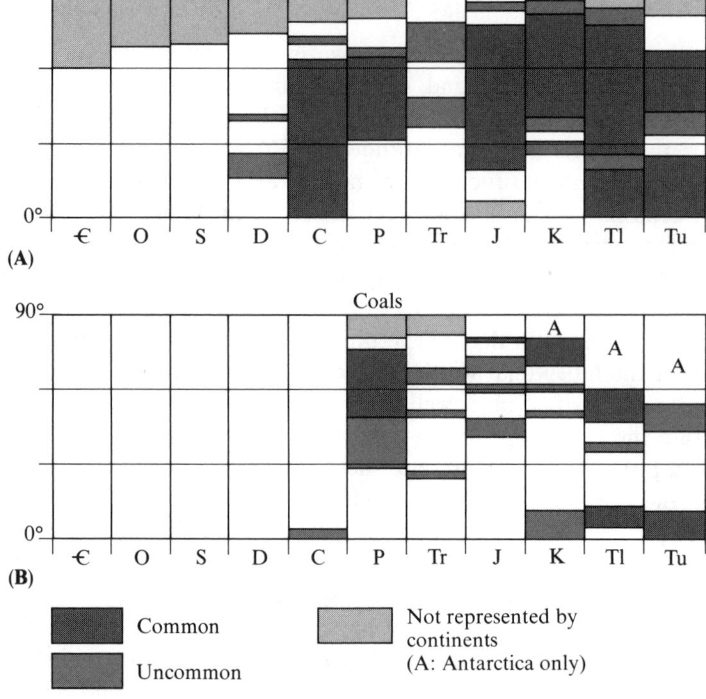

Figure 13–17
Relationship among abundance of coal, paleolatitude, and geologic time. (**A**) Northern Hemisphere. (**B**) Southern Hemisphere. Coal is more abundant in Northern Hemisphere not only because most of Earth's land area is located there, but also because southern continents were near polar region during most of Paleozoic time. Coals are less abundant in cooler climates. [J. K. A. Habicht, 1979, Amer. Assoc. Petroleum Geol. Studies in Geol. No. 9.]

The Fruitland Formation consists of interbedded, fine-grained, immature to submature subarkosic sandstone (sometimes containing terrestrial vertebrate fossils), siltstone, shale, and thin micritic limestone that contains lacustrine pelecypods. Many of the detrital rocks are carbonaceous, and coal beds form 2.7% of the formation. The coal underlies an area of more than 16,000 km^2. Nearly all the rock units in the Fruitland are discontinuous and lenticular, and most pinch out laterally within 100 m. The coal beds are the most continuous units and in places can be traced laterally for about 5 km along strike.

Coal Beds

Although the coal beds are well exposed in the outcrop belt of the Fruitland Formation, the bulk of the coal is located in the subsurface within the basin. The position and thickness of the coal beds in the subsurface were determined by interpretation of

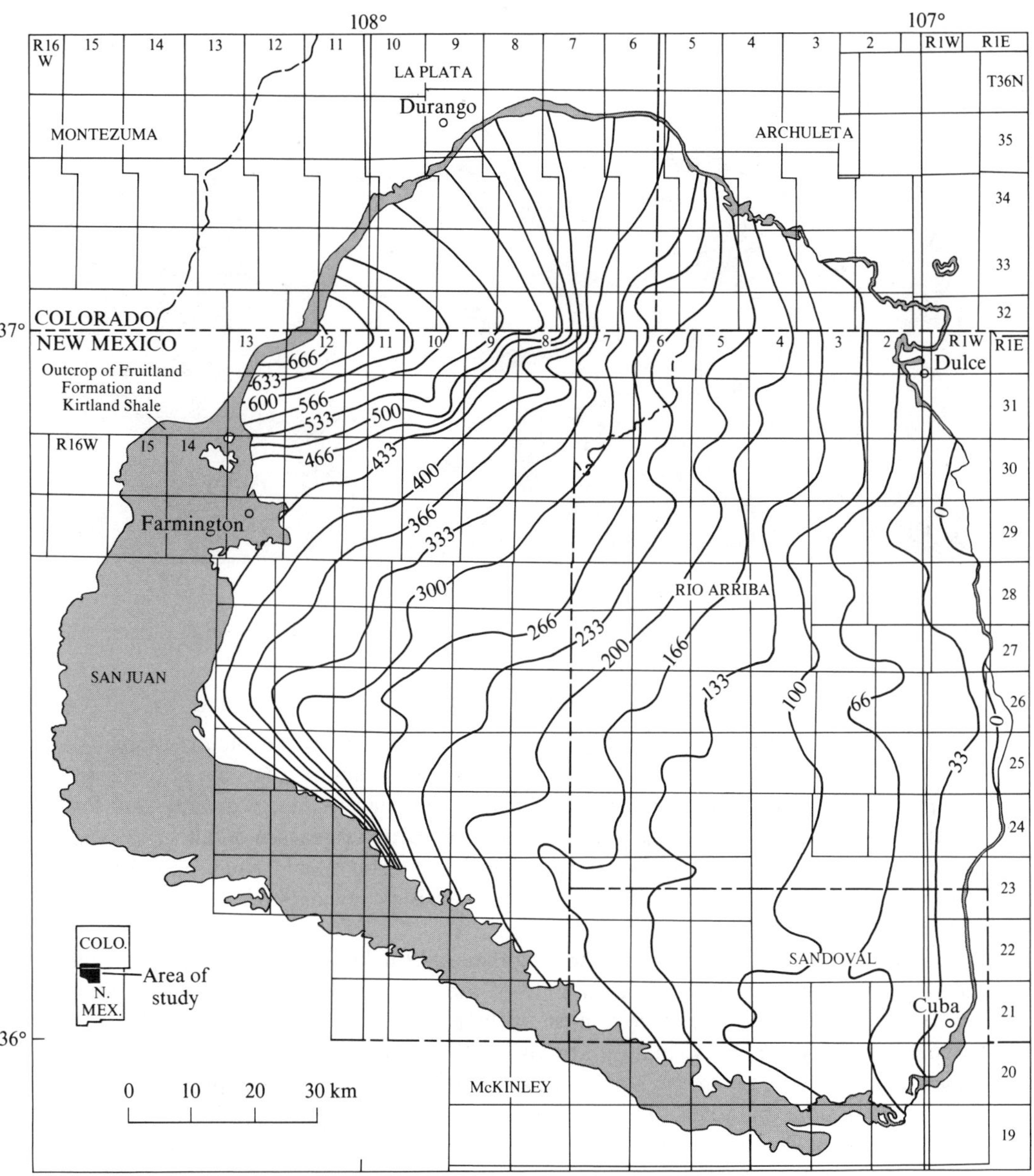

Figure 13–18
Isopach map of combined Fruitland Formation and Kirtland Shale. The two units cannot be separated in eastern half of area. Contour interval in meters. [Fassett and Hinds, 1971.]

Figure 13–19
Outcrop of Fruitland Formation and underlying and overlying units, 18 km southwest of Cuba, New Mexico. [Fassett and Hinds, 1971.]

the electric-log records and drilling-rate charts generated by petroleum companies during drilling in the San Juan Basin. Figure 13–20 shows the character of the response of the Fruitland coal beds on an induction-electric log, a drilling-rate log, and a gamma ray–neutron log. The coal has a distinctive signature on all three types of logs. Other types of well logs, particularly acoustic (sonic) logs and focused varieties of electric logs can delineate the boundaries of the coal beds to within 15 cm or less.

Figure 13–21 is an isopach map showing the total thickness of coal in the Fruitland Formation. The map shows that, in spite of the extreme lenticularity of individual beds and the random occurrence of most of them, clearly defined patterns of distribution of total coal in the formation are visible. Most prominent is the band of thick coal trending NW–SE in the northern half of the map area, with a maximum coal thickness of more than 25 m. The northeastern limits of this thick buildup are defined by the trend of the northeasternmost and largest of two major stratigraphic rises of the underlying Pictured Cliffs Sandstone. In the southeast, the extensive swamps that produced the thick coal zone appear to have ended near the base of a well-drained highland, based on regional stratigraphic relationships. At the southwestern edge of the map area is a smaller zone of relatively thick coal, also trending NW–SE. This area is adjacent to the southwesternmost of the two major stratigraphic rises in the Pictured Cliffs Sandstone. The two NW–SE zones of coal concentration contain not

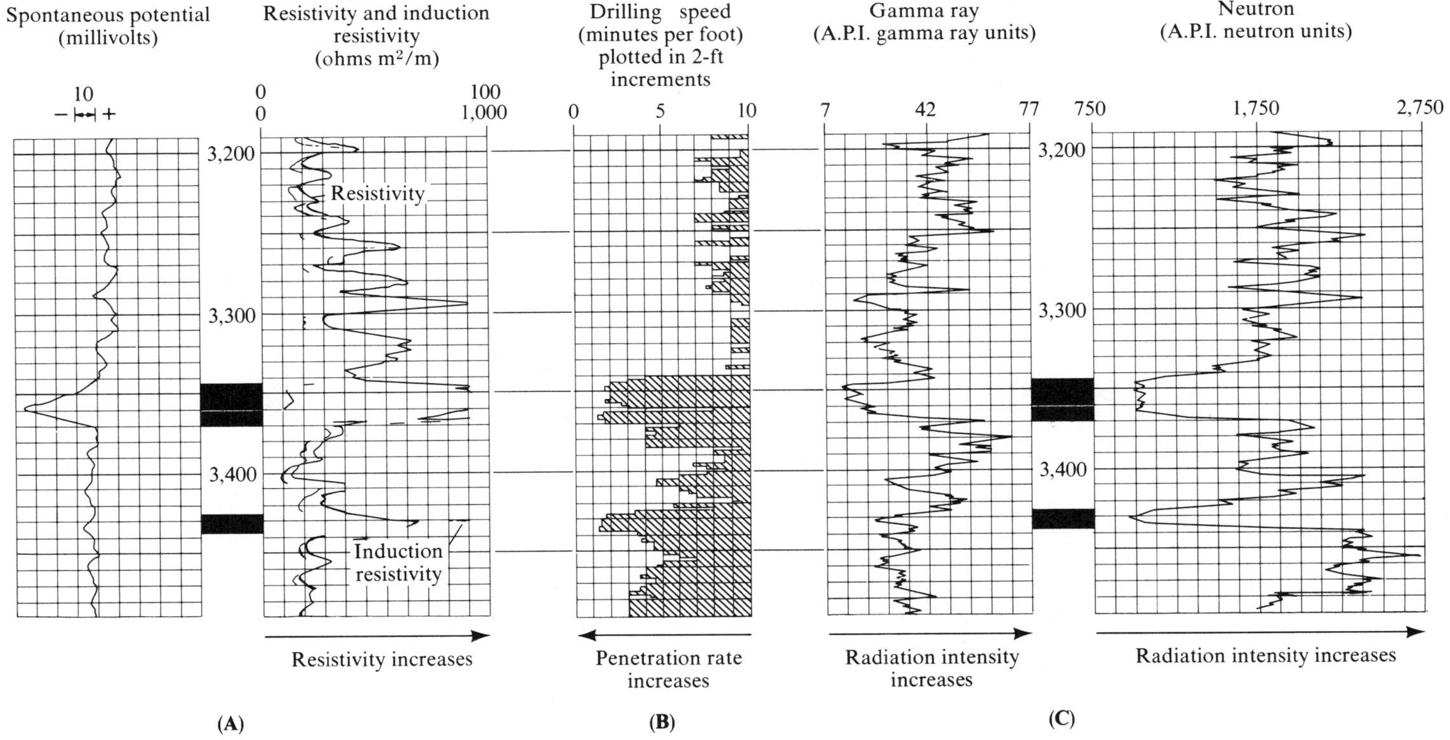

Figure 13–20
(**A**) Induction-electric, (**B**) drilling-rate, and (**C**) radioactivity logs through lower part of Fruitland Formation at typical well site, showing characteristics of response to coal (black bars). [Fassett and Hinds, 1971.]

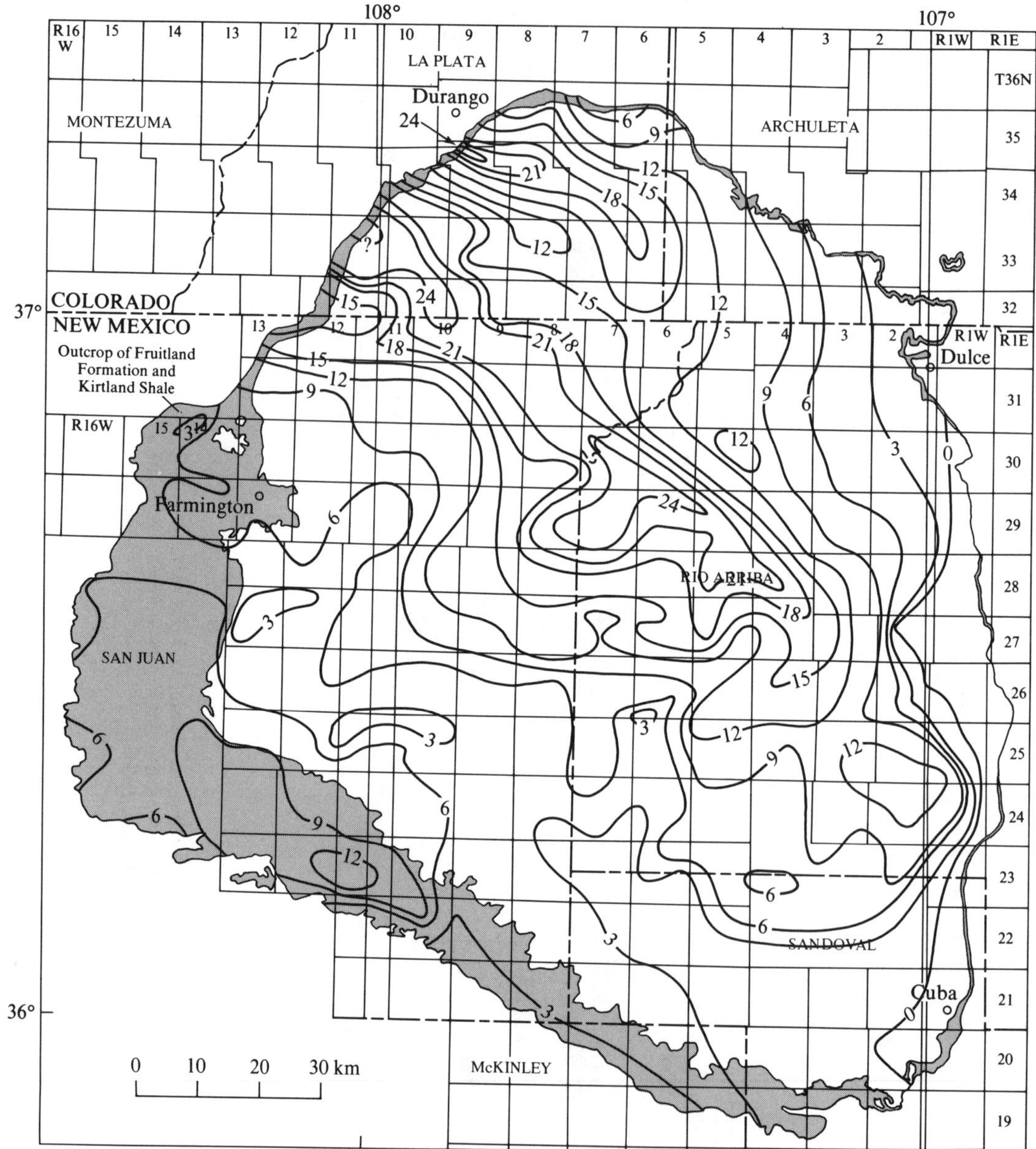

Figure 13–21
Isopach map of total thickness of coal in Fruitland Formation. Contours in meters. [Fassett and Hinds, 1971.]

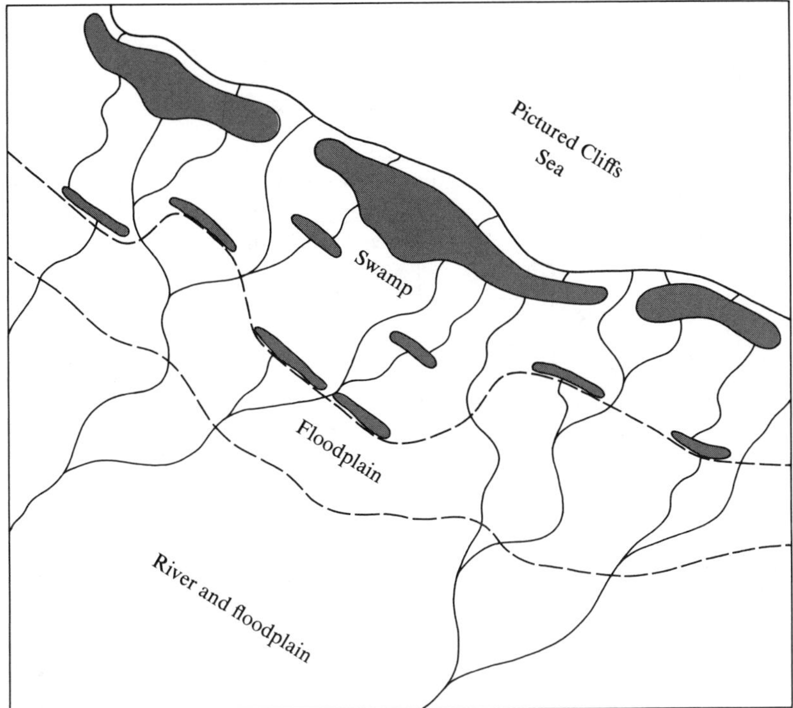

Figure 13–22
Diagrammatic paleogeographic map, showing environments of deposition of rocks that now compose Fruitland Formation. Ruled areas are peat bogs. [Fassett and Hinds, 1971.]

only the major total thicknesses of coal but also the thickest individual coal beds. The environmental setting in which the Fruitland Formation coal accumulated is shown in Figure 13–22.

Figure 13–23 shows the distribution of the fixed-carbon percentage of Fruitland Formation coals on a dry, ash-free basis. This map reflects the depositional zones of Figure 13–22. In the zone of coal buildup in the southwestern part of the basin, fixed-carbon percentages are in excess of 55%. Across the zone of thin coal to the northeast, the values are 50–55%. Further to the northeast, in the zone of thickest coal accumulation, the fixed-carbon percentages rise to a maximum of more than 70% before declining again toward the Pictured Cliffs sea. Fixed-carbon percentages of 50–70% correspond to subbituminous or bituminous coal.

Dekoven Seam (Carboniferous), Southern Illinois

The approach of Fassett and Hinds (1971) was a broad-scale field and stratigraphic study of coals in relation to coal abundance and depositional environment. A complementary approach is illustrated by Cameron (1978), who emphasized detailed mega-

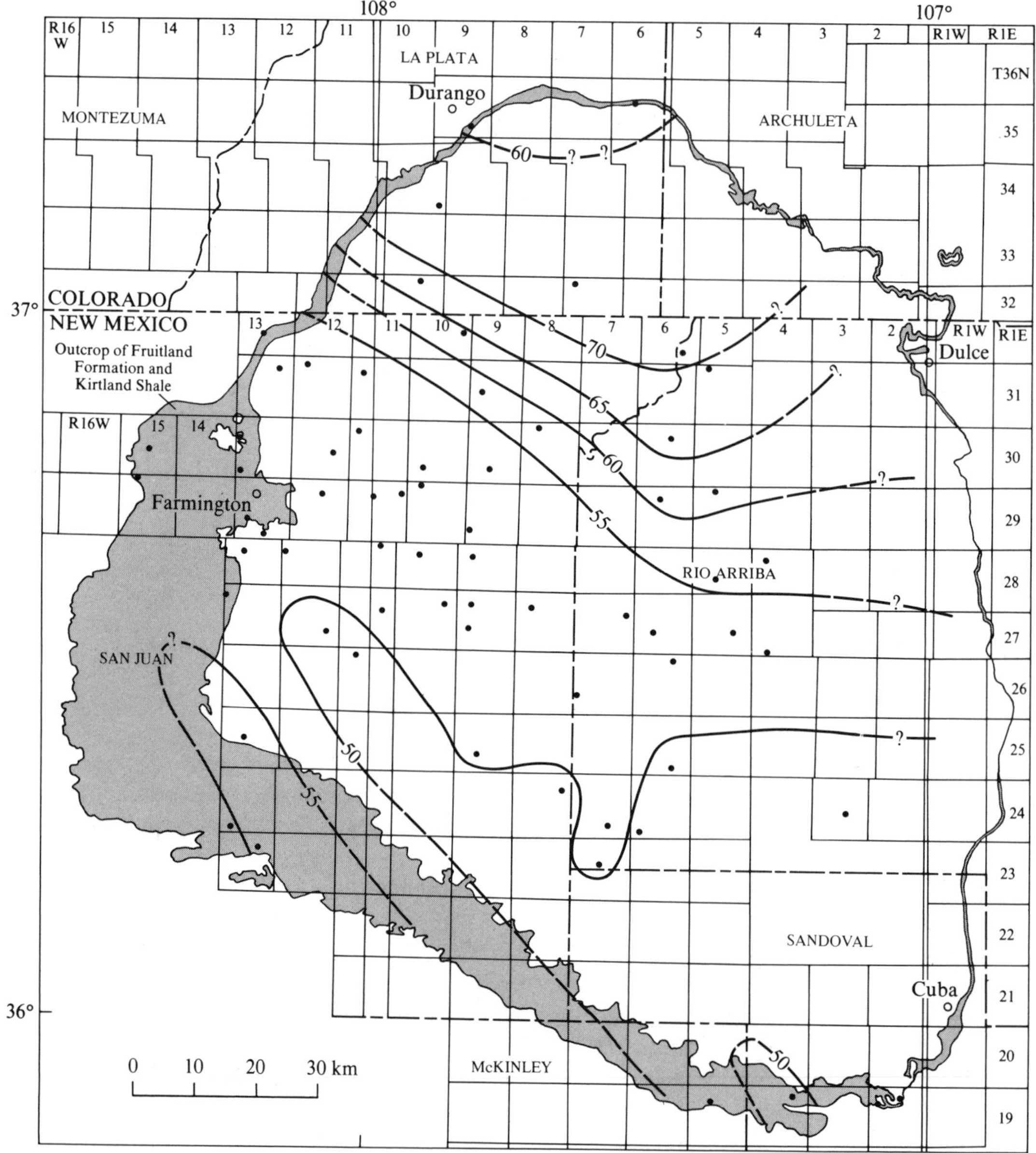

Figure 13–23
Fixed-carbon percentage of coal samples (dry, ash-free) from Fruitland Formation. Contours in percent; interval varies. [Fassett and Hinds, 1971.]

scopic description of individual coal beds supplemented by microscopic studies. At the outcrop, the coal beds are collected as a stratigraphically labeled series of small blocks, whose widths are set by the spacing of vertical joints ("cleats") and whose heights are set by surfaces of easy breakage parallel to banding in the coal. The blocks of coal are sequentially labeled and subsequently mounted in an epoxy resin as a stratigraphic column and polished to obtain clean, unweathered surfaces. Fresh surfaces are necessary for accurate description of lithotypes. Laminae as thin as 1 mm were described, and an attempt was made to subdivide Stopes' (1919) clarain category into "bright" and "dull" because the bulk of the seams was clarain.

Cameron studied four seams; one of them, the Dekoven seam, can serve as an example of his results. He divided this coal seam into nine major petrographic zones characterized as follows:

Interval IX. From the top of the seam to 7.7 cm; a zone characterized by an abundance of dull clarain with subordinate amounts of bright clarain and vitrain.

Interval VIII. From 7.7 to 12.5 cm; bright clarain is the dominant constituent.

Interval VII. From 12.5 to 28.5 cm; dull clarain is most abundant.

Interval VI. From 28.5 to 32.4 cm; mainly bright clarain.

Interval V. From 32.4 to 34.7 cm; dull clarain is the most abundant lithotype with some intercalated layers of bright clarain, especially near the base.

Interval IV. From 34.7 to 41.7 cm; a bright zone with a concentration of vitrain layers.

Interval III. From 41.7 to 45.0 cm; mainly dull with two layers identified as durain.

Interval II. From 45.0 to 50.3 cm; shale parting.

Interval I. From 50.3 to 77.8 cm; the base of the seam; the most prominent lithotype is bright clarain.

The lithotype contents for these intervals are summarized in Figure 13–24. The vertical ordering of lithotypes in a seam has genetic significance. It is interpreted as reflecting sequential changes in swamp conditions, although the exact nature of the changes is not always clear. Differences among the intervals are substantial. For example, Interval V is dominated by dull clarain, but the overlying Interval VI is largely bright clarain. Cameron noted that the Dekoven seam appears, from the lithotypic column, to contain only three fusain layers, but it is likely that this is an underestimate. When blocks of the coal seam were collected, they tended to part at the fusain layers because they were less compact (i.e., more friable). Despite an effort to protect these surfaces, some of the friable fusain was nearly always lost during sampling.

Cameron's study of Illinois coal included microscopic analyses of the megascopic lithotypes for two of the four seams he studied. A total of 376 bands within the seams was examined to relate the various lithotypes to their contents of vitrinite, inertinite,

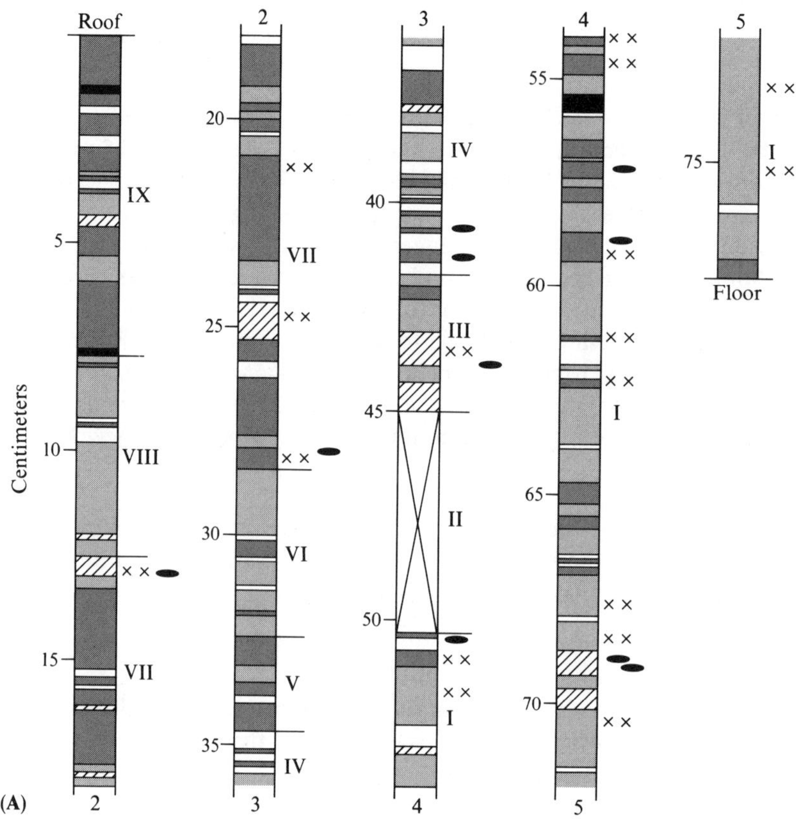

Roof
2
3
4
5
Floor
Centimeters
IX
VIII
VII
VI
V
IV
III
II
I
(A)

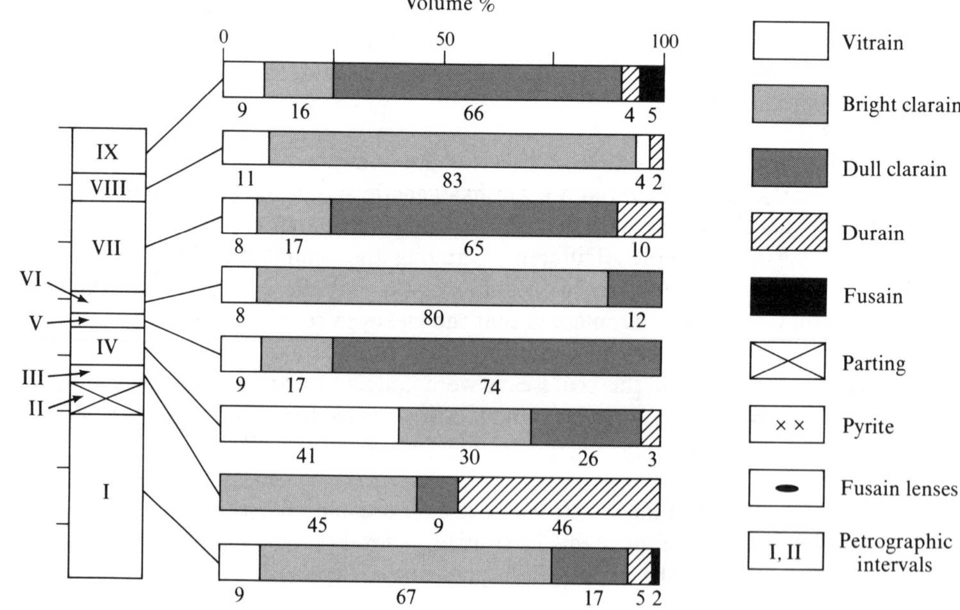

Volume %
0
50
100
IX
VIII
VII
VI
V
IV
III
II
I
9 16 66 4 5
11 83 4 2
8 17 65 10
8 80 12
9 17 74
41 30 26 3
45 9 46
9 67 17 5 2
Vitrain
Bright clarain
Dull clarain
Durain
Fusain
Parting
Pyrite
Fusain lenses
I, II
Petrographic intervals
(B)

exinite, and mineral matter (see Figure 13–25). For example, Figure 13–25A shows that of the 109 vitrain bands studied 95% or 103 contained 90–100% vitrinite. For the bright clarains, approximately 48% of the 131 bands studied contained 80–90% vitrinite. Clearly, glossiness or brightness is highly correlated with content of vitrinite; the powdery, charcoal-like fusain has a median vitrinite content of only 10%. The most sharply defined lithotypes in terms of maceral content are vitrain and fusain; they are almost monomaceralic. Vitrain consists of vitrinite; fusain consists of inertinite (fusinite and semifusinite) (Figure 13–25B). The maceral contents of clarain and durain are more variable.

Whether or not good correlations exist between lithotypes and macerals, lithotypes are still useful to describe the banding sequence within a coal seam, and the sequence can be used for seam identification and correlation. The objectives of a characterization of banding are the same as the objectives of a stratigrapher studying inorganic sedimentary rocks: age relationships, determination of depositional environments, and diagenetic history of the unit.

MODERN COAL ENVIRONMENTS

The initiation and maintenance of areally extensive peat-forming environments are relatively uncommon geologic events. This is suggested by a survey of modern sedimentary environments but is even more evident from the infrequency with which we encounter coal seams in ancient sedimentary rocks. The explanation for the scarcity of these carbonaceous deposits is well known: the fact that organic matter decomposes rapidly to carbon dioxide plus water when free oxygen is present, the abundance of oxygen in the atmosphere, the equilibrium generally maintained between the oxygen in the atmosphere and the oxygen dissolved in surface waters, and the typical free circulation of these waters at and near the Earth's surface. Bacteria and fungi are prominent in most surficial environments and rapidly convert dead organic matter ultimately into carbon dioxide and water.

The development of a coal seam, however, requires more than the inhibition of the decay process. The formation of laterally extensive coal seams requires that anaerobic conditions occur in a subsiding basin, so that a thick accumulation of plant debris is possible. It requires 3–20 m of plant debris to yield 1 m of coal (depending on the type of plant and the degree of degradation) because of the high initial porosity and easy compactibility of fresh plant attritus. A corollary of this fact is that high organic productivity is required if thick coal seams are to result. The relationship among

---

Figure 13–24
(**A**) Detailed megascopic profile of Dekoven coal seam. (**B**) Summary of proportions of various lithotypes within each of nine petrographic intervals. [Cameron, 1978. Copyright, American Society for Testing and Materials, 1916 Race Street, Philadelphia, PA 19103. Reprinted with permission.]

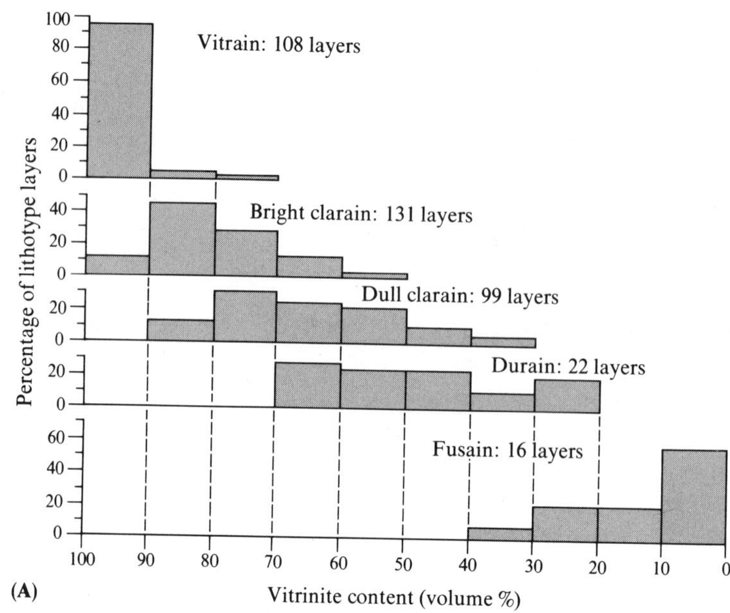

Vitrain: 108 layers
Bright clarain: 131 layers
Dull clarain: 99 layers
Durain: 22 layers
Fusain: 16 layers
Percentage of lithotype layers
100 80 60 40 20 0
100 90 80 70 60 50 40 30 20 10 0
(A)
Vitrinite content (volume %)

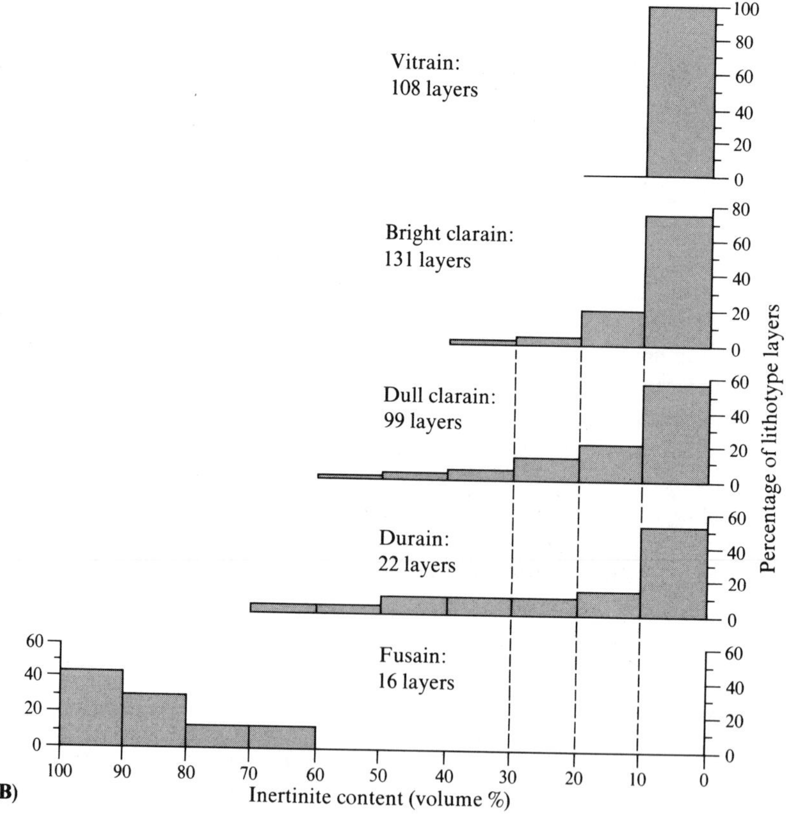

Vitrain:
108 layers
Bright clarain:
131 layers
Dull clarain:
99 layers
Durain:
22 layers
Fusain:
16 layers
Percentage of lithotype layers
100 80 60 40 20 0
100 90 80 70 60 50 40 30 20 10 0
(B)
Inertinite content (volume %)

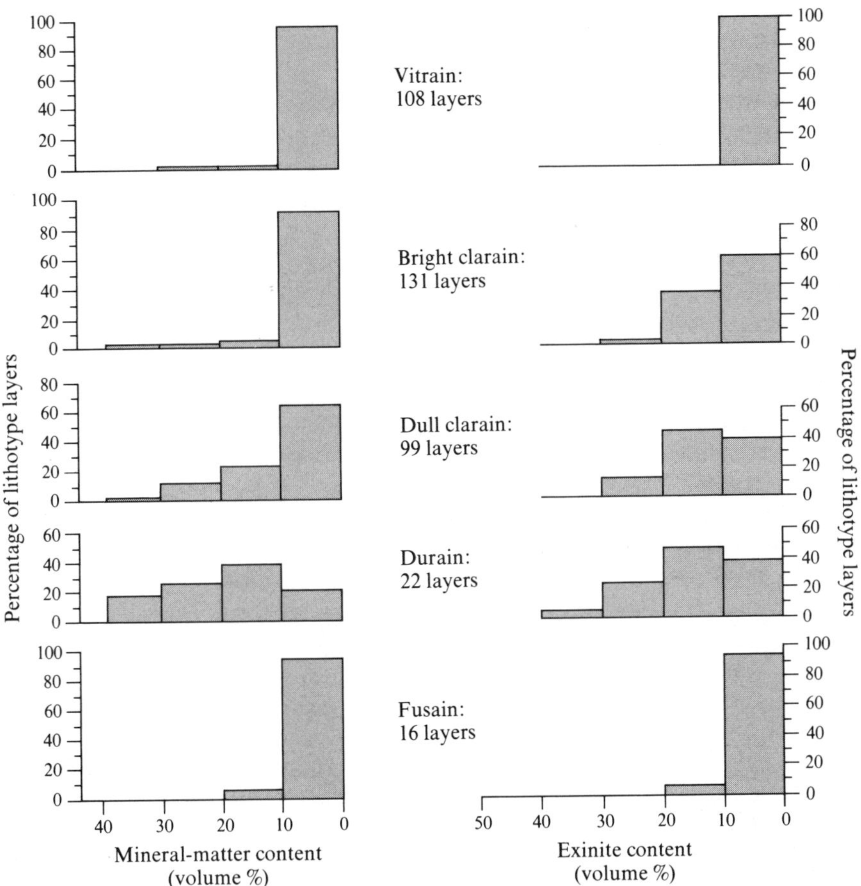

Figure 13–25
Frequency distribution of lithotype bands according to (**A**) vitrinite content, (**B**) inertinite content, (**C**) exinite and mineral-matter content. [Cameron, 1978. Copyright, American Society for Testing and Materials, 1916 Race Street, Philadelphia, PA 19103. Adapted with permission.]

organic growth, decay, and mean annual temperature is shown in Figure 13–26. Although organic accumulations in swamps are most significant at 20–25°C, peats may accumulate in cool temperate and subarctic climates as well; the rate of organic growth is slowed, but the cool temperatures retard the decomposition process even more. Examples of modern peat accumulations in humid temperate to subtropical climates are numerous, particularly in coastal lowlands such as the southern and southeastern United States, from Louisiana to Virginia. Among those that have been studied in some detail are the Louisiana coastal plain (Frazier and Osanik, 1969), the Everglades swamp–marsh complex of southern Florida (Spackman et al., 1966;

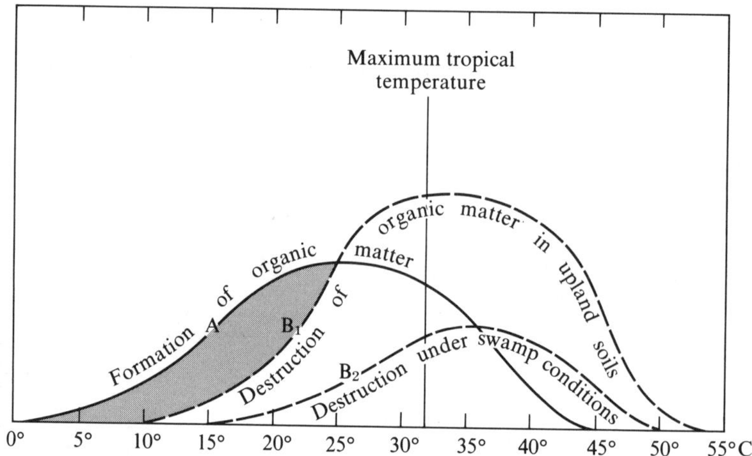

Figure 13–26
Relationships among temperature, rate of organic growth, and destruction of organic matter by microflora. Zone between curves of formation of organic matter and its destruction under swamp (anaerobic) conditions is substantial at most surface temperatures, assuming adequate moisture. Under aerobic conditions, cool temperatures are required if organic matter is to accumulate. At temperatures in excess of 25°C in presence of free oxygen, rate of organic decomposition always exceeds rate of production. [M. Gordon et al., 1958, U.S. Geol. Surv. Prof. Paper No. 299.]

Spackman et al., 1969; Scholl, 1969), the Okefenokee swamp–marsh complex of southern Georgia (Cohen, 1973, 1974), and the Snuggedy Swamp of South Carolina (Staub and Cohen, 1979). Accumulations of freshwater peat in humid, cool to temperate climates are also very common in the Northern Hemisphere, although they have not been as well studied as those in warmer climates. Many examples are known, however, in Siberia and in Canada. The Canadian deposits are known as *muskeg* and cover an area of 1.3 million square kilometers (Radforth, 1969).

The beds of peat usually rest on sediments representing the soil surface in which the initial swamp vegetation rooted itself. These ancient soils, leached of alkalis by plant growth, and still containing traces of plant roots, may be recognized in the "seat-earths" or underclays that underlie coal seams.

In those parts of the swamp where the water is too deep for land plants to take root, aquatic plants and animals flourish, contributing their remains to the sapropelic deposits. These deposits may eventually accumulate to sufficient thickness for swamp and terrestrial plants to establish themselves and for normal peat formation to commence. On the other hand, sapropelic deposits are sometimes overlain by sands and muds, which represent incursions of sediment-bearing currents of water into stagnant pools and swampy lagoons. These strong currents may also remove parts of previously formed organic deposits; their activity is shown by the deeply scoured channels, filled with inorganic sediments, which are found in the "roofs" of many coal seams.

Thus, the requirements for the development of thick peat deposits and consequently for the formation of coal seams are:

1. A slow, continuous rise of the groundwater table, which keeps abreast of peat formation.

2. Protection of the swamp (by beaches, sandbars, spits, etc.) against major inundations by the sea and (by natural levees) against river floodwaters.

3. An extensive area of low relief landward of the swamp, so that the input of fluvial sediments is minimal, to prevent interruption of peat formation.

If the groundwater table rises too high, generally because of rapid subsidence, the swamps will drown and lacustrine or marine sediments will be deposited. If subsidence is too slow, the plant material on the surface will rot and the peat that has already formed will be eroded. Alternatively, fire may consume existing peat deposits. A slow oscillation of the water level, in conjunction with a general regional subsidence, may give rise to a considerable thickness of alternating layers of peat and inorganic sediments. With the retreat of the invading waters, the emerging land surfaces rapidly reacquire a cover of vegetation, and the conditions for the formation of peat are again resumed.

The very early diagenesis of peat in modern swamp environments includes both microbial and chemical changes. The most severe alteration occurs at the peat surface and immediately below it to a depth of approximately 0.5 m, the zone in which aerobic bacteria, actinomyces, and fungi are active. Values of pH are 3.5–6.5. With increasing depth these organisms are replaced by anaerobic bacteria, which can function to depths of about 10 m. Below this depth, microbial life is extinguished, and only chemical changes occur—primarily condensation, polymerization, and reducing reactions.

Okefenokee Swamp–Marsh Complex, Georgia

The Okefenokee Swamp (Cohen, 1973, 1974) is a region of both swamps and marshes covering more than 1,600 km^2 of southeastern Georgia and extending slightly into northeastern Florida (see Figure 13–27). The forested portions of the Okefenokee (i.e., the swamps proper) are composed primarily of cypresses (*Taxodium* spp.), bays (*Persea, Magnolia,* and *Gordonia* spp.), gums (*Nyssa* spp.), or shrubs (e.g., *Cyrilla* and *Ilex* spp.). The marshlands are regions of standing water occupied by floating or herbaceous aquatic plants such as *Nymphaea odorata* Ait. (white water lily), *Panicum hemitomon* Schultes (maidencane), or *Woodwardia virginica* (L.) Smith (chainfern). These open, unforested areas are sometimes called "prairies."

The Okefenokee Swamp originated as a filling of a lagoon on the landward (western) side of a Pleistocene offshore barrier island or bar, now evident as Trail Ridge.

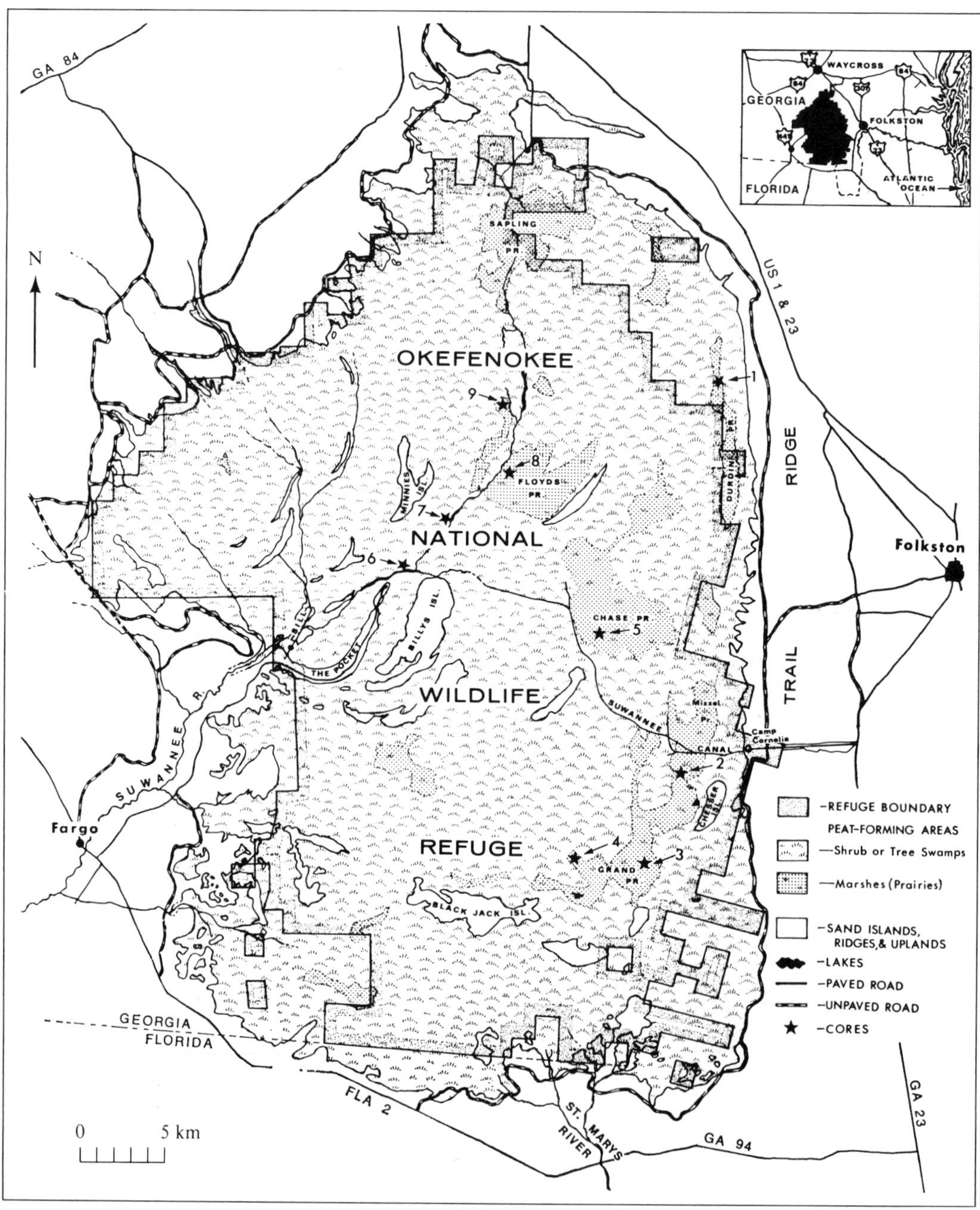

Figure 13–27
Map of Okefenokee Swamp, showing topography and coring locations. [Cohen, 1974.]

The islands within the swamp are composed of quartz sand and are believed to be remnants of bars on the landward side of Trail Ridge. The maximum thickness of the peat accumulation within the swamp is about 4 m and C^{14} analyses of basal peats indicate that the accumulation began 5,000–6,500 years ago. The basal peats contain only plants that are not tolerant of salt and also contain abundant freshwater sponge spicules, suggesting a period of erosion of the underlying Pleistocene surface and resulting flushing out of the marine pore and surface waters before the basal peats were deposited. The basal peats also contain large amounts of fine-grained charcoal, the result of burning of previously deposited phytogenic sediments. Figure 13–28 illustrates a typical core of peat from the surface down to the underlying carbonaceous sandy clay of the Pleistocene substrate. The variation in plant species with time can be interpreted in terms of either climatic change or, more likely, laterally shifting positions of environments within the swamp (Table 13–2). The typical succession in the Okefenokee ranges from open-water marshes (dominated by floating aquatics), through glade marshes and island fringes (dominated by grasses, sedges, or ferns), to tree islands and swamps (dominated by cypress, bays, gums, or other trees).

FROM MODERN BOTANY TO COAL PETROLOGY

The types of data contained in Table 13–2 are useful in characterizing the various peat types with respect to their original vegetational and depositional environments. As sedimentary petrologists, however, we also need quantitative data in a form more useful for predicting the eventual petrography of the coal that could result from each type of peat. Table 13–3 gives the average percentages of cell filling (and/or secretions), structurally intact cell walls (or cell-wall fragments), fine, granular (unstructured) materials including mineral matter, fungal remains, and charcoal. The cell fillings and secretions will probably become the resinite macerals of lignites or bituminous coals. Cell-wall materials will probably become the vitrinites. Much of the fine, granular debris will probably become the micrinite macerals, although some small portion may result in very fine grains of vitrinite. The fungal remains will become sclerotinite, and the pieces of charcoal and partially burned materials will constitute the fusinitic and semifusinitic portions of the coal. Of course, this type of analysis should be taken only as a rough approximation of the eventual coal composition because, at this early stage of peatification, many of the structured plant tissues have yet to be altered in a way that would be characteristic of these tissues with depth. This is especially true of the roots, which make up a significant portion of some of these samples. Roots tend to alter in such a manner as to produce appreciable amounts of premicrinitic materials. This would especially affect the predictions concerning the open-marsh and glades peats, because these have a high N/S ratio (Table 13–2).

The herbaceous peats have very little preresinitic, fusinitic, or sclerotinitic material, and considerable amounts of fine, granular (premicrinitic) materials. It is very likely

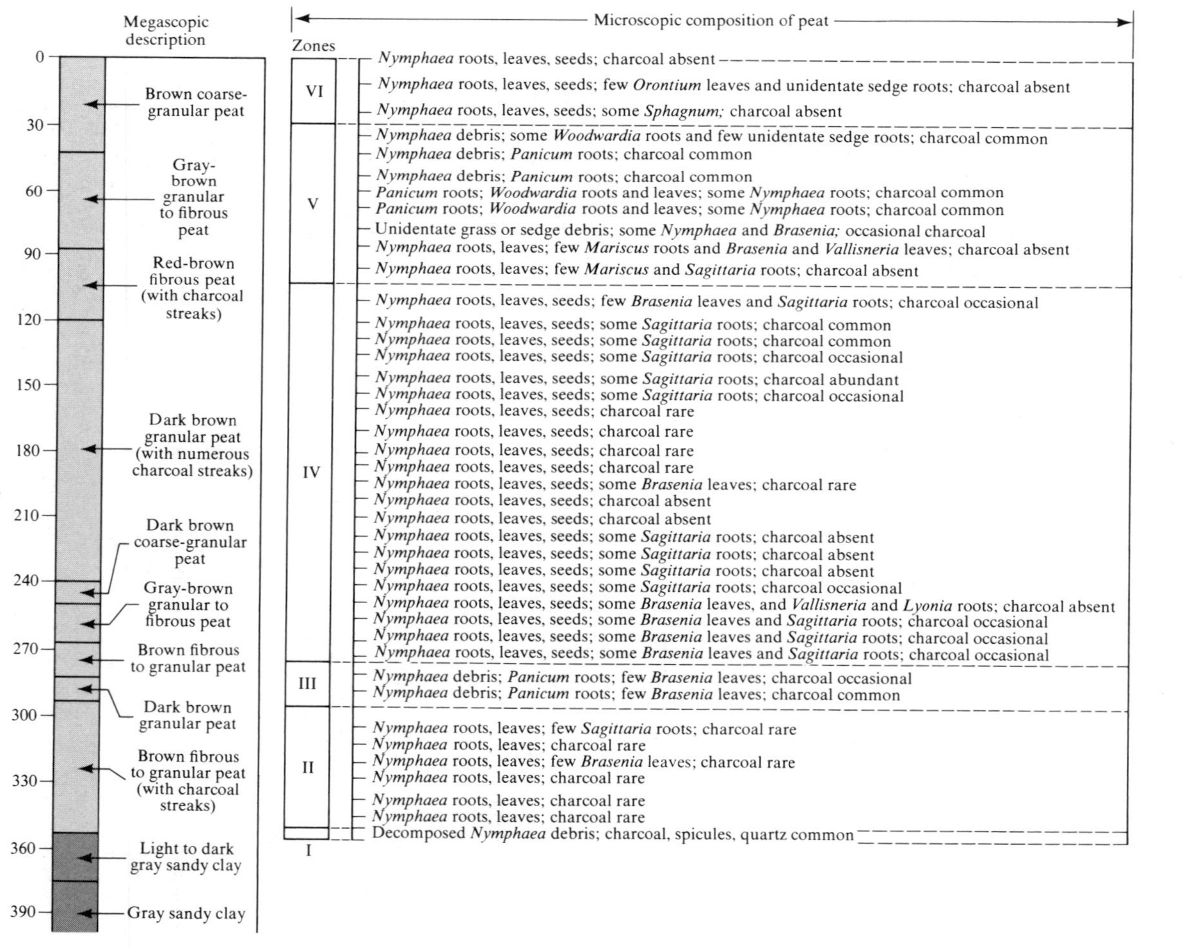

Figure 13–28
Petrographic and paleobotanical analysis of Core 4 in southeastern part of Okefenokee Swamp. [Cohen, 1974.]

Table 13–2
Characteristics of Some Surface Peat Sediments from the Okefenokee Swamp

	Dominant plant species	Megascopic characteristics		Microscopic characteristics			
		Color	Texture	Color (30X)	Avg ratio (F/M)	Avg ratio (N/S)	Characteristic plant fragments
Open-water marshes	*Nymphaea odorata* (white water lily)	Reddish brown	Fine fibrous to granular	Light yellow	44/56	75/25	*Nymphaea* leaf and root sclereids; vesicular red cell fillings of rhizome
	Carex hyalinolepis (sedge)	Reddish brown to brown	Fine fibrous to granular	Light yellow to light brown	46/54	60/40	*Nymphaea* leaf and root sclereids
Shallow "glades" and island fringes	*Panicum hemitomon* (maidencane)	Reddish brown to dark brown	Coarse fibrous	Light yellow	52/48	75/25	*Panicum* phytoliths
	Woodwardia virginica (chain fern)	Reddish brown to dark brown	Fibrous to granular	Red-brown	49/51	60/40	Root hairs and inner cortical cells of *Woodwardia*
"Tree islands" and swamps	*Cyrilla racemiflora* (titi)	Reddish brown to dark brown	Coarse granular to woody	Light brown	66/34	35/65	Red-brown to yellow cell fillings (abundant)
	Taxodium distichum (swamp cypress)	Dark brown	Coarse granular to woody	Red-brown	50/50	40/60	Orange to yellow "vitreous" spheres

[a]rare <0.5%; common 0.5–2%; abundant >2%.
[b]rare <2%; common 2–10%; abundant >10%.
[c]rare <1%; common 1–3%; abundant >3%.
Note: F = framework; M = matrix; N = nonsedimentary debris; S = sedimentary debris.
Source: Cohen, 1973.

Table 13-2 (*continued*)

Microscopic characteristics							
Fungal remains[a]	Fecal pellets[b]	Diatoms[a]	Insect parts[a]	Sponge spicules[c]	Quartz[a]	Charcoal[b]	Microscopically identifiable plant species
Absent to common	Rare to abundant	Rare to common	Rare to common	Common	Rare (fine-grained, angular, pitted)	Rare	*Nymphaea* roots, leaves, stems, and seeds; other species (e.g., *Sagittaria, Orontium*) less common
Rare to common	Absent to rare	Rare to common	Absent to common	Rare to common	Rare (fine-grained, angular, pitted)	Absent to common	*Carex* leaves and roots not esp. common; mostly debris from *Nymphaea, Sphagnum,* and *Vallisneria*
Absent to common	Absent to common	Common	Rare to common	Common	Rare (fine-grained, angular, pitted)	Rare to common	*Panicum* leaves and roots common; *Nymphaea* debris also common
Absent to common	Absent to common	Common	Rare to common	Rare to common	Rare (fine-grained, angular, pitted)	Absent to common	*Woodwardia* leaves, roots, rhizomes; *Sphagnum, Nymphaea, Carex, Taxodium, Panicum* common
Common (esp. in leaves and wood)	Common	Common	Common	Common	Absent to rare	Rare to abundant	*Cyrilla* leaves, roots, and wood; *Lyonia, Ilex, Persea, Leucothoe, Carex, Sagittaria, Magnolia,* and *Sphagnum* common
Common (esp. in leaves and wood)	Common	Common	Absent to common	Common to abundant	Rare (fine-grained, angular, pitted)	Common to abundant	*Taxodium* leaves, twigs, roots, wood; occasional hardwood debris, *Tillandsia,* and *Sphagnum*

Table 13–3
Quantitative Petrographic Data Applicable to Origin of Coal Types

Coal type	Peat type	Average percent[a] cell fillings and secretions (preresinites and prephlobophenites)	Average percent cell wall materials (previtrinites)	Average percent fine granular debris (predominantly premicrinites with some previtrinites and mineral matter)	Average percent fungal remains (presclerotinites)	Average percent charcoal (fusinites and semifusinites)
Herbaceous	*Nymphaea*	1	49	49	0.3	1
	Carex	1	65	32	0.3	2
	Panicum	1	62	35	0.3	2
	Woodwardia	2	46	49	0.3	3
Arboreous	*Cyrilla*	15	56	25	1	4
	Taxodium	10	60	24	1	5

[a]Data based on point counts of 5 vertically oriented microtome sections of each peat type (1,000 points at 125×; pore space disregarded).
Source: Cohen, 1973.

that these peats will increase in premicrinitic and decrease in previtrinitic materials as peatification progresses. The coal that might be predicted from these peats would be fine-grained, massive (i.e., relatively unlaminated on a megascopic or microscopic scale), and very dull in appearance.

On the other hand, the peats derived from tree vegetation (i.e., *Cyrilla* and *Taxodium*) have significantly greater percentages of preresinites, fusinites, and sclerotinites than the herbaceous peats and considerably less premicrinitic debris. The resulting coal should therefore be somewhat brighter and more laminated (due to the higher proportion of woody or leafy debris) than the coals from herbaceous sources. In fact, these predicted coals would seem to be somewhat similar in composition to some duroclarainic bituminous coals of Pennsylvanian age.

SUMMARY

Coal is a widely distributed rock type in Late Paleozoic and younger rocks but has received very little study by sedimentary petrologists. Within the past 10 years, however, interest in coal petrology has increased significantly, in part as a result of increased economic interest in coal as a partial replacement for petroleum.

The formation of extensive coal deposits requires a moist climate but can occur at both tropical and subarctic temperatures. Large accumulations of terrestrial plant debris require not only a suitable climate but also an anaerobic depositional environment and a slowly subsiding basin—conditions that can be satisfied in a wide variety of tectonic and paleogeographic settings. Modern accumulations exist adjacent to rapidly subsiding basins such as the Gulf Coast region of the United States, the trailing plate margin of the eastern coast of North America, and the stable cratonic areas of Siberia and eastern Canada.

Most coal beds are banded or contain layers that differ in luster. These bands are termed lithotypes; the most common categories of lithotypes are vitrain, clarain, durain, and fusain, from most lustrous to least lustrous. The individual constituents other than minerals seen in polished sections of coal are termed macerals. They are analogous to minerals in inorganic rocks. A large number of macerals has been named, but the nomenclature is still evolving because of the relative newness of the field of coal petrography. The three major groups of macerals are vitrinite, exinite or liptinite, and inertinite, each of which has numerous varieties.

Coal rank refers to the position of a coal in the metamorphic series from peat to lignite to subbituminous coal, bituminous coal, anthracite coal, and meta-anthracite coal. Estimation of coal rank can be made on several criteria, among which the most commonly used are percentage of fixed carbon or percentage of volatile matter, reflectance of vitrinite macerals, and calorific value (Btu). Temperature increase is the most important cause of increase in rank.

FURTHER READING

American Society for Testing and Materials. 1980. *Annual Book of Standards.* Part 26: *Coal and Coke; Atmospheric Analysis,* 937 pp.

Bostick, N. H. 1979. Microscopic measurement of the level of catagenesis of solid organic matter in sedimentary rocks to aid exploration for petroleum and to determine former burial temperatures—a review. In P. A. Scholle and P. R. Schluger (eds.), *Aspects of Diagenesis.* Soc. Econ. Paleontol. Mineral. Spec. Pub. No. 26, pp. 17–43.

Cameron, A. R. 1978. Megascopic description of coal with particular reference to seams in southern Illinois. In R. R. Dutcher (ed.), *Field Description of Coal.* Amer. Soc. for Testing and Materials Spec. Tech. Pub. No. 661, pp. 9–32.

Cohen, A. D. 1973. Petrology of some Holocene sediments from the Okefenokee swamp–marsh complex of southern Georgia. *Geol. Soc. Amer. Bull., 84,* 3867–3878.

Cohen, A. D. 1974. Petrography and paleoecology of Holocene peats from the Okefenokee swamp–marsh complex of Georgia. *Jour. Sed. Petrology, 44,* 716–726.

Crelling, J. C., and R. R. Dutcher. 1980. *Principles and Applications of Coal Petrology.* Soc. Econ. Paleontol. Mineral. Short Course Notes No. 8, 127 pp.

Dapples, E. C., and M. E. Hopkins (eds.). 1969. *Environments of Coal Deposition.* Geol. Soc. Amer. Spec. Paper No. 114, 204 pp.

Davis, A. 1978. The reflectance of coal. In C. Karr, Jr. (ed.), *Analytical Methods for Coal and Coal Products.* New York: Academic Press, vol. 1, pp. 27–81.

Dutcher, R. R., et al. 1974. *Carbonaceous Materials as Indicators of Metamorphism.* Geol. Soc. Amer. Spec. Paper No. 153, 108 pp.

Dutcher, R. R. (ed.). 1978. *Field Description of Coal.* Amer. Soc. for Testing and Materials Spec. Tech. Pub. No. 661, 71 pp.

Fassett, J. E. 1977. Stratigraphy of the coals of the San Juan Basin. In D. K. Murray (ed.), *Geology of Rocky Mountain Coal.* Colo. Geol. Surv. Resource Series No. 1, pp. 61–71.

Fassett, J. E., and J. S. Hinds. 1971. *Geology and Fuel Resources of the Fruitland Formation and the Kirtland Shale of the San Juan Basin, New Mexico and Colorado.* U.S. Geol. Surv. Prof. Paper No. 676, 76 pp.

Fettweis, G. B. 1979. *World Coal Resources.* New York: Elsevier, 415 pp.

Frazier, D. A., and A. Osanik. 1969. Recent peat deposits—Louisiana coastal plain. In E. C. Dapples and M. E. Hopkins (eds.), *Environments of Coal Deposition.* Geol. Soc. Amer. Spec. Paper No. 114, pp. 63–85.

Geol. Soc. Amer. Bull. 1981, *92,* 519–577. Nine articles concerned with coal resources in North America.

Hower, J. C., and A. Davis. 1981. Application of vitrinite reflectance anisotropy in the evaluation of coal metamorphism. *Geol. Soc. Amer. Bull., 92,* 350–366.

Hubbert, M. K. 1971. The energy resources of the earth. *Scientific American, 224,* 60–70.

International Committee for Coal Petrology. 1963. *International Handbook of Coal Petrography,* 2nd ed. Paris: Centre National de la Recherche Scientifique. Supplements published in 1971, 1975.

Kottlowski, F. E., A. T. Cross, and A. A. Meyerhoff. 1978. *Coal Resources of the Americas.* Geol. Soc. Amer. Spec. Paper No. 179, 90 pp.

McKinney, J. S. 1959. Petrographic analysis of the Croweburg coal and its associated sediments. M.Sc. thesis, University of Oklahoma, 124 pp.

Marchioni, D. L. 1980. Petrography and depositional environment of the Liddell seam, upper Hunter Valley, New South Wales. *Internat. Jour. Coal Geol., 1,* 35–61.

Murchison, D., and T. S. Westoll (eds.). 1968. *Coal and Coal-Bearing Strata.* New York: Elsevier, 418 pp.

Petrakis, L., and D. W. Grandy. 1980. Coal analysis, characterization and petrography. *Jour. Chem. Educ., 57,* 689–694.

Radforth, N. W. 1969. Environmental and structural differentials in peatland development. In E. C. Dapples and M. E. Hopkins (eds.), *Environments of Coal Deposition.* Geol. Soc. Amer. Spec. Paper No. 114, pp. 87–104.

Ryer, T. A., and A. W. Langer. 1980. Thickness change involved in the peat-to-coal transformation for a bituminous coal of Cretaceous age in central Utah. *Jour. Sed. Petrology, 50,* 987–992.

Scholle, D. W. 1969. Modern coastal mangrove swamp stratigraphy and the ideal cyclothem. In E. C. Dapples and M. E. Hopkins (eds.), *Environments of Coal Deposition.* Geol. Soc. Amer. Spec. Paper No. 114, pp. 37–61.

Schopf, J. F. 1960. *Field Description and Sampling of Coal Beds.* U.S. Geol. Surv. Bull. 1111-B, 70 pp.

Spackman, W. 1958. The maceral concept and the study of modern environments as a means of understanding the nature of coal. *Trans. N.Y. Acad. Sci., 20,* 411–423.

Spackman, W., C. P. Dolsen, and W. L. Riegel. 1966. Phytogenic organic sediments and sedimentary environments in the Everglades–mangrove complex. *Palaeontographica,* B 117, pp. 135–152.

Spackman, W., W. L. Riegel, and C. P. Dolsen. 1969. Geological and biological interactions in the swamp–marsh complex of southern Florida. In E. C. Dapples and M. E. Hopkins (eds.), *Environments of Coal Deposition.* Geol. Soc. Amer. Spec. Paper No. 114, pp. 1–35.

Stach, E. 1975. *Handbook of Coal Petrology,* 2nd ed. Berlin: Gebrüder Borntraeger, 428 pp.

Staub, J. R., and A. D. Cohen. 1979. The Snuggedy Swamp of South Carolina: a back-barrier estuarine coal-forming environment. *Jour. Sed. Petrology, 49,* 133–144.

Stopes, M. C. 1919. On the four visible ingredients in banded bituminous coal. *Proc. Royal Soc. London, 90,* 69–87.

Stopes, M. C. 1935. On the petrology of banded bituminous coals. *Fuel, 14,* 4–13.

Teichmüller, M., and R. Teichmüller. 1968. Geological aspects of coal metamorphism. In D. Murchison and T. S. Westoll (eds.), *Coal and Coal-Bearing Strata.* New York: Elsevier, pp. 233–267.

Teichmüller, M. and R. Teichmüller. 1979. Diagenesis of coal (coalification). In G. Larsen and G. V. Chilingar (eds.), *Diagenesis in Sediments and Sedimentary Rocks.* New York: Elsevier, pp. 207–246.

Trumbull, J. 1960. *Coal Fields of the United States.* U.S. Geol. Surv. Map, 1:5,000,000.

Van Krevelen, D. W. 1961. *Coal.* New York: Elsevier, 514 pp.

Vinopal, R. 1979. X-ray radiography of coal. *Jour. Sed. Petrology, 49,* 483–486.

14

The Development of a Research Project

Blessed is he who expects nothing,
for he shall never be disappointed.

ALEXANDER POPE

As we observed earlier, there is no substitute for a good idea. Ideas are what separate the prosaic from the exciting, the mundane from the saintly. Psychologists are uncertain about the origin of new ideas, but it is clear that there are at least two essential requirements: (1) a wide background of facts about the general subject of interest; and (2) time to think about the facts, perhaps mentally reorganizing them so that gaps in understanding become more apparent.

BACKGROUND INFORMATION

With regard to sedimentary petrology, certain types of subject matter are clearly prerequisite to logical thought. Areas such as inorganic, organic, and physical chemistry, mineralogy, igneous and metamorphic petrology, and stratigraphy would appear on most lists of important background subjects. Fluid mechanics, paleontology, and ecology might be helpful as well, depending on the nature of the subject matter to be investigated. A large and growing proportion of the advances in our understanding of sedimentary petrogenesis during the past 20 years has been made by geoscientists with unusually strong backgrounds in appropriate mathematics, hydraulic engineering, and the sister sciences of chemistry, physics, and biology. Geology consists of the applications of other fields of science to a particular object: the Earth. In this respect, geologists are more closely allied to the engineering fields than to the "basic" sciences. We deal with applications rather than with the development of new fundamental principles (Kitts, 1974). From my point of view, you are doing yourself a great disservice if you enroll in only the minimum number of "outside science" courses required of you during your years in college. The shortest road to professional obsolescence is through deficiencies in training in the sciences that form the underpinnings of geology.

TIME FOR THOUGHT

It takes time to generate high-quality research ideas—time to assemble background information and time to consider its meaning. There is no substitute for intense thought about a subject over a considerable length of time. Although the source of "inspiration" commonly cannot be traced, few would doubt that it is typically preceded by mental effort—perhaps not during the time immediately before the flash of insight, but certainly at some time in the past. Nothing in science springs full-blown from a vacuum.

The first step in the development of a research project is to identify the major problems faced by workers interested in the same types of things. Useful descriptions of these problems are explicitly presented in the widely used textbooks that deal with sedimentary petrology, such as Bathurst (1975), Blatt et al. (1980), Pettijohn (1975), Pettijohn et al. (1973), and Potter et al. (1980). Clues to the initial identification of important problems may also be obtained by reading recent issues of the major publications in the field, such as *Clays and Clay Minerals, Journal of Sedimentary Petrology, Sedimentary Geology,* and *Sedimentology.* For those particularly interested in the geochemical aspects of sedimentary petrology, *Geochimica et Cosmochimica Acta* is a must. In addition, many journals of a more general character sometimes publish important papers dealing with major sedimentary petrological problems, for example, *Bulletin of the American Association of Petroleum Geologists, Bulletin of the Geological Society of America, Journal of Geology,* and *Marine Geology.* The research of others can be a useful initial guide when learning to separate the wheat from the chaff in research.

With experience and repeated exposure to the scientific literature, you will soon develop your own list of journals most useful for your particular interests. Each journal of significance should be examined on a regular schedule. This means at least scanning the titles of all articles, reading the abstracts of those that seem, from the titles, to be of some importance to your personal interests (perhaps 25% of the articles), and reading in full those that seem most pertinent. It would, of course, be better to read everything, but the number of publications in sedimentary geology is now so large and the size of each issue so thick that the ideal in reading habits is no longer attainable. It is becoming easy to imagine the time when a sizable proportion of published work will largely duplicate previously published material, simply because it will have become too time-consuming to search for, read, and digest earlier work. Perhaps progress in science is a self-limiting process.

THE EVOLUTION OF A PROJECT: "THE MINERAL COMPOSITION OF MUDROCKS"

How do you select a specific research project? There is no unique method for selection, and the examples we discuss in the remainder of this chapter were chosen simply because I am very familiar with the reasons they were selected and the way in which they developed.

Mudrocks are the most abundant type of sedimentary rock, forming about two-thirds of the stratigraphic column; and from field observations, it is obvious that the group of minerals we call clays is an important part of these rocks. There are no clay minerals in igneous and metamorphic rocks. Primary source rocks of sediments consist mostly of feldspars, with lesser amounts of quartz and other nonclay, silicate minerals. It seems reasonable to hypothesize, therefore, that the average composition of mudrocks might tell us a great deal about trends in mineralogic maturation caused by sedimentary processes. On a more local level, it seems possible that the mineral composition of sedimentary rocks might differ among different sedimentary environments. Why not investigate these questions by determining the mineral composition of a "world-average mudrock" obtained from an analysis of mudrocks of different ages from different geographic areas and from different sedimentary environments? It seemed to Daniel B. Shaw and Charles E. Weaver that this would be a worthwhile project and so, in the early 1960s, they decided to undertake the project. At the time, Mr. Shaw was studying for a master's degree at the University of Houston. Dr. Weaver, then employed in the research laboratory of the Shell Oil Company in Houston, was a specialist in clay mineralogy and agreed to supervise the research, which would form the master's thesis project for Mr. Shaw. The study was conducted in 1962–1963 and published in 1965 in the *Journal of Sedimentary Petrology,* a leading journal in the field of sedimentary mineralogy and petrology.

Search of the Literature

The first step in an investigation is to determine exactly what is known about the topic of study. So, before we consider the methodology and reasoning that Shaw and Weaver used in their work, it is useful to discuss the general principles of a literature search. There nearly always exists a small amount of published data relevant to a topic, and commonly you will find in seemingly unrelated articles bits and pieces of information that will turn out to be useful to you.

How do you make a literature search? One good place to start is with the references listed in major textbooks published within the past five years or so, for example, Potter et al. (1980), Blatt et al. (1980), and Friedman and Sanders (1978). Such books are essentially summaries of the knowledge existing at the time the books were published; each of the authors can be considered knowledgeable and familiar with the literature; and each book contains an extensive treatment of mudrocks. Read the sections dealing with mudrocks in each book, taking notes on 3 × 5 index cards about points that seem significant to you. When references of apparent importance are cited in the text, make a separate 3 × 5 card for each. Then scan the list at the end of the section for additional references that seem significant to you but that may not have seemed to the author of the book worth citing in the body of the text.

At the end of this phase of the search, you may have perhaps 50 references. Recall, however, that the books from which they came are not current. It takes a couple of years from the time an author submits a manuscript to a publisher until it appears for sale, so that the reference lists may be somewhat out-of-date when the book appears. To remedy this deficiency in your literature search, you need to examine the journals in which articles dealing with mudrock mineralogy are published. Look at the titles of articles in all issues of relevant journals published within three years of the date of publication of the most recent textbook you have examined. For example, if you were investigating the topic in 1982 and the most recent text was that of Blatt et al. in 1980, then you would need to examine journal issues from 1977 onward. Read all pertinent articles you find, makes notes on 3 × 5 cards, and examine the reference lists in each article—just as you did with the textbooks. You may have 70 cards at this point, perhaps 60 of which are for articles you have not yet read. But do not panic. Possibly half of these will turn out to be of no significance to you, given the approach to the topic that you wish to pursue, and this will be revealed by a rapid scanning of the article. However, for each article that *is* significant, you must examine its reference lists for new candidates. You will discover that many of those cited are titles you have previously encountered so you can start to breathe easier. You have passed the "expansion stage" in the search, and the number of duplicates will soon outnumber the number of new titles. Eventually, no new titles will appear. Study all your index cards and integrate in your mind the information on them, so that you have a good grasp of the current level of understanding of the subject.

How do you recognize the journals that you must examine during the literature search? The best way is to see which are cited most frequently in the textbooks you examined at the start of the search. For mudrock mineralogy, the leading journals may be *Journal of Sedimentary Petrology, Sedimentology, Bulletin of the Geological Society of America, Clays and Clay Minerals,* and *Geochimica et Cosmochimica Acta.* But many other publications will contain articles of interest, and it is well worth your time to examine recent issues of as many of them as possible. In this regard, there is no substitute for a good library. Items not present in your library can be obtained on interlibrary loan through the science librarian, but the process normally requires two to four weeks and can be very frustrating, depending on the presumed significance of the reference requested.

The search of the literature conducted by Shaw and Weaver in the early 1960s resulted in only 10 references worth citing in their 1965 paper, although they examined a much larger number during their search. Two of these were of particular significance. They were estimates of the mineral composition of the average "shale," made by Leith and Mead (1915, p. 76) and Clarke (1924, p. 33) based on interpretations of chemical analyses of 78 "carefully selected and weighted" shales analyzed in the laboratories of the U.S. Geological Survey in Washington, D.C., by Clarke prior to 1908 (see Table 14–1). It is noteworthy that, although the same chemical data were used in the two publications, the mineralogic interpretations ("modes") were strikingly different (see Table 14–2). Leith and Mead inferred only 34.80% micaceous minerals (kaolinite, muscovite, and chlorite) in the average "shale"; Clarke, even less. Leith and Mead inferred 17.60% feldspar; Clarke, 30.0%. These percentages of phyllosilicate minerals and feldspar seem very strange. Is it really possible that the average fissile mudrock (shale) contains only one-third clays and micas? Is it possible that the average shale contains 50–100% more feldspar than does the average sandstone? Probably not.

As Shaw and Weaver realized, probably most of the sodium and potassium that Leith and Mead and Clarke attributed to feldspar actually resided in montmorillonitic and illitic clays. In other words, it was apparent that the best one can hope to do in inferring minerals from a chemical analysis of a shale is to lump clays and feldspars together as inseparable without petrographic or X-ray diffraction studies. On this basis, Leith and Mead found 52.4% clay and feldspar; Clarke, 55–60%. Although we might suspect that these percentages still are a bit low for fissile mudrocks, they do seem "in the ballpark" of what we might expect.

Table 14–1
Composite Oxide Analysis of 78 Shales

Oxide	Percent
SiO_2	58.38
Al_2O_3	15.47
Fe_2O_3	4.03
FeO	2.46
MgO	2.45
CaO	3.12
Na_2O	1.31
K_2O	3.25
H_2O	5.02
TiO_2	0.65
CO_2	2.64
P_2O_5	0.17
SO_3	0.65
BaO	0.05
C, organic	0.81
Total	100.46[a]

[a]Numbers do not total to exactly 100.00% because of minor inaccuracies in the analyses.

Source: Clarke, 1924.

Table 14–2
Mineralogic Interpretation of Oxide Analyses Made by Clarke in 1908 as Made by Leith and Mead (1915) and Clarke (1924)

Mineral	Leith and Mead (1915)	Clarke (1924)
Quartz	31.91	22.3
Orthoclase	12.05	30.0 (feldspar)
Albite	5.55	
Kaolinite	10.00	25.0 (clay)
Muscovite	18.40	
Limonite	4.75	5.6
Dolomite	7.90	5.7 (carbonates)
Chlorite	6.40	11.4 (others)
Gypsum	1.17	
Rutile	0.66	
Apatite	0.40	
Carbon	0.81	
Total	100.00	100.0

Techniques of Rock Analysis

In order for Shaw and Weaver to analyze accurately the mineral composition of rocks as fine-grained as mudrocks, it was necessary to develop reliable and rapid methods of qualitative and quantitative analysis. Petrographic methods were of little help because clays conceal a large proportion of the quartz and feldspar in thin sections of shales. An X-ray technique seemed like a good method. As we have seen in Figure 3–8, each crystalline constituent in a diffractogram produces a characteristic X-ray diffraction pattern that can be distinguished readily from the patterns of other minerals in the powdered mixture. Diffraction patterns are an excellent tool with which to determine the presence or absence of a mineral in a mixture. However, they possess some inherent difficulties when quantification of mineral percentages is desired.

Shaw and Weaver had to develop a method for obtaining accurate and reproducible results for mineral percentages in polymineralic mixtures. Fortunately, Dr. Weaver was (and is) a clay mineralogist with extensive experience in X-ray diffraction studies and was able to develop a technique based on the comparison of the diffraction tracing in each sample with artificial standards. After extensive experimentation, he developed a method in which the deviation of the percentages in the unknowns from the true percentages was within 10% of the amount present. This degree of accuracy was deemed satisfactory for the purpose of the investigation.

Procedure for Sampling

The objective of Shaw and Weaver was to characterize "the world of mudrocks" from a mineralogic point of view. How does one sample "the world"? True random selection is clearly not possible because of the usual limitations of time and funds for traveling around the world, to say nothing of the political difficulties associated with an American citizen getting permission to sample mudrocks in Afghanistan, Albania, Cambodia, and similar places. The solution to the sampling problem adopted by Shaw and Weaver was to restrict the study to American rocks and to assume that these would be a good representation of the world's mudrocks. This would seem to be a reasonable assumption, as the United States is large enough to assure that rocks of all geologic ages and depositional environments are represented. Certainly, ancient rocks and depositional environments were not restricted by the political boundaries that exist today.

The problem of sampling the United States was alleviated by the fact that Dr. Weaver was employed by a major oil company that has a large collection of surface samples and core samples from all parts of the country. Mr. Shaw's access to these materials was arranged and, although some surface samples were used, most samples were from subsurface cores. The major sampling areas were:

1. Tertiary of southeastern Texas Gulf Coast.
2. Mesozoic of western Canada and west-central United States.
3. Paleozoic of Marathon Uplift, western Texas.
4. Paleozoic of Bend Arch area, north-central Texas.
5. Paleozoic of Ouachita and Arbuckle Mountains, Ardmore and McAlester Basin, Oklahoma and Arkansas.
6. Paleozoic of Appalachian Province, Pennsylvania, West Virginia, and Tennessee.
7. Miscellaneous shales.

A total of 400 samples was analyzed. Samples that were obviously silty or calcareous were arbitrarily excluded by Shaw and Weaver. Therefore, their conclusions concerning mudrocks do not apply to siltstones and silt-shales, as we defined them in Chapter 3 (see Table 3–1). This leads us to suspect that their results would be biased toward mudrocks with higher percentages of clay minerals and lesser amounts of quartz and feldspar. Probably this bias was also present in the samples analyzed chemically by Clarke in the early part of this century. He stated that he analyzed "shales;" and to nearly all students of sedimentary rocks, the word "shale" implies fissility. Few rocks that contain large percentages of silt are fissile.

Results of the Research

The average composition of the samples from each of the seven geographic–stratigraphic sediment volumes studied by Shaw and Weaver is shown in Table 14–3. With the exceptions of the high percentage of quartz in sample group 3 and the high percentage of feldspar in group 7, there are no striking differences in gross composition among the seven groups. We note also that, although only mudrocks lacking visible silt were analyzed, the percentage of quartz is higher than we might have suspected: 31%. If a more randomly chosen group of mudrocks was analyzed, the amount of quartz might well be as much as 40% because coarser-grained rocks (excluding conglomerates) contain more quartz than do finer-grained ones. The same statement would apply to feldspar. The amount of clay minerals would decrease correspondingly to perhaps 50%.

How do these data compare with the estimates of Leith and Mead and of Clarke, which were based on inferences from chemical analyses? The earlier workers inferred too much feldspar and too little clay, as we suspected. Clarke, in particular, overestimated feldspar by a factor of 7, which led him to significantly underestimate the percentage of quartz. Alkali feldspars are silica-rich minerals. Clarke used so much silica to make feldspar in his estimate of mineral percentages that he had little left over to use for quartz! It is clear that potassium-rich clay (illite) must be very abun-

Table 14–3
Average Percentages of Quartz, Feldspar, Carbonate, and Clay Minerals Based on 305 Shale Samples from United States and Canada as Determined with X-Ray Diffraction-Absorption Technique

		Nonclay minerals (±10%)			Estimated
Area	No. of samples	Quartz	Feldspar	Carbonate	% Clay
1	21	26.5	6	3.5	64
2	39	30	5	7	58
3	27	47	2	1	50
4	38	31	2	3	64
5	103	29	2.5	—	68.5
6	20	23	3	8	66
7	57	29	11	3	56
Unweighted average		30.8	4.5	3.7	60.9

Source: Shaw and Weaver, 1965.

dant among the clays in shales. (This inference is supported by many studies of the clay mineralogy of shales, as we observed in Chapter 3.)

What other things of possible interest might be done with the data obtained by Shaw and Weaver? Well, we have determined the percentages of minerals in several hundred shales. We might construct frequency curves to show how often different values appear. Perhaps shales tend to have either no feldspar or large percentages, resulting in the calculated average of 4.5%. Is a shale as likely to contain 2.5% feldspar as 6.5%? Examination of cumulative frequency curves for each mineral constituent can answer questions such as these. This led Shaw and Weaver to construct the series of cumulative frequency curves shown in Figure 14–1. (Note that the averages given in the figure differ slightly from those in Table 14–3 because of the different method of calculating the averages.) From the figure we see that the distribution of quartz percentages is aproximately Normal, with a mean of 28% and a standard deviation of 9%. Feldspar, in contrast, reveals a much different frequency distribution, not only in the mean percentage (hardly a surprise) but in the shape of the curve as well. A very strong positive skewness is obvious and is calculated to be +0.79—exceptionally high for any frequency distribution. (Recall that the mathematical limits of this parameter are from −1 to +1.) What does this indicate? Why do about 50% of the samples contain percentages of feldspar so markedly different (the sharp change in slope of the frequency curve) from the other 50%? Might the shales that contain large amounts of feldspar be those that are interbedded with

Figure 14–1
Cumulative frequency curves on arithmetic probability graph paper of percentages of (**A**) quartz, (**B**) feldspar, (**C**) carbonate, and (**D**) clay minerals in shales with less than 50% quartz studied by Shaw and Weaver (1965). Parameters calculated using formulas on pages 522 and 524.

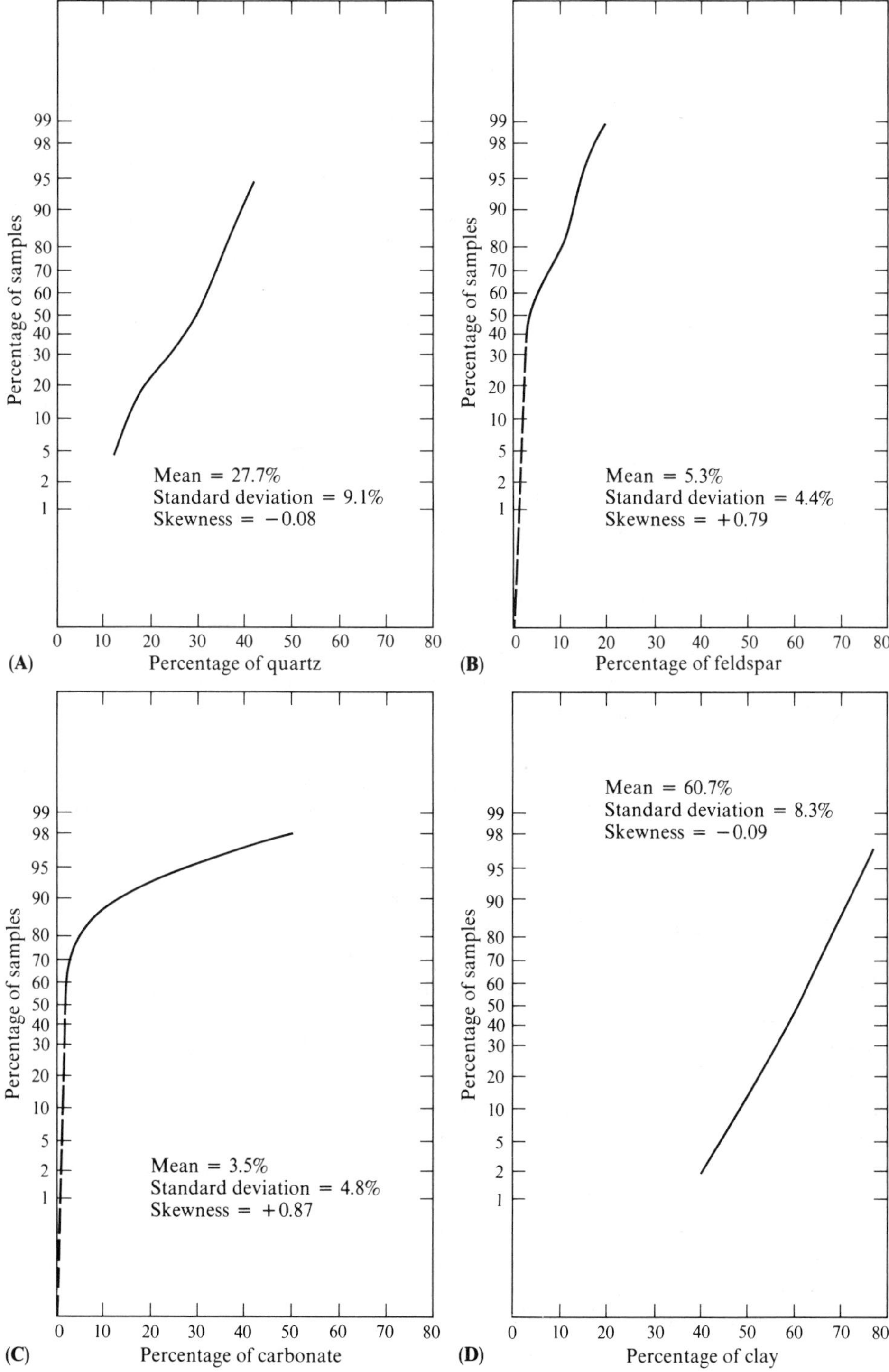

Percentage of samples
99
98
95
90
80
70
60
50
40
30
20
10
5
2
1
Mean = 27.7%
Standard deviation = 9.1%
Skewness = −0.08
0 10 20 30 40 50 60 70 80
(A)
Percentage of quartz
Mean = 5.3%
Standard deviation = 4.4%
Skewness = +0.79
(B)
Percentage of feldspar
Mean = 3.5%
Standard deviation = 4.8%
Skewness = +0.87
(C)
Percentage of carbonate
Mean = 60.7%
Standard deviation = 8.3%
Skewness = −0.09
(D)
Percentage of clay

arkosic sandstones? If so, we may be able to use the nonclay fraction of shales for provenance studies, just as we use the nonclay fraction of sandstones for this purpose.

Shaw and Weaver tried to assess this possibility by analyzing a collection of shales with known sandstone associations. They divided the shales into three groups:

1. One with 89–95% quartz, 0–5% feldspar, and a quartz/feldspar ratio of 15.
2. A second group that contained 50–75% quartz, 10% feldspar, and a quartz/feldspar ratio of 8.
3. A final group with approximately equal amounts of quartz and feldspar (35%) and a quartz/feldspar ratio of 1.

For the associated sandstones in the first group, the analogous numbers were 31 Q, 3.5 F, and a Q/F of 8.6; in the second group, 32, 4.3, and 7.5; and in the third group, 23, 19, and 1.2. From Figure 14–2 it is apparent that, when the percentage of feldspar in the rocks is low, the Q/F ratio is the same in interbedded sandstones and shales. But when the ratio exceeds about 8, the sandstones seem to lose their feldspar; the Q/F ratio rises much faster in the sandstones. What might be the explanation for this occurrence? Perhaps the more feldspathic sandstones are also more permeable, and their contained feldspars are being altered to clay or dissolved by diagenetic waters. Such speculations lead to new ideas for research into both the accuracy of provenance studies of sandstones and the importance of diagenesis in modifying the depositional mineral composition of sedimentary rocks.

The frequency curve for the percentage of carbonate minerals in shales is, like that for feldspars, extremely positively skewed. This time, about 70% of the samples contained a "normal" amount of carbonate (less than 2%), and 30% seemed peculiarly rich in these minerals. Could the natural separation of shales into a "high" carbonate group and a "low" carbonate group be a result of provenance variation, as we inferred for the feldspars? Is carbonate in detrital rocks normally detrital? Our experience with rocks suggests not. Carbonate minerals result from either fossil shells or diagenetic cement in nearly all detrital rocks. Which is more likely in shales? There are no data with which to decide, but perhaps fossils are the better guess because of the low permeabilities to diagenetic waters associated with such fine-grained rocks. Once again, new research projects are suggested. In addition, both calcite and dolomite are common minerals in carbonate-bearing rocks, based on investigations of sandstones and conglomerates. And the ratio of dolomite to calcite is known to increase in carbonate rocks with increasing geologic age (see Chapter 6). Is this true of detrital mudrocks as well? If it is, what might we infer about the processes that form dolomite from precursor calcite?

The frequency distribution of percentages of clay minerals (What kinds?) in shales is, like the distribution for quartz, approximately Normal. (Values of skewness between -0.1 and $+0.1$ generally are considered to be "unskewed"; i.e., the departure from perfect Normality is too small to be interpretable.) Shales are as likely to contain 69% clay as 52.4% clay, as likely to contain 77.3% as 44.1%. There seem to be no

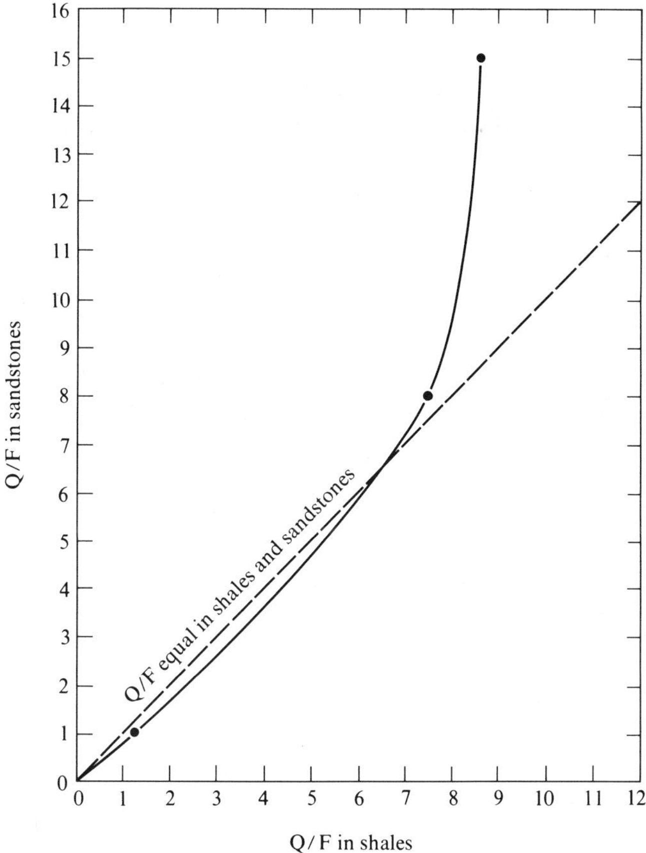

Figure 14–2
Relationship between quartz/feldspar ratios in three groups of interbedded shales and sandstones studied by Shaw and Weaver (1965).

natural peculiarities present of the types we found for feldspars and carbonate minerals. What do you think would happen to the shape of the frequency curve for clays if the more coarsely grained mudrocks had been included in the analyses made by Shaw and Weaver?

The study of mudrocks by Shaw and Weaver filled an obvious gap in our understanding of mudrocks (more exactly, shales). Because of their work, we now have a much clearer understanding of the mineral composition of these abundant sedimentary rocks, particularly of the nonclay fraction. As with most investigations, however, their results raised as many new questions as they answered. How closely does grain size control the mineral composition of the nonclay fraction of detrital rocks? What is the importance of diagenesis as a control of the mineral composition of an ancient rock? What degree of accuracy can we reasonably expect in a provenance interpretation based on the mineral composition of an ancient sandstone? The first of these questions was investigated for quartz by Blatt and Schultz (1976).

ONE GOOD PROJECT LEADS TO ANOTHER: "THE SIZE DISTRIBUTION OF QUARTZ IN MUDROCKS"

As we observed in Chapter 3, the sedimented size distribution of clay minerals in fine-grained detrital rocks is not determinable because of uncertainties in the degree of original flocculation, the inherent imprecision of peptizing techniques, and the effects of demonstrable recrystallization on original crystal size with the creation of new clay-mineral particles. It may, however, be possible to recover the sedimented size distribution of quartz and feldspar, which form at least 35% of the average mudrock. To determine the grain-size distribution of these grains, it is obvious that we must free them from the enclosing mass of clay minerals. How might this be done? One possibility is the peptizing method used to disaggregate the clays. This would have two drawbacks, however. First, the presence of undisaggregated clay floccules would cause problems because the floccules probably enclose fine silt- and clay-size quartz within them. Second, how do we isolate the quartz and feldspar from the clay minerals in the disaggregated sediment we have produced from the mudrock? The common clay minerals have specific gravities very close to those of quartz and feldspars, so that heavy-liquid separation would not be effective. It is commonly noted in "technique books" dealing with sediments that micas can be separated from less platy minerals by sliding the loose grains down an inclined and vibrating sheet of paper, but it seems unlikely that this technique would be very effective on particles as small as clays and on invisible particles such as flaky quartz and feldspar only a few microns in diameter. What we need is a chemical technique that would magically dissolve all the minerals except the quartz and feldspar. Fortunately, exactly such a technique was devised in 1969 by Chapman et al., as noted in Chapters 3 and 15.

In 1973, at the University of Oklahoma, Douglas J. Schultz was looking for a research topic for his master's thesis and decided to follow up on the work of Shaw and Weaver by using the technique of Chapman et al. to determine the grain-size distribution of quartz in a "world-average mudrock." Once again, a sampling problem needed to be solved before the study could begin. The sampling problem was made easier by the results of Shaw and Weaver, which indicated that the percentage of quartz in mudrocks differed very little among different geographic areas (see Table 14–3). Blatt and Schultz made the assumption that the size distribution of the mineral would be similarly independent of geographic location. Consequently, they conducted the sampling of mudrocks mostly in the Oklahoma–New Mexico–Colorado area. A few additional samples were used from existing rock collections at the University of Oklahoma. In order to extend the data from the mudrocks into coarser sizes, Blatt and Schultz also collected samples from the fine-grained sandstones interbedded with the mudrocks. These samples were subjected to the same chemical treatment as the mudrocks to assure that the results for the two groups of rocks would not be affected by differences in the chemical treatment they received. Schultz analyzed 16 mudrock–sandstone pairs (32 rock samples) ranging in age from

Devonian through Cretaceous, although there was no reason to believe that geologic age would affect the results. The range in age was simply a byproduct of sampling over a geographic area covering tens of thousands of square kilometers. The significant variable in a grain-size project such as Schultz was attempting should be kinetic energy levels in the depositional environment, and these are unrelated to geologic age.

The results obtained by Blatt and Schultz are shown in Table 14–4 and Figure 14–3, and the correspondence between their results and those of Shaw and Weaver is striking. Shaw and Weaver estimated the graphic mean percentage and standard deviation of quartz in mudrocks to be 27.7% and 9.1%, respectively, based on 400 samples. The estimates by Blatt and Schultz, based on 16 samples, were 27.6% and 10.7%. From Table 14–4 it is evident that about three-fourths of the quartz grains in mudrocks are silt size, with the remainder equally divided between sand and clay sizes. The mean size of the quartz is barely into the fine silt range at 6.1 ϕ. The small size of the quartz fragments in mudrocks, which form two-thirds of all sedimentary rocks, raises a new significant question. How are such small grains produced? Certainly, there is very little quartz 6.1 ϕ in size falling from freshly disintegrated granites, gneisses, and schists. Are slates and phyllites abundant enough to supply all this very fine-grained quartz, or has the quartz been produced by sedimentary processes acting on the coarser grains? How might this question be investigated?

Figure 14–3 shows the relationship between the percentage of quartz (including chert) and the mean grain size of the quartz in the mudrocks and sandstones examined by Blatt and Schultz. A high degree of correlation is obvious between the two variables. Statistical analysis indicates that there is less than 1 chance in 100 that the correlation could be spurious. The finest mean size in the 16 mudrocks examined was 7.3 ϕ; but if we extrapolate the best-fit line to zero percent quartz, we can determine

Table 14–4
Percentage of Quartz in Mudrocks and Sandstones Studied by Blatt and Schultz (1976) and Grain-Size Distribution of Quartz in These Rocks

Amount of quartz	
Mudrocks	**Sandstones**
Mean = 27.6% quartz	Mean = 61.8% quartz
Standard deviation = 10.7%	Standard deviation = 11.4%
Size frequency distribution of quartz	
Mean size = 6.1 ϕ (14μm)	4.0 ϕ (62 μm)
Standard deviation = 1.5 ϕ	1.4 ϕ
Skewness = 0.0	+.3
Sand = 11.6%	55.4%
Silt = 73.0%	42.1%
Clay = 15.4%	2.5%

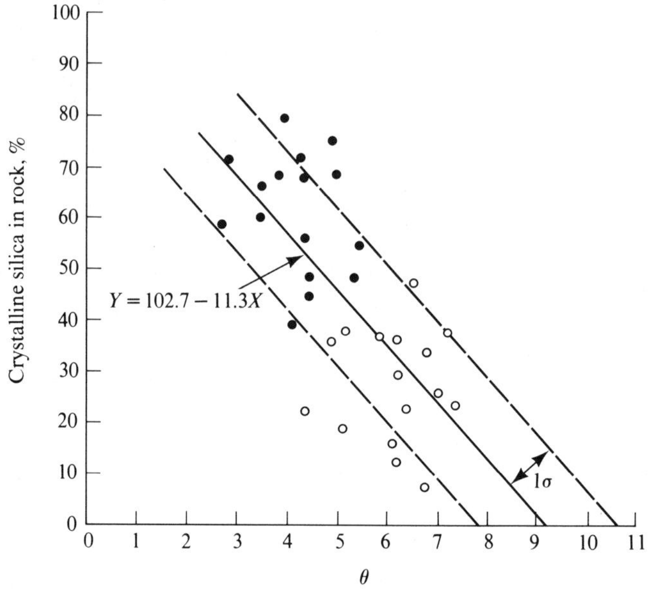

Figure 14–3
Relationship between percentage of quartz and mean grain size of quartz in mudrocks and sandstones examined by Blatt and Schultz (1976). Open circles = mudrock. Solid circles = sandstone.

the average grain size at which quartz disappears from mudrocks. This size is 9.1 ϕ (1.8 μm). Note that this is the *average* size of the quartz grains, not the size of the smallest grain of quartz. The value of 1.8 μm agrees well with observations made by clay mineralogists. They routinely conduct clay-mineral analyses of the size fraction smaller than 2 μm because the presence of quartz generates diffractogram peaks that overlap with and conceal diagnostic clay-mineral peaks. The clay mineralogists' experience tells them that the quartz peaks are greatly diminished or absent in the fraction less than 2 μm. The value of 2 μm is also supported by the work of Moss and Green (1975), who examined quartz from plutonic rocks with the scanning electron microscope. They found that the quartz crystals in the plutons are almost always cut by partially healed microfractures separated by distances of 2–20 μm. This planar fracture system may set the lower limit of detrital quartz to be expected in fine-grained rocks.

It is clear that grain size exercises a major control on the mineral assemblage that can reasonably be expected in a detrital rock. Granitic rock fragments do not appear in the silt sizes; clay minerals do not form a large percentage of the sediment in most cobble conglomerates. Further, it is noticeable from Figure 14–3 that mudrocks and sandstones have overlapping grain sizes. Both mudrocks and their interbedded sandstones have abundant quartz in the size range 4–6 ϕ. Possibly, this would be equally true of feldspar grains in interbedded muds and sands.

AND ANOTHER: "FELDSPAR GRAINS AND DIAGENESIS"

The fact that (1) grain size is an important control on sediment mineralogy and that (2) muds and sands have overlapping grain-size distributions generated a new insight into mudrock mineralogy. It is suspected by many sedimentary petrologists that feldspar grains are commonly dissolved by percolating underground waters during diagenesis. How might we determine whether this has happened in a particular sediment we are interested in? Clearly, to determine how much (if any) feldspar is missing, we need to know the amount originally present. It occurred to Blatt, following the results of Schultz, that the existence of overlapping size distributions in muds and sands might provide the key to determining the amount of feldspar deposited in a sediment. It seemed reasonable to hypothesize that a size fraction common to both a fresh mud and a fresh sand would contain an identical mineral composition. Consider the 4.0–4.5 ϕ fraction in a newly deposited river mud and adjacent fine sand. A feldspar grain does not know or care whether it comes to rest among mostly clay grains or mostly quartz grains. Its deposition results from a decrease in current velocity, which may occur in the water overlying mud or overlying sand. If this is true, then the proportion of feldspar in the 4.0–4.5 ϕ size fraction of the mud and adjacent sand should be identical at the time of deposition. Muds, however, compact rapidly and efficiently and are soon impermeable, whereas sands can maintain their permeability to water for very long periods of time. Therefore, ancient mudrocks might well preserve the feldspar/quartz ratio they had when deposited, but sandstones might not. They probably would contain less feldspar because of removal by percolating underground waters during diagenesis over the tens or hundreds of millions of years since they were deposited.

In the fall of 1975, R. G. Charles enrolled in the graduate program in geology at the University of Oklahoma to work toward a master's degree in sedimentary petrology. Blatt suggested that Charles investigate the hypothesis that identical size fractions from a mud and a sand would have the same feldspar/quartz ratio. As was true for the study by Schultz, the location from which the samples came was of no significance. The only sampling requirement was to obtain mud–sand sample pairs that ranged widely in feldspar content so that the final x–y plot comparing the feldspar content in the muds with that in the sands would show a scattering of points extensive enough to permit a significant line to be drawn through them.

This requirement was satisfied by sampling 20 mud–sand pairs (40 samples) from rivers in Oklahoma, Arkansas, Missouri, Louisiana, Texas, and Mississippi during the summer of 1976. The 40 samples were fused in sodium bisulfate to remove all constituents except quartz and feldspar, the grains were sieved to isolate the 4.0–4.5 ϕ size fraction (and other fractions as well), and atomic absorption spectrophotometry was used to obtain both the amount of feldspar and the normative percentages of orthoclase, albite, anorthite, and quartz. The amounts of each of these normative minerals were compared for each mud–sand pair. Some of the results are shown in Figure 14–4. Nearly identical results were obtained for the other normative

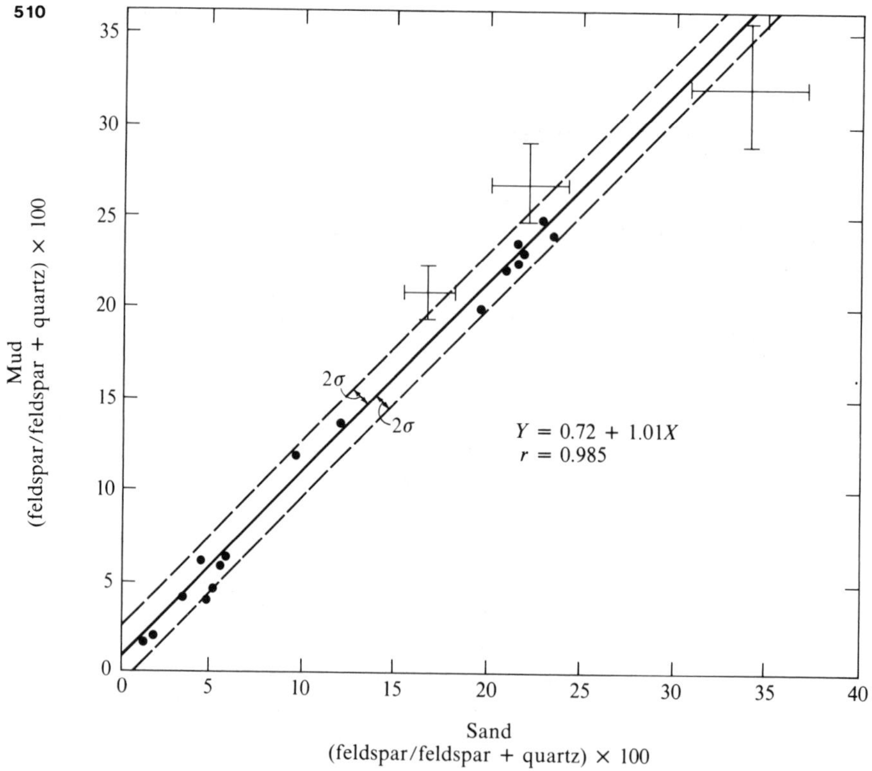

Figure 14–4
Percentage of feldspar in feldspar/feldspar plus quartz fraction of 4.0–4.5 ϕ size fraction of 20 mud–sand pairs examined by Charles and Blatt (1978). Error bars around the four data points furthest from best-fit line indicate one standard deviation for atomic absorption data, which is 10% of amount of feldspar present in a sample. Line of best fit to data points is at 45° angle, indicating that ratio of feldspar to feldspar plus quartz is the same for muds and sands.

minerals and for the 10–20 μm size fraction isolated by the use of micromesh sieves (see Chapter 3). It is clear from Figure 14–4 that muds and sands do indeed have identical feldspar/quartz ratios at the time of deposition if we examine a narrowly sized subfraction of the sediments. The original hypothesis now proven, a method had been devised for determining the extent of removal of feldspar from permeable sandstones by diagenetic processes. The results were published jointly by Charles and Blatt (1978). Investigations of ancient mudrock–sandstone pairs are now in progress to evaluate the general importance of diagenetic dissolution of feldspar from sandstones. Knowing that it *can* occur is one thing; demonstrating that it is a geologically widespread process remains to be proven.

It is worth noting that at the start of Charles' research, neither he nor his supervisor knew how to use an atomic absorption spectrophotometer. They knew in general what the machine was designed to do, and they obtained permission to use the one belonging to the Oklahoma Geological Survey. The chemists in the Survey's laboratory offered to teach them how to use the spectrophotometer. The point is that it can be very useful to understand the types of things that can be done by chemists and geochemists. Learning how to do them is always possible if the desire is present. Do not be afraid to use what seems to you to be "exotic" equipment.

SUMMARY

The purpose of this chapter has been to illustrate that ideas do not spring from a vacuum and that each investigator stands on the shoulders of those who preceded him or her. The examples used were chosen because they are familiar to the author of this textbook; he was intimately involved with them. Other researchers could cite analogous examples from their own experience. Everyone's thought processes are different, however, so that different tracks are followed by different investigators. The key element common to all is a consuming interest in and commitment to the investigative process. If we assume that everyone who spends four or more years of study in the fundamentals of geology has such a commitment, we must conclude that good research can be accomplished by most of us.

FURTHER READING

Bathurst, R. G. C. 1975. *Carbonate Sediments and Their Diagenesis,* 2nd ed. New York: Elsevier, 658 pp.

Blatt, H., G. V. Middleton, and R. C. Murray. 1980. *Origin of Sedimentary Rocks,* 2nd ed. Englewood Cliffs, NJ: Prentice-Hall, 782 pp.

Blatt, H., and D. J. Schultz. 1976. Size distribution of quartz in mudrocks. *Sedimentology, 23,* 857–866.

Chapman, S. L., J. K. Syers, and M. L. Jackson. 1969. Quantitative determination of quartz in soils, sediments, and rocks by pyrosulfate fusion and hydrofluosilicic acid treatment. *Soil Sci., 107,* 348–355.

Charles, R. G., and H. Blatt. 1978. Quartz, chert, and feldspars in modern fluvial muds and sands. *Jour. Sed. Petrology, 48,* 427–432.

Clarke, F. W. 1924. *The Data of Geochemistry.* U.S. Geol. Surv. Bull., *770,* 841 pp.

Friedman, G. M., and J. E. Sanders. 1978. *Principles of Sedimentology.* New York: Wiley, 792 pp.

Kitts, D. B. 1974. Physical theory and geological knowledge. *Jour. Geol., 82,* 1–23.

Leith, C. K., and W. J. Mead. 1915. *Metamorphic Geology.* New York: Holt, 337 pp.

Moss, A. J., and P. Green. 1975. Sand and silt grains: predetermination of their formation and properties by microfractures in quartz. *Jour. Geol. Soc. Austral., 22,* 485–495.

Pettijohn, F. J. 1975. *Sedimentary Rocks,* 3rd ed. New York: Harper, 628 pp.

Pettijohn, F. J., P. E. Potter, and R. Siever. 1973. *Sand and Sandstone.* New York: Springer-Verlag, 618 pp.

Potter, P. E., J. B. Maynard, and W. A. Pryor. 1980. *Sedimentology of Shale.* New York: Springer-Verlag, 306 pp.

Shaw, D. B., and C. E. Weaver. 1965. The mineralogical composition of shales. *Jour. Sed. Petrology, 35,* 213–222.

15

The Practice of Sedimentary Petrology

If some great catastrophe is not announced every morning, we feel a certain void.

PAUL VALÉRY

HISTORICAL PERSPECTIVE

The interest of people in sedimentary materials predates recorded history but no doubt originated because of the need for tools and weapons and the need to understand naturally occurring dangers such as floods and landslides. With the advent of agrarian communities, considerations of sediment variations increased because of questions concerned with the nourishment of plants. Which crops grow best in certain types of soil, and why? Is an argillaceous soil better than one that is not? Is a soil developed on a volcanic slope superior to one on a white sandstone? Practical answers to such questions could often be gained simply by observation and trial and error. Scientifically satisfying answers, however, had to await the development of the microscope, particularly the polarizing microscope, and the knowledge of X-rays.

The polarizing prism was described by William Nicol in 1829, and during the succeeding two decades many thin sections were made of objects of interest to a modern sedimentary petrologist, such as chert, chalcedony, molluscan shells, petrified wood, coal, concretions, and limestones. It was not until 1849, however, that the fundamental geologic importance of rock thin sections was recognized in a public forum. In that year, an untrained high school graduate in Sheffield, England, presented a talk to the local Literary and Philosophical Society. The man was Henry Clifton Sorby, now universally regarded as the father of petrography. His first article on the subject appeared in 1851; it was a paper on a cherty limestone in which he distinguished among calcite, detrital quartz, and authigenic chalcedony. In 1877 and 1880 he described the different appearances in polarized light of quartz in and from igneous rocks, metamorphic rocks, and sandstones; and he showed how quartz could be used in provenance studies. In 1879 Sorby published the first paper dealing with the petrography of carbonate rocks, delineating the microscopic shell structure and mineralogy of fossils and modern invertebrates, investigating their selective abrasion, discussing the replacement of aragonitic shells by calcite and the mechanism by which it occurs, and describing coccoliths, calcite cementation, and dolomitization. Sorby's papers are valuable reading even today; a bibliography of the more important ones is given by Folk (1965).

X-rays were discovered in 1895 by Wilhelm Roentgen in Germany, and the power of these short wavelengths for the identification of minerals and the determination of their crystal structures was recognized in 1912 by W. Friedrich, C. M. P. Knipping, and M. T. F. von Laue. Succeeding decades saw the development of intense interest in such work. From the viewpoint of sedimentary petrography, the most important characteristic of X-radiation is that it can be used to recognize and distinguish among the various clay minerals—particles too small to be identified firmly with a polarizing microscope. X-ray techniques are used routinely by sedimentary petrologists. More recent developments in analytical techniques have proven nearly as powerful as X-rays in the study of sedimentary rocks. Tools such as the electron microprobe (1950s) and cathodoluminescence petrography (1960s) have greatly increased our understanding of sedimentary mineralogy and diagenesis. The ion microprobe (1970s) promises to increase even more our insights into the petrogenesis of sedimentary rocks.

THE CHOICE OF A PROJECT

It is apparent that there exists an extremely large number of objects of interest to geologists; for practical purposes the number can be considered infinite. These objects include formations, members, faults, brachiopods, quartz grains, cross beds, and topographic irregularities of one kind or another. Life, however, is not infinite despite the extraordinary achievements of Late Holocene medicine. Therefore, it is necessary for those who wish to advance our understanding of sedimentary rocks to choose projects carefully. The importance of care in choosing projects has been succinctly stated by Nobel laureate P. B. Medawar in his excellent collection of essays *Advice to a Young Scientist* (1979, p. 13):

> It can be said with complete confidence that any scientist of any age who *wants to make important discoveries must study important problems.* Dull or piffling problems yield dull or piffling answers. It is not enough that a problem should be "interesting"—almost any problem is interesting if it is studied in sufficient depth.

There are two requirements that must be satisfied to achieve the highest order of success in research: inspiration and perspiration.

The Inspiration of an Idea

The ability to generate ideas that lead to new insights is uncommon in all types of human endeavors; sedimentary petrology is no exception. Since the beginning of the field 130 years ago, only a few published papers stand out as classics; Sorby (1880) on quartz, Rubey (1929) on the Mowry Shale, Russell (1937) on Mississippi River sands, Pettijohn (1941) on intrastratal solution, Bramlette (1946) on the Monterey cherts, Krynine (1950) on arkoses, and Folk (1959) on limestones are among those generally cited as standards.

Is there a common characteristic among these classics? If there is, it is the nature of the question each investigator asked before choosing his project. The question may be expressed as: What is an important problem in my field of interest? For Sorby it was the relationship between crystalline source rocks and the characteristics of detrital quartz grains, the most abundant mineral in sandstones. For Russell it was the effect of fluvial transport on the mineral composition of sands. For Pettijohn it was the effect of diagenetic waters on the dissolution of heavy minerals. For Folk it was the use of textures to determine the depositional history and diagenesis of limestones. It is noteworthy that none of these studies originated with the statement: No one has examined the Makework Formation yet; I will study it. The gathering of data from randomly chosen formations is useful and may lead to new insights, but it is not the surest path to this objective. The most productive hours a scientist can spend may consist of staring at a blank wall and thinking. The most productive person may be the one whose hands are still at frequent intervals.

The Design of the Project

The choice of a project forms the first half of the early part of research. The second half consists of designing the work in such a way that as many variables as possible are controlled and that as unique a solution to the problem is obtained as possible. This may involve care in choosing the geographic or geologic region in which the field study is to be done. It may involve prudent choice of laboratory techniques or possibly comparison of surface with subsurface samples to make sure that the lack of cement in outcrop samples is not due simply to surface weathering rather than to ancient diagenetic history.

Projects have an inherent tendency to expand beyond the bounds originally set; so if you wish to reach a useful conclusion in a reasonable length of time, you must define the project very carefully. No matter how limited and compact it appears at the outset, it is almost certain to grow uncontrollably if not weeded regularly. It may be nice to analyze 20 additional samples, but is it really worth the time and effort? Masses of excess numbers are impressive, but commonly they are of questionable scientific benefit. A mere 20 samples may yield as useful a conclusion as 50; the precision of a quartz/feldspar ratio in a sandstone is typically satisfactory when based on a count of 100 grains, despite the impressiveness of a 300-grain count. An elementary understanding of mathematical statistics goes a long way toward preventing a time-consuming excess of data collection.

The object of data collection is to obtain the maximum amount of important information in the shortest period of time. Probably, 80% of "The Truth" can be obtained in 20% of the time required to reach 95% of such truth; 100% is impossible. Do not become sidetracked by minor details, regardless of their psychological appeal or esoteric character. Keep your eye on the ball.

Sampling Procedures

A sampling program can be simple or complex, but its design must be always guided by the question: Exactly what question am I trying to answer in my work? As obvious as this may seem, it is commonly lost sight of during the choice of samples for subsequent laboratory analysis. For example, surface samples may be a poor choice for analysis in a study aimed at determining the variability in type and degree of cementation of a sandstone. The effects on cementation of the past 20 years of weathering since the exposure was created by the state highway department may overshadow completely the cementation pattern of the previous 200 million years.

In the determination of the provenance of a sequence of sandstones and mudrocks, it may be better to sample the mudrocks rather than the sandstones because of the likelihood of the removal of diagnostic minerals from the sandstone by intrastratal solution. The less permeable mudrocks commonly retain unstable minerals no longer present in the sandstones. Remember, the fact that the sandstone is tightly cemented and impermeable to water at present does not necessarily mean that it has always been so. It may have been cemented and decemented many times during its 200-million-year history. A mudrock, on the other hand, is not susceptible to such changes. Once lithified, mummification of the sand and silt grains in it is much more likely than in interbedded sandstones.

The geologist, like other scientists, finds the Earth too vast and complex to be studied as a whole. It is necessary, therefore, to treat a part of it—to select individuals for observation that are characteristic of the larger population from which they are drawn. The sample is used to obtain measurements from which inferences are made about the population.

In the choice of a sample, it is essential that preferences and prejudices be excluded. Moreover, the method of selection must not be correlated with the attribute under study. It is easy to prove that all odd numbers are prime if the sample of odd numbers that are studied is selected to contain only prime numbers. To circumvent the natural tendency toward subjective bias (usually undesirable), the concept of the random sample has evolved. The term *random* may be defined as the circumstance that every individual in the population has an equal chance of being chosen for a sample. This implies the absence of pattern, which could be used to predict what the next position might be.

Randomness is foreign to human processes, so that some mechanical means must be relied on to secure a true random sample. Random numbers are available in tables of the Rand Corporation and in some statistics texts and can be utilized for map work. Randomness is characterized by clustering. If clusters are not present, then there must, by necessity, exist a series of points whose density of occurrence is identical in all directions. Clearly, such a situation is the opposite of our definition, because it requires perfect symmetry. Meditate for a few moments on these last few sentences. Somewhat paradoxical but true.

Once the population of interest is defined, a sampling plan (a plan to insure randomness?) is set up in terms of the objectives of the study. Several sampling plans in use for geologic problems are (1) simple random, (2) systematic, (3) nested, (4) stratified, and (5) dumbbell.

In most petrological projects, some modifications must be made in the five sampling plans listed because we do not have complete freedom of sampling. Sometimes a sampling point is located under a talus slope, or perhaps the layer of interest has been eroded by a modern stream at the intended sampling point. In such cases, adjustments must be made in the original sampling plan, so that a sample can be obtained and randomness preserved. The design of a sampling plan is a key part of an investigation. Conclusions can be no better than the samples on which they are based.

Simple Random Sampling

Samples are taken from an area or a line of traverse in terms of random positions taken from random-number tables. This is applicable to an area that appears homogeneous, e.g., same bedrock, drainage pattern.

Systematic Sampling

Samples are taken at fixed intervals along a line or on a grid. This is an ideal way of sampling if a trend or gradient is suspected. However, hidden "sine waves," e.g., ripple marks, may affect results.

Nested Sampling

Major units are first sampled at random, e.g., a township. Then, within these, smaller units are sampled at random for several levels. A similar system may be used in linear fashion along a stream or an outcropping bed.

Stratified Sampling

A line or an area is divided into natural segments, e.g., a beach (foreshore, backshore, etc.), along a barrier, or sand lenses in shale. Randomized samples are taken separately in each stratum.

Dumbbell Sampling

A base line is set up at random points (1,000 x apart), and stations established at additional points perpendicular to the ends of this line (100 x), until the last position, say x, is spun off mechanically. This gives good areal coverage with fewer samples than a grid pattern.

Laboratory Procedures

All scientific theories and generalizations are based on the interpretation of qualitative or quantitative data. Because of the small grain and crystal sizes of nearly all sedimentary rocks, a key element in most petrologic work is laboratory study of field samples. In the ideal working situation, the routine and techniques should be designed with only one purpose: to obtain as accurate and precise a picture of the truth as possible. Under conditions of industrial pressure, however, some allowance must usually be made for inadequate time, so that the quality of the final result is a compromise between the ideal and the possible. In basic research work, no such compromise is permitted.

During the course of the laboratory work, the mature judgment of an experienced research worker may suggest which operations appear to be more fundamentally important and, hence, should be emphasized at the expense of others that seem to be less important and can be treated less exhaustively. However, when in uncharted seas, even mature judgment may not be adequate to establish such priorities, and equal attention must be given to all methods and procedures.

A very large number of measurement techniques is available to the sedimentary petrologist. Some can be applied to individual grains or crystals; others require a sample weight larger than a single sand grain. In order to choose the best techniques to use during the course of laboratory work, it is necessary to answer two important questions at the outset:

1. What is your objective? That is, what specific parameter or property of the grain or rock is this particular technique supposed to measure?

2. How precisely does the technique measure the variable? Is the precision adequate for the objective? For example, examine the data in Figure 15–1 that show the result of two counts totaling 8,300 grains of heavy minerals separated from a modern stream sand. It is clear that the precision (reproducibility) of the data is no better than about 1%, despite the exceptionally large number of grains counted. In most investigations, only 100–300 grains are counted. Do not report results with pseudo-precision (e.g., percentages of heavy minerals to the nearest 0.1%) unless justified by your procedure. Figure 15–2 is a graph that enables you to determine the precision of your results.

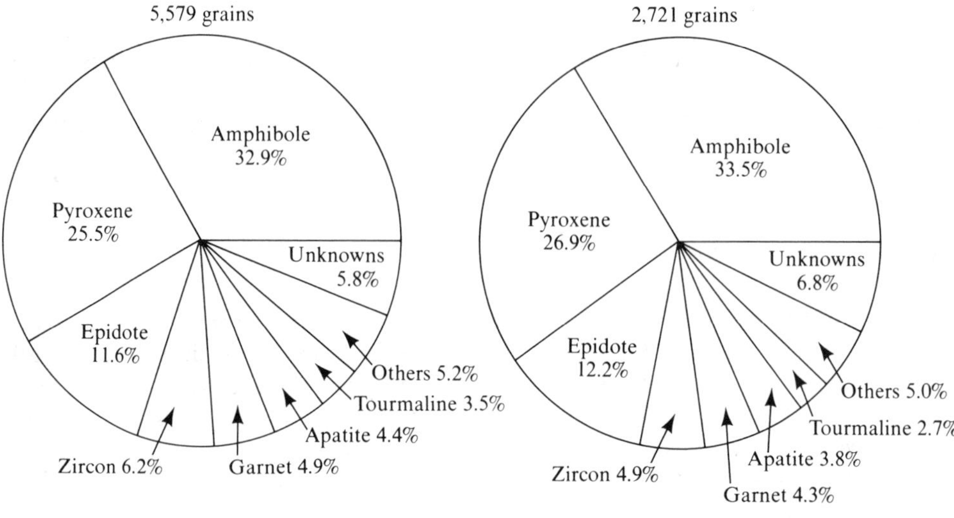

Figure 15–1
Results of a count of 8,300 grains of heavy minerals from Rio Grande, New Mexico. [Blatt et al., 1980. Copyright © 1980. Reprinted by permission of Prentice-Hall, Inc., Englewood Cliffs, New Jersey.]

Grain-Size Determination

The distribution of grain sizes in a detrital rock is one of its most fundamental properties and is determined routinely in petrologic studies. This is done for two reasons. The first is that grain-size distributions can supply valuable clues to the strength of the currents that deposited the sediment and the environment of deposition (see Figure 4–7). For example, desert dunes, beaches, and marine bars are characterized by sand with a very narrow range of grain size; fluvial sands typically contain a small amount of mud-size sediment; sands deposited by turbidity currents characteristically contain 10–20% mud-size material.

The second reason petrologists are interested in grain-size distributions of detrital sediment is that the mineral composition of a rock is often closely controlled by the sizes of the detrital particles. For example, a very fine-grained sandstone does not contain fragments of granite because the crystals of feldspar and quartz in a granite are so large that small detrital fragments derived from a granite are almost certain to be formed only of parts of single crystals of a mineral; coarser grains from the granite are rich in fragments of the rock that contain crystals of both feldspar and quartz (see Chapter 4). The relationship between detrital grain size and the composition of the detrital fraction of conglomerates, sandstones, and mudrocks is illustrated in Figure 15–3. Keep in mind, however, that the graph shows only a "world-average rock"; specific sediments can depart widely from the average. Sediment derived from

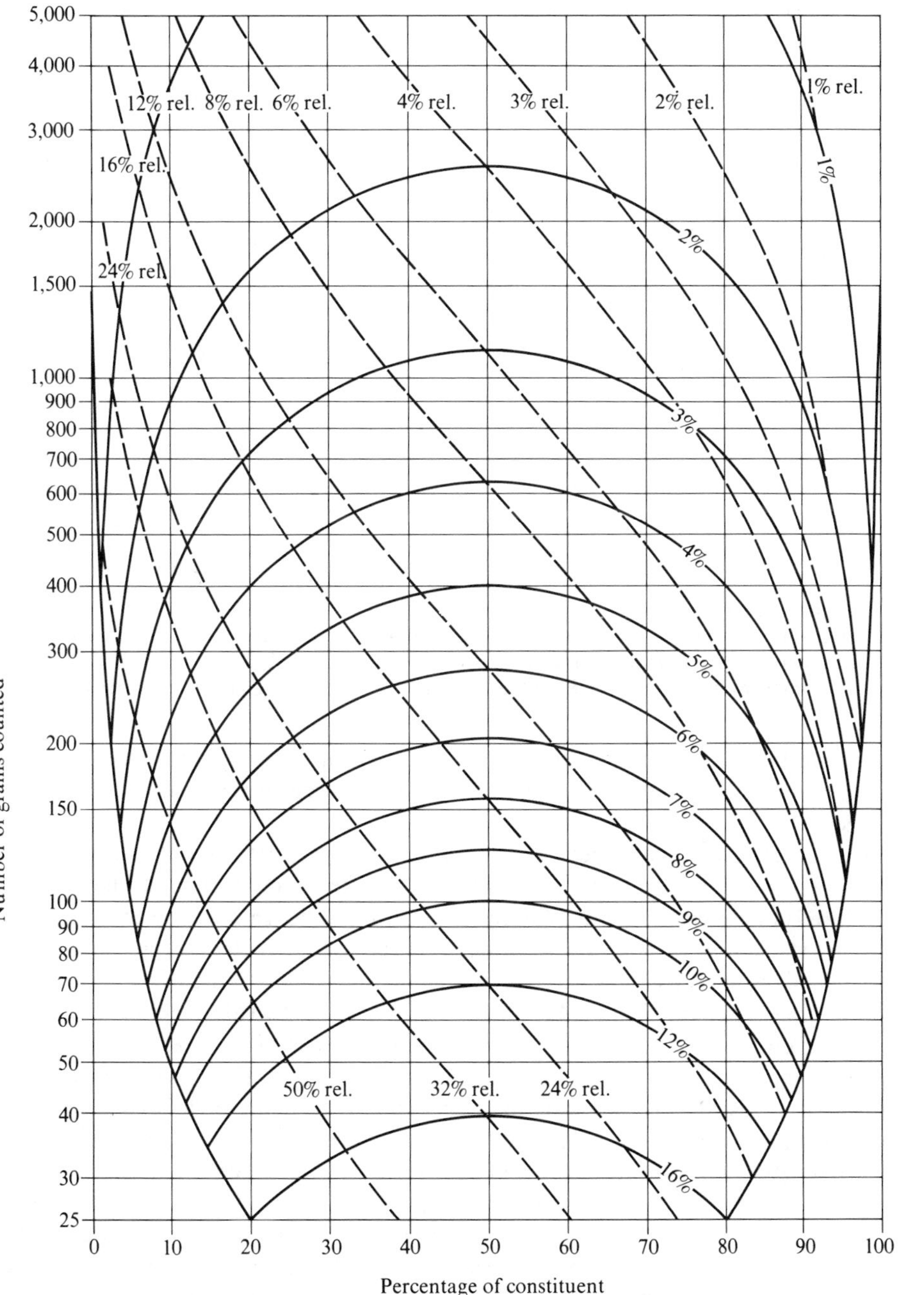

Figure 15–2
Graph for determining precision of grain counts. Unbroken curves show 95% certainty of result. For example, based on a count of 900 grains, you have determined that the unit contains 10% feldspar. Graph indicates that you can be 95% certain that true percentage lies within 2% of that value, between 8% and 12%. Broken curves show relative percentage certainty; for your 10% estimate relative certainty is 20% because 2% is 20% of 10%. [L. Van Der Plas and A. C. Tobi, 1965, *Amer. Jour. Sci., 263.*]

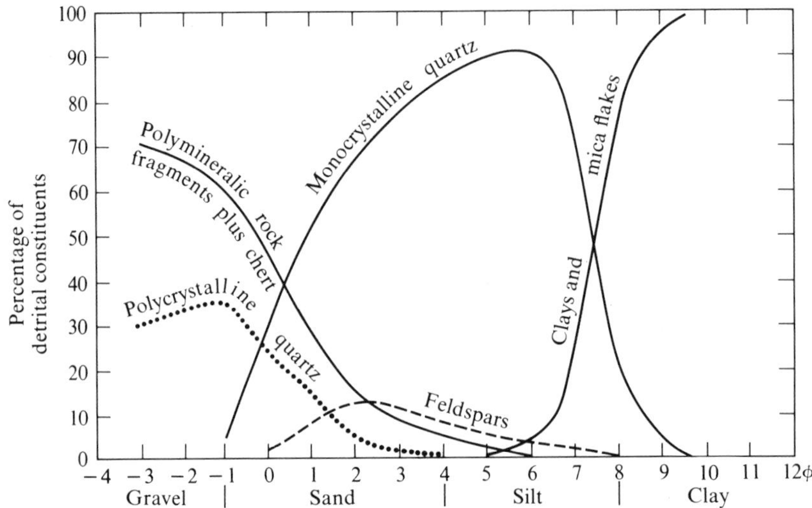

Figure 15–3
Average relationship between detrital grain size and mineral composition in fine conglomerates, sandstones, and mudrocks. [Blatt et al., 1980. Copyright © 1980. Reprinted by permission of Prentice-Hall, Inc., Englewood Cliffs, New Jersey.]

a metaquartzite contains only quartz, irrespective of detrital grain size; sediment from a rhyolite may consist only of rock fragments even in the coarse and medium silt sizes.

Several techniques are in general use for the determination of detrital grain size. When one is dealing with loose sediment such as a beach or river sand, the standard procedure is to use either a nest of sieves or a settling tube. In the sieve technique, the sieves are stacked in sequence, with those of larger hole diameter on top, smaller hole diameter below, and a pan at the base of the stack. If the sediment contains a wide range of grain sizes (more than 4 ϕ units), sieves at 1 ϕ intervals generate a sufficient number of data points for subsequent statistical analysis of the size distribution. If the size range is less, sieves at 0.5 ϕ intervals should be used. In cases of exceptionally well-sorted sands, 0.25 ϕ intervals may be required to achieve satisfactory results. Standard sieves with square holes and formed of woven wire are commercially available from pebble sizes down to 4.5 ϕ (44 μm) and special "micromesh" sieves can be purchased with hole diameters of 40, 30, 20, 10, and 5 μm. The latter sieve screens are made of plastic, have round, die-punched holes, and are used while immersed in a shallow pan of water. The tip of an ultrasonic vibrator is placed between the water surface and the sieve surface to cause the vibration of the sediment needed to pass it through the sieve holes (Nuckolls and Fuller, 1966).

When standard sieves are used, approximately 20–30 g of the sample is poured onto the top sieve, the sieve is capped with a fitted cover, and the stack is set on a shaking apparatus (see Figure 15–4) for 10–15 minutes. The stack is then removed,

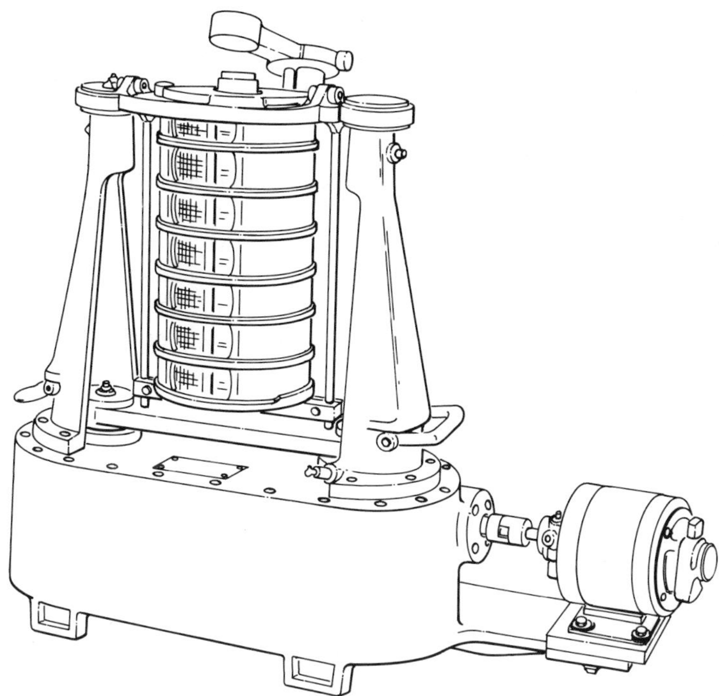

Figure 15–4
A Ro-Tap mechanical sieve shaker, one of the machines in common use for determination of grain-size distribution. Metal bar at base of stack of sieves rotates while hammer at top pounds vertically. [Carver, 1971.]

the cover lifted off, and the sediment on the upper sieve emptied onto a piece of glazed paper (so that grains do not stick in fibers of the paper). A brush, such as a toothbrush, is rubbed gently over the bottom of the screen on the paper to dislodge grains partially stuck in the screen holes. The collection of sediment from the screen is then weighed to the nearest 0.1 g and stored in a small vial for possible future use in mineralogic studies. This procedure is repeated with the grains on each sieve and pan; the total of the weights of grains on each sieve should equal the weight of sediment originally poured into the nest of sieves.

It is also possible to determine the grain-size distribution of sediments with a settling tube—a water-filled cylinder into which sediment is dropped and the size distribution determined by the rate of settling. Using a settling tube can produce data more rapidly than sieving but, unfortunately, the method has many limitations in regard to both the significance of the property measured and the precision of the measuring technique (Blatt et al., 1980, pp. 59–66). From the viewpoint of mineralogic analysis, the method of size determination by settling tube has an additional disadvantage. At the conclusion of the analysis, we are left with a pile of sediment at the base of a column of water and cannot separate the pile into its various size fractions to study changes in mineral composition as a function of grain size. In contrast, sieve analysis provides not only the frequency distribution of grain sizes but groups of grains separated into size fractions as narrowly defined as desired for subsequent mineralogic analyses.

Grain-size data can be illustrated in many ways, the most common of which are as a bar graph called a histogram (see Figure 15–5A), as a noncumulative frequency curve (see Figure 15–5B), as a cumulative frequency curve on standard arithmetic graph paper (see Figure 15–5C), and as a cumulative frequency curve on arithmetic probability graph paper (see Figure 15–5D). Most frequency distributions of grain size plot as nearly "bell-shaped curves" when shown as noncumulative plots (see Figure 15–5B). They appear as S-shaped curves when shown as cumulative curves on standard arithmetic graph paper (see Figure 15–5C). Arithmetic probability graph paper is designed to convert a perfect bell-shaped or S-shaped curve into a straight line, so that it is easier to identify deviations from the perfect curve. It is typically these deviations that supply the most valuable clues to the hydrodynamic environment prevailing at the time of sediment accumulation.

Numerous parameters are in common use to specify characteristics of a grain-size frequency distribution. Those in most common use are modal size, mean size, standard deviation of grain sizes (sorting), and skewness (lopsidedness) of the frequency distribution. The modal grain size is defined as the most frequently occurring size class and is normally determined simply by visual inspection of the graph. It is the highest bar on a histogram, the highest point on a noncumulative curve, the steepest part of a cumulative curve on standard arithmetic graph paper, and the sieve size that contains the highest percentage of the sediment as seen on a cumulative curve on probability graph paper.

Mean size, standard deviation, and skewness are determined through calculation by picking numerical values at selected points on a cumulative frequency curve. The data points used are the phi values at 5%, 16%, 50%, 84% and 95%.

For *mean size* the formula is

$$\text{Mean size} = \frac{16\% + 50\% + 84\%}{3}$$

where 16% is the phi value at 16% on the cumulative curve, etc. The phi value at 16% represents the average size of the coarsest one-third of the grains; the 50% value, the central one-third; and the 84% value, the finest one-third.

Figure 15–5
Results of sieve analysis displayed in various ways. (**A**) Histogram. (**B**) Noncumulative size frequency curve. (**C**) Cumulative curve on standard arithmetic graph paper. (**D**) Cumulative curve on arithmetic probability graph paper. Modal class is 0–1 ϕ; mean is 1.0 ϕ; standard deviation is 1.2 ϕ; skewness is +0.11 (slightly positively skewed).

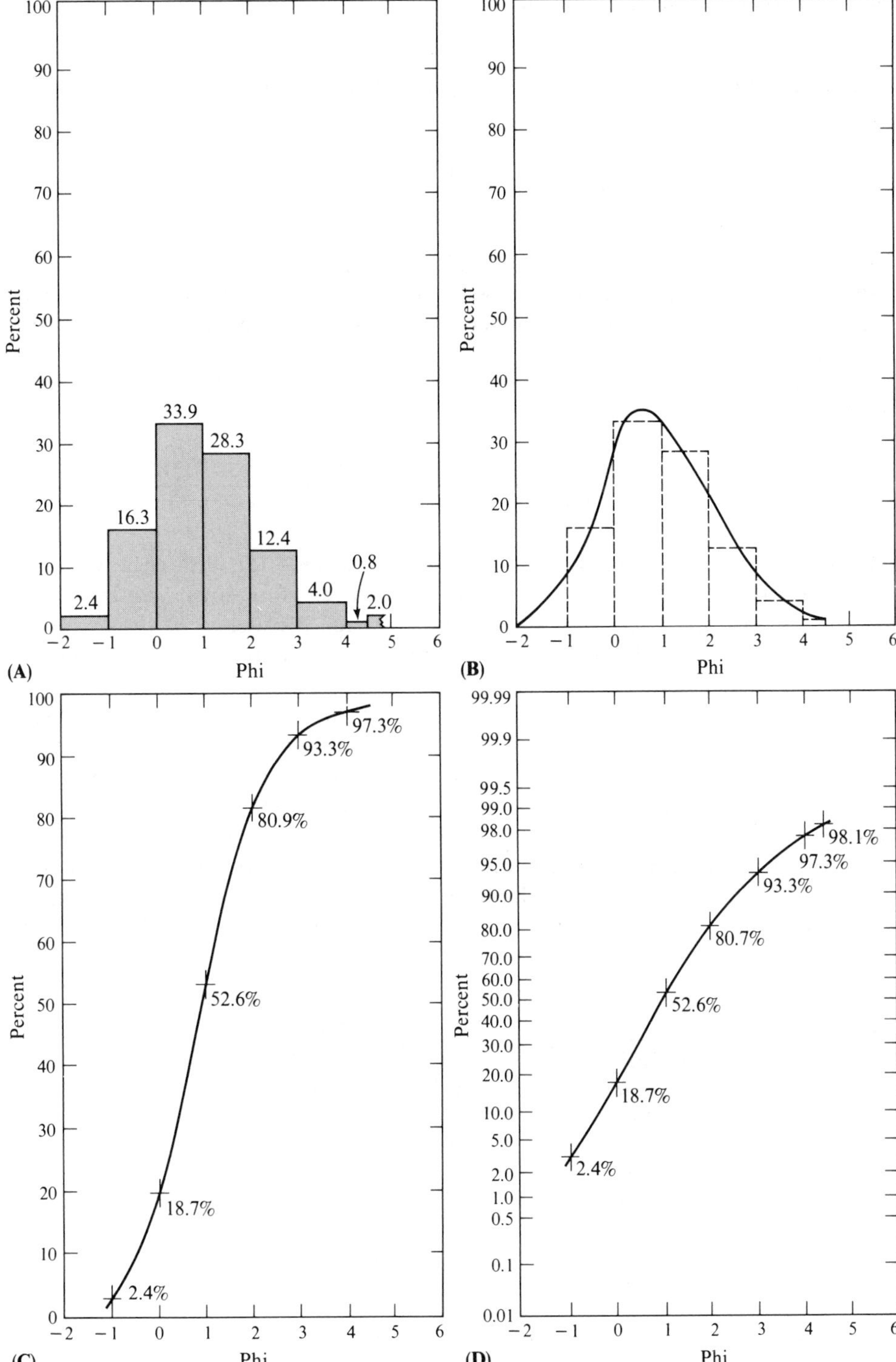
Percent
Phi
2.4
16.3
33.9
28.3
12.4
4.0
0.8
2.0
(A)
(B)
2.4%
18.7%
52.6%
80.9%
93.3%
97.3%
(C)
2.4%
18.7%
52.6%
80.7%
93.3%
97.3%
98.1%
(D)

Standard deviation is the accepted measure of the sorting of the sediment and is defined by

$$\text{Standard deviation} = \frac{84\% - 16\%}{4} + \frac{95\% - 5\%}{6.6}$$

In Figure 4–3 are shown drawings of sediment sorting in relation to the numerical values of standard deviation obtained with this formula. On a cumulative frequency curve, the steeper the curve, the better the sorting (i.e., the smaller the standard deviation).

Skewness is a measure of the symmetry of a frequency distribution. The formula used is

$$\text{Skewness} = \frac{16\% + 84\% - 2(50\%)}{2(84\% - 16\%)} + \frac{5\% + 95\% - 2(50\%)}{2(95\% - 5\%)}$$

The mathematical limits of skewness are from -1.0 to $+1.0$, with negative values indicating skewness (lopsidedness) toward the coarser grain sizes; positive values, skewness toward the finer grain sizes. More thorough discussions of the derivation and sedimentologic significance of these three parameters are given by Folk (1974) and most sedimentology texts.

It commonly happens that a sandstone is lithified by quartz and cannot be disaggregated. In these circumstances, it is necessary to determine the mean grain size and standard deviation in thin section. The procedure is to use a microscope eyepiece that contains a micrometer and to measure the longest dimension of 500 grains encountered in a linear traverse (see Figure 15–6). Experiment has shown that the mean grain size determined in this way is within 0.3 ϕ of the value obtained by sieving the same sand sample. Typically, the thin-section size is the coarser of the two because sand grains are ellipsoidal in shape and the thin section commonly slices through

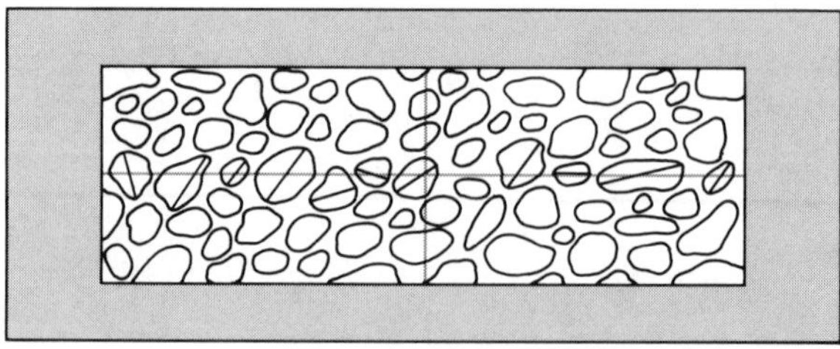

Figure 15–6
Diagrammatic sketch of thin section of sandstone, showing 11 grains that would be encountered along an east–west traverse in a point count (cement omitted). Length that should be measured in each grain is shown.

grains to expose a measurement length longer than the intermediate grain diameter measured by the sieve hole. The standard deviation obtained by thin-section measurements is also close to the value obtained by sieving (Friedman, 1958). When clay is present, however, measurement of grain diameter in thin section is not possible; and if the rock cannot be disaggregated, the size-frequency distribution of the grains in the rock cannot be determined accurately.

Mudrock Disaggregation

The majority of sandstones is cemented by calcite, dolomite, or hematite, and these minerals are dissolved easily by acetic or hydrochloric acid to free the detrital grains. Sieve analysis or mineralogic analyses as a function of grain size can then be made as desired. Mudrocks, however, pose a more difficult problem in disaggregation. Even though complete disaggregation of the clay fraction of the rock into individual clay flakes is impossible, the mudrock can be dispersed effectively enough to free nearly the entire weight of quartz, feldspar, lithic fragments, and accessory heavy minerals. (The same technique can be used for sandstones cemented by clay, either detrital or authigenic.)

The method of disaggregation of a mudrock involves the use of a solution called a peptizer, which penetrates the mudrock and causes the clay minerals in it to disperse. The most effective peptizers are salts of a strong base and a weak acid, such as sodium hexametaphosphate or sodium oxalate. Prepare a 0.01 N solution of the peptizer and put about 200 ml of the solution into a 400 ml beaker. Fragment a sample of the mudrock into pea-sized pieces with a mortar and pestle. The purpose of the fragmentation is to speed the disaggregation. Pea size is sufficiently coarse so that none of the silt or fine-grained sand in the rock will be crushed by the fragmentation. The time required for thorough disaggregation of the pieces of mudrock ranges from a few hours to perhaps six weeks, depending on fissility, clay content and type, permeability, and other factors. When the research must be completed by a certain time, you have two choices: (1) allow excessive time for the completion of the lab procedure, or (2) pray that your mudrocks are responsive to your needs. Success is marked by the increasing muddiness of the solution as clays go into suspension and the number of fragments on the bottom of the beaker goes to zero. At this point the muddy solution must be further dispersed with an ultrasonic vibrator. The tip of the vibrator is placed into the beaker, and the suspension is agitated for 5–10 minutes, then washed into sieves as needed to remove the clay and to concentrate the sand and silt grains.

Heavy-Mineral Concentration

Disaggregation of lithified rocks is useful for grain-size determination, but absolutely necessary for the examination of accessory minerals. The essential parts of an apparatus for separating light from heavy minerals are illustrated in Figure 15–7. Ap-

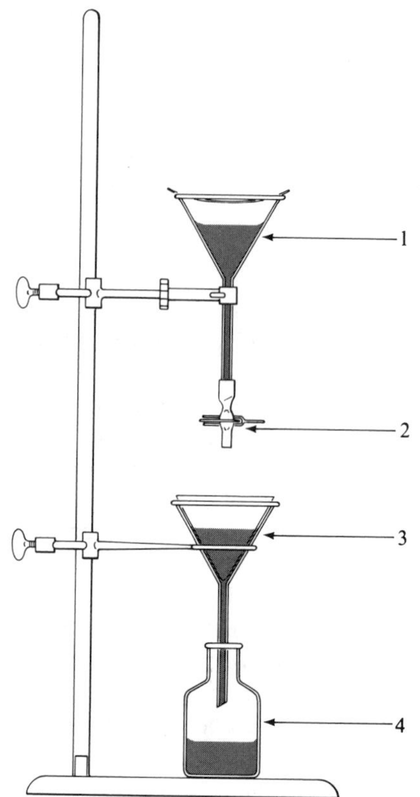

Figure 15–7
Simple apparatus for separation of heavy minerals. (1) Funnel containing heavy liquid. (2) Rubber tube and clamp. (3) Funnel fitted with filter paper. (4) Bottle to collect used heavy liquid. [W. C. Krumbein and F. J. Pettijohn, 1938, *Manual of Sedimentary Petrography* (New York: Plenum).]

proximately 30 g of sediment is poured into a heavy liquid in the upper funnel. Settling of the heavy minerals toward the clamp begins immediately and is normally complete after about 5 minutes. The clamp is opened momentarily to release the heavy minerals into the funnel and filter paper below, and the heavy liquid seeps through the paper into a bottle and is recovered to be used again. The filter paper is removed, inverted into a small porcelain dish that contains a solvent such as acetone, and recovered. About 30 g of sand will typically yield 0.1–0.3 g of heavy minerals—an amount adequate for point counting many hundreds of grains. The rate of settling of a heavy mineral in the upper funnel depends on the mass of the grain (its volume and specific gravity) and on the specific gravity of the liquid. Therefore, it is useful to separate the heavy minerals with the aid of a centrifuge to shorten the time and improve the accuracy of the separation. This is of particular importance when one is dealing with silt-size heavy minerals. The theory and practice of heavy-mineral separation are discussed in detail by Carver (1971, pp. 427–452).

All heavy liquids are toxic and can damage human lungs. These liquids should be used only in a fume hood that is adequately ventilated. Several heavy liquids are in common use by sedimentary petrologists, two favorites being bromoform ($CHBr_3$, sp. gr. 2.89) and 1,1,2,2-tetrabromoethane ($C_2H_2Br_4$, sp. gr. 2.97). The only common minerals that occur in the range of specific gravity 2.9–3.0 are muscovite and biotite; these minerals are not claimed by either light- or heavy-mineral enthusiasts and are not included in counts of heavy minerals. Other heavy liquids that have been used include methylene iodide—diiodomethane (CH_2I_2, sp. gr. 3.32)—which is about 15 times more expensive than bromoform or tetrabromoethane, and Clerici solution, a mixture of thallium compounds that has a specific gravity of 4.2 but is used only rarely because of its extreme toxicity (and cost). The common heavy liquids are miscible in all proportions with available organic solvents such as benzene (sp. gr. 0.9) so that specific gravities of 0.9–3.0 can be produced as needed.

Following separation, the heavy minerals are examined with a polarizing microscope, either as whole grains embedded in epoxy cement or as thin-sectioned grains on a glass slide. A thin section is preferred because of the difficulty of identifying whole grains and because of the need to examine internal features of the grains in diagenetic studies.

Magnetic separation. Some elements, because of the arrangement of electrons around their nuclei, are attracted toward a magnet with greater intensity than are other elements. Examples of magnetic elements include iron, nickel, cobalt, manganese, and chromium; weakly magnetic or nonmagnetic elements include silicon, zinc, copper, sulfur, and phosphorous. Minerals that contain magnetic elements and are themselves magnetic are termed *paramagnetic;* minerals that are less magnetic are termed *diamagnetic*. Among the abundant elements in the Earth's crust, iron is the most important cause of paramagnetic minerals.

For purposes of separating minerals according to their magnetic properties, minerals may be classed as highly magnetic (magnetite and pyrrhotite), moderately magnetic (ilmenite, chromite, and almandite), weakly magnetic (monazite and tourmaline), and almost nonmagnetic (quartz and zircon). Many minerals have a range of compositions, however, and the degree of magnetic susceptibility varies with mineral composition. This fact is the main deterrent to the effective use of magnetism for the separation of heavy minerals (see Figure 15–8). For example, the amount of iron varies significantly in minerals such as the epidote group, garnet group, and hornblendes. In addition, inclusions in minerals can cause problems in separating sand-size grains. In some heavy-mineral suites, magnetic separation can be quite effective; in others, it is almost useless.

Magnetite is relatively constant in composition and so highly magnetic that it can be removed from a group of loose grains with a small hand magnet purchased at what used to be called a "five and ten cent store." For all other minerals the standard machine used for separation is the Frantz isodynamic magnetic separator (see Figure 15–9). The electromagnet is mounted on a universal joint and may be tilted both transversely and longitudinally; the intensity of magnetic flux is controlled by a rheostat. Sediment is fed into a cone-shaped hopper to the chute, whose transverse slope is adjusted so that the gravitational and magnetic forces are opposed. By a

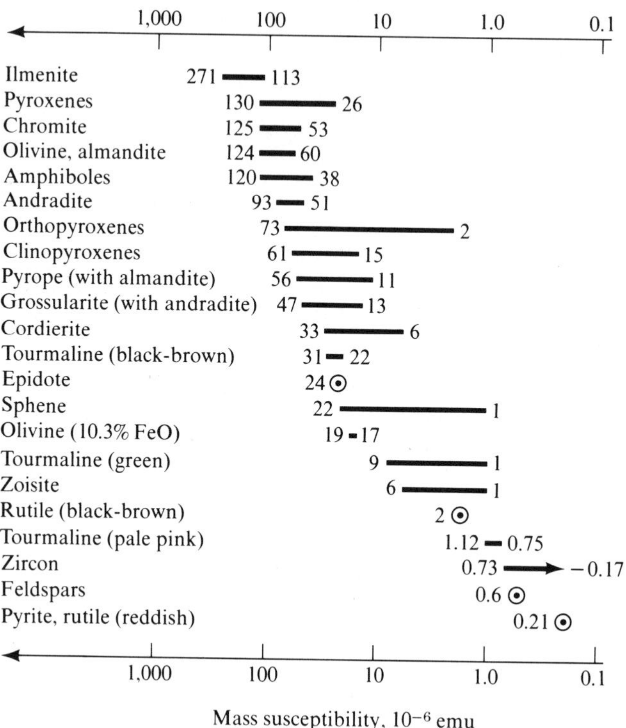

Figure 15–8
Mass susceptibilities of common paramagnetic minerals, showing that iron-rich minerals are the most paramagnetic. Pyrite is an exception because of arrangement of electrons between iron and sulfur atoms. [G. Müller, 1967, *Methods in Sedimentary Petrology* (New York: Hafner Publishing Co.).]

suitable balance of these forces, separation can be obtained—the more magnetic minerals being moved against the chute slope to be discharged into one of two small collector buckets.

Insoluble Residues

Concentrates of "insoluble" minerals from sedimentary rocks have, until recently, been made almost entirely from carbonate rocks. In most cases, these insolubles form less than 5% of the rock. A sample of the rock about 10 g in weight is placed in a beaker that contains either a 10% solution of commercial hydrochloric acid or a 20% solution of acetic acid, and acid is added as needed until frothing (evolution of carbon dioxide gas) stops. Acetic acid does not attack phosphatic materials such as detrital apatite, conodonts, or fragments of linguloid brachipods, and thus is preferred by

Figure 15–9
Frantz isodynamic magnetic separator. Vertical tilt screw is on far side of magnet and is not visible. [Photo courtesy S. G. Frantz Co.]

paleontologists. Petrologists are in the habit of ignoring apatite as a detrital mineral.

The typical residue from a carbonate rock consists of powdery to microgranular chert, clay minerals, organic matter, some silt-size and fine sand-size detrital quartz and chert grains, perhaps dolomite rhombs (if the acid was not heated during digestion of the rock), and occasionally pyrite, silicified fossils, arenaceous foraminifera, or other materials. These grains can be subdivided according to any useful criteria. Examples include rounded quartz, euhedral quartz (secondary growths), various colors or textures of chert (oolitic, chalcedonic, etc.), or types of clay minerals. Insoluble residues from carbonate rocks are useful not only for a more complete characterization of the rock than is possible in a thin section, but also for lithologic correlation (Ireland, 1971).

In recent years, a technique has been developed for freeing quartz and feldspar from mudrocks (Chapman et al., 1969). This is necessary because although these two mineral groups form an average of about 35% of a typical mudrock, the grains are largely concealed in thin section because of the high birefringence and platy shape of the common clay minerals. The technique involves digestion of the mudrock in sodium bisulfate and other reagents, which completely dissolve all constituents of the rock except quartz, chert, and feldspars (Chapman et al., 1969). These grains can then be analyzed with either a polarizing microscope, an electron microscope, a microprobe, or one of several geochemical techniques such as atomic absorption spectrophotometry. Feldspar grains in mudrocks are less susceptible to alteration and removal during diagenesis than are feldspar grains in sandstones. For this reason, coarse silt and fine sand grains in mudrocks can be very useful in studies of provenance as well as diagenesis.

Acetate Peels

The texture, fabric, packing, mineral composition, and mineral distribution in a hand specimen of a carbonate rock can be studied effectively using acetate peels (McCrone, 1963). The manufacture of the peel takes only a few minutes, and the technique is inexpensive. Some features of carbonate rocks, such as delicate textures, are seen more clearly in peels than in thin sections. Peels and thin sections of carbonate rocks therefore complement each other. Peels can also be stored easily and projected on a screen.

Acetate peels of carbonate rocks are made by cutting and polishing a rock slab and etching the polished face in dilute hydrochloric acid for 10 seconds (longer for dolomite). The slab is then placed in a container filled with sand, with the etched surface face-up and horizontal. The etched surface is wetted with acetone, and a piece of cellulose acetate is placed dull side down over the etched surface. The film adheres easily to the surface as the acetone dissolves film along the preferential relief produced by the etching. The peel dries within 15 minutes and is peeled from the rock slab. It is then mounted between two glass slides and bound. As is evident in Figure 15–10, the slide produced looks in almost every way like a thin section, the major difference being that crossed polarizers will cause extinction of light over the entire peel because the acetate is isotropic.

Grain Staining

Frequently, it is difficult to distinguish in thin section betwen calcite and dolomite or between orthoclase and untwinned plagioclase feldspar. To overcome these difficulties, selective staining techniques have been developed (Friedman, 1971, for carbonates; Houghton, 1980, for feldspars). Thin sections or hand specimens are etched briefly in hydrochloric acid (for carbonate) or hydrofluoric acid (for silicates) and stained with various dyes. The procedure is rapid and effective; for orthoclase and plagioclase, the percentage stained is within 10% of the true value in artificial mixtures 90% of the time. The accuracy of staining for feldspars decreases when microperthite grains are present and when the grain size of feldspar is smaller than sand.

The caution concerning the need for an adequately ventilated fume hood when working with heavy liquids is even more applicable when using hydrofluoric acid. The acid reacts rapidly with human tissue, but pain and other overt signs of deep burns may not be noticed for several hours. To avoid painful burns, gloves and safety glasses should be worn, and hood ventilation is imperative.

Grain Point Counts

The glass slide on which the grains (or rock chip) are located is placed in a "mechanical stage" that is mounted on the rotating stage of a polarizing microscope. The slide is moved an arbitrarily set distance (e.g., 0.3 mm) and, if a grain lands at the inter-

Figure 15–10
Negative photographic print of acetate peel of limestone slab from supratidal facies of Manlius Formation (Devonian), New York. Peel shows a massive burrowed dolomicrite overlain by intraclastic dolomicrite. Flat clasts were produced by erosion of high-energy waves as they moved across dolomicritic supratidal flats. [L. F. Laporte, 1967, *Amer. Assoc. Petroleum Geol. Bull., 51*. Photo courtesy L. F. Laporte.]

section of the cross hairs in the eyepiece, it is identified and described. If a grain is not intersected, the slide is moved another unit of distance. This procedure is known as point counting and results in an estimate of the volume percentage of each mineral in the slide. The method used when working with accessory minerals is to count all grains along the linear traverse, whether or not they land at the intersection of the cross hairs. This is necessitated by the small number of grains recovered in a heavy-mineral separation relative to the number of grains in a thin section of a rock chip. The procedure gives a percentage by number of grains, not by volume or weight.

Opaque heavy-mineral grains cannot be identified mineralogically in a standard thin section. Special techniques—a polished thin section and a reflecting microscope—are required and, as a result, very little is known about the types of opaque grains in sedimentary rocks. Such grains, however, typically form 50% of all heavy-mineral grains.

THE ART OF OBSERVATION

A thorough description of a rock is an essential prerequisite to an adequate interpretation. Ideally, the description of a thin section of a rock or grain mount should be so complete that another petrologist who reads it could draw a perfect picture of what you saw on the stage of the microscope. In practice, of course, there is neither the time nor the perceptiveness to write such an extensive description. Choices must be made and it is useful, at least when learning to observe, to have a checklist of important features to note (see Tables 15–1 and 15–2).

Observation is an art that must be learned and practiced repeatedly to attain proficiency. A conscious effort must be made initially, but the more frequently it is done, the more automatic it becomes. If you believe you are already observant, answer the following questions:

1. What were the colors of the shirt and shoes of the last male student you spoke to?

2. What were the colors of the last out-of-state license plate you saw?

3. How many and what types of rings were on the fingers of the last person you passed in the hallway?

4. What is the copyright date of this book? What is the name of the publisher? What page are you reading at the moment?

THE INTERPRETATION OF DATA

When interpreting data you have collected, be conservative in your conclusions. There is an almost overpowering tendency, particularly among beginners, to overextrapolate on the basis of insufficient evidence. Remember, although it took you many weeks (or months or years) to accumulate your data, your information can be only a minute sample of the amount available or potentially available. Do not be carried away by your own "brilliance" or "insight." A moderate amount of ego is a useful stimulant; an excess of ego is an almost certain road to scientific disaster. As expressed by Medawar (1979, p. 39):

> I cannot give any scientist of any age better advice than this: the intensity of the conviction that a hypothesis is true has no bearing on whether it is true or not.

There are numerous historical examples of excellent data that led to justified but erroneous conclusions. A familiar example from antiquity is the belief of Aristotle (384–322 B.C.) and his disciples that all matter consisted of four "elements"—earth, water, air, and fire—a belief that prevailed for about 2,000 years until the beginnings of modern experimental science in the sixteenth century. Lest you think modern experimental science did away with Aristotelian errors of interpretation, consider the famous study of J. B. van Helmont (1577–1644), who conducted the first quantitative experiments to determine the source of the materials of which plants are composed (Gabriel and Fogel, 1955, p. 155):

> That all vegetable [matter] immediately and materially arises from the element of water alone I learned from this experiment. I took an earthenware pot, placed in it 200 lb of

Table 15–1
Checklist for Petrographic Description of Thin Sections of Sandstones

I. Name of formation or member, geographic and stratigraphic location, geologic age
II. Texture
   A. Percentages of gravel, sand, and mud
   B. Mean grain size and sorting
      1. Is the sediment polymodal?
      2. What are the modal sizes?
   C. Give the rock a textural name, e.g., poorly sorted, muddy, medium sandstone
   D. Grain shape: idiomorphism, roundness, relation of roundness to grain size or mineralogy, roundness sorting
   E. Stage of textural maturity
   F. Fabric
      1. Is the slide homogeneous or laminated? Cause and scale of lamination? Evidence of bioturbation? Geopetal structures?
      2. Orientations of grains (lineation, imbrication)? Parallelism of fossils?
      3. Packing: closeness, presence of deformed ductile grains, sutured grain contacts, relation of compaction effects to mineralogy
      4. Precement porosity, i.e., percentage of pore-filling cement; etching of grains by cement
III. Mineral composition
   A. Percentage of quartz
      1. Monocrystalline vs. polycrystalline
      2. Undulatory vs. nonundulatory
      3. Size and texture of crystals in polycrystalline grains, e.g., sutured, elongated
   B. Percentage of chert
   C. Percentage of feldspar
      1. Orthoclase vs. plagioclase vs. microcline vs. sanidine vs. perthite
      2. Twinned vs. untwinned; type of twinning
      3. Degree and type of alteration, relation of alteration to mineralogy
      4. Relation between mineralogy and grain size, e.g., orthoclase coarser than plagioclase
   D. Percentage of lithic fragments
      1. Types of fragments and relative abundances, describing fragment mineralogy accurately, e.g., sillimanite schist rather than metamorphic rock fragment; basalt and biomicrite rather than volcanic rock fragment and limestone fragment
      2. Relation between fragment type and grain size
   E. Micas
   F. Other terrigenous minerals, such as tourmaline and zircon
   G. Clay matrix: most abundant clay based on optical properties (an X-ray diffractogram is essential as a supplement to the rock description)
IV. Classification
V. Interpretation and paragenesis
   A. Probable major source areas, e.g., mostly granite with minor older sandstones
   B. Kinetic energy of depositional environment, based on grain size and sorting
   C. Permissible depositional environments
   D. Diagenetic changes
      1. Age relations of authigenic constituents
      2. Effects of intrastratal solution
      3. Effects of compaction
VI. Economic significance and implications

Table 15–2
Checklist for Petrographic Description of Thin Sections of Limestones and Dolomites

I. Name of formation or member, geographic and stratigraphic location, geologic age
II. Texture
   A. Types and percentages of allochems
      1. Fossils: types, state of articulation of valves, whole or broken and/or abraded, fragment size, degree of preservation of internal structure
      2. Peloids: size
      3. Ooliths: types of nuclei, thickness and crystal orientation of oolithic coating; spherical, ellipsoidal, or deformed outlines
      4. Intraclasts: size, degree of rounding, internal character
   B. Types and percentages of orthochems
      1. Micrite
      2. Microspar: distribution in slide, relation to other features such as allochems or lamination
      3. Sparry calcite: crystal size, growth habit and relation to other features
   C. Fabric
      1. Is the slide homogeneous or laminated? Cause and scale of lamination? Evidence of bioturbation?
      2. Grain-supported or mud-supported?
      3. Other features such as stromatolites, cross bedding, biologic growth textures, geopetal structures, stylolites
III. Mineral composition
   A. Percentages of calcite and aragonite
      1. Allochems or orthochems, e.g., aragonitic cement around calcitic fossils
      2. Relation to fabric elements, e.g., aragonitic biologic growth textures on calcitic allochems
   B. Percentage of dolomite
      1. Crystal size and relation to texture
      2. Presence of zoning
   C. Terrigenous constituents
      1. Types and percentages
      2. Distribution in slide, e.g., quartz silt along laminae
IV. Classification
V. Interpretation and paragenesis
   A. Kinetic energy of depositional environment: based on presence or absence of micrite, abraded allochems, cross bedding, and presence of ooliths
   B. Fossils: planktonic or benthonic, shallow or deep, sessile or vagrant
   C. Terrigenous sediment
      1. Grain size indicates strength of currents entering basin
      2. Mineral composition indicates nature of rocks surrounding basin
   D. Diagenetic changes
      1. Age relation of authigenic constituents, e.g., dolomite, quartz
      2. Time of stylolitization in relation to other diagenetic changes
      3. Compaction effects
VI. Economic significance and implications

> earth dried in an oven, soaked this with water, and planted in it a willow shoot weighing 5 lb. After five years had passed the tree grown thereform weighed 169 lb and about 3 oz. But the earthenware pot was constantly wet only with rain or (when necessary) distilled water; and it was ample (in size) and imbedded in the ground; and, to prevent dust flying around from mixing with the earth, the rim of the pot was kept covered with an iron plate coated with tin and pierced with many holes. I did not compute the weight of the deciduous leaves of the four autumns. Finally, I again dried the earth of the pot, and it was found to be the same 200 lb minus about 2 oz. Therefore, 164 lb of wood, bark, and root had arisen from the water alone.

This was an excellent experiment—well planned, carefully done, and accurately described. The conclusion, of course, was totally wrong, but was quite consistent with the scientific knowledge of the time.

Still unconvinced about the need for humility in scientific inference? If so, simply examine the literature dealing with the origin of dolomite. Prior to 1964, when dolomite was discovered forming in a number of modern environments, it was generally believed that dolomite formed from preexisting limestone almost entirely after burial to hundreds of meters. Current ideas relate the origin of most dolomite to replacement at the surface within a few hundred or a thousand years after limestone deposition (see Chapter 8), that is, penecontemporaneously with the formation of the limestone. You might also read articles dealing with the past locations of the continents written before the concept of plate tectonics was widely accepted. Some of the ideas used to explain structural patterns and faunal distributions centered around "connecting land bridges" between continents and "the oceanization" of continental crust lying above some present oceanic areas. As strange as some of these ideas seem to us from the vantage point of the 1980s, they seemed quite plausible to many geologists of the 1950s and 1960s. Today's "remarkable insights" may be tomorrow's lunacies.

SUMMARY

Sedimentary petrology is an important and fascinating part of geology that was born about 100 years ago, underwent a great leap forward with the application of X-ray techniques 70 years ago, and is currently in another period of expansion.

The single most important part of an investigation is the idea on which it is based. There is no substitute for imagination. However, imagination must be accompanied by a well-thought-out plan of sampling and laboratory procedures to achieve the best results. A wide variety of laboratory techniques is available for the separation and identification of minerals; new techniques are being devised continually.

As in other fields of scientific investigation, careful observation and description are an essential part of a study. The number and quality of observations may be fairly low at first but are certain to increase with practice. Learning takes time and involves hard work, but its fruits are delicious. A successful scientist is a happy scientist.

FURTHER READING

Blatt, H., G. V. Middleton, and R. C. Murray. 1980. *Origin of Sedimentary Rocks,* 2nd ed. Englewood Cliffs, NJ: Prentice-Hall, 782 pp.

Bramlette, M. N. 1946. *The Monterey Formation of California and the Origin of Its Siliceous Rocks.* U.S. Geol. Surv. Prof. Paper No. 212, 57 pp.

Carver, R. E. (ed.). 1971. *Procedures in Sedimentary Petrology.* New York: Wiley, 653 pp.

Chapman, S. L., J. K. Syers, and M. L. Jackson. 1969. Quantitative determination of quartz in soils, sediments, and rocks by pyrosulfate fusion and hydrofluosilicic acid treatment. *Soil Sci., 107,* 348–355.

Devismes, P. 1978. *Photographic Altas of Detrital Minerals.* Mémoire du Bureau de Recherches Géologiques et Minières No. 95, 203 pp.

Folk, R. L. 1959. Practical petrographic classification of limestones. *Amer. Assoc. Petroleum Geol. Bull., 43,* 1–38.

Folk, R. L. 1965. Henry Clifton Sorby (1826–1908), the founder of petrography. *Jour. Geol. Educ., 13,* 43–47.

Folk, R. L. 1974. *Petrology of Sedimentary Rocks.* Austin, TX: Hemphill, 182 pp.

Friedman, G. M. 1958. Determination of sieve-size distribution from thin-section data for sedimentary petrological studies. *Jour. Geol., 66,* 394–416.

Friedman, G. M. 1971. Staining. In R. E. Carver (ed.), *Procedures in Sedimentary Petrology.* New York: Wiley, pp. 511–530.

Gabriel, M. L., and S. Fogel. 1955. *Great Experiments in Biology.* Englewood Cliffs, NJ: Prentice-Hall, 317 pp.

Houghton, H. F. 1980. Refined techniques for staining plagioclase and alkali feldspars in thin section. *Jour. Sed. Petrology, 50,* 629–631.

Ireland, H. A. 1971. Insoluble residues. In R. E. Carver (ed.), *Procedures in Sedimentary Petrology.* New York: Wiley, pp. 479–498.

Krynine, P. D. 1950. *Petrology, Stratigraphy and Origin of the Triassic Sedimentary Rocks of Connecticut.* Conn. State Geol. Nat. Hist. Surv. Bull., *73,* 239 pp.

McCrone, A. W. 1963. Quick preparation of peel-prints for sedimentary petrography. *Jour. Sed. Petrology, 33,* 228–230.

Medawar, P. B. 1979. *Advice to a Young Scientist.* New York: Harper, 109 pp.

Norman, M. B. 1974. Improved techniques for selective staining of feldspar and other minerals using amaranth. *Jour. Res.* (U.S. Geol. Surv.), *2,* 73–79.

Nuckolls, M. J., and R. K. Fuller. 1966. Sieve analysis of particles smaller than 44 microns in diameter. *Soil Sci., 102,* 292–295.

Pettijohn, F. J. 1941. Persistence of heavy minerals and geologic age. *Jour. Geol., 49,* 610–625.

Picot, P., and Z. Johan. 1977. *Atlas of Metallic Minerals.* Mémoire du Bureau de Recherches Géologiques et Minières No. 90, 403 pp.

Ramdohr, P. 1980. *The Ore Minerals and Their Intergrowths,* 2nd ed. 2 vols. New York: Pergamon, 1205 pp.

Rubey, W. W. 1929. Origin of the siliceous Mowry Shale of the Black Hills region. U. S. Geol. Surv. Prof. Paper No. 154–D, pp. 153–170.

Russell, R. D. 1937. Mineral composition of Mississippi River sands. *Geol. Soc. Amer. Bull., 48,* 1307–1348.

Sorby, H. C. 1880. On the structure and origin of non-calcareous stratified rocks. *Proc. Geol. Soc. London, 36,* 46–92.

Glossary

"When I *use a word," Humpty Dumpty said . . . ,*
"it means just what I choose it to mean—
neither more nor less."

LEWIS CARROLL

This glossary is intended to provide definitions of the terms more frequently used in sedimentary petrology. The definitions are consistent with those in the *Glossary of Geology,* 2nd ed., 1980, published by the American Geological Institute. The *Glossary of Geology* contains more than 35,000 geological terms and is the generally accepted arbiter for geologists.

Abyssal Referring to the ocean environment or depth zone deeper than 1,000 m.

Accessory Minerals Those minerals in a sedimentary rock that occur in relatively small quantities, typically less than 1% of the rock. Synonym of **Heavy Minerals.**

Allochem The carbonate particles that form the framework in mechanically deposited limestones, as distinguished from micrite matrix and sparry cement. Important allochems include silt-, sand-, and gravel-size intraclasts, ooliths, fossils, and pellets.

Allochthonous Formed in a location other than where it is presently found, e.g., sand grains on a beach.

Amorphous Referring to a noncrystalline solid.

Anhedral Referring to a mineral grain or crystal that shows no crystal faces, because of either abrasion or interference by other crystals during growth.

Aphanitic See **Microcrystalline.**

Arc-Trench System An elongate zone in the marginal parts of an oceanic basin that includes a deep-sea trench and the adjoining group of volcanic islands, e.g., the Aleutian Trench and Aleutian Islands.

Arenite A detrital rock whose grains are mostly 0.06–2.0 mm in diameter. Synonym of **Sandstone.**

Argillaceous Referring to a rock having an appreciable percentage of clay minerals. Synonym of **Clayey.**

Argillite A nonfissile mudrock that is very highly indurated, perhaps weakly metamorphosed, and lacking slaty cleavage.

Arkose A detrital rock that contains an appreciable percentage of feldspar grains, typically at least 20%.

Aulacogen A tectonic trough on a craton, bounded by convergent normal faults and oriented normal to the cratonic boundary.

Authigenic Referring to rock constituents that are formed in place and not transported; they are formed either at the time of sediment deposition or afterward, e.g., authigenic quartz cement.

Autochthonous Referring to rock constituents that are formed in place. Similar to **Authigenic.**

Backarc Basin A basinal area between an island arc and the continental landmass.

Bathyal Referring to the ocean environment or depth zone 200–1,000 m.

Bauxite A rock or soil composed mostly of amorphous or crystalline hydrous aluminum oxides and aluminum hydroxides, typically gibbsite; aluminous laterite.

Beachrock Limestone formed in the intertidal zone, typically a calcarenite or biosparite.

Bed A layer of rock distinguishable in the field; a subdivision of a formation; thicker than 1 cm. See also **Lamina.**

Benthonic Bottom-dwelling, e.g., clams and crinoids, as contrasted to sharks and jellyfish. See also **Nektonic** and **Planktonic.**

Bentonite Claystone produced by alteration of glassy igneous material; composed essentially of the montmorillonite group of clays and colloidal silica; originally a tuff or volcanic ash.

Bimodality A frequency distribution characterized by two peaks, e.g., sediment formed of sand and clay but lacking silt.

Bioclastic Referring to a rock consisting of fragments of the hard parts of organisms, as in bioclastic limestone; the polar bioclastic limestone is coquina.

Bioherm A mound-shaped structure built by the *in situ* growth of sedentary organisms. See also **Reef.**

Biolithite Limestone constructed by organisms that grew and remained in place, characterized by a rigid framework; typical of reef cores.

Biomicrite Limestone composed of skeletal debris and carbonate mud.

Biosparite Limestone composed of skeletal debris cemented by clear calcite or aragonite whose crystal size exceeds about 20 μm. See also **Microspar.**

Biostrome A sheetlike accumulation of skeletal debris, generally calcium carbonate.

Bioturbation The churning and stirring of a sediment by organisms.

Bird's-Eye Structure A distinctive pattern of small (typically 1–3 mm) voids in limestones; may be filled with minerals, commonly calcite or anhydrite.

Bitterns The liquid remaining after seawater has been concentrated past the point of halite precipitation; also the very soluble minerals that may precipitate following halite, e.g., magnesium chloride and borates.

Boudinage A process in which an originally continuous rock layer has been transformed by stretching or shrinking into elongate segments resembling sausages.

Boulder A sediment particle with a diameter greater than 256 mm.

Boundstone Limestone whose original constituents were bound together during deposition and remained substantially in the position of growth; may be biohermal or biostromal. Not an exact synonym of **Biolithite.**

Breccia Coarse-grained clastic rock composed of angular fragments cemented in a fine-grained matrix.

Brine Surface or subsurface water containing more than 35 ppt of dissolved substances; more saline than normal seawater. Synonym of **Hypersaline.**

Brown Clay Pelagic sediment containing less than 30% skeletal debris. Synonym of **Red Clay.**

Calcarenite A limestone consisting of sand-size particles; typically a **Biosparite** or **Grainstone.**

Calcilutite A limestone consisting of mud-size particles; typically a **Micrite** or **Mudstone.**

Calcirudite A limestone consisting of allochemical gravel-size particles.

Calcrete A limestone precipitated as surface or near-surface crusts and nodules by the evaporation of soil moisture in semiarid climates. Synonym of **Caliche;** a variety of **Duricrust.**

Caliche See **Calcrete.**

Carbonate Bank A large shoal underlain by carbonate sediment.

Carbonate Buildup A nongenetic term for any feature composed of carbonate that (1) differs in character from laterally equivalent and/or overlying strata, (2) is typically thicker than laterally equivalent carbonate deposits, and (3) probably stood topographically higher than surrounding sediment during its depositional history.

Carbonate-Compensation Depth The depth in a body of water at which the rate of dissolution of solid calcium carbonate equals the rate of supply; the depth below which no calcium carbonate can accumulate.

Carbonate Rocks Rocks composed mostly of carbonate minerals.

Cementation Solidification of sediment, usually caused by the introduction of pore-filling material or by recrystallization of fine-grained materials. Synonym of **Lithification.**

Chalcedony A rock composed of microcrystalline quartz crystals of fibrous habit, as seen with a polarizing microscope. See also **chert.**

Chalk A soft, pure micritic limestone composed of the shells of planktonic foraminifera and/or coccoliths.

Chelation, Chelate The binding of a metallic ion by two or more atoms of an organic molecule.

Chert A rock composed of microcrystalline quartz crystals of subequant habit, typically with diameters less than 20 μm.

Chicken-Wire Structure Anhydrite (or gypsum) rock composed of "nodules" of anhydrite in a matrix of microcrystalline anhydrite. Synonym of **Nodular Anhydrite.**

Clastic Referring to a rock or sediment composed of pieces of preexisting materials.

Clay Used to indicate (1) fragmental material less than 4 μm in size and also (2) a group of minerals of phyllosilicate structure and particular composition; i.e., clay size versus clay mineral. The clay-size fraction of most silicate sediments is composed mostly of clay minerals.

Clay Membrane An assemblage of clay-mineral flakes that act as a selective filter for dissolved substances in water passing through them.

Claystone A lithified, nonfissile rock composed mostly of clay-size silicate materials.

Cobble A sediment particle with a diameter of 64–256 mm.

Colloform Referring to the rounded, globular texture of a colloidal mineral deposit or organic growth.

Colloid A particle less than 0.24 μm in diameter that has the ability to scatter visible light when suspended in a column of water; colloids do not settle in the gravitational field.

Collophane Cryptocrystalline to X-ray amorphous phosphate sediment.

Compaction Reduction in bulk volume and decrease in porosity as a result of the increasing weight of overburden as deposition continues; tectonic compaction may occur subsequently.

Compensation Depth See **Carbonate-Compensation Depth.**

Concretion A clearly bounded volume of rock within one or more beds of enclosing rock; commonly formed by early lithification.

Conglomerate A coarse-grained clastic rock composed of rounded gravel, particles greater than 2 mm in diameter. Synonym of **Rudite.**

Connate Water Water trapped in the pore space of sediment at the time of deposition.

Continental Borderland That area of the continental margin between the shoreline and the continental slope that is topographically more complex than the continental shelf; characterized by ridges and basins.

Continental Crust That type of the Earth's crust that underlies the continents and continental shelves; composed of sial above and sima below and is 35–60 km thick.

Continental Rise That part of the continental margin that is between the continental slope and the abyssal plain, except where an oceanic trench is present.

Continental Shelf That part of the continental margin that is between the shoreline and the continental slope (or, in the absence of a noticeable continental slope, a depth of 200 m). The shelf is characterized by its very gentle slope of about 0.1°.

Continental Slope That part of the continental margin that is between the continental shelf and the continental rise (or oceanic trench). It is characterized by its relatively steep slope of 3–6°.

Coquina A weakly indurated clastic or detrital limestone composed of mechanically sorted fossil debris that experienced transport and abrasion before reaching the depositional site.

Craton A part of the Earth's crust that has attained stability and has not undergone significant orogenic activity for a prolonged period; located in the central parts of continental masses.

Cross Beds Layers within a stratified unit that are oriented at an angle to the dominant stratification.

Cryptocrystalline Referring to crystals too small to be visible with either the naked eye or a polarizing microscope and detectable only with X-ray techniques.

Crystalline Limestone An obsolete and nonsensical term formerly used to indicate a coarse-grained limestone, a phaneritic limestone. All limestones must be crystalline, as calcium carbonate cannot be amorphous.

Darcy's Law An empirical formula applicable to the laminar flow of a fluid through a permeable rock. The equation relates apparent velocity of flow, fluid discharge, cross-sectional area of the rock, permeability, fluid viscosity, distance of fluid flow, and the pressure gradient during the flow.

Dedolomitization The process in which dolomite is replaced by another mineral, usually calcite.

Deep-Sea Fan A terrigenous, fan-shaped deposit located seaward of large rivers or submarine canyons. Synonym of **Submarine Fan.**

Deep-Sea Trench A narrow, elongate, steep-sided, rock-walled depression that generally is deeper than the adjacent seafloor by 200 m or more.

Depocenter The area in which the thickest accumulation of sediments occurs within a depositional basin.

Detrital Referring to sediment composed of pieces of preexisting materials derived from outside the depositional basin.

Diachronous Referring to a rock unit that is of noticeably different ages in different areas.

Diagenesis All of the chemical, physical, and biological changes in the characteristics of a sediment accumulation from the time the grains are deposited until they are metamorphosed or melted, excluding weathering.

Diapir A body of rock that has moved upward by plastic flow, piercing the overlying beds, e.g., salt diapirs in the Gulf Coast region.

Diastem A brief interruption in sediment deposition during which no erosion occurs, e.g., a bedding plane.

Diatomite Indurated diatom ooze. Synonym of **Diatomaceous Earth.**

Distal Outer; referring to a sedimentary deposit that formed farthest from the source area. See also **Proximal.**

Dolomicrite Mud-size dolomite crystals in a carbonate rock.

Dolomite A mineral with the formula $CaMg(CO_3)_2$ but also used to refer to the sedimentary rock composed largely or entirely of the mineral. See also **Dolostone.**

Dolostone A term used by some petrologists to designate a sedimentary rock composed largely or entirely of the mineral dolomite.

Dorag Dolomitization Replacement of limestone by dolomite through the mechanism of a mixture of fresh water (about 95%) and seawater (about 5%), presumably along coastlines in subtropical and tropical areas.

Drainage Basin A region bounded by a drainage divide and occupied by a stream system; the provenance of sediment in the lower part of the trunk stream.

Drusy Mosaic A mosaic of equant anhedral crystals whose sizes increase outward from a grain surface into a pore space.

Duricrust The hard crust on the surface of a soil, formed by the evaporation of soil moisture; calcrete and silcrete are varieties of duricrust.

Effective Porosity The percentage of a rock volume that consists of interconnected void spaces.

Eh A measure, usually expressed in millivolts, of the relative intensity of oxidation or reduction in a solution; values range from -1.0 volt (most reducing) to $+1.0$ volt (most oxidizing).

Enterolithic Folding A sedimentary structure consisting of ribbons of intestine-like folds that originate through chemical changes involving an increase or a decrease in the volume of a rock, as when gypsum dehydrates to anhydrite.

Epeiric Sea A sea on the continental shelf or within a continent. Synonym of **Epicontinental Sea.**

Epeirogenic Uplift A broad, vertical movement of the Earth's crust unaccompanied by crumpling of the strata involved in the uplift.

Epicontinental Sea See **Epeiric Sea.**

Eugeosyncline A geosyncline in which volcanism is associated with detrital sedimentation.

Euhedral Referring to a crystal that has well-developed crystal outlines, such as a rhomb of dolomite in a limestone.

Eustatic Change in Sea Level Worldwide change in sea level caused by changes in the volume of seawater, as by the melting of glaciers.

Euxinic Referring to a state of an environment characterized by a lack of circulation of water and consequent depletion of dissolved oxygen, resulting in anaerobic conditions.

Evaporite A sedimentary deposit composed of minerals more soluble than dolomite, such as gypsum, halite, or sylvite.

Expandable Clay A clay mineral whose layers can be readily separated by the absorption of water; swelling clay.

Fabric The orientation, or lack of it, of the crystals or grains of which a sedimentary rock is composed.

Facies A characteristic or group of characteristics of a rock that distinguish it from other rocks, e.g., redbed facies, shale facies, sandy facies.

Fanglomerate Referring to a conglomerate formed as part of an alluvial fan.

Fenestral Fabric Fabric in a carbonate rock characterized by discontinuities in the rock framework that are larger than grain-supported voids. Fenestra may be empty or filled. Examples are bird's-eye and stromatactis structures.

First-Cycle Grain A detrital particle in a sediment that was derived directly from an igneous or a metamorphic rock and has not previously been part of a sedimentary rock.

Fissility The ability of a rock to split easily into thin sheets parallel to the bedding of the unit; caused by parallelism of orientation of clay minerals in the unit.

Flint Dark-gray to black chert.

Flocculation A process by which individual flakes of clay minerals are aggregated into clumps (floccules) because of electrostatic charges on clay surfaces; the process is more rapid in saline water than in fresh water.

Flute Cast A spoon-shaped sole mark in which the deep part of the spoon is upcurrent.

Flysch Marine sedimentary facies characterized by a thick sequence of poorly fossiliferous, thinly bedded, graded deposits consisting of rhythmically interbedded mudstones and sandstones; formed during the later stages of filling of a geosynclinal trough by rapid erosion of a rising mountain belt.

Forearc Basin A basin lying between a deep-sea trench and its adjoining island arc.

Foreland Basin A basin in a backarc area that is floored by continental crust. Synonym of **Backarc Basin** and **Retroarc Basin.**

Forereef Facies Coarse sediment that faces the prevailing wind direction; commonly composed of reef talus.

Formation An extensive, mappable lithologic unit in a sedimentary sequence.

Formation Waters Waters present in the pores of a sedimentary unit.

Friable Referring to the state of incomplete cementation in which a rock is coherent but crumbles easily under a small amount of pressure, such as that applied by the fingers.

Geopetal Structure Any rock feature that permits distinctions between the top and bottom of a bed at the time of deposition, e.g., cross bedding or stromatactis.

Geosyncline A large area of the Earth's crust that has subsided more or less continuously over an extensive period of time (one or more geologic periods) and has accumulated a filling of sediment much thicker than is present in surrounding rocks of the same age.

Geothermal Gradient The rate of increase of temperature with depth in the Earth.

Graded Bed A layer of sedimentary rock in which the particle sizes change systematically in a direction normal to the bedding surface, from coarser at the base to finer at the top.

Grainstone A clastic limestone in which the grains are resting on each other (grain-supported) and no micrite is present.

Granule A sediment particle with a diameter of 2–4 mm.

Grapestone A composite particle consisting of two or more sand-size calcium carbonate particles of various kinds that have become bound together by cementation shortly after deposition.

Gravel All rounded particles with a diameter greater than 2 mm.

Graywacke A type of sandstone typical of flysch sequences and characterized by grayish-green color, clay matrix, poor sorting of the sand grains, and abundant lithic fragments; the word has been redefined many times since its origin 200 years ago.

Groove Cast A long, narrow sole mark usually a few millimeters in height and width but many centimeters in length.

Groundwater That part of the subsurface water that is below the groundwater table, in the zone of saturation. Synonym of **Phreatic Water.**

Grus An accumulation of sediment resulting from granular disintegration of crystalline rocks, especially granitic ones; untransported debris of weathering.

Halmyrolysis The geochemical reaction between seawater and sediments in an area of little or no sedimentation, e.g., modification of clay minerals on the seafloor or the authigenesis of glauconite from preexisting biotite.

Heavy Minerals Those minerals in a sediment or sedimentary rock that have specific gravities greater than about 2.9, defined by the specific gravity of the liquid used to separate them from the more abundant quartz and feldspar. Synonym of **Accessory Minerals.**

Hemipelagic Deposits Deep-sea sediments containing a small amount of terrigenous material as well as remains of pelagic organisms.

High-Energy Environment A depositional environment characterized by relatively high kinetic energies, e.g., beach or sand dune. See also **Low-Energy Environment.**

High-Magnesian Calcite A calcite containing more than 4% $MgCO_3$ in solid solution. See also **Low-Magnesian Calcite.**

High-Strontian Aragonite An aragonite containing more than 7 ppt of strontium. See also **Low-Strontian Aragonite.**

Hypersaline Referring to water that contains more than 35 ppt of dissolved solids. Synonym of **Brine.**

Ichnofossil A sedimentary structure consisting of a fossilized track, trail, or burrow formed by the life activities of an animal, such as feeding or moving in soft sediment. Synonym of **Trace Fossil.**

Imbrication A sedimentary structure consisting of detrital grains (typically gravel) stacked with their flat surfaces at an angle to the main bedding plane; the flat surfaces dip upstream.

Immature Sediment Refers to either the texture or the mineral composition of a coarse detrital rock, especially sandstone; textural immaturity means the presence of at least 5% clay; mineralogic immaturity means an abundance of feldspar and lithic fragments. See also **Textural Maturity** and **Mineralogic Maturity.**

Inclusions Gas, liquid, or mineral crystals contained within a larger crystal.

Induration The hardening of a rock by the action of heat, pressure, or the introduction of some pore filling. Synonym of **Lithification.**

Interarc Basin A basin in a backarc area that is floored with oceanic crust.

Interstitial Referring to location in pore spaces, between the detrital grains.

Intra-Arc Basin A basin located within the volcanic island arc.

Intrabasinal Rocks Sedimentary rocks whose constituent particles were derived from within the basin of deposition, e.g., most limestones.

Intraclast A carbonate fragment composed of penecontemporaneous limestone or dolomite formed within the basin of deposition or on its fringes, e.g., dolomite from a supratidal flat or shallow seafloor carbonate torn up by a storm.

Intracontinental Basin Synonym of **Intracratonic Basin.**

Intracratonic Basin A tectonic basin within a continent and underlain by continental crust, e.g., Williston Basin, Illinois Basin.

Intraformational Conglomerate A conglomerate in which the clasts are penecontemporaneous in origin with the matrix; developed by the fragmentation and rounding of newly formed rock and its rapid incorporation into new sedimentary deposits.

Intramicrite Limestone consisting of intraclasts in a micrite matrix.

Intrasparite Limestone consisting of intraclasts and sparry calcite cement.

Intrastratal Solution Dissolution of minerals after they have been deposited in a body of sediment and buried.

Iron Formation A rock unit consisting of alternating bands of quartz (including chert) and iron-bearing minerals, totaling at least 15% Fe.

Ironstone A sedimentary rock containing at least 15% Fe, typically unbanded and lacking chert; typically of Phanerozoic age.

Island Arc A chain of islands, e.g., the Aleutians, rising from the deep seafloor and located near a continent; the chain generally is curved and convex toward the open ocean.

Isostasy A condition of gravitational equilibrium, comparable to floating, of the brittle crust above the more plastic mantle.

Jasper Red chert; the color results from hematite impurity.

Juvenile Water Water derived directly from magma and that is coming to the Earth's surface for the first time.

Labile Referring to minerals or lithic fragments that are mechanically or chemically unstable, e.g., schist fragments or calcic feldspars.

Lamina A layer of sediment or sedimentary rock less than 1 cm thick. See also **Bed.**

Lamination Stratification on a scale less than 1 cm.

Laterite A surficially formed deposit consisting largely or entirely of iron and/or aluminum oxides and hydroxides.

Leaching The selective removal of soluble constituents from minerals and rocks by percolating waters; used mostly in reference to surficial processes.

Lee Downstream

Light Minerals Those minerals in a sediment or sedimentary rock that have specific gravities less than about 2.9.

Limeclast A carbonate rock fragment of clastic or detrital origin contained within a carbonate rock; the fragment may be intrabasinal (intraclastic) or extrabasinal (terrigenous).

Limestone A sedimentary rock consisting largely or entirely of calcium carbonate.

Litharenite A sandstone containing an appreciable amount of undisaggregated rock (lithic) fragments, typically at least 20%.

Lithic Fragment A detrital fragment of undisaggregated rock, e.g., a schist fragment or basalt fragment. Synonym of **Rock Fragment.**

Lithification The conversion of unconsolidated sediment into a coherent aggregate. Synonym of **Cementation** and **Induration.**

Loess A wind-deposited sediment consisting mostly of silt.

Low-Energy Environment A depositional environment characterized by relatively low kinetic energies, e.g., lakes. See also **High-Energy Environment.**

Low-Magnesian Calcite A calcite containing less than 4% $MgCO_3$ in solid solution. See also **High-Magnesian Calcite.**

Low-Strontian Aragonite An aragonite containing less than 2 ppt of strontium. See also **High-Strontian Aragonite.**

Luminescence Petrography The examination of a thin section during bombardment by a broad beam of low-energy electrons.

Lutite A sedimentary rock consisting of particles of silt and clay sizes; generally restricted to silicate rocks. Synonym of **Pelite.**

Macrocrystalline Referring to crystals distinguishable with the naked eye. Synonym of **Phaneritic.**

Marl A friable mixture of subequal amounts of micrite and clay minerals.

Matrix The finer-grained material in a sediment with a conspicuous range in grain size, e.g., the clay in a texturally immature sandstone.

Mature Sediment Refers to either the texture or the mineral composition of a coarse detrital rock, especially sandstone; textural maturity means no clay is present and the sand grains are well sorted but not round; mineralogic maturity means the sandstone is composed mostly of quartz. See also **Textural Maturity** and **Mineralogic Maturity.**

Mélange A mappable body of rock composed of a tectonically chaotic mixture of very large fragments (perhaps several kilometers in length) of preexisting rock set in a pelitic matrix.

Mean Size A measure of average size calculated by summing all of the size determinations and dividing this sum by the number of determinations.

Member A body of rock, not necessarily mappable, that is a subdivision of a formation; may consist of one of more beds.

Meteoric Water Water of recent atmospheric origin.

Micrite Microcrystalline calcite or aragonite mud, normally of clay size, in a carbonate rock. See also **Dolomicrite.**

Microcrystalline Referring to crystals distinguishable with a polarizing microscope but not resolvable with the naked eye; smaller than about 10 μm; e.g., micrite or shale. Synonym of **Aphanitic.**

Microspar Calcite matrix in limestones occurring as uniformly sized crystals 5–20 μm in diameter; originates by recrystallization of micrite.

Mineralogic Maturity Refers to a detrital sediment composed mostly of quartz.

Miogeosyncline A linear geosyncline, formed on the continental shelf, that is filled with shallow-water sediments lacking volcanic detritus.

Modal Analysis Determination of the actual mineral composition of a rock, typically by counting grains in thin section with a polarizing microscope. See also **Normative Analysis.**

Mode That value or group of values that occur with the greatest frequency in a set of data.

Molasse A continental, deltaic, and/or marine sedimentary facies consisting of a thick sequence of cross-bedded, fossiliferous detrital rocks with minor coal and carbonate rocks; a postorogenic accumulation.

Monocrystalline Referring to a clastic grain consisting of a single crystal.

Mud Sediment whose particles are less than 0.06 mm in size.

Mudrock Detrital rock whose grains are mostly less than 0.06 mm in size.

Mudstone Nonfissile mudrock containing subequal amounts of silt- and clay-size silicate detritus; also a mud-supported limestone that contains less than 10% sand- or gravel-size particles.

Mud-Supported Referring to limestone in which the sand- and gravel-size grains are not sufficiently abundant to form the supporting framework.

Nannofossil A calcareous planktonic microfossil smaller than about 5 μm, such as coccoliths and discoasters; common in Mesozoic chalks.

Nektonic Referring to aquatic pelagic organisms that are free-swimming as contrasted to **Planktonic.** See also **Benthonic.**

Neritic Referring to the depth zone in the ocean between the low-tide line and 200 m or between the low-tide line and the edge of the continental shelf.

Nodular Anhydrite Synonym of **Chicken-Wire Structure.**

Nodule A small, hard, round, irregularly shaped body of a different composition from the rock in which it is located, e.g., a phosphate nodule in a shale.

Nonundulatory Extinction Referring to a crystal that has not been plastically deformed and therefore extinguishes as a unit in thin section on rotation of the microscope stage. See also **Undulatory Extinction.**

Normative Analysis Chemical analysis of a rock, typically a microcrystalline one, followed by a prescribed calculation of the proportions of a set of minerals (normative minerals) that may or may not actually be present in the rock; useful for comparative purposes. See also **Modal Analysis.**

Novaculite Milky-white chert.

Obduction The welding of oceanic ophiolite and/or sediment onto the edge of the continental block as the ocean floor moves into a subduction zone.

Oligomictic Referring to conglomerates whose particles consist of only one kind of rock.

Olistostrome A chaotic mass of fragmental rock that originated by submarine slumping or sliding; it is mappable, is lens-shaped, and lacks true bedding. Nontectonic equivalent of **Mélange.**

Oncolith Synonym of algal **Pisolith.**

Oolith A spherical or an elliptical particle less than 2 mm in diameter, typically composed of calcium carbonate (but also hematite or quartz) and having a nucleus of mineral matter surrounded by a multilayered rim with a concentric or radial fabric.

Oomicrite Limestone composed of ooliths in a micrite matrix.

Oomoldic Porosity Referring to porosity created by the dissolution of ooliths from a rock.

Oosparite Limestone composed of ooliths cemented by sparry calcite.

Ooze Pelagic sediment containing at least 30% skeletal remains of microorganisms, e.g., diatom ooze.

Opaque Mineral A mineral that does not transmit visible light, e.g., magnetite.

Ophiolite A group of mafic and ultramafic rocks, often albitized, that are produced on the ocean floor and sometimes obducted onto the edge of the continent as the ocean floor moves into a trench.

Orogenic Referring to mountain building.

Orthochem The carbonate material that binds the allochems of limestones, including micrite and sparry cement.

Orthogeosyncline A geosyncline located at the juncture of a continental craton and the oceanic floor, including both a **Miogeosyncline** and a **Eugeosyncline.**

Orthoquartzite Sandstone or conglomerate composed of more than 90% detrital quartz and cemented by quartz.

Overgrowth Secondary material precipitated from solution in crystallographic continuity around a crystal of the same composition; enlargement during diagenesis.

Packing The spatial density of grains in a sedimentary deposit.

Packing Density A measure of the extent to which the allochemical or terrigenous particles of a sedimentary rock occupy the rock volume.

Packing Proximity An estimate of the proportion of allochemical or terrigenous particles that are in contact with their neighbors.

Packstone A clastic limestone in which the grains are resting on each other (grain-supported) but some micrite is present.

Paleocurrent Indicator Any feature of a sedimentary rock that indicates the direction in which the sediment was transported.

Paleogeography The study and description of the physical geography of the geologic past, e.g., topographic relief.

Paleogeology The study and description of the geology of the geologic past, e.g., its tectonics and outcrop patterns of rocks.

Paleoslope The direction of slope of a former land surface, particularly one of regional extent.

Parting Lineation Parallel ridges and grooves a few millimeters wide and many centimeters long found on the bedding surfaces of thinly bedded sandstones.

Pebble A sediment particle with a diameter of 2–64 mm.

Pelagic Deposit Deep-sea sediment lacking terrigenous material, e.g., brown clay and organic oozes.

Pelagic Realm The oceanic area lying seaward of the low-tide line.

Pelite See **Lutite.**

Pellet A small, round, unstructured aggregate of micrite, normally containing organic matter and thought to be of fecal origin.

Pelmicrite Limestone composed of peloids in a micrite matrix.

Peloid Similar to a **Pellet** but of uncertain origin.

Pelsparite Limestone composed of peloids cemented by sparry calcite.

Peritidal Referring to a sea-margin area subject to the effects of tidal fluctuations.

Permeability The capacity of a porous material to transmit a fluid; depends on the characteristics of the material, the fluid, and the physical conditions of the flow.

Petrogenesis The branch of petrology that deals with the origin and formation of rocks.

Petrography The branch of geology dealing with the description and systematic classification of rocks, conventionally centering around the use of a polarizing microscope.

Petrology The branch of geology dealing with the origin, occurrence, structure, and history of rocks; it is broader in scope than **Petrography.**

pH The negative logarithm to the base 10 of the hydrogen ion activity of a solution.

Phaneritic See **Macrocrystalline.**

Phi Scale A geometric scale of the sizes of sedimentary particles using class boundaries based on negative logarithms to the base 2. The phi scale relates to the millimeter scale as follows: $\phi = -\log_2$ size in millimeters. It is used extensively in sedimentology and sedimentary petrology.

Phosphorite A sedimentary rock composed principally of phosphate minerals, typically a carbonate fluorapatite.

Photic Zone That part of the ocean in which there is sufficient penetration of light to support photosynthesis; in clear ocean water, about 30 m.

Phreatic Water Water in the zone of saturation. Synonym of **Groundwater.**

Phytoplankton Unicellular, floating aquatic plants.

Pisolith A small, round accretionary body, typically composed of calcium carbonate, with a diameter greater than 2 mm.

Placer Deposit An accumulation of sediment formed by mechanical concentration of mineral particles, typically those of specific gravity greater than most minerals, e.g., gold or magnetite.

Planktonic Referring to aquatic organisms that are not free-swimming and are transported solely by the prevailing currents, as contrasted to **Nektonic.** See also **Benthonic.**

Platform Facies A sedimentary facies that contains sediments produced in the neritic environment of shelf seas marginal to a craton.

Polycrystalline Referring to a sedimentary particle composed of two or more crystals.

Polycyclic Sediment See **Recycled Sediment.**

Polymictic Referring to conglomerates whose particles consist of more than one type of rock.

Porcellanite An impure chert having the texture, dull luster, or general appearance of unglazed porcelain.

Porosity The percentage of rock volume that consists of void space.

Pressure Solution Solution occurring preferentially at the contact surfaces of crystals where the external pressure exceeds the hydrostatic pressure of the interstitial fluid.

Provenance The area from which the constituent materials of a sediment are derived; the source area.

Proximal Upfan; referring to a sedimentary deposit that formed nearest the source area. See also **Distal.**

Proximate Source The most recent source rock of a sedimentary particle, e.g., an older sandstone, as contrasted to the **Ultimate Source,** which may have been a granite.

Recycled Sediment Detrital grains that have been part of a previous sedimentary rock since their release from the parent igneous or metamorphic rock. Synonym of **Polycyclic Sediment.**

Red Clay See **Brown Clay.**

Reef A rigid, wave-resistant **Bioherm.**

Reflux A process in which the salinity and the specific gravity of surface water are increased by evaporation; the water sinks and flows through underlying rock. Commonly the process results in the replacement of minerals in the underlying rock by new minerals.

Retroarc Basin A basin in a backarc area that is floored by continental crust and separated from the arc by a system of folds and thrusts.

Reworked Sediment See **Recycled Sediment.**

Rifted Continental Margin A long, narrow continental trough bounded by normal faults; a graben of regional extent with faults extending through the entire thickness of the crust.

Ripple Marks Small ridges of sediment resembling a ripple of water and formed on the bedding surface of a sediment.

Rock Fragment See **Lithic Fragment.**

Roundness The degree of abrasion of a particle as expressed by the angularity of its corners; often expressed by the rho (ρ) scale as values ranging from zero (most angular) to six (most rounded).

Rudite See **Conglomerate.**

Sabkha A supratidal salt flat in an arid environment.

Salina A playa in which the sediment consists predominantly of evaporite minerals.

Salinity The total amount of dissolved salts in water, typically reported as parts per thousand or milligrams per liter.

Salt Sieving The process in which subsurface water is forced through an assemblage of clay-mineral flakes (a "clay membrane") with the result that ions are selectively passed through or retained behind the membrane.

Sand A sediment particle with a diameter of 0.06–2.0 mm.

Sandstone A sedimentary rock in which sand-size particles predominate; normally restricted to silicate rocks. See also **Arenite.**

Sapropel An unconsolidated sludge composed of plant remains putrefying in an anaerobic environment on the shallow bottoms of lakes and seas.

Sedimentary Structure A megascopic feature, generally best studied in outcrop rather than in hand specimen or thin section, and representing a discontinuity or major inhomogeneity, e.g., cross bedding.

Sedimentation Unit That thickness of sediment that was deposited under essentially constant physical conditions.

Sessile Benthos Aquatic organisms that attach themselves to the sediment bottom and remain in one spot throughout their adult lives, e.g., crinoids.

Shale A fissile, terrigenous, sedimentary silicate rock whose particles are mostly of silt and clay size.

Shard A fragment of glass, usually of volcanic origin, in a sediment; characteristically the fragment has curved surfaces of fracture.

Silcrete A siliceous crust of sand and gravel cemented by opal, chert, and/or quartz; formed by evaporation of water in a semiarid climate; like **Calcrete,** a variety of **Duricrust.**

Silt A sediment particle 0.004–0.062 mm.

Siltstone A nonfissile sedimentary rock in which silt-size particles predominate; normally restricted to silicate rocks.

Skeletal Particle Rigid material, originally secreted within the soft parts of a living organism, that has been freed from the organic matrix.

Skewness The asymmetry of a frequency distribution.

Sole Mark Any sedimentary structure found on the bottom surface of a bed.

Sorting The selection during transport of particles according to their sizes, specific gravities, and shapes; a well-sorted sediment has only a small amount of variability among the diameters of its particles. See also **Standard Deviation.**

Sparry Calcite A pore-filling cement in either sandstones or carbonate rocks that consists of a mosaic of calcite crystals larger than micrite.

Sphericity The approach to equality of three mutually perpendicular axes passing through a particle.

Standard Deviation. A departure from the middle of a normal distribution that includes 34% of the total population of individual values; the perfection of sorting is given by the value of the standard deviation.

Stoss Upstream.

Stromatactis Open-space sedimentary structures in carbonate rocks that are characterized by horizontal flat bottoms and irregular upper surfaces; the cavities may be filled with internal sediment and/or sparry calcite cement.

Stromatolite A laminated and lithified calcareous sedimentary structure formed by sediment-binding blue-green algae that trap silty detritus suspended in the water washing over the algal filaments; may have several gross outward topographic forms depending on interactions with currents.

Stylolite A thin seam or a surface of contact between two beds marked by an irregular and interlocking or mutual interpenetration of the two sides, the columns and projections on one side fitting into their counterparts on the other; relatively insoluble constituents may be concentrated along the seam; common in limestones.

Subduction The process of one crustal block descending beneath another, as along convergent plate margins.

Submarine Fan See **Deep-Sea Fan.**

Sucrosic Texture A "sugary" appearance in dolomites; commonly such rocks are formed of euhedral crystals.

Supratidal Zone An area lying above but close to the mean high-tide level but not higher than the level of the highest storm tide.

Survival Potential The probability that a detrital sedimentary particle will remain after the rigors of abrasion and chemical attack; e.g., quartz has a high survival potential, schist fragments a low one.

Suture Belt A zone in the Earth's crust marking the juncture of formerly separate crustal segments.

Sutured Contact An irregularly interlocking boundary; intense suturing that affects a large number of grains produces a stylolite.

Syntaxial Cement Cement formed by crystals that have grown as overgrowths on clastic particles, typically in limestones and quartz sandstones.

Taconite A local term used for banded iron ore in the Lake Superior region.

Tephra A general term for all particles exploded from volcanoes.

Terrigenous Referring to material eroded from the land surface.

Textural Maturity A sequential series of changes in texture that occurs during the transportation of sediment, from the presence of clay (immaturity) to loss of clay (submaturity) to sorting of the nonclay fraction (maturity) to rounding of the sand grains (supermaturity).

Texture The size, shape, and arrangement of the crystals or grains of which a rock is composed.

Thermocline A thin layer of water within which the temperature changes rapidly with depth.

Thin Section A fragment of rock mechanically ground to a standard thickness of 0.03 mm, polished, and mounted between two pieces of glass as a microscope slide; used for textural, mineralogic, and diagenetic studies.

Tonstein A kaolinitic claystone commonly occurring as a thin seam in a coal seam.

Trace Fossil See **Ichnofossil.**

Travertine A light-colored, micritic limestone, often having a fibrous or concentric structure, formed by rapid precipitation in surface water and groundwater.

Trench A long, narrow depression of the deep-sea floor, oriented parallel to a continental margin and adjacent to it.

Tripoli A porous, friable, light-colored, siliceous rock resulting from the weathering of chert or siliceous limestone.

Tufa A calcium carbonate precipitate formed by evaporation as a thin crust around a spring or along the margin of a stream.

Turbidite A deposit of a turbidity current.

Turbidity Current A current that consists of a suspension of detritus and water that flows down a submarine slope.

Ultimate Source The igneous or metamorphic parent rock from which a detrital grain was derived, irrespective of whether the grain has been a part of an earlier sedimentary rock. See also **Proximate Source.**

Undulatory Extinction Referring to a crystal that has been deformed plastically and therefore does not extinguish as a unit in thin section on rotation of the microscope stage. See also **Nonundulatory Extinction.**

Upwelling The rising of cold subsurface water in an ocean or a lake to replace warmer waters moved away from the site by prevailing winds; upwelling waters are rich in nutrients, which are required in large amounts by growing reefs.

Vadose Water Water in the zone of aeration, above the groundwater table.

Vagrant Benthos Benthonic organisms able to move around at the bottom of a body of water, e.g., clams and trilobites.

Varve A rhythmically bedded sediment in which the time span of the rhythm is one year.

Vug A large void in a rock, commonly not joined to other voids, e.g., the spherical cavity formed by dissolution of an oolith in a limestone.

Wacke A texturally (and typically mineralogically) immature sandstone.

Wackestone A mud-supported limestone that contains more than 10% sand- or gravel-size particles.

Wentworth Scale See **Phi Scale.**

Whiting A milky-white body of water made white by a dense suspension of calcium carbonate sediment.

Zoned Crystal A crystal showing systematic variation in chemical composition between the center and the rim, e.g., rhomb-shaped zones of iron concentration in a dolomite crystal.

Zooxanthellae Unicellular algae that live symbiotically within the tissues of reef-building types of corals.

ZTR Index The percentage of zircon, tourmaline, and rutile among the nonopaque heavy-mineral grains of a detrital rock.

Index of Names

An italic page number indicates an illustration. A lowercase letter "t" after a page number indicates a table.

Index of Topics

An italic page number indicates an illustration. A lowercase letter "t" after a page number indicates a table.